Title 14—Aeronautics and Space

(This book contains parts 1 to 59)

CHAPTER I—FEDERAL AVIATION ADMINISTRATION, DEPARTMENT OF TRANSPORTATION

SUBCHAPTER A—DEFINITIONS AND GENERAL REQUIREMENTS

PART 1—DEFINITIONS AND ABBREVIATIONS

AUTHORITY: 49 U.S.C. 106(f), 106(g), 40113, 44701.

§ 1.1 General definitions.

As used in Subchapters A through K of this chapter, unless the context requires otherwise:

Administrator means the Federal Aviation Administrator or any person to whom he has delegated his authority in the matter concerned.

Aerodynamic coefficients means nondimensional coefficients for aerodynamic forces and moments.

Air carrier means a person who undertakes directly by lease, or other arrangement, to engage in air transportation.

Air commerce means interstate, overseas, or foreign air commerce or the transportation of mail by aircraft or any operation or navigation of aircraft within the limits of any Federal airway or any operation or navigation of aircraft which directly affects, or which may endanger safety in, interstate, overseas, or foreign air commerce.

Aircraft means a device that is used or intended to be used for flight in the air.

Aircraft engine means an engine that is used or intended to be used for propelling aircraft. It includes turbosuperchargers, appurtenances, and accessories necessary for its functioning, but does not include propellers.

Airframe means the fuselage, booms, nacelles, cowlings, fairings, airfoil surfaces (including rotors but excluding propellers and rotating airfoils of engines), and landing gear of an aircraft and their accessories and controls.

Airplane means an engine-driven fixed-wing aircraft heavier than air, that is supported in flight by the dynamic reaction of the air against its wings.

Airport means an area of land or water that is used or intended to be used for the landing and takeoff of aircraft, and includes its buildings and facilities, if any.

Airship means an engine-driven lighter-than-air aircraft that can be steered.

Air traffic means aircraft operating in the air or on an airport surface, exclusive of loading ramps and parking areas.

Air traffic clearance means an authorization by air traffic control, for the purpose of preventing collision between known aircraft, for an aircraft to proceed under specified traffic conditions within controlled airspace.

Air traffic control means a service operated by appropriate authority to promote the safe, orderly, and expeditious flow of air traffic.

Air Traffic Service (ATS) route is a specified route designated for channeling the flow of traffic as necessary for the provision of air traffic services. The term "ATS route" refers to a variety of airways, including jet routes, area navigation (RNAV) routes, and arrival and departure routes. An ATS route is defined by route specifications, which may include:

(1) An ATS route designator;

(2) The path to or from significant points;

(3) Distance between significant points;

(4) Reporting requirements; and

(5) The lowest safe altitude determined by the appropriate authority.

Air transportation means interstate, overseas, or foreign air transportation or the transportation of mail by aircraft.

Alert Area. An alert area is established to inform pilots of a specific area wherein a high volume of pilot training or an unusual type of aeronautical activity is conducted.

Alternate airport means an airport at which an aircraft may land if a landing at the intended airport becomes inadvisable.

Altitude engine means a reciprocating aircraft engine having a rated takeoff power that is producible from sea level to an established higher altitude.

Amateur rocket means an unmanned rocket that:

(1) Is propelled by a motor or motors having a combined total impulse of 889,600 Newton-seconds (200,000 pound-seconds) or less; and

(2) Cannot reach an altitude greater than 150 kilometers (93.2 statute miles) above the earth's surface.

Appliance means any instrument, mechanism, equipment, part, apparatus, appurtenance, or accessory, including communications equipment, that is used or intended to be used in operating or controlling an aircraft in flight, is installed in or attached to the aircraft, and is not part of an airframe, engine, or propeller.

Approved, unless used with reference to another person, means approved by the FAA or any person to whom the FAA has delegated its authority in the matter concerned, or approved under the provisions of a bilateral agreement between the United States and a foreign country or jurisdiction.

Area navigation (RNAV) is a method of navigation that permits aircraft operations on any desired flight path.

Area navigation (RNAV) route is an ATS route based on RNAV that can be used by suitably equipped aircraft.

Armed Forces means the Army, Navy, Air Force, Marine Corps, and Coast Guard, including their regular and reserve components and members serving without component status.

Autorotation means a rotorcraft flight condition in which the lifting rotor is driven entirely by action of the air when the rotorcraft is in motion.

Auxiliary rotor means a rotor that serves either to counteract the effect of the main rotor torque on a rotorcraft or to maneuver the rotorcraft about one or more of its three principal axes.

Balloon means a lighter-than-air aircraft that is not engine driven, and that sustains flight through the use of either gas buoyancy or an airborne heater.

Brake horsepower means the power delivered at the propeller shaft (main drive or main output) of an aircraft engine.

Calibrated airspeed means the indicated airspeed of an aircraft, corrected for position and instrument error. Calibrated airspeed is equal to true airspeed in standard atmosphere at sea level.

Canard means the forward wing of a canard configuration and may be a fixed, movable, or variable geometry surface, with or without control surfaces.

Canard configuration means a configuration in which the span of the forward wing is substantially less than that of the main wing.

Category:

(1) As used with respect to the certification, ratings, privileges, and limitations of airmen, means a broad classification of aircraft. Examples include: airplane; rotorcraft; glider; and lighter-than-air; and

(2) As used with respect to the certification of aircraft, means a grouping of aircraft based upon intended use or operating limitations. Examples include: transport, normal, utility, acrobatic, limited, restricted, and provisional.

Category A, with respect to transport category rotorcraft, means multiengine rotorcraft designed with engine and system isolation features specified in Part 29 and utilizing scheduled takeoff and landing operations under a critical engine failure concept which assures adequate designated surface area and adequate performance capability for continued safe flight in the event of engine failure.

Category B, with respect to transport category rotorcraft, means single-engine or multiengine rotorcraft which do not fully meet all Category A standards. Category B rotorcraft have no guaranteed stay-up ability in the event of engine failure and unscheduled landing is assumed.

Category II operations, with respect to the operation of aircraft, means a straight-in ILS approach to the runway of an airport under a Category II ILS instrument approach procedure issued by the Administrator or other appropriate authority.

Category III operations, with respect to the operation of aircraft, means an ILS approach to, and landing on, the runway of an airport using a Category III ILS instrument approach procedure issued by the Administrator or other appropriate authority.

Ceiling means the height above the earth's surface of the lowest layer of clouds or obscuring phenomena that is reported as "broken", "overcast", or "obscuration", and not classified as "thin" or "partial".

Civil aircraft means aircraft other than public aircraft.

Class:

(1) As used with respect to the certification, ratings, privileges, and limitations of airmen, means a classification of aircraft within a category having similar operating characteristics. Examples include: single engine; multiengine; land; water; gyroplane; helicopter; airship; and free balloon; and

(2) As used with respect to the certification of aircraft, means a broad grouping of aircraft having similar characteristics of propulsion, flight, or landing. Examples include: airplane; rotorcraft; glider; balloon; landplane; and seaplane.

Clearway means:

(1) For turbine engine powered airplanes certificated after August 29, 1959, an area beyond the runway, not less than 500 feet wide, centrally located about the extended centerline of the runway, and under the control of the airport authorities. The clearway is expressed in terms of a clearway plane, extending from the end of the runway with an upward slope not exceeding 1.25 percent, above which no object nor any terrain protrudes. However, threshold lights may protrude above the plane if their height above the end of the runway is 26 inches or less and if they are located to each side of the runway.

(2) For turbine engine powered airplanes certificated after September 30, 1958, but before August 30, 1959, an area beyond the takeoff runway extending no less than 300 feet on either side of the extended centerline of the runway, at an elevation no higher than the elevation of the end of the runway, clear of all fixed obstacles, and under the control of the airport authorities.

Climbout speed, with respect to rotorcraft, means a referenced airspeed which results in a flight path clear of the height-velocity envelope during initial climbout.

Commercial operator means a person who, for compensation or hire, engages in the carriage by aircraft in air commerce of persons or property, other than as an air carrier or foreign air carrier or under the authority of Part 375 of this title. Where it is doubtful that an operation is for "compensation or hire", the test applied is whether the carriage by air is merely incidental to the person's other business or is, in itself, a major enterprise for profit.

Configuration, Maintenance, and Procedures (CMP) document means a document approved by the FAA that contains minimum configuration, operating, and maintenance requirements, hardware life-limits, and Master Minimum Equipment List (MMEL) constraints necessary for an airplane-engine combination to meet ETOPS type design approval requirements.

Consensus standard means, for the purpose of certificating light-sport aircraft, an industry-developed consensus standard that applies to aircraft design, production, and airworthiness. It includes, but is not limited to, standards for aircraft design and performance, required equipment, manufacturer quality assurance systems, production acceptance test procedures, operating instructions, maintenance and inspection procedures, identification and recording of major repairs and major alterations, and continued airworthiness.

Controlled airspace means an airspace of defined dimensions within which air traffic control service is provided to IFR flights and to VFR flights in accordance with the airspace classification.

NOTE: Controlled airspace is a generic term that covers Class A, Class B, Class C, Class D, and Class E airspace.

Controlled Firing Area. A controlled firing area is established to contain activities, which if not conducted in a controlled environment, would be hazardous to nonparticipating aircraft.

Crewmember means a person assigned to perform duty in an aircraft during flight time.

Critical altitude means the maximum altitude at which, in standard atmosphere, it is possible to maintain, at a specified rotational speed, a specified power or a specified manifold pressure.

Unless otherwise stated, the critical altitude is the maximum altitude at which it is possible to maintain, at the maximum continuous rotational speed, one of the following:

(1) The maximum continuous power, in the case of engines for which this power rating is the same at sea level and at the rated altitude.

(2) The maximum continuous rated manifold pressure, in the case of engines, the maximum continuous power of which is governed by a constant manifold pressure.

Critical engine means the engine whose failure would most adversely affect the performance or handling qualities of an aircraft.

Decision altitude (DA) is a specified altitude in an instrument approach procedure at which the pilot must decide whether to initiate an immediate missed approach if the pilot does not see the required visual reference, or to continue the approach. Decision altitude is expressed in feet above mean sea level.

Decision height (DH) is a specified height above the ground in an instrument approach procedure at which the pilot must decide whether to initiate an immediate missed approach if the pilot does not see the required visual reference, or to continue the approach. Decision height is expressed in feet above ground level.

Early ETOPS means ETOPS type design approval obtained without gaining non-ETOPS service experience on the candidate airplane-engine combination certified for ETOPS.

EFVS operation means an operation in which visibility conditions require an EFVS to be used in lieu of natural vision to perform an approach or landing, determine enhanced flight visibility, identify required visual references, or conduct a rollout.

Enhanced flight visibility (EFV) means the average forward horizontal distance, from the cockpit of an aircraft in flight, at which prominent topographical objects may be clearly distinguished and identified by day or night by a pilot using an enhanced flight vision system.

Enhanced flight vision system (EFVS) means an installed aircraft system which uses an electronic means to provide a display of the forward external scene topography (the natural or man-made features of a place or region especially in a way to show their relative positions and elevation) through the use of imaging sensors, including but not limited to forward-looking infrared, millimeter wave radiometry, millimeter wave radar, or low-light level image intensification. An EFVS includes the display element, sensors, computers and power supplies, indications, and controls.

Equivalent airspeed means the calibrated airspeed of an aircraft corrected for adiabatic compressible flow for the particular altitude. Equivalent airspeed is equal to calibrated airspeed in standard atmosphere at sea level.

ETOPS Significant System means an airplane system, including the propulsion system, the failure or malfunctioning of which could adversely affect the safety of an ETOPS flight, or the continued safe flight and landing of an airplane during an ETOPS diversion. Each ETOPS significant system is either an ETOPS group 1 significant system or an ETOPS group 2 significant system.

(1) An ETOPS group 1 Significant System—

(i) Has fail-safe characteristics directly linked to the degree of redundancy provided by the number of engines on the airplane.

(ii) Is a system, the failure or malfunction of which could result in an IFSD, loss of thrust control, or other power loss.

(iii) Contributes significantly to the safety of an ETOPS diversion by providing additional redundancy for any system power source lost as a result of an inoperative engine.

(iv) Is essential for prolonged operation of an airplane at engine inoperative altitudes.

(2) An ETOPS group 2 significant system is an ETOPS significant system that is not an ETOPS group 1 significant system.

Extended Operations (ETOPS) means an airplane flight operation, other than an all-cargo operation in an airplane with more than two engines, during

8

which a portion of the flight is conducted beyond a time threshold identified in part 121 or part 135 of this chapter that is determined using an approved one-engine-inoperative cruise speed under standard atmospheric conditions in still air.

Extended over-water operation means—

(1) With respect to aircraft other than helicopters, an operation over water at a horizontal distance of more than 50 nautical miles from the nearest shoreline; and

(2) With respect to helicopters, an operation over water at a horizontal distance of more than 50 nautical miles from the nearest shoreline and more than 50 nautical miles from an offshore heliport structure.

External load means a load that is carried, or extends, outside of the aircraft fuselage.

External-load attaching means means the structural components used to attach an external load to an aircraft, including external-load containers, the backup structure at the attachment points, and any quick-release device used to jettison the external load.

Final approach fix (FAF) defines the beginning of the final approach segment and the point where final segment descent may begin.

Final takeoff speed means the speed of the airplane that exists at the end of the takeoff path in the en route configuration with one engine inoperative.

Fireproof—

(1) With respect to materials and parts used to confine fire in a designated fire zone, means the capacity to withstand at least as well as steel in dimensions appropriate for the purpose for which they are used, the heat produced when there is a severe fire of extended duration in that zone; and

(2) With respect to other materials and parts, means the capacity to withstand the heat associated with fire at least as well as steel in dimensions appropriate for the purpose for which they are used.

Fire resistant—

(1) With respect to sheet or structural members means the capacity to withstand the heat associated with fire at least as well as aluminum alloy in dimensions appropriate for the purpose for which they are used; and

(2) With respect to fluid-carrying lines, fluid system parts, wiring, air ducts, fittings, and powerplant controls, means the capacity to perform the intended functions under the heat and other conditions likely to occur when there is a fire at the place concerned.

Flame resistant means not susceptible to combustion to the point of propagating a flame, beyond safe limits, after the ignition source is removed.

Flammable, with respect to a fluid or gas, means susceptible to igniting readily or to exploding.

Flap extended speed means the highest speed permissible with wing flaps in a prescribed extended position.

Flash resistant means not susceptible to burning violently when ignited.

Flightcrew member means a pilot, flight engineer, or flight navigator assigned to duty in an aircraft during flight time.

Flight level means a level of constant atmospheric pressure related to a reference datum of 29.92 inches of mercury. Each is stated in three digits that represent hundreds of feet. For example, flight level 250 represents a barometric altimeter indication of 25,000 feet; flight level 255, an indication of 25,500 feet.

Flight plan means specified information, relating to the intended flight of an aircraft, that is filed orally or in writing with air traffic control.

Flight simulation training device (FSTD) means a full flight simulator or a flight training device.

Flight time means:

(1) Pilot time that commences when an aircraft moves under its own power for the purpose of flight and ends when the aircraft comes to rest after landing; or

(2) For a glider without self-launch capability, pilot time that commences when the glider is towed for the purpose of flight and ends when the glider comes to rest after landing.

Flight training device (FTD) means a replica of aircraft instruments, equipment, panels, and controls in an open flight deck area or an enclosed aircraft cockpit replica. It includes the equipment and computer programs necessary to represent aircraft (or set of aircraft) operations in ground and

9

flight conditions having the full range of capabilities of the systems installed in the device as described in part 60 of this chapter and the qualification performance standard (QPS) for a specific FTD qualification level.

Flight visibility means the average forward horizontal distance, from the cockpit of an aircraft in flight, at which prominent unlighted objects may be seen and identified by day and prominent lighted objects may be seen and identified by night.

Foreign air carrier means any person other than a citizen of the United States, who undertakes directly, by lease or other arrangement, to engage in air transportation.

Foreign air commerce means the carriage by aircraft of persons or property for compensation or hire, or the carriage of mail by aircraft, or the operation or navigation of aircraft in the conduct or furtherance of a business or vocation, in commerce between a place in the United States and any place outside thereof; whether such commerce moves wholly by aircraft or partly by aircraft and partly by other forms of transportation.

Foreign air transportation means the carriage by aircraft of persons or property as a common carrier for compensation or hire, or the carriage of mail by aircraft, in commerce between a place in the United States and any place outside of the United States, whether that commerce moves wholly by aircraft or partly by aircraft and partly by other forms of transportation.

Forward wing means a forward lifting surface of a canard configuration or tandem-wing configuration airplane. The surface may be a fixed, movable, or variable geometry surface, with or without control surfaces.

Full flight simulator (FFS) means a replica of a specific type; or make, model, and series aircraft cockpit. It includes the assemblage of equipment and computer programs necessary to represent aircraft operations in ground and flight conditions, a visual system providing an out-of-the-cockpit view, a system that provides cues at least equivalent to those of a three-degree-of-freedom motion system, and has the full range of capabilities of the systems installed in the device as described in part 60 of this chapter and the qualification performance standards (QPS) for a specific FFS qualification level.

Glider means a heavier-than-air aircraft, that is supported in flight by the dynamic reaction of the air against its lifting surfaces and whose free flight does not depend principally on an engine.

Ground visibility means prevailing horizontal visibility near the earth's surface as reported by the United States National Weather Service or an accredited observer.

Go-around power or thrust setting means the maximum allowable inflight power or thrust setting identified in the performance data.

Gyrodyne means a rotorcraft whose rotors are normally engine-driven for takeoff, hovering, and landing, and for forward flight through part of its speed range, and whose means of propulsion, consisting usually of conventional propellers, is independent of the rotor system.

Gyroplane means a rotorcraft whose rotors are not engine-driven, except for initial starting, but are made to rotate by action of the air when the rotorcraft is moving; and whose means of propulsion, consisting usually of conventional propellers, is independent of the rotor system.

Helicopter means a rotorcraft that, for its horizontal motion, depends principally on its engine-driven rotors.

Heliport means an area of land, water, or structure used or intended to be used for the landing and takeoff of helicopters.

Idle thrust means the jet thrust obtained with the engine power control level set at the stop for the least thrust position at which it can be placed.

IFR conditions means weather conditions below the minimum for flight under visual flight rules.

IFR over-the-top, with respect to the operation of aircraft, means the operation of an aircraft over-the-top on an IFR flight plan when cleared by air traffic control to maintain "VFR conditions" or "VFR conditions on top".

Indicated airspeed means the speed of an aircraft as shown on its pitot static airspeed indicator calibrated to reflect standard atmosphere adiabatic compressible flow at sea level uncorrected for airspeed system errors.

In-flight shutdown (IFSD) means, for ETOPS only, when an engine ceases to function (when the airplane is airborne) and is shutdown, whether self induced, flightcrew initiated or caused by an external influence. The FAA considers IFSD for all causes: for example, flameout, internal failure, flightcrew initiated shutdown, foreign object ingestion, icing, inability to obtain or control desired thrust or power, and cycling of the start control, however briefly, even if the engine operates normally for the remainder of the flight. This definition excludes the airborne cessation of the functioning of an engine when immediately followed by an automatic engine relight and when an engine does not achieve desired thrust or power but is not shutdown.

Instrument means a device using an internal mechanism to show visually or aurally the attitude, altitude, or operation of an aircraft or aircraft part. It includes electronic devices for automatically controlling an aircraft in flight.

Instrument approach procedure (IAP) is a series of predetermined maneuvers by reference to flight instruments with specified protection from obstacles and assurance of navigation signal reception capability. It begins from the initial approach fix, or where applicable, from the beginning of a defined arrival route to a point:

(1) From which a landing can be completed; or

(2) If a landing is not completed, to a position at which holding or en route obstacle clearance criteria apply.

Interstate air commerce means the carriage by aircraft of persons or property for compensation or hire, or the carriage of mail by aircraft, or the operation or navigation of aircraft in the conduct or furtherance of a business or vocation, in commerce between a place in any State of the United States, or the District of Columbia, and a place in any other State of the United States, or the District of Columbia; or between places in the same State of the United States through the airspace over any place outside thereof; or between places in the same territory or posses-sion of the United States, or the District of Columbia.

Interstate air transportation means the carriage by aircraft of persons or property as a common carrier for compensation or hire, or the carriage of mail by aircraft in commerce:

(1) Between a place in a State or the District of Columbia and another place in another State or the District of Columbia;

(2) Between places in the same State through the airspace over any place outside that State; or

(3) Between places in the same possession of the United States;

Whether that commerce moves wholly by aircraft of partly by aircraft and partly by other forms of transportation.

Intrastate air transportation means the carriage of persons or property as a common carrier for compensation or hire, by turbojet-powered aircraft capable of carrying thirty or more persons, wholly within the same State of the United States.

Kite means a framework, covered with paper, cloth, metal, or other material, intended to be flown at the end of a rope or cable, and having as its only support the force of the wind moving past its surfaces.

Landing gear extended speed means the maximum speed at which an aircraft can be safely flown with the landing gear extended.

Landing gear operating speed means the maximum speed at which the landing gear can be safely extended or retracted.

Large aircraft means aircraft of more than 12,500 pounds, maximum certificated takeoff weight.

Light-sport aircraft means an aircraft, other than a helicopter or powered-lift that, since its original certification, has continued to meet the following:

(1) A maximum takeoff weight of not more than—

(i) 1,320 pounds (600 kilograms) for aircraft not intended for operation on water; or

(ii) 1,430 pounds (650 kilograms) for an aircraft intended for operation on water.

(2) A maximum airspeed in level flight with maximum continuous power (V_{H}) of not more than 120 knots CAS

under standard atmospheric conditions at sea level.

(3) A maximum never-exceed speed (V_{NE}) of not more than 120 knots CAS for a glider.

(4) A maximum stalling speed or minimum steady flight speed without the use of lift-enhancing devices (V_{S1}) of not more than 45 knots CAS at the aircraft's maximum certificated takeoff weight and most critical center of gravity.

(5) A maximum seating capacity of no more than two persons, including the pilot.

(6) A single, reciprocating engine, if powered.

(7) A fixed or ground-adjustable propeller if a powered aircraft other than a powered glider.

(8) A fixed or feathering propeller system if a powered glider.

(9) A fixed-pitch, semi-rigid, teetering, two-blade rotor system, if a gyroplane.

(10) A nonpressurized cabin, if equipped with a cabin.

(11) Fixed landing gear, except for an aircraft intended for operation on water or a glider.

(12) Fixed or retractable landing gear, or a hull, for an aircraft intended for operation on water.

(13) Fixed or retractable landing gear for a glider.

Lighter-than-air aircraft means aircraft that can rise and remain suspended by using contained gas weighing less than the air that is displaced by the gas.

Load factor means the ratio of a specified load to the total weight of the aircraft. The specified load is expressed in terms of any of the following: aerodynamic forces, inertia forces, or ground or water reactions.

Long-range communication system (LRCS). A system that uses satellite relay, data link, high frequency, or another approved communication system which extends beyond line of sight.

Long-range navigation system (LRNS). An electronic navigation unit that is approved for use under instrument flight rules as a primary means of navigation, and has at least one source of navigational input, such as inertial navigation system or global positioning system.

Mach number means the ratio of true airspeed to the speed of sound.

Main rotor means the rotor that supplies the principal lift to a rotorcraft.

Maintenance means inspection, overhaul, repair, preservation, and the replacement of parts, but excludes preventive maintenance.

Major alteration means an alteration not listed in the aircraft, aircraft engine, or propeller specifications—

(1) That might appreciably affect weight, balance, structural strength, performance, powerplant operation, flight characteristics, or other qualities affecting airworthiness; or

(2) That is not done according to accepted practices or cannot be done by elementary operations.

Major repair means a repair:

(1) That, if improperly done, might appreciably affect weight, balance, structural strength, performance, powerplant operation, flight characteristics, or other qualities affecting airworthiness; or

(2) That is not done according to accepted practices or cannot be done by elementary operations.

Manifold pressure means absolute pressure as measured at the appropriate point in the induction system and usually expressed in inches of mercury.

Maximum engine overtorque, as it applies to turbopropeller and turboshaft engines incorporating free power turbines for all ratings except one engine inoperative (OEI) ratings of two minutes or less, means the maximum torque of the free power turbine rotor assembly, the inadvertent occurrence of which, for periods of up to 20 seconds, will not require rejection of the engine from service, or any maintenance action other than to correct the cause.

Maximum speed for stability characteristics, V_{FC}/M_{FC} means a speed that may not be less than a speed midway between maximum operating limit speed (V_{MO}/M_{MO}) and demonstrated flight diving speed (V_{DF}/M_{DF}), except that, for altitudes where the Mach number is the limiting factor, M_{FC} need not exceed the Mach number at which effective speed warning occurs.

Medical certificate means acceptable evidence of physical fitness on a form prescribed by the Administrator.

Military operations area. A military operations area (MOA) is airspace established outside Class A airspace to separate or segregate certain nonhazardous military activities from IFR Traffic and to identify for VFR traffic where theses activities are conducted.

Minimum descent altitude (MDA) is the lowest altitude specified in an instrument approach procedure, expressed in feet above mean sea level, to which descent is authorized on final approach or during circle-to-land maneuvering until the pilot sees the required visual references for the heliport or runway of intended landing.

Minor alteration means an alteration other than a major alteration.

Minor repair means a repair other than a major repair.

National defense airspace means airspace established by a regulation prescribed, or an order issued under, 49 U.S.C. 40103(b)(3).

Navigable airspace means airspace at and above the minimum flight altitudes prescribed by or under this chapter, including airspace needed for safe takeoff and landing.

Night means the time between the end of evening civil twilight and the beginning of morning civil twilight, as published in the Air Almanac, converted to local time.

Nonprecision approach procedure means a standard instrument approach procedure in which no electronic glide slope is provided.

Operate, with respect to aircraft, means use, cause to use or authorize to use aircraft, for the purpose (except as provided in §91.13 of this chapter) of air navigation including the piloting of aircraft, with or without the right of legal control (as owner, lessee, or otherwise).

Operational control, with respect to a flight, means the exercise of authority over initiating, conducting or terminating a flight.

Overseas air commerce means the carriage by aircraft of persons or property for compensation or hire, or the carriage of mail by aircraft, or the operation or navigation of aircraft in the conduct or furtherance of a business or vocation, in commerce between a place in any State of the United States, or the District of Columbia, and any place in a territory or possession of the United States; or between a place in a territory or possession of the United States, and a place in any other territory or possession of the United States.

Overseas air transportation means the carriage by aircraft of persons or property as a common carrier for compensation or hire, or the carriage of mail by aircraft, in commerce:

(1) Between a place in a State or the District of Columbia and a place in a possession of the United States; or

(2) Between a place in a possession of the United States and a place in another possession of the United States; whether that commerce moves wholly by aircraft or partly by aircraft and partly by other forms of transportation.

Over-the-top means above the layer of clouds or other obscuring phenomena forming the ceiling.

Parachute means a device used or intended to be used to retard the fall of a body or object through the air.

Person means an individual, firm, partnership, corporation, company, association, joint-stock association, or governmental entity. It includes a trustee, receiver, assignee, or similar representative of any of them.

Pilotage means navigation by visual reference to landmarks.

Pilot in command means the person who:

(1) Has final authority and responsibility for the operation and safety of the flight;

(2) Has been designated as pilot in command before or during the flight; and

(3) Holds the appropriate category, class, and type rating, if appropriate, for the conduct of the flight.

Pitch setting means the propeller blade setting as determined by the blade angle measured in a manner, and at a radius, specified by the instruction manual for the propeller.

Portable oxygen concentrator means a medical device that separates oxygen from other gasses in ambient air and dispenses this concentrated oxygen to the user.

13

Positive control means control of all air traffic, within designated airspace, by air traffic control.

Powered parachute means a powered aircraft comprised of a flexible or semi-rigid wing connected to a fuselage so that the wing is not in position for flight until the aircraft is in motion. The fuselage of a powered parachute contains the aircraft engine, a seat for each occupant and is attached to the aircraft's landing gear.

Powered-lift means a heavier-than-air aircraft capable of vertical takeoff, vertical landing, and low speed flight that depends principally on engine-driven lift devices or engine thrust for lift during these flight regimes and on nonrotating airfoil(s) for lift during horizontal flight.

Precision approach procedure means a standard instrument approach procedure in which an electronic glide slope is provided, such as ILS and PAR.

Preventive maintenance means simple or minor preservation operations and the replacement of small standard parts not involving complex assembly operations.

Prohibited area. A prohibited area is airspace designated under part 73 within which no person may operate an aircraft without the permission of the using agency.

Propeller means a device for propelling an aircraft that has blades on an engine-driven shaft and that, when rotated, produces by its action on the air, a thrust approximately perpendicular to its plane of rotation. It includes control components normally supplied by its manufacturer, but does not include main and auxiliary rotors or rotating airfoils of engines.

Public aircraft means any of the following aircraft when not being used for a commercial purpose or to carry an individual other than a crewmember or qualified non-crewmenber:

(1) An aircraft used only for the United States Government; an aircraft owned by the Government and operated by any person for purposes related to crew training, equipment development, or demonstration; an aircraft owned and operated by the government of a State, the District of Columbia, or a territory or possession of the United States or a political subdivision of one of these governments; or an aircraft exclusively leased for at least 90 continuous days by the government of a State, the District of Columbia, or a territory or possession of the United States or a political subdivision of one of these governments.

(i) For the sole purpose of determining public aircraft status, *commercial purposes* means the transportation of persons or property for compensation or hire, but does not include the operation of an aircraft by the armed forces for reimbursement when that reimbursement is required by any Federal statute, regulation, or directive, in effect on November 1, 1999, or by one government on behalf of another government under a cost reimbursement agreement if the government on whose behalf the operation is conducted certifies to the Administrator of the Federal Aviation Administration that the operation is necessary to respond to a significant and imminent threat to life or property (including natural resources) and that no service by a private operator is reasonably available to meet the threat.

(ii) For the sole purpose of determining public aircraft status, *governmental function* means an activity undertaken by a government, such as national defense, intelligence missions, firefighting, search and rescue, law enforcement (including transport of prisoners, detainees, and illegal aliens), aeronautical research, or biological or geological resource management.

(iii) For the sole purpose of determining public aircraft status, *qualified non-crewmember* means an individual, other than a member of the crew, aboard an aircraft operated by the armed forces or an intelligence agency of the United States Government, or whose presence is required to perform, or is associated with the performance of, a governmental function.

(2) An aircraft owned or operated by the armed forces or chartered to provide transportation to the armed forces if—

(i) The aircraft is operated in accordance with title 10 of the United States Code;

(ii) The aircraft is operated in the performance of a governmental function under title 14, 31, 32, or 50 of the

United States Code and the aircraft is not used for commercial purposes; or

(iii) The aircraft is chartered to provide transportation to the armed forces and the Secretary of Defense (or the Secretary of the department in which the Coast Guard is operating) designates the operation of the aircraft as being required in the national interest.

(3) An aircraft owned or operated by the National Guard of a State, the District of Columbia, or any territory or possession of the United States, and that meets the criteria of paragraph (2) of this definition, qualifies as a public aircraft only to the extent that it is operated under the direct control of the Department of Defense.

Rated 30-second OEI Power, with respect to rotorcraft turbine engines, means the approved brake horsepower developed under static conditions at specified altitudes and temperatures within the operating limitations established for the engine under part 33 of this chapter, for continuation of one flight operation after the failure or shutdown of one engine in multiengine rotorcraft, for up to three periods of use no longer than 30 seconds each in any one flight, and followed by mandatory inspection and prescribed maintenance action.

Rated 2-minute OEI Power, with respect to rotorcraft turbine engines, means the approved brake horsepower developed under static conditions at specified altitudes and temperatures within the operating limitations established for the engine under part 33 of this chapter, for continuation of one flight operation after the failure or shutdown of one engine in multiengine rotorcraft, for up to three periods of use no longer than 2 minutes each in any one flight, and followed by mandatory inspection and prescribed maintenance action.

Rated continuous OEI power, with respect to rotorcraft turbine engines, means the approved brake horsepower developed under static conditions at specified altitudes and temperatures within the operating limitations established for the engine under part 33 of this chapter, and limited in use to the time required to complete the flight after the failure or shutdown of one engine of a multiengine rotorcraft.

Rated maximum continuous augmented thrust, with respect to turbojet engine type certification, means the approved jet thrust that is developed statically or in flight, in standard atmosphere at a specified altitude, with fluid injection or with the burning of fuel in a separate combustion chamber, within the engine operating limitations established under Part 33 of this chapter, and approved for unrestricted periods of use.

Rated maximum continuous power, with respect to reciprocating, turbopropeller, and turboshaft engines, means the approved brake horsepower that is developed statically or in flight, in standard atmosphere at a specified altitude, within the engine operating limitations established under part 33, and approved for unrestricted periods of use.

Rated maximum continuous thrust, with respect to turbojet engine type certification, means the approved jet thrust that is developed statically or in flight, in standard atmosphere at a specified altitude, without fluid injection and without the burning of fuel in a separate combustion chamber, within the engine operating limitations established under part 33 of this chapter, and approved for unrestricted periods of use.

Rated takeoff augmented thrust, with respect to turbojet engine type certification, means the approved jet thrust that is developed statically under standard sea level conditions, with fluid injection or with the burning of fuel in a separate combustion chamber, within the engine operating limitations established under part 33 of this chapter, and limited in use to periods of not over 5 minutes for takeoff operation.

Rated takeoff power, with respect to reciprocating, turbopropeller, and turboshaft engine type certification, means the approved brake horsepower that is developed statically under standard sea level conditions, within the engine operating limitations established under part 33, and limited in use to periods of not over 5 minutes for takeoff operation.

Rated takeoff thrust, with respect to turbojet engine type certification, means the approved jet thrust that is

15

developed statically under standard sea level conditions, without fluid injection and without the burning of fuel in a separate combustion chamber, within the engine operating limitations established under part 33 of this chapter, and limited in use to periods of not over 5 minutes for takeoff operation.

Rated 30-minute OEI power, with respect to rotorcraft turbine engines, means the approved brake horsepower developed under static conditions at specified altitudes and temperatures within the operating limitations established for the engine under part 33 of this chapter, and limited in use to one period of use no longer than 30 minutes after the failure or shutdown of one engine of a multiengine rotorcraft.

Rated 2½-minute OEI power, with respect to rotorcraft turbine engines, means the approved brake horsepower developed under static conditions at specified altitudes and temperatures within the operating limitations established for the engine under part 33 of this chapter for periods of use no longer than 2½ minutes each after the failure or shutdown of one engine of a multiengine rotorcraft.

Rating means a statement that, as a part of a certificate, sets forth special conditions, privileges, or limitations.

Reference landing speed means the speed of the airplane, in a specified landing configuration, at the point where it descends through the 50 foot height in the determination of the landing distance.

Reporting point means a geographical location in relation to which the position of an aircraft is reported.

Restricted area. A restricted area is airspace designated under Part 73 within which the flight of aircraft, while not wholly prohibited, is subject to restriction.

Rocket means an aircraft propelled by ejected expanding gases generated in the engine from self-contained propellants and not dependent on the intake of outside substances. It includes any part which becomes separated during the operation.

Rotorcraft means a heavier-than-air aircraft that depends principally for its support in flight on the lift generated by one or more rotors.

Rotorcraft-load combination means the combination of a rotorcraft and an external-load, including the external-load attaching means. Rotorcraft-load combinations are designated as Class A, Class B, Class C, and Class D, as follows:

(1) *Class A rotorcraft-load combination* means one in which the external load cannot move freely, cannot be jettisoned, and does not extend below the landing gear.

(2) *Class B rotorcraft-load combination* means one in which the external load is jettisonable and is lifted free of land or water during the rotorcraft operation.

(3) *Class C rotorcraft-load combination* means one in which the external load is jettisonable and remains in contact with land or water during the rotorcraft operation.

(4) *Class D rotorcraft-load combination* means one in which the external-load is other than a Class A, B, or C and has been specifically approved by the Administrator for that operation.

Route segment is a portion of a route bounded on each end by a fix or navigation aid (NAVAID).

Sea level engine means a reciprocating aircraft engine having a rated takeoff power that is producible only at sea level.

Second in command means a pilot who is designated to be second in command of an aircraft during flight time.

Show, unless the context otherwise requires, means to show to the satisfaction of the Administrator.

Small aircraft means aircraft of 12,500 pounds or less, maximum certificated takeoff weight.

Small unmanned aircraft means an unmanned aircraft weighing less than 55 pounds on takeoff, including everything that is on board or otherwise attached to the aircraft.

Small unmanned aircraft system (small UAS) means a small unmanned aircraft and its associated elements (including communication links and the components that control the small unmanned aircraft) that are required for the safe and efficient operation of the small unmanned aircraft in the national airspace system.

Special VFR conditions mean meteorological conditions that are less than those required for basic VFR flight in controlled airspace and in which some aircraft are permitted flight under visual flight rules.

Special VFR operations means aircraft operating in accordance with clearances within controlled airspace in meteorological conditions less than the basic VFR weather minima. Such operations must be requested by the pilot and approved by ATC.

Standard atmosphere means the atmosphere defined in U.S. Standard Atmosphere, 1962 (Geopotential altitude tables).

Stopway means an area beyond the takeoff runway, no less wide than the runway and centered upon the extended centerline of the runway, able to support the airplane during an aborted takeoff, without causing structural damage to the airplane, and designated by the airport authorities for use in decelerating the airplane during an aborted takeoff.

Suitable RNAV system is an RNAV system that meets the required performance established for a type of operation, e.g. IFR; and is suitable for operation over the route to be flown in terms of any performance criteria (including accuracy) established by the air navigation service provider for certain routes (e.g. oceanic, ATS routes, and IAPs). An RNAV system's suitability is dependent upon the availability of ground and/or satellite navigation aids that are needed to meet any route performance criteria that may be prescribed in route specifications to navigate the aircraft along the route to be flown. Information on suitable RNAV systems is published in FAA guidance material.

Synthetic vision means a computer-generated image of the external scene topography from the perspective of the flight deck that is derived from aircraft attitude, high-precision navigation solution, and database of terrain, obstacles and relevant cultural features.

Synthetic vision system means an electronic means to display a synthetic vision image of the external scene topography to the flight crew.

Takeoff power:

(1) With respect to reciprocating engines, means the brake horsepower that is developed under standard sea level conditions, and under the maximum conditions of crankshaft rotational speed and engine manifold pressure approved for the normal takeoff, and limited in continuous use to the period of time shown in the approved engine specification; and

(2) With respect to turbine engines, means the brake horsepower that is developed under static conditions at a specified altitude and atmospheric temperature, and under the maximum conditions of rotor shaft rotational speed and gas temperature approved for the normal takeoff, and limited in continuous use to the period of time shown in the approved engine specification.

Takeoff safety speed means a referenced airspeed obtained after lift-off at which the required one-engine-inoperative climb performance can be achieved.

Takeoff thrust, with respect to turbine engines, means the jet thrust that is developed under static conditions at a specific altitude and atmospheric temperature under the maximum conditions of rotorshaft rotational speed and gas temperature approved for the normal takeoff, and limited in continuous use to the period of time shown in the approved engine specification.

Tandem wing configuration means a configuration having two wings of similar span, mounted in tandem.

TCAS I means a TCAS that utilizes interrogations of, and replies from, airborne radar beacon transponders and provides traffic advisories to the pilot.

TCAS II means a TCAS that utilizes interrogations of, and replies from, airborne radar beacon transponders and provides traffic advisories and resolution advisories in the vertical plane.

TCAS III means a TCAS that utilizes interrogation of, and replies from, airborne radar beacon transponders and provides traffic advisories and resolution advisories in the vertical and horizontal planes to the pilot.

Time in service, with respect to maintenance time records, means the time from the moment an aircraft leaves the surface of the earth until it touches it at the next point of landing.

Traffic pattern means the traffic flow that is prescribed for aircraft landing at, taxiing on, or taking off from, an airport.

True airspeed means the airspeed of an aircraft relative to undisturbed air. True airspeed is equal to equivalent airspeed multiplied by $(\rho 0/\rho)^{1/2}$.

Type:

(1) As used with respect to the certification, ratings, privileges, and limitations of airmen, means a specific make and basic model of aircraft, including modifications thereto that do not change its handling or flight characteristics. Examples include: DC–7, 1049, and F–27; and

(2) As used with respect to the certification of aircraft, means those aircraft which are similar in design. Examples include: DC–7 and DC–7C; 1049G and 1049H; and F–27 and F–27F.

(3) As used with respect to the certification of aircraft engines means those engines which are similar in design. For example, JT8D and JT8D–7 are engines of the same type, and JT9D–3A and JT9D–7 are engines of the same type.

United States, in a geographical sense, means (1) the States, the District of Columbia, Puerto Rico, and the possessions, including the territorial waters, and (2) the airspace of those areas.

United States air carrier means a citizen of the United States who undertakes directly by lease, or other arrangement, to engage in air transportation.

Unmanned aircraft means an aircraft operated without the possibility of direct human intervention from within or on the aircraft.

Unmanned aircraft system means an unmanned aircraft and its associated elements (including communication links and the components that control the unmanned aircraft) that are required for the safe and efficient operation of the unmanned aircraft in the airspace of the United States.

VFR over-the-top, with respect to the operation of aircraft, means the operation of an aircraft over-the-top under VFR when it is not being operated on an IFR flight plan.

Warning area. A warning area is airspace of defined dimensions, extending from 3 nautical miles outward from the coast of the United States, that contains activity that may be hazardous to nonparticipating aircraft. The purpose of such warning areas is to warn nonparticipating pilots of the potential danger. A warning area may be located over domestic or international waters or both.

Weight-shift-control aircraft means a powered aircraft with a framed pivoting wing and a fuselage controllable only in pitch and roll by the pilot's ability to change the aircraft's center of gravity with respect to the wing. Flight control of the aircraft depends on the wing's ability to flexibly deform rather than the use of control surfaces.

Winglet or tip fin means an out-of-plane surface extending from a lifting surface. The surface may or may not have control surfaces.

[Doc. No. 1150, 27 FR 4588, May 15, 1962]

EDITORIAL NOTE: For FEDERAL REGISTER citations affecting §1.1, see the List of CFR Sections Affected, which appears in the Finding Aids section of the printed volume and at *www.govinfo.gov*.

§ 1.2   Abbreviations and symbols.

In Subchapters A through K of this chapter:

AFM means airplane flight manual.

AGL means above ground level.

ALS means approach light system.

APU means auxiliary power unit.

ASR means airport surveillance radar.

ATC means air traffic control.

ATS means Air Traffic Service.

CAMP means continuous airworthiness maintenance program.

CAS means calibrated airspeed.

CAT II means Category II.

CMP means configuration, maintenance, and procedures.

DH means decision height.

DME means distance measuring equipment compatible with TACAN.

EAS means equivalent airspeed.

EFVS means enhanced flight vision system.

Equi-Time Point means a point on the route of flight where the flight time, considering wind, to each of two selected airports is equal.

ETOPS means extended operations.

EWIS, as defined by §25.1701 of this chapter, means electrical wiring interconnection system.

FAA means Federal Aviation Administration.

FFS means full flight simulator.

FM means fan marker.

FSTD means flight simulation training device.

FTD means flight training device.

GS means glide slope.

HIRL means high-intensity runway light system.

IAS means indicated airspeed.

ICAO means International Civil Aviation Organization.

IFR means instrument flight rules.

IFSD means in-flight shutdown.

ILS means instrument landing system.

IM means ILS inner marker.

INT means intersection.

LDA means localizer-type directional aid.

LFR means low-frequency radio range.

LMM means compass locator at middle marker.

LOC means ILS localizer.

LOM means compass locator at outer marker.

M means mach number.

MAA means maximum authorized IFR altitude.

MALS means medium intensity approach light system.

MALSR means medium intensity approach light system with runway alignment indicator lights.

MCA means minimum crossing altitude.

MDA means minimum descent altitude.

MEA means minimum en route IFR altitude.

MEL means minimum equipment list.

MM means ILS middle marker.

MOCA means minimum obstruction clearance altitude.

MRA means minimum reception altitude.

MSL means mean sea level.

NDB (ADF) means nondirectional beacon (automatic direction finder).

NM means nautical mile.

NOPAC means North Pacific area of operation.

NOPT means no procedure turn required.

OEI means one engine inoperative.

OM means ILS outer marker.

OPSPECS means operations specifications.

PACOTS means Pacific Organized Track System.

PAR means precision approach radar.

PMA means parts manufacturer approval.

POC means portable oxygen concentrator.

PTRS means Performance Tracking and Reporting System.

RAIL means runway alignment indicator light system.

RBN means radio beacon.

RCLM means runway centerline marking.

RCLS means runway centerline light system.

REIL means runway end identification lights.

RFFS means rescue and firefighting services.

RNAV means area navigation.

RR means low or medium frequency radio range station.

RVR means runway visual range as measured in the touchdown zone area.

SALS means short approach light system.

SATCOM means satellite communications.

SSALS means simplified short approach light system.

SSALSR means simplified short approach light system with runway alignment indicator lights.

TACAN means ultra-high frequency tactical air navigational aid.

TAS means true airspeed.

TCAS means a traffic alert and collision avoidance system.

TDZL means touchdown zone lights.

TSO means technical standard order.

TVOR means very high frequency terminal omnirange station.

V_A means design maneuvering speed.

V_B means design speed for maximum gust intensity.

V_C means design cruising speed.

V_D means design diving speed.

V_{DF}/M_{DF} means demonstrated flight diving speed.

V_{EF} means the speed at which the critical engine is assumed to fail during takeoff.

V_F means design flap speed.

V_{FC}/M_{FC} means maximum speed for stability characteristics.

V_{FE} means maximum flap extended speed.

V_{FTO} means final takeoff speed.

V_H means maximum speed in level flight with maximum continuous power.

V_{LE} means maximum landing gear extended speed.

V_{LO} means maximum landing gear operating speed.

V_{LOF} means lift-off speed.

V_{MC} means minimum control speed with the critical engine inoperative.

V_{MO}/M_{MO} means maximum operating limit speed.

V_{MU} means minimum unstick speed.

V_{NE} means never-exceed speed.

V_{NO} means maximum structural cruising speed.

V_R means rotation speed.

V_{REF} means reference landing speed.

V_S means the stalling speed or the minimum steady flight speed at which the airplane is controllable.

V_{S0} means the stalling speed or the minimum steady flight speed in the landing configuration.

V_{S1} means the stalling speed or the minimum steady flight speed obtained in a specific configuration.

V_{SR} means reference stall speed.

V_{SR0} means reference stall speed in the landing configuration.

V_{SR1} means reference stall speed in a specific configuration.

V_{SW} means speed at which onset of natural or artificial stall warning occurs.

V_{TOSS} means takeoff safety speed for Category A rotorcraft.

V_X means speed for best angle of climb.

V_Y means speed for best rate of climb.

V_1 means the maximum speed in the takeoff at which the pilot must take the first action (e.g., apply brakes, reduce thrust, deploy speed brakes) to stop the airplane within the accelerate-stop distance. V_1 also means the minimum speed in the takeoff, following a failure of the critical engine at V_{EF}, at which the pilot can continue the takeoff and achieve the required height above the takeoff surface within the takeoff distance.

V_2 means takeoff safety speed.

V_{2min} means minimum takeoff safety speed.

VFR means visual flight rules.

$VGSI$ means visual glide slope indicator.

VHF means very high frequency.

VOR means very high frequency omnirange station.

$VORTAC$ means collocated VOR and TACAN.

[Doc. No. 1150, 27 FR 4590, May 15, 1962]

EDITORIAL NOTE: For FEDERAL REGISTER citations affecting §1.2, see the List of CFR Sections Affected, which appears in the Finding Aids section of the printed volume and at *www.govinfo.gov*.

§ 1.3 Rules of construction.

(a) In Subchapters A through K of this chapter, unless the context requires otherwise:

(1) Words importing the singular include the plural;

(2) Words importing the plural include the singular; and

(3) Words importing the masculine gender include the feminine.

(b) In Subchapters A through K of this chapter, the word:

(1) *Shall* is used in an imperative sense;

(2) *May* is used in a permissive sense to state authority or permission to do the act prescribed, and the words "no person may * * *" or "a person may not * * *" mean that no person is required, authorized, or permitted to do the act prescribed; and

(3) *Includes* means "includes but is not limited to".

[Doc. No. 1150, 27 FR 4590, May 15, 1962, as amended by Amdt. 1–10, 31 FR 5055, Mar. 29, 1966]

PART 3—GENERAL REQUIREMENTS

Subpart A—General Requirements Concerning Type Certificated Products or Products, Parts, Appliances, or Materials That May Be Used on Type-Certificated Products

Subpart B—Security Threat Disqualification

by individuals for a certificate or any part of a certificate.

AUTHORITY: 49 U.S.C. 106(g), 40113, 44701, 44704, and 46111.

SOURCE: 70 FR 54832, Sept. 16, 2005, unless otherwise noted.

Subpart A—General Requirements Concerning Type Certificated Products or Products, Parts, Appliances, or Materials That May Be Used on Type-Certificated Products

§3.1 Applicability.

(a) This part applies to any person who makes a record regarding:

(1) A type-certificated product, or

(2) A product, part, appliance or material that may be used on a type-certificated product.

(b) Section 3.5(b) does not apply to records made under part 43 of this chapter.

§3.5 Statements about products, parts, appliances and materials.

(a) *Definitions.* The following terms will have the stated meanings when used in this section:

Airworthy means the aircraft conforms to its type design and is in a condition for safe operation.

Product means an aircraft, aircraft engine, or aircraft propeller.

Record means any writing, drawing, map, recording, tape, film, photograph or other documentary material by which information is preserved or conveyed in any format, including, but not limited to, paper, microfilm, identification plates, stamped marks, bar codes or electronic format, and can either be separate from, attached to or inscribed on any product, part, appliance or material.

(b) *Prohibition against fraudulent and intentionally false statements.* When conveying information related to an advertisement or sales transaction, no person may make or cause to be made:

(1) Any fraudulent or intentionally false statement in any record about the airworthiness of a type-certificated product, or the acceptability of any product, part, appliance, or material for installation on a type-certificated product.

(2) Any fraudulent or intentionally false reproduction or alteration of any record about the airworthiness of any type-certificated product, or the acceptability of any product, part, appliance, or material for installation on a type-certificated product.

(c) *Prohibition against intentionally misleading statements.* (1) When conveying information related to an advertisement or sales transaction, no person may make, or cause to be made, a material representation that a type-certificated product is airworthy, or that a product, part, appliance, or material is acceptable for installation on a type-certificated product in any record if that representation is likely to mislead a consumer acting reasonably under the circumstances.

(2) When conveying information related to an advertisement or sales transaction, no person may make, or cause to be made, through the omission of material information, a representation that a type-certificated product is airworthy, or that a product, part, appliance, or material is acceptable for installation on a type-certificated product in any record if that representation is likely to mislead a consumer acting reasonably under the circumstances.

(d) The provisions of §3.5(b) and §3.5(c) shall not apply if a person can show that the product is airworthy or that the product, part, appliance or material is acceptable for installation on a type-certificated product.

Subpart B—Security Threat Disqualification

SOURCE: 84 FR 42803, Aug. 19, 2019, FAA–2018–0656; Amendment No. 3–2, unless otherwise noted.

§3.200 Effect of Transportation Security Administration notification on a certificate or any part of a certificate held by an individual.

When the Transportation Security Administration (TSA) notifies the FAA that an individual holding a certificate or part of a certificate issued by the FAA poses, or is suspected of posing, a risk of air piracy or terrorism or a threat to airline or passenger safety, the FAA will issue an order amending, modifying, suspending, or revoking any certificate or part of a certificate issued by the FAA.

§ 3.205 **Effect of Transportation Security Administration notification on applications by individuals for a certificate or any part of a certificate.**

(a) When the TSA notifies the FAA that an individual who has applied for a certificate or any part of a certificate issued by the FAA poses, or is suspected of posing, a risk of air piracy or terrorism or a threat to airline or passenger safety, the FAA will hold the individual's certificate applications in abeyance pending further notification from the TSA.

(b) When the TSA notifies the FAA that the TSA has made a final security threat determination regarding an individual, the FAA will deny all the individual's certificate applications. Alternatively, if the TSA notifies the FAA that it has withdrawn its security threat determination, the FAA will continue processing the individual's applications.

PART 5—SAFETY MANAGEMENT SYSTEMS

Subpart A—General

Sec.
5.1   Applicability.
5.3   General requirements.
5.5   Definitions.

Subpart B—Safety Policy

5.21   Safety policy.
5.23   Safety accountability and authority.
5.25   Designation and responsibilities of required safety management personnel.
5.27   Coordination of emergency response planning.

Subpart C—Safety Risk Management

5.51   Applicability.
5.53   System analysis and hazard identification.
5.55   Safety risk assessment and control.

Subpart D—Safety Assurance

5.71   Safety performance monitoring and measurement.
5.73   Safety performance assessment.
5.75   Continuous improvement.

Subpart E—Safety Promotion

5.91   Competencies and training.
5.93   Safety communication.

Subpart F—SMS Documentation and Recordkeeping

5.95   SMS documentation.
5.97   SMS records.

AUTHORITY: Pub. L. 111–216, sec. 215 (Aug. 1, 2010); 49 U.S.C. 106(f), 106(g), 40101, 40113, 40119, 41706, 44101, 44701–44702, 44705, 44709–44711, 44713, 44716–44717, 44722, 46105.

SOURCE: 80 FR 1326, Jan. 8, 2015, unless otherwise noted.

Subpart A—General

§ 5.1 **Applicability.**

(a) A certificate holder under part 119 of this chapter authorized to conduct operations in accordance with the requirements of part 121 of this chapter must have a Safety Management System that meets the requirements of this part and is acceptable to the Administrator by March 9, 2018.

(b) A certificate holder must submit an implementation plan to the FAA Administrator for review no later than September 9, 2015. The implementation plan must be approved no later than March 9, 2016.

(c) The implementation plan may include any of the certificate holder's existing programs, policies, or procedures that it intends to use to meet the requirements of this part, including components of an existing SMS.

[80 FR 1326, Jan. 8, 2015, as amended at 80 FR 1584, Jan. 13, 2015]

§ 5.3 **General requirements.**

(a) Any certificate holder required to have a Safety Management System under this part must submit the Safety Management System to the Administrator for acceptance. The SMS must be appropriate to the size, scope, and complexity of the certificate holder's operation and include at least the following components:

(1) Safety policy in accordance with the requirements of subpart B of this part;

(2) Safety risk management in accordance with the requirements of subpart C of this part;

(3) Safety assurance in accordance with the requirements of subpart D of this part; and

(4) Safety promotion in accordance with the requirements of subpart E of this part.

(b) The Safety Management System must be maintained in accordance with the recordkeeping requirements in subpart F of this part.

(c) The Safety Management System must ensure compliance with the relevant regulatory standards in chapter I of Title 14 of the Code of Federal Regulations.

§ 5.5 Definitions.

Hazard means a condition that could foreseeably cause or contribute to an aircraft accident as defined in 49 CFR 830.2.

Risk means the composite of predicted severity and likelihood of the potential effect of a hazard.

Risk control means a means to reduce or eliminate the effects of hazards.

Safety assurance means processes within the SMS that function systematically to ensure the performance and effectiveness of safety risk controls and that the organization meets or exceeds its safety objectives through the collection, analysis, and assessment of information.

Safety Management System (SMS) means the formal, top-down, organization-wide approach to managing safety risk and assuring the effectiveness of safety risk controls. It includes systematic procedures, practices, and policies for the management of safety risk.

Safety objective means a measurable goal or desirable outcome related to safety.

Safety performance means realized or actual safety accomplishment relative to the organization's safety objectives.

Safety policy means the certificate holder's documented commitment to safety, which defines its safety objectives and the accountabilities and responsibilities of its employees in regards to safety.

Safety promotion means a combination of training and communication of safety information to support the implementation and operation of an SMS in an organization.

Safety Risk Management means a process within the SMS composed of describing the system, identifying the hazards, and analyzing, assessing and controlling risk.

Subpart B—Safety Policy

§ 5.21 Safety policy.

(a) The certificate holder must have a safety policy that includes at least the following:

(1) The safety objectives of the certificate holder.

(2) A commitment of the certificate holder to fulfill the organization's safety objectives.

(3) A clear statement about the provision of the necessary resources for the implementation of the SMS.

(4) A safety reporting policy that defines requirements for employee reporting of safety hazards or issues.

(5) A policy that defines unacceptable behavior and conditions for disciplinary action.

(6) An emergency response plan that provides for the safe transition from normal to emergency operations in accordance with the requirements of § 5.27.

(b) The safety policy must be signed by the accountable executive described in § 5.25.

(c) The safety policy must be documented and communicated throughout the certificate holder's organization.

(d) The safety policy must be regularly reviewed by the accountable executive to ensure it remains relevant and appropriate to the certificate holder.

§ 5.23 Safety accountability and authority.

(a) The certificate holder must define accountability for safety within the organization's safety policy for the following individuals:

(1) Accountable executive, as described in § 5.25.

(2) All members of management in regard to developing, implementing, and maintaining SMS processes within their area of responsibility, including, but not limited to:

(i) Hazard identification and safety risk assessment.

(ii) Assuring the effectiveness of safety risk controls.

(iii) Promoting safety as required in subpart E of this part.

(iv) Advising the accountable executive on the performance of the SMS and on any need for improvement.

(3) Employees relative to the certificate holder's safety performance.

(b) The certificate holder must identify the levels of management with the authority to make decisions regarding safety risk acceptance.

§ 5.25 **Designation and responsibilities of required safety management personnel.**

(a) *Designation of the accountable executive.* The certificate holder must identify an accountable executive who, irrespective of other functions, satisfies the following:

(1) Is the final authority over operations authorized to be conducted under the certificate holder's certificate(s).

(2) Controls the financial resources required for the operations to be conducted under the certificate holder's certificate(s).

(3) Controls the human resources required for the operations authorized to be conducted under the certificate holder's certificate(s).

(4) Retains ultimate responsibility for the safety performance of the operations conducted under the certificate holder's certificate.

(b) *Responsibilities of the accountable executive.* The accountable executive must accomplish the following:

(1) Ensure that the SMS is properly implemented and performing in all areas of the certificate holder's organization.

(2) Develop and sign the safety policy of the certificate holder.

(3) Communicate the safety policy throughout the certificate holder's organization.

(4) Regularly review the certificate holder's safety policy to ensure it remains relevant and appropriate to the certificate holder.

(5) Regularly review the safety performance of the certificate holder's organization and direct actions necessary to address substandard safety performance in accordance with § 5.75.

(c) *Designation of management personnel.* The accountable executive must designate sufficient management personnel who, on behalf of the accountable executive, are responsible for the following:

(1) Coordinate implementation, maintenance, and integration of the SMS throughout the certificate holder's organization.

(2) Facilitate hazard identification and safety risk analysis.

(3) Monitor the effectiveness of safety risk controls.

(4) Ensure safety promotion throughout the certificate holder's organization as required in subpart E of this part.

(5) Regularly report to the accountable executive on the performance of the SMS and on any need for improvement.

§ 5.27 **Coordination of emergency response planning.**

Where emergency response procedures are necessary, the certificate holder must develop and the accountable executive must approve as part of the safety policy, an emergency response plan that addresses at least the following:

(a) Delegation of emergency authority throughout the certificate holder's organization;

(b) Assignment of employee responsibilities during the emergency; and

(c) Coordination of the certificate holder's emergency response plans with the emergency response plans of other organizations it must interface with during the provision of its services.

Subpart C—Safety Risk Management

§ 5.51 **Applicability.**

A certificate holder must apply safety risk management to the following:

(a) Implementation of new systems.

(b) Revision of existing systems.

(c) Development of operational procedures.

(d) Identification of hazards or ineffective risk controls through the safety assurance processes in subpart D of this part.

§**5.53 System analysis and hazard identification.**

(a) When applying safety risk management, the certificate holder must analyze the systems identified in §5.51. Those system analyses must be used to identify hazards under paragraph (c) of this section, and in developing and implementing risk controls related to the system under §5.55(c).

(b) In conducting the system analysis, the following information must be considered:

(1) Function and purpose of the system.

(2) The system's operating environment.

(3) An outline of the system's processes and procedures.

(4) The personnel, equipment, and facilities necessary for operation of the system.

(c) The certificate holder must develop and maintain processes to identify hazards within the context of the system analysis.

§**5.55 Safety risk assessment and control.**

(a) The certificate holder must develop and maintain processes to analyze safety risk associated with the hazards identified in §5.53(c).

(b) The certificate holder must define a process for conducting risk assessment that allows for the determination of acceptable safety risk.

(c) The certificate holder must develop and maintain processes to develop safety risk controls that are necessary as a result of the safety risk assessment process under paragraph (b) of this section.

(d) The certificate holder must evaluate whether the risk will be acceptable with the proposed safety risk control applied, before the safety risk control is implemented.

Subpart D—Safety Assurance

§**5.71 Safety performance monitoring and measurement.**

(a) The certificate holder must develop and maintain processes and systems to acquire data with respect to its operations, products, and services to monitor the safety performance of the organization. These processes and systems must include, at a minimum, the following:

(1) Monitoring of operational processes.

(2) Monitoring of the operational environment to detect changes.

(3) Auditing of operational processes and systems.

(4) Evaluations of the SMS and operational processes and systems.

(5) Investigations of incidents and accidents.

(6) Investigations of reports regarding potential non-compliance with regulatory standards or other safety risk controls established by the certificate holder through the safety risk management process established in subpart C of this part.

(7) A confidential employee reporting system in which employees can report hazards, issues, concerns, occurrences, incidents, as well as propose solutions and safety improvements.

(b) The certificate holder must develop and maintain processes that analyze the data acquired through the processes and systems identified under paragraph (a) of this section and any other relevant data with respect to its operations, products, and services.

[80 FR 1326, Jan. 8, 2015, as amended at 82 FR 24010, May 25, 2017]

§**5.73 Safety performance assessment.**

(a) The certificate holder must conduct assessments of its safety performance against its safety objectives, which include reviews by the accountable executive, to:

(1) Ensure compliance with the safety risk controls established by the certificate holder.

(2) Evaluate the performance of the SMS.

(3) Evaluate the effectiveness of the safety risk controls established under §5.55(c) and identify any ineffective controls.

(4) Identify changes in the operational environment that may introduce new hazards.

(5) Identify new hazards.

(b) Upon completion of the assessment, if ineffective controls or new hazards are identified under paragraphs (a)(2) through (5) of this section, the certificate holder must use the safety risk management process described in subpart C of this part.

§ 5.75 Continuous improvement.

The certificate holder must establish and implement processes to correct safety performance deficiencies identified in the assessments conducted under § 5.73.

Subpart E—Safety Promotion

§ 5.91 Competencies and training.

The certificate holder must provide training to each individual identified in § 5.23 to ensure the individuals attain and maintain the competencies necessary to perform their duties relevant to the operation and performance of the SMS.

§ 5.93 Safety communication.

The certificate holder must develop and maintain means for communicating safety information that, at a minimum:

(a) Ensures that employees are aware of the SMS policies, processes, and tools that are relevant to their responsibilities.

(b) Conveys hazard information relevant to the employee's responsibilities.

(c) Explains why safety actions have been taken.

(d) Explains why safety procedures are introduced or changed.

Subpart F—SMS Documentation and Recordkeeping

§ 5.95 SMS documentation.

The certificate holder must develop and maintain SMS documentation that describes the certificate holder's:

(a) Safety policy.

(b) SMS processes and procedures.

§ 5.97 SMS records.

(a) The certificate holder must maintain records of outputs of safety risk management processes as described in subpart C of this part. Such records must be retained for as long as the control remains relevant to the operation.

(b) The certificate holder must maintain records of outputs of safety assurance processes as described in subpart D of this part. Such records must be retained for a minimum of 5 years.

(c) The certificate holder must maintain a record of all training provided under § 5.91 for each individual. Such records must be retained for as long as the individual is employed by the certificate holder.

(d) The certificate holder must retain records of all communications provided under § 5.93 for a minimum of 24 consecutive calendar months.

SUBCHAPTER B—PROCEDURAL RULES

PART 11—GENERAL RULEMAKING PROCEDURES

Subpart A—Rulemaking Procedures

Subpart B—Paperwork Reduction Act Control Numbers

AUTHORITY: 49 U.S.C. 106(f), 106(g), 40101, 40103, 40105, 40109, 40113, 44110, 44502, 44701–44702, 44711, 46102, and 51 U.S.C. 50901–50923.

SOURCE: Docket No. FAA–1999–6622, 65 FR 50863, Aug. 21, 2000, unless otherwise noted.

EDITORIAL NOTE: Nomenclature changes to part 11 appear at 61 FR 18052, April 24, 1996.

Subpart A—Rulemaking Procedures

§ 11.1 To what does this part apply?

This part applies to the issuance, amendment, and repeal of any regulation for which FAA ("we") follows public rulemaking procedures under the Administrative Procedure Act ("APA") (5 U.S.C. 553).

DEFINITION OF TERMS

§ 11.3 What is an advance notice of proposed rulemaking?

An advance notice of proposed rulemaking (ANPRM) tells the public that FAA is considering an area for rulemaking and requests written comments on the appropriate scope of the rulemaking or on specific topics. An advance notice of proposed rulemaking may or may not include the text of potential changes to a regulation.

§ 11.5 What is a notice of proposed rulemaking?

A notice of proposed rulemaking (NPRM) proposes FAA's specific regulatory changes for public comment and contains supporting information. It includes proposed regulatory text.

§ 11.7 What is a supplemental notice of proposed rulemaking?

On occasion, FAA may decide that it needs more information on an issue, or that we should take a different approach than we proposed. Also, we may want to follow a commenter's suggestion that goes beyond the scope of the original proposed rule. In these cases, FAA may issue a supplemental notice of proposed rulemaking (SNPRM) to give the public an opportunity to comment further or to give us more information.

§ 11.9 What is a final rule?

A final rule sets out new or revised requirements and their effective date. It also may remove requirements. When preceded by an NPRM, a final rule will also identify significant substantive issues raised by commenters in response to the NPRM and will give the agency's response.

§ 11.11 What is a final rule with request for comments?

A final rule with request for comment is a rule that the FAA issues in final (with an effective date) that invites public comment on the rule. We usually do this when we have not first issued an ANPRM or NPRM, because we have found that doing so would be impracticable, unnecessary, or contrary to the public interest. We give our reasons for our determination in the preamble. The comment period often ends after the effective date of the rule. A final rule not preceded by an ANPRM or NPRM is commonly called an "immediately adopted final rule." We invite comments on these rules only if we think that we will receive useful information. For example, we would not invite comments when we are just making an editorial clarification or correction.

§ 11.13 What is a direct final rule?

A direct final rule is a type of final rule with request for comments. Our reason for issuing a direct final rule without an NPRM is that we would not expect to receive any adverse comments, and so an NPRM is unnecessary. However, to be certain that we are correct, we set the comment period to end before the effective date. If we receive an adverse comment, we will either publish a document withdrawing the direct final rule before it becomes effective and may issue an NPRM, or proceed by any other means permitted under the Administrative Procedure Act, 5 U.S.C. 551 *et seq.*, consistent with procedures at 49 CFR 5.13(l).

[Docket No. FAA–1999–6622, 65 FR 50863, Aug. 21, 2000, as amended at 84 FR 71717, Dec. 27, 2019]

§ 11.15 What is a petition for exemption?

A petition for exemption is a request to the FAA by an individual or entity asking for relief from the requirements of a current regulation. For petitions for waiver of commercial space transportation regulations, see part 404 of this title.

[FAA–2016–6761, Amdt. No. 11–62, 83 FR 28534, June 20, 2016]

§ 11.17 What is a petition for rule-making?

A petition for rulemaking is a request to FAA by an individual or entity asking the FAA to adopt, amend, or repeal a regulation.

§ 11.19 What is a special condition?

A special condition is a regulation that applies to a particular aircraft design. The FAA issues special conditions when we find that the airworthiness regulations for an aircraft, aircraft engine, or propeller design do not contain adequate or appropriate safety standards, because of a novel or unusual design feature.

GENERAL

§ 11.21 What are the most common kinds of rulemaking actions for which FAA follows the Administrative Procedure Act?

FAA follows the Administrative Procedure Act (APA) procedures for these common types of rules:

(a) Rules found in the Code of Federal Regulations;

(b) Airworthiness directives issued under part 39 of this chapter; and

(c) Airspace Designations issued under various parts of this chapter.

§ 11.23 Does FAA follow the same procedures in issuing all types of rules?

Yes, in general, FAA follows the same procedures for all rule types. There are some differences as to which FAA official has authority to issue each type, and where you send petitions for FAA to adopt, amend, or repeal each type. Assume that the procedures in this subpart apply to all rules, except where we specify otherwise.

§ 11.25 How does FAA issue rules?

(a) The FAA uses APA rulemaking procedures to adopt, amend, or repeal regulations. To propose or adopt a new regulation, or to change a current regulation, FAA will issue one or more of the following documents. We publish these rulemaking documents in the FEDERAL REGISTER unless we name and personally serve a copy of a rule on every person subject to it. We also make all documents available to the public by posting them in the Federal Docket Management System at *http://www.regulations.gov.*

(1) An advance notice of proposed rulemaking (ANPRM).

(2) A notice of proposed rulemaking (NPRM).

(3) A supplemental notice of proposed rulemaking (SNPRM).

(4) A final rule.

(5) A final rule with request for comments.

(6) A direct final rule.

(b) Each of the rulemaking documents in paragraph (a) of this section generally contains the following information:

(1) The topic involved in the rulemaking document.

(2) FAA's legal authority for issuing the rulemaking document.

(3) How interested persons may participate in the rulemaking proceeding (for example, by filing written comments or making oral presentations at a public meeting).

(4) Whom to call if you have questions about the rulemaking document.

(5) The date, time, and place of any public meetings FAA will hold to discuss the rulemaking document.

(6) The docket number and regulation identifier number (RIN) for the rulemaking proceeding.

[Doc. No. 1999–6622, 65 FR 50863, Aug. 21, 2000, as amended at 72 FR 68474, Dec. 5, 2007]

§ 11.27 Are there other ways FAA collects specific rulemaking recommendations before we issue an NPRM?

Yes, the FAA obtains advice and recommendations from rulemaking advisory committees. One of these committees is the Aviation Rulemaking Advisory Committee (ARAC), which is a formal standing committee comprised of representatives of aviation associations and industry, consumer groups, and interested individuals. In conducting its activities, ARAC complies with the Federal Advisory Committee Act and the direction of FAA. We task ARAC with providing us with recommended rulemaking actions dealing with specific areas and problems. If we accept an ARAC recommendation to change an FAA rule, we ordinarily publish an NPRM using the procedures in this part. The FAA may establish other rulemaking advisory committees as needed to focus on specific issues for a limited period of time.

§ 11.29 May FAA change its regulations without first issuing an ANPRM or NPRM?

The FAA normally adds or changes a regulation by issuing a final rule after an NPRM. However, FAA may adopt, amend, or repeal regulations without first issuing an ANPRM or NPRM in the following situations:

(a) We may issue a final rule without first requesting public comment if, for good cause, we find that an NPRM is impracticable, unnecessary, or contrary to the public interest. We place that finding and a brief statement of the reasons for it in the final rule. For example, we may issue a final rule in response to a safety emergency.

(b) If an NPRM would be unnecessary because we do not expect to receive adverse comment, we may issue a direct final rule.

§ 11.31 How does FAA process direct final rules?

(a) A direct final rule will take effect on a specified date unless FAA receives an adverse comment within the comment period—generally 60 days after the direct final rule is published in the FEDERAL REGISTER. An adverse comment explains why a rule would be inappropriate, or would be ineffective or unacceptable without a change. It may challenge the rule's underlying premise or approach. Under the direct final rule process, we do not consider the following types of comments to be adverse:

(1) A comment recommending another rule change, in addition to the change in the direct final rule at issue. We consider the comment adverse, however, if the commenter states why the direct final rule would be ineffective without the change.

(2) A frivolous or insubstantial comment.

(b) If FAA has not received an adverse comment, we will publish a confirmation document in the FEDERAL REGISTER, generally within 15 days after the comment period closes. The confirmation document tells the public the effective date of the rule.

(c) If we receive an adverse comment, we will advise the public by publishing a document in the FEDERAL REGISTER before the effective date of the direct final rule. This document may withdraw the direct final rule in whole or in part. If we withdraw a direct final rule because of an adverse comment, we may incorporate the commenter's recommendation into another direct final rule or may publish a notice of proposed rulemaking.

[Docket No. FAA–1999–6622, 65 FR 50863, Aug. 21, 2000, as amended at 84 FR 71717, Dec. 27, 2019]

§ 11.33 How can I track FAA's rulemaking activities?

The best ways to track FAA's rulemaking activities are with the docket number or the regulation identifier number.

(a) *Docket ID.* We assign a docket ID to each rulemaking document proceeding. Each rulemaking document FAA issues in a particular rulemaking proceeding, as well as public comments on the proceeding, will display the same docket ID. This ID allows you to search the Federal Docket Management System (FDMS) for information on most rulemaking proceedings. You can view and copy docket materials during regular business hours at the U.S. Department of Transportation, Docket Operations, West Building Ground Floor, Room W12–140, 1200 New Jersey Avenue, SE., Washington, DC 20590. Or you can view and download docketed materials through the Internet at *http://www.regulations.gov.* If you can't find the material in the electronic docket, contact the person listed under **FOR FURTHER INFORMATION CONTACT** in the document you are interested in.

(b) *Regulation identifier number.* DOT publishes a semiannual agenda of all current and projected DOT rulemakings, reviews of existing regulations, and completed actions. This semiannual agenda appears in the Unified Agenda of Federal Regulations, published in the FEDERAL REGISTER in April and October of each year. The semiannual agenda tells the public about DOT's—including FAA's—regulatory activities. DOT assigns a regulation identifier number (RIN) to each individual rulemaking proceeding in the semiannual agenda. This number appears on all rulemaking documents published in the FEDERAL REGISTER and makes it easy for you to track those rulemaking proceedings in both the FEDERAL REGISTER and the semiannual regulatory agenda.

[Doc. No. 1999–6622, 65 FR 50863, Aug. 21, 2000, as amended at 72 FR 68474, Dec. 5, 2007]

§ 11.35 Does FAA include sensitive security information and proprietary information in the Federal Docket Management System (FDMS)?

(a) *Sensitive security information.* You should not submit sensitive security information to the rulemaking docket, unless you are invited to do so in our request for comments. If we ask for this information, we will tell you in the specific document how to submit this information, and we will provide a separate non-public docket for it. For all proposed rule changes involving civil aviation security, we review comments as we receive them, before they are placed in the docket. If we find that a comment contains sensitive security information, we remove that information before placing the comment in the general docket.

(b) *Proprietary information.* When we are aware of proprietary information filed with a comment, we do not place it in the docket. We hold it in a separate file to which the public does not have access, and place a note in the docket that we have received it. If we receive a request to examine or copy this information, we treat it as any other request under the Freedom of Information Act (5 U.S.C. 552). We process such a request under the DOT procedures found in 49 CFR part 7.

[Doc. No. 1999–6622, 65 FR 50863, Aug. 21, 2000, as amended at 72 FR 68474, Dec. 5, 2007]

§ 11.37 Where can I find information about an Airworthiness Directive, an airspace designation, or a petition handled in a region?

The FAA includes most documents concerning Airworthiness Directives, airspace designations, or petitions handled in a region in the electronic dock-et. If the information isn't in the docket, contact the person listed under **FOR FURTHER INFORMATION CONTACT** in the FEDERAL REGISTER document about the action.

§ 11.38 What public comment procedures does the FAA follow for Special Conditions?

Even though the Administrative Procedure Act does not require notice and comment for rules of particular applicability, FAA does publish proposed special conditions for comment. In the following circumstances we may not invite comment before we issue a special condition. If we don't, we will invite comment when we publish the final special condition.

(a) The FAA considers prior notice to be impracticable if issuing a design approval would significantly delay delivery of the affected aircraft. We consider such a delay to be contrary to the public interest.

(b) The FAA considers prior notice to be unnecessary if we have provided previous opportunities to comment on substantially identical proposed special conditions, and we are satisfied that new comments are unlikely.

§ 11.39 How may I participate in FAA's rulemaking process?

You may participate in FAA's rulemaking process by doing any of the following:

(a) File written comments on any rulemaking document that asks for comments, including an ANPRM, NPRM, SNPRM, a final rule with request for comments, or a direct final rule. Follow the directions for commenting found in each rulemaking document.

(b) Ask that we hold a public meeting on any rulemaking, and participate in any public meeting that we hold.

(c) File a petition for rulemaking that asks us to adopt, amend, or repeal a regulation.

§ 11.40 Can I get more information about a rulemaking?

You can contact the person listed under **FOR FURTHER INFORMATION CONTACT** in the preamble of a rule. That person can explain the meaning and intent of a proposed rule, the technical aspects of a document, the terminology in a document, and can tell you our published schedule for the rulemaking process. We cannot give you information that is not already available to other members of the public. The Department of Transportation policy regarding public contacts during rulemaking appears at 49 CFR 5.19.

[Docket No. FAA–1999–6622, 65 FR 50863, Aug. 21, 2000, as amended at 84 FR 71717, Dec. 27, 2019]

WRITTEN COMMENTS

§ 11.41 Who may file comments?

Anyone may file written comments about proposals and final rules that request public comments.

§ 11.43 What information must I put in my written comments?

(a) Your written comments must be in English and must contain the following:

(1) The docket number of the rulemaking document you are commenting on, clearly set out at the beginning of your comments.

(2) Your name and mailing address, and, if you wish, other contact information, such as a fax number, telephone number, or e-mail address.

(3) Your information, views, or arguments, following the instructions for participation in the rulemaking document on which you are commenting.

(b) You should also include all material relevant to any statement of fact or argument in your comments, to the extent that the material is available to you and reasonable for you to submit. Include a copy of the title page of the document. Whether or not you submit a copy of the material to which you refer, you should indicate specific places in the material that support your position.

§ 11.45 Where and when do I file my comments?

(a) Send your comments to the location specified in the rulemaking document on which you are commenting. If you are asked to send your comments to the Federal Document Management System, you may send them in either of the following ways:

(1) By mail to: U.S. Department of Transportation, Docket Operations, West Building Ground Floor, Room W12–140, 1200 New Jersey Avenue, SE., Washington, DC 20590.

(2) Through the Internet to *http:// www.regulations.gov.*

(3) In any other manner designated by FAA.

(b) Make sure that your comments reach us by the deadline set out in the rulemaking document on which you are commenting. We will consider late-filed comments to the extent possible only if they do not significantly delay the rulemaking process.

(c) We may reject your paper or electronic comments if they are frivolous, abusive, or repetitious. We may reject comments you file electronically if you do not follow the electronic filing instructions at the Federal Docket Management System Web site.

[Doc. No. 1999–6622, 65 FR 50863, Aug. 21, 2000, as amended at 72 FR 68474, Dec. 5, 2007]

§ 11.47 May I ask for more time to file my comments?

Yes, if FAA grants your request for more time to file comments, we grant all persons the same amount of time. We will notify the public of the extension by a document in the FEDERAL REGISTER. If FAA denies your request, we will notify you of the denial. To ask for more time, you must file a written or electronic request for extension at least 10 days before the end of the comment period. Your letter or message must—

(a) Show the docket number of the rule at the top of the first page;

(b) State, at the beginning, that you are requesting an extension of the comment period;

(c) Show that you have good cause for the extension and that an extension is in the public interest;

(d) Be sent to the address specified for comments in the rulemaking document on which you are commenting.

§ 11.51 May I request that FAA hold a public meeting on a rulemaking action?

Yes, you may request that we hold a public meeting. FAA holds a public meeting when we need more than written comments to make a fully informed decision. Submit your written request to the address specified in the rulemaking document on which you are commenting. Specify at the top of your letter or message that you are requesting that the agency hold a public meeting. Submit your request no later than 30 days after our rulemaking notice. If we find good cause for a meeting, we will notify you and publish a notice of the meeting in the FEDERAL REGISTER.

§ 11.53 What takes place at a public meeting?

A public meeting is a non-adversarial, fact-finding proceeding conducted by an FAA representative. Public meetings are announced in the FEDERAL REGISTER. We invite interested persons to attend and to present their views to the agency on specific issues. There are no formal pleadings and no adverse parties, and any regulation issued afterward is not necessarily based exclusively on the record of the meeting.

PETITIONS FOR RULEMAKING AND FOR EXEMPTION

§ 11.61 May I ask FAA to adopt, amend, or repeal a regulation, or grant relief from the requirements of a current regulation?

(a) Using a petition for rulemaking, you may ask FAA to add a new regulation to title 14 of the Code of Federal Regulations (14 CFR) or ask FAA to amend or repeal a current regulation in 14 CFR.

(b) Using a petition for exemption, you may ask FAA to grant you relief from current regulations in 14 CFR.

§ 11.63 How and to whom do I submit my petition for rulemaking or petition for exemption?

(a) To submit a petition for rulemaking or exemption—

(1) By electronic submission, submit your petition for rulemaking or exemption to the FAA through the internet at *http://www.regulations.gov*, the Federal Docket Management System website. For additional instructions, you may visit *http://www.faa.gov*, and navigate to the Rulemaking home page.

(2) By paper submission, send the original signed copy of your petition for rulemaking or exemption to this address: U.S. Department of Transportation, Docket Operations, West Building Ground Floor, Room W12–140, 1200 New Jersey Avenue, SE., Washington, DC 20590.

(b) Submit a petition for rulemaking or exemption from part 139 of this chapter—

(1) To the appropriate FAA airport field office in whose area your airport is, or will be, established; and

(2) To the U.S. Department of Transportation, Docket Operations, West Building Ground Floor, Room W12–140, 1200 New Jersey Avenue, SE., Washington, DC 20590 or by electronic submission to this Internet address: *http://www.regulations.gov*.

(c) The FAA may designate other means by which you can submit petitions in the future.

(d) Submit your petition for exemption 120 days before you need the exemption to take effect.

[Amdt. 11–50, 69 FR 22386, Apr. 26, 2004, as amended at 72 FR 68474, Dec. 5, 2007; Amdt. 11–55, 74 FR 202, Jan. 5, 2009; FAA–2016–6761, Amdt. No. 11–62, 83 FR 28534, June 20, 2016]

§ 11.71 What information must I include in my petition for rulemaking?

(a) You must include the following information in your petition for rulemaking:

(1) Your name and mailing address and, if you wish, other contact information such as a fax number, telephone number, or e-mail address.

(2) An explanation of your proposed action and its purpose.

(3) The language you propose for a new or amended rule, or the language you would remove from a current rule.

(4) An explanation of why your proposed action would be in the public interest.

(5) Information and arguments that support your proposed action, including relevant technical and scientific data available to you.

(6) Any specific facts or circumstances that support or demonstrate the need for the action you propose.

(b) In the process of considering your petition, we may ask that you provide information or data available to you about the following:

(1) The costs and benefits of your proposed action to society in general, and identifiable groups within society in particular.

(2) The regulatory burden of your proposed action on small businesses, small organizations, small governmental jurisdictions, and Indian tribes.

(3) The recordkeeping and reporting burdens of your proposed action and whom the burdens would affect.

(4) The effect of your proposed action on the quality of the natural and social environments.

§ 11.73 How does FAA process petitions for rulemaking?

After we have determined the disposition of your petition, we will contact you in writing about our decision. The FAA may respond to your petition for rulemaking in one of the following ways:

(a) If we determine that your petition justifies our taking the action you suggest, we may issue an NPRM or ANPRM. We will do so no later than 6 months after the date we receive your petition. In making our decision, we consider:

(1) The immediacy of the safety or security concerns you raise;

(2) The priority of other issues the FAA must deal with; and

(3) The resources we have available to address these issues.

(b) If we have issued an ANPRM or NPRM on the subject matter of your petition, we will consider your arguments for a rule change as a comment in connection with the rulemaking proceeding. We will not treat your petition as a separate action.

(c) If we have begun a rulemaking project in the subject area of your petition, we will consider your comments and arguments for a rule change as part of that project. We will not treat your petition as a separate action.

(d) If we have tasked ARAC to study the general subject area of your petition, we will ask ARAC to review and evaluate your proposed action. We will not treat your petition as a separate action.

(e) If we determine that the issues you identify in your petition may have merit, but do not address an immediate safety concern or cannot be addressed because of other priorities and resource constraints, we may dismiss your petition. Your comments and arguments for a rule change will be placed in a database, which we will examine when we consider future rulemaking.

§ 11.75 Does FAA invite public comment on petitions for rulemaking?

Generally, FAA does not invite public comment on petitions for rulemaking.

§ 11.77 Is there any additional information I must include in my petition for designating airspace?

In petitions asking FAA to establish, amend, or repeal a designation of airspace, including special use airspace, you must include all the information specified by § 11.71 and also:

(a) The location and a description of the airspace you want assigned or designated;

(b) A complete description of the activity or use to be made of that airspace, including a detailed description of the type, volume, duration, time, and place of the operations to be conducted in the area;

(c) A description of the air navigation, air traffic control, surveillance, and communication facilities available and to be provided if we grant the designation; and

(d) The name and location of the agency, office, facility, or person who would have authority to permit the use of the airspace when it was not in use for the purpose to which you want it assigned.

§ 11.81 What information must I include in my petition for an exemption?

You must include the following information in your petition for an exemption and submit it to FAA as soon as you know you need an exemption.

(a) Your name and mailing address and, if you wish, other contact information such as a fax number, telephone number, or e-mail address;

(b) The specific section or sections of 14 CFR from which you seek an exemption;

(c) The extent of relief you seek, and the reason you seek the relief;

(d) The reasons why granting your request would be in the public interest; that is, how it would benefit the public as a whole;

(e) The reasons why granting the exemption would not adversely affect safety, or how the exemption would provide a level of safety at least equal to that provided by the rule from which you seek the exemption;

(f) A summary we can publish in the FEDERAL REGISTER, stating:

(1) The rule from which you seek the exemption; and

(2) A brief description of the nature of the exemption you seek;

(g) Any additional information, views or arguments available to support your request; and

(h) If you want to exercise the privileges of your exemption outside the United States, the reason why you need to do so.

§ 11.83 How can I operate under an exemption outside the United States?

If you want to be able to operate under your exemption outside the United States, you must request this when you petition for relief and give us the reason for this use. If you do not provide your reason or we determine that it does not justify this relief, we will limit your exemption to use within the United States. Before we extend your exemption for use outside the United States, we will verify that the exemption would be in compliance with the Standards of the International Civil Aviation Organization (ICAO). If it would not, but we still believe it would be in the public interest to allow you to do so, we will file a difference with ICAO. However, a foreign country still may not allow you to operate in that country without meeting the ICAO standard.

§ 11.85 Does FAA invite public comment on petitions for exemption?

Yes, FAA publishes information about petitions for exemption in the FEDERAL REGISTER. The information includes—

(a) The docket number of the petition;

(b) The citation to the rule or rules from which the petitioner requested relief;

(c) The name of the petitioner;

(d) The petitioner's summary of the action requested and the reasons for requesting it; and

(e) A request for comments to assist FAA in evaluating the petition.

§ 11.87 Are there circumstances in which FAA may decide not to publish a summary of my petition for exemption?

The FAA may not publish a summary of your petition for exemption and request comments if you present or we find good cause why we should not delay action on your petition. The factors we consider in deciding not to request comment include:

(a) Whether granting your petition would set a precedent.

(b) Whether the relief requested is identical to exemptions granted previously.

(c) Whether our delaying action on your petition would affect you adversely.

(d) Whether you filed your petition in a timely manner.

§ 11.89 How much time do I have to submit comments to FAA on a petition for exemption?

The FAA states the specific time allowed for comments in the FEDERAL REGISTER notice about the petition. We usually allow 20 days to comment on a petition for exemption.

§ 11.91 How does FAA inform me of its decision on my petition for exemption?

The FAA will notify you in writing about its decision on your petition. A copy of this decision is also placed in the public docket. We will include the docket number associated with your petition in our letter to you.

[Doc. No. FAA–2005–22982, 71 FR 1485, Jan. 10, 2006]

§ 11.101 **May I ask FAA to reconsider my petition for rulemaking or petition for exemption if it is denied?**

Yes, you may petition FAA to reconsider your petition denial. You must submit your request to the address to which you sent your original petition, and FAA must receive it within 60 days after we issued the denial. For us to accept your petition, show the following:

(a) That you have a significant additional fact and why you did not present it in your original petition;

(b) That we made an important factual error in our denial of your original petition; or

(c) That we did not correctly interpret a law, regulation, or precedent.

§ 11.103 **What exemption relief may be available to federal, state, and local governments when operating aircraft that are not public aircraft?**

The Federal Aviation Administration may grant a federal, state, or local government an exemption from part A of subtitle VII of title 49 United States Code, and any regulation issued under that authority that is applicable to an aircraft as a result of the Independent Safety Board Act Amendments of 1994, Public Law 103–411, if—

(a) The Administrator finds that granting the exemption is necessary to prevent an undue economic burden on the unit of government; and

(b) The Administrator certifies that the aviation safety program of the unit of government is effective and appropriate to ensure safe operations of the type of aircraft operated by the unit of government.

[68 FR 25488, May 13, 2003]

Subpart B—Paperwork Reduction Act Control Numbers

§ 11.201 **Office of Management and Budget (OMB) control numbers assigned under the Paperwork Reduction Act.**

(a) The Paperwork Reduction Act of 1995 (44 U.S.C. 3501–3520) requires FAA to get approval from OMB for our information collection activities, and to list a record of those approvals in the FEDERAL REGISTER. This subpart lists the control numbers OMB assigned to FAA's information collection activities.

(b) The table listing OMB control numbers assigned to FAA's information collection activities follows:

14 CFR part or section identified and described	Current OMB control number
Part 13.5	2120–0795
Part 14	2120–0539
Part 17	2120–0632
Part 21	2120–0018, 2120–0552
Part 34	2120–0508
Part 39	2120–0056
Part 43	2120–0020
Part 45	2120–0508
Part 47	2120–0024, 2120–0042
Part 48	2120–0765
Part 49	2120–0043
Part 60	2120–0680
Part 61	2120–0021, 2120–0034, 2120–0543, 2120–0571
Part 63	2120–0007
Part 65	2120–0022, 2120–0535, 2120–0571, 2120–0648
Part 67	2120–0034, 2120–0543
Part 77	2120–0001
Part 89	2120–0781, 2120–0782, 2120–0783, 2120–0785.
Part 91	2120–0005, 2120–0026, 2120–0027, 2120–0573, 2120–0606, 2120–0620, 2120–0631, 2120–0651
Part 93	2120–0524, 2120–0606, 2120–0639
Part 101	2120–0027
Part 105	2120–0027, 2120–0641
Part 107	2120–0005, 2120–0021, 2120–0027, 2120–0767, 2120–0768, 2120–0775.
Part 111	2120–0607
Part 119	2120–0593
Part 121	2120–0008, 2120–0028, 2120–0535, 2120–0571, 2120–0600, 2120–0606, 2120–0614, 2120–0616, 2120–0631, 2120–0651, 2120–0653, 2120–0691, 2120–0739, 2120–0760, 2120–0766.
Part 125	2120–0028, 2120–0085, 2120–0616, 2120–0651
Part 129	2120–0028, 2120–0536, 2120–0616, 2120–0638
Part 133	2120–0044
Part 135	2120–0003, 2120–0028, 2120–0039, 2120–0535, 2120–0571, 2120–0600, 2120–0606, 2120–0614, 2120–0616, 2120–0620, 2120–0631, 2120–0653, 2120–0766.
Part 137	2120–0049
Part 139	2120–0045, 2120–0063
Part 141	2120–0009
Part 142	2120–0570
Part 145	2120–0003, 2120–0010, 2120–0571
Part 147	2120–0040
Part 150	2120–0517
Part 157	2120–0036
Part 158	2120–0557
Part 161	2120–0563
Part 171	2120–0014
Part 183	2120–0033, 2120–0604
Part 193	2120–0646

14 CFR part or section identified and described	Current OMB control number
Part 198	2120–0514
Part 400	2120–0643, 2120–0644, 0649
Part 401	2120–0608
Part 440	2120–0601
SFAR 36	2120–0507
SFAR 71	2120–0620

[Doc. No. 1999–6622, 65 FR 50863, Aug. 21, 2000, as amended by Amdt. 11–47, 67 FR 9553, Mar. 1, 2002; Amdt. 11–49, 68 FR 61321, Oct. 27, 2003; Amdt. 11–49, 68 FR 70132, Dec. 17, 2003; 70 FR 40163, July 12, 2005; 71 FR 63426, Oct. 30, 2006; 72 FR 59599, Oct. 22, 2007; Amdt. 11–56, 79 FR 12937, Mar. 7, 2014; Amdt. 11–57, 80 FR 58586, Sept. 30, 2015; Doc. FAA–2015–7396, Amdt. 11–58, 80 FR 79255, Dec. 21, 2015; Doc. FAA–2011–1136, Amdt. 11–59, 81 FR 13969, Mar. 16, 2016; Doc. FAA–2014–0554, Amdt. 11–60, 81 FR 33117, May 24, 2016; 81 FR 38573, June 14, 2016; Doc. FAA–2016–9064, Amdt. 11–61, 81 FR 59129, Aug. 29, 2016; FAA–2018–1087, Amdt. 11–64, 86 FR 4381, Jan. 15, 2021; Docket No. FAA–2019–1100, Amdt. 11–63, 86 FR 4503, Jan. 15, 2021; FAA–2020–0246, Amdt. 11–65, 86 FR 31060, June 10, 2021; FAA–2018–1051, Amdt. 13–40A, 87 FR 61233, Oct. 11, 2022]

PART 13—INVESTIGATIVE AND ENFORCEMENT PROCEDURES

Subpart A—General Authority to Re-Delegate and Investigative Procedures

Sec.
13.1 Re-delegation.
13.2 Reports of violations.
13.3 Investigations (general).
13.5 Formal complaints.
13.7 Records, documents, and reports.

Subpart B—Administrative Actions

13.11 Administrative disposition of certain violations.

Subpart C—Legal Enforcement Actions

13.13 Consent orders.
13.14 [Reserved]
13.15 Civil penalties: Other than by administrative assessment.
13.16 Civil penalties: Administrative assessment against a person other than an individual acting as a pilot, flight engineer, mechanic, or repairman; administrative assessment against all persons for hazardous materials violations.
13.17 Seizure of aircraft.
13.18 Civil penalties: Administrative assessment against an individual acting as a pilot, flight engineer, mechanic, or repairman.

13.19 Certificate actions appealable to the National Transportation Safety Board.
13.20 Orders of compliance, cease and desist orders, orders of denial, and other orders.
13.21–13.29 [Reserved]

Subpart D—Rules of Practice for FAA Hearings

13.31 Applicability.
13.33 Parties, representatives, and notice of appearance.
13.35 Request for hearing, complaint, and answer.
13.37 Hearing officer: Assignment and powers.
13.39 Disqualification of hearing officer.
13.41 Separation of functions and prohibition on ex parte communications.
13.43 Service and filing of pleadings, motions, and documents.
13.44 [Reserved]
13.45 Computation of time and extension of time.
13.47 Withdrawal or amendment of the complaint, answer, or other filings.
13.49 Motions.
13.51 Intervention.
13.53 Discovery.
13.55 Notice of hearing.
13.57 Subpoenas and witness fees.
13.59 Evidence.
13.61 Argument and submittals.
13.63 Record, decision, and aircraft registration proceedings.
13.65 Appeal to the Administrator, reconsideration, and judicial review.
13.67 Procedures for expedited proceedings.
13.69 Other matters: Alternative dispute resolution, standing orders, and forms.

Subpart E—Orders of Compliance Under the Hazardous Materials Transportation Act

13.70 Delegation of authority.
13.71 Applicability.
13.73 Notice of proposed order of compliance.
13.75 Reply or request for hearing.
13.77 Consent order of compliance.
13.79 [Reserved]
13.81 Emergency orders.
13.83–13.87 [Reserved]

Subpart F—Formal Fact-Finding Investigation Under an Order of Investigation

13.101 Applicability.
13.103 Order of investigation.
13.105 Notification.
13.107 Designation of additional parties.
13.109 Convening the investigation.
13.111 Subpoenas.
13.113 Noncompliance with the investigative process.
13.115 Public proceedings.
13.117 Conduct of investigative proceeding or deposition.

AUTHORITY: 18 U.S.C. 6002; 28 U.S.C. 2461 (note); 49 U.S.C. 106(g), 5121–5124, 40113–40114, 44103–44106, 44701–44704, 44709–44710, 44713, 44725, 44742, 44802 (note), 46101–46111, 46301, 46302 (for a violation of 49 U.S.C. 46504), 46304–46316, 46318–46320, 46501–46502, 46504, 46507, 47106, 47107, 47111, 47122, 47306, 47531–47532; 49 CFR 1.83.

SOURCE: Docket No. 18884, 44 FR 63723, Nov. 5, 1979, unless otherwise noted.

Subpart A—General Authority to Re-Delegate and Investigative Procedures

SOURCE: Docket No. FAA-2018-1051; Amdt. No. 13-40, 86 FR 54526, Oct. 1, 2021, unless otherwise noted.

§ 13.1 Re-delegation.

Unless otherwise specified, the Chief Counsel, each Deputy Chief Counsel, and the Assistant Chief Counsel for Enforcement may re-delegate the authority delegated to them under this part.

§ 13.2 Reports of violations.

(a) Any person who knows of any violation of 49 U.S.C. subtitle VII, 49 U.S.C. chapter 51, or any rule, regulation, or order issued under those statutes, should report the violation to FAA personnel.

(b) FAA personnel will review each report made under this section to determine whether any additional investigation or action is warranted.

§ 13.3 Investigations (general).

(a) The Administrator may conduct investigations; hold hearings; issue subpoenas; require the production of relevant documents, records, and property; and take evidence and depositions.

(b) The Administrator has delegated the authority to conduct investigations to the various services and offices for matters within their respective areas.

(c) The Administrator delegates to the Chief Counsel, each Deputy Chief Counsel, and the Assistant Chief Counsel for Enforcement the authority to:

(1) Issue orders;

(2) Conduct formal investigations;

(3) Subpoena witnesses and records in conducting a hearing or investigation;

(4) Order depositions and production of records in a proceeding or investigation; and

(5) Petition a court of the United States to enforce a subpoena or order described in paragraphs (c)(3) and (4) of this section.

(d) A complaint against the sponsor, proprietor, or operator of a federally assisted airport involving violations of the legal authorities listed in § 16.1 of this chapter must be filed in accordance with the provisions of part 16 of this chapter.

§ 13.5 Formal complaints.

(a) Any person may file a complaint with the Administrator with respect to a violation by a person of any requirement under 49 U.S.C. subtitle VII, 49 U.S.C. chapter 51, or any rule, regulation, or order issued under those statutes, as to matters within the jurisdiction of the Administrator. This section does not apply to complaints against the Administrator or employees of the FAA acting within the scope of their employment.

(b) Complaints filed under this section must—

(1) Be submitted in writing and identified as a complaint seeking an appropriate order or other enforcement action;

(2) Be submitted to the Federal Aviation Administration, Office of the Chief Counsel, Attention: Formal Complaint Clerk (AGC–300), 800 Independence Avenue SW, Washington, DC 20591;

(3) Set forth the name and address, if known, of each person who is the subject of the complaint and, with respect to each person, the specific provisions of the statute, rule, regulation, or order that the complainant believes were violated;

(4) Contain a concise but complete statement of the facts relied upon to substantiate each allegation;

(5) State the name, address, telephone number, and email of the person filing the complaint; and

(6) Be signed by the person filing the complaint or an authorized representative.

(c) A complaint that does not meet the requirements of paragraph (b) of this section will be considered a report under § 13.2.

(d) The FAA will send a copy of a complaint that meets the requirements of paragraph (b) of this section to the subject(s) of the complaint by certified mail.

(e) A subject of the complaint may serve a written answer to the complaint to the Formal Complaint Clerk at the address specified in paragraph (b)(2) of this section no later than 20 days after service of a copy of the complaint. For purposes of this paragraph (e), the date of service is the date on which the FAA mailed a copy of the complaint to the subject of the complaint.

(f) After the subject(s) of the complaint have served a written answer or after the allotted time to serve an answer has expired, the Administrator will determine if there are reasonable grounds for investigating the complaint, and—

(1) If the Administrator determines that a complaint does not state facts that warrant an investigation or action, the complaint may be dismissed without a hearing and the reason for the dismissal will be given, in writing, to the person who filed the complaint and the subject(s) of the complaint; or

(2) If the Administrator determines that reasonable grounds exist, an informal investigation may be initiated or an order of investigation may be issued in accordance with subpart F of this part, or both. The subject(s) of a complaint will be advised which official has been delegated the responsibility under § 13.3(b) or (c), as applicable, for conducting the investigation.

(g) If the investigation substantiates the allegations set forth in the complaint, the Administrator may take action in accordance with applicable law and FAA policy.

(h) The complaint and other records relating to the disposition of the complaint are maintained in the Formal Complaint Docket (AGC–300), Office of the Chief Counsel, Federal Aviation Administration, 800 Independence Avenue SW, Washington, DC 20591. Any interested person may examine any docketed material at that office at any time after the docket is established, except material that is required to be withheld from the public under applicable law, and may obtain a copy upon paying the cost of the copy.

§ 13.7 Records, documents, and reports.

Each record, document, and report that FAA regulations require to be maintained, exhibited, or submitted to the Administrator may be used in any investigation conducted by the Administrator; and, except to the extent the use may be specifically limited or prohibited by the section which imposes the requirement, the records, documents, and reports may be used in any civil penalty action, certificate action, or other legal proceeding.

Subpart B—Administrative Actions

SOURCE: Docket No. FAA-2018-1051; Amdt. No. 13-40, 86 FR 54527, Oct. 1, 2021, unless otherwise noted.

§ 13.11 Administrative disposition of certain violations.

(a) If, after an investigation, FAA personnel determine that an apparent violation of 49 U.S.C. subtitle VII, 49 U.S.C. chapter 51, or any rule, regulation, or order issued under those statutes, does not require legal enforcement action, an appropriate FAA official may take administrative action to address the apparent violation.

(b) An administrative action under this section does not constitute a formal adjudication of the matter, and may take the form of—

(1) A Warning Notice that recites available facts and information about the incident or condition and indicates that it may have been a violation; or

(2) A Letter of Correction that states the corrective action the apparent violator has taken or agrees to take. If the apparent violator does not complete the agreed corrective action, the FAA may take legal enforcement action.

Subpart C—Legal Enforcement Actions

SOURCE: Docket No. FAA-2018-1051; Amdt. No. 13-40, 86 FR 54527, Oct. 1, 2021, unless otherwise noted.

§ 13.13 Consent orders.

(a) The Chief Counsel, each Deputy Chief Counsel, and the Assistant Chief Counsel for Enforcement may issue a consent order to resolve any matter with a person that may be subject to legal enforcement action.

(b) A person that may be subject to legal enforcement action may propose a consent order. The proposed consent order must include—

(1) An admission of all jurisdictional facts;

(2) An express waiver of the right to further procedural steps and of all rights to legal review in any forum;

(3) An express waiver of attorney's fees and costs;

(4) If a notice or order has been issued prior to the proposed consent order, an incorporation by reference of the notice or order and an acknowledgment that the notice or order may be used to construe the terms of the consent order; and

(5) If a request for hearing or appeal is pending in any forum, a provision that the person will withdraw the request for hearing or notice of appeal.

§ 13.14 [Reserved]

§ 13.15 Civil penalties: Other than by administrative assessment.

(a) The FAA uses the procedures in this section when it seeks a civil penalty other than by the administrative assessment procedures in § 13.16 or § 13.18.

(b) The authority of the Administrator to seek a civil penalty, and the ability to refer cases to the United States Attorney General, or the delegate of the Attorney General, for prosecution of civil penalty actions sought by the Administrator is delegated to the Chief Counsel, each Deputy Chief Counsel, and the Assistant Chief Counsel for Enforcement. This delegation applies to cases involving one or more of the following:

(1) An amount in controversy in excess of:

(i) $400,000, if the violation was committed by a person other than an individual or small business concern; or

(ii) $50,000, if the violation was committed by an individual or small business concern.

(2) An in rem action, seizure of aircraft subject to lien, suit for injunctive relief, or for collection of an assessed civil penalty.

(c) The Administrator may compromise any civil penalty proposed under this section, before referral to the United States Attorney General, or the delegate of the Attorney General, for prosecution.

(1) The Administrator, through the Chief Counsel, a Deputy Chief Counsel, or the Assistant Chief Counsel for Enforcement sends a civil penalty letter to the person charged with a violation. The civil penalty letter contains a statement of the charges; the applicable law, rule, regulation, or order; and the amount of civil penalty that the Administrator will accept in full settlement of the action or an offer to compromise the civil penalty.

(2) Not later than 30 days after receipt of the civil penalty letter, the person cited with an alleged violation may respond to the civil penalty letter by—

(i) Submitting electronic payment, a certified check, or money order in the amount offered by the Administrator in the civil penalty letter. The agency attorney will send a letter to the person charged with the violation stating that payment is accepted in full settlement of the civil penalty action; or

(ii) Submitting one of the following to the agency attorney:

(A) Written material or information that may explain, mitigate, or deny the violation or that may show extenuating circumstances; or

(B) A written request for an informal conference to discuss the matter with the agency attorney and to submit any relevant information or documents that may explain, mitigate, or deny the violation; or that may show extenuating circumstances.

(3) The documents, material, or information submitted under paragraph (c)(2)(ii) of this section may include support for any claim of inability to pay the civil penalty in whole or in part, or for any claim of small business status as defined in 49 U.S.C. 46301(i).

(4) The Administrator will consider any material or information submitted under paragraph (c)(2)(ii) of this section to determine whether the person is subject to a civil penalty or to determine the amount for which the Administrator will compromise the action.

(5) If the parties cannot agree to compromise the civil penalty, the Administrator may refer the civil penalty action to the United States Attorney General, or the delegate of the Attorney General, to begin proceedings in a U.S. district court to prosecute and collect a civil penalty.

§13.16 Civil penalties: Administrative assessment against a person other than an individual acting as a pilot, flight engineer, mechanic, or repairman; administrative assessment against all persons for hazardous materials violations.

(a) *General.* The FAA uses the procedures in this section when it assesses a civil penalty against a person other than an individual acting as a pilot, flight engineer, mechanic, or repairman for a violation cited in the first sentence of 49 U.S.C. 46301(d)(2), or in 49 U.S.C. 47531, or any implementing rule, regulation, or order, except when the U.S. district courts have exclusive jurisdiction.

(b) *District court jurisdiction.* The U.S. district courts have exclusive jurisdiction of any civil penalty action initiated by the FAA for violations described in paragraph (a) of this section if—

(1) The amount in controversy is more than $400,000 for a violation committed by a person other than an individual or small business concern;

(2) The amount in controversy is more than $50,000 for a violation committed by an individual or a small business concern;

(3) The action is in rem or another action in rem based on the same violation has been brought;

(4) The action involves an aircraft subject to a lien that has been seized by the Government; or

(5) Another action has been brought for an injunction based on the same violation.

(c) *Hazardous materials violations.* An order assessing a civil penalty for a violation under 49 U.S.C. chapter 51, or a rule, regulation, or order issued under 49 U.S.C. chapter 51, is issued only after the following factors have been considered:

41

(1) The nature, circumstances, extent, and gravity of the violation;

(2) With respect to the violator, the degree of culpability, any history of prior violations, the ability to pay, and any effect on the ability to continue to do business; and

(3) Other matters that justice requires.

(d) *Delegation of authority.* The authority of the Administrator is delegated to each Deputy Chief Counsel and the Assistant Chief Counsel for Enforcement, as follows:

(1) Under 49 U.S.C. 46301(d), 47531, and 5123, and 49 CFR 1.83, to initiate and assess civil penalties for a violation of those statutes or a rule, regulation, or order issued under those provisions;

(2) Under 49 U.S.C. 5123, 49 CFR 1.83, 49 U.S.C. 46301(d), and 49 U.S.C. 46305, to refer cases to the Attorney General of the United States or a delegate of the Attorney General for collection of civil penalties;

(3) Under 49 U.S.C. 46301(f), to compromise the amount of a civil penalty imposed; and

(4) Under 49 U.S.C. 5123(e) and (f) and 49 CFR 1.83, to compromise the amount of a civil penalty imposed.

(e) *Order assessing civil penalty.* (1) An order assessing civil penalty may be issued for a violation described in paragraph (a) or (c) of this section, or as otherwise provided by statute, after notice and opportunity for a hearing, when:

(i) A person charged with a violation agrees to pay a civil penalty for a violation; or

(ii) A person charged with a violation does not request a hearing under paragraph (g)(2)(ii) of this section within 15 days after receipt of a final notice of proposed civil penalty.

(2) The following also serve as an order assessing civil penalty:

(i) An initial decision or order issued by an administrative law judge as described in § 13.232(e).

(ii) A decision or order issued by the FAA decisionmaker as described in § 13.233(j).

(f) *Notice of proposed civil penalty.* A civil penalty action is initiated by sending a notice of proposed civil penalty to the person charged with a violation, the designated agent for the person, or if there is no such designated agent, the president of the company charged with a violation. In response to a notice of proposed civil penalty, a company may designate in writing another person to receive documents in that civil penalty action. The notice of proposed civil penalty contains a statement of the charges and the amount of the proposed civil penalty. Not later than 30 days after receipt of the notice of proposed civil penalty, the person charged with a violation may—

(1) Submit the amount of the proposed civil penalty or an agreed-upon amount, in which case either an order assessing civil penalty or compromise order under paragraph (n) of this section may be issued in that amount;

(2) Submit to the agency attorney one of the following:

(i) Written information, including documents and witness statements, demonstrating that a violation of the regulations did not occur or that a penalty or the amount of the penalty is not warranted by the circumstances.

(ii) A written request to reduce the proposed civil penalty, stating the amount of reduction and the reasons and providing any documents supporting a reduction of the proposed civil penalty, including records indicating a financial inability to pay or records showing that payment of the proposed civil penalty would prevent the person from continuing in business.

(iii) A written request for an informal conference to discuss the matter with the agency attorney and to submit relevant information or documents; or

(3) Request a hearing conducted in accordance with subpart G of this part.

(g) *Final notice of proposed civil penalty.* A final notice of proposed civil penalty will be sent to the person charged with a violation, the designated agent for the person, the designated agent named in accordance with paragraph (f) of this section, or the president of the company charged with a violation. The final notice of proposed civil penalty contains a statement of the charges and the amount of the proposed civil penalty and, as a result of information submitted to the agency attorney during informal procedures, may modify an allegation or a proposed civil penalty contained in a notice of proposed civil penalty.

(1) A final notice of proposed civil penalty may be issued—

(i) If the person charged with a violation fails to respond to the notice of proposed civil penalty within 30 days after receipt of that notice; or

(ii) If the parties participated in any procedures under paragraph (f)(2) of this section and the parties have not agreed to compromise the action or the agency attorney has not agreed to withdraw the notice of proposed civil penalty.

(2) Not later than 15 days after receipt of the final notice of proposed civil penalty, the person charged with a violation may do one of the following:

(i) Submit the amount of the proposed civil penalty or an agreed-upon amount, in which case either an order assessing civil penalty or a compromise order under paragraph (n) of this section may be issued in that amount; or

(ii) Request a hearing conducted in accordance with subpart G of this part.

(h) *Request for a hearing.* Any person requesting a hearing, under paragraph (f)(3) or (g)(2)(ii) of this section must file the request with the FAA Hearing Docket Clerk and serve the request on the agency attorney in accordance with the requirements in subpart G of this part.

(i) *Hearing.* The procedural rules in subpart G of this part apply to the hearing.

(j) *Appeal.* Either party may appeal the administrative law judge's initial decision to the FAA decisionmaker under the procedures in subpart G of this part. The procedural rules in subpart G of this part apply to the appeal.

(k) *Judicial review.* A person may seek judicial review only of a final decision and order of the FAA decisionmaker in accordance with §13.235.

(l) *Payment.* (1) A person must pay a civil penalty by:

(i) Sending a certified check or money order, payable to the Federal Aviation Administration, to the FAA office identified in the notice of proposed civil penalty, the final notice of proposed civil penalty, or the order assessing civil penalty; or

(ii) Making an electronic payment according to the directions specified in the notice of proposed civil penalty, the final notice of proposed civil penalty, or the order assessing civil penalty.

(2) The civil penalty must be paid within 30 days after service of the order assessing civil penalty, unless otherwise agreed to by the parties. In cases where a hearing is requested, an appeal to the FAA decisionmaker is filed, or a petition for review of the FAA decisionmaker's decision is filed in a U.S. court of appeals, the civil penalty must be paid within 30 days after all litigation in the matter is completed and the civil penalty is affirmed in whole or in part.

(m) *Collection of civil penalties.* If an individual does not pay a civil penalty imposed by an order assessing civil penalty or other final order, the Administrator may take action to collect the penalty.

(n) *Compromise.* The FAA may compromise the amount of any civil penalty imposed under this section under 49 U.S.C. 5123(e), 46301(f), or 46318 at any time before referring the action to the United States Attorney General, or the delegate of the Attorney General, for collection.

(1) When a civil penalty is compromised with a finding of violation, an agency attorney issues an order assessing civil penalty.

(2) When a civil penalty is compromised without a finding of violation, the agency attorney issues a compromise order that states the following:

(i) The person has paid a civil penalty or has signed a promissory note providing for installment payments.

(ii) The FAA makes no finding of a violation.

(iii) The compromise order will not be used as evidence of a prior violation in any subsequent civil penalty proceeding or certificate action proceeding.

§13.17 Seizure of aircraft.

(a) The Chief Counsel, or a Regional Administrator for an aircraft within the region, may issue an order authorizing a State or Federal law enforcement officer or a Federal Aviation Administration safety inspector to seize an aircraft that is involved in a violation for which a civil penalty may be imposed on its owner or the individual commanding the aircraft.

(b) Each person seizing an aircraft under this section places it in the nearest available and adequate public storage facility in the judicial district in which it was seized.

(c) The Regional Administrator or Chief Counsel, without delay, sends a written notice and a copy of this section to the registered owner of the seized aircraft and to each other person shown by FAA records to have an interest in it, stating the—

(1) Time, date, and place of seizure;

(2) Name and address of the custodian of the aircraft;

(3) Reasons for the seizure, including the violations alleged or proven to have been committed; and

(4) Amount that may be tendered as—

(i) A compromise of a civil penalty for the alleged violation; or

(ii) Payment for a civil penalty imposed for a proven violation.

(d) The Chief Counsel or Assistant Chief Counsel for Enforcement immediately sends a report to the United States Attorney for the judicial district in which it was seized, requesting the United States Attorney to institute proceedings to enforce a lien against the aircraft.

(e) The Regional Administrator or Chief Counsel directs the release of a seized aircraft when—

(1) The alleged violator pays a civil penalty or an amount agreed upon in compromise, and the costs of seizing, storing, and maintaining the aircraft;

(2) The aircraft is seized under an order of a court of the United States in proceedings in rem initiated under 49 U.S.C. 46305 to enforce a lien against the aircraft;

(3) The United States Attorney General, or the delegate of the Attorney General, notifies the FAA that the United States Attorney General, or the delegate of the Attorney General, refuses to institute proceedings in rem under 49 U.S.C. 46305 to enforce a lien against the aircraft; or

(4) A bond in the amount and with the sureties prescribed by the Chief Counsel or the Assistant Chief Counsel for Enforcement is deposited, conditioned on payment of the penalty or the compromise amount, and the costs of seizing, storing, and maintaining the aircraft.

§ 13.18 Civil penalties: Administrative assessment against an individual acting as a pilot, flight engineer, mechanic, or repairman.

(a) *General.* (1) This section applies to each action in which the FAA seeks to assess a civil penalty by administrative procedures against an individual acting as a pilot, flight engineer, mechanic, or repairman under 49 U.S.C. 46301(d)(5) for a violation listed in 49 U.S.C. 46301(d)(2). This section does not apply to a civil penalty assessed for a violation of 49 U.S.C. chapter 51, or a rule, regulation, or order issued thereunder.

(2) Notwithstanding the provisions of paragraph (a)(1) of this section, the U.S. district courts have exclusive jurisdiction of any civil penalty action involving an individual acting as a pilot, flight engineer, mechanic, or repairman for violations described in paragraph (a)(1), or under 49 U.S.C. 46301(d)(4), if:

(i) The amount in controversy is more than $50,000;

(ii) The action involves an aircraft subject to a lien that has been seized by the government; or

(iii) Another action has been brought for an injunction based on the same violation.

(b) *Definitions.* As used in this part, the following definitions apply:

(1) *Flight engineer* means an individual who holds a flight engineer certificate issued under part 63 of this chapter.

(2) *Individual acting as a pilot, flight engineer, mechanic, or repairman* means an individual acting in such capacity, whether or not that individual holds the respective airman certificate issued by the FAA.

(3) *Mechanic* means an individual who holds a mechanic certificate issued under part 65 of this chapter.

(4) *Pilot* means an individual who holds a pilot certificate issued under part 61 of this chapter.

(5) *Repairman* means an individual who holds a repairman certificate issued under part 65 of this chapter.

(c) *Delegation of authority.* The authority of the Administrator is delegated to the Chief Counsel and each Deputy Chief Counsel, and the Assistant Chief Counsel for Enforcement, as follows:

(1) To initiate and assess civil penalties under 49 U.S.C. 46301(d)(5);

(2) To refer cases to the Attorney General of the United States, or the delegate of the Attorney General, for collection of civil penalties; and

(3) To compromise the amount of a civil penalty under 49 U.S.C. 46301(f).

(d) *Notice of proposed assessment.* A civil penalty action is initiated by sending a notice of proposed assessment to the individual charged with a violation specified in paragraph (a) of this section. The notice of proposed assessment contains a statement of the charges and the amount of the proposed civil penalty. The individual charged with a violation may do the following:

(1) Submit the amount of the proposed civil penalty or an agreed-upon amount, in which case either an order of assessment or a compromise order will be issued in that amount.

(2) Answer the charges in writing by submitting information, including documents and witness statements, demonstrating that a violation of the regulations did not occur or that a penalty, or the amount of the penalty, is not warranted by the circumstances.

(3) Submit a written request to reduce the proposed civil penalty, stating the amount of reduction and the reasons, and providing any documents supporting a reduction of the proposed civil penalty, including records indicating a financial inability to pay.

(4) Submit a written request for an informal conference to discuss the matter with an agency attorney and submit relevant information or documents.

(5) Request that an order of assessment be issued so that the individual charged may appeal to the National Transportation Safety Board.

(e) *Failure to respond to notice of proposed assessment.* An order of assessment may be issued if the individual charged with a violation fails to respond to the notice of proposed assessment within 15 days after receipt of that notice.

(f) *Order of assessment.* An order of assessment, which imposes a civil penalty, may be issued for a violation described in paragraph (a) of this section after notice and an opportunity to answer any charges and be heard as to why such order should not be issued.

(g) *Appeal.* Any individual who receives an order of assessment issued under this section may appeal the order to the National Transportation Safety Board. The appeal stays the effectiveness of the Administrator's order.

(h) *Judicial review.* A party may seek judicial review only of a final decision and order of the National Transportation Safety Board under 49 U.S.C. 46301(d)(6) and 46110. Neither an initial decision, nor an order issued by an administrative law judge that has not been appealed to the National Transportation Safety Board, nor an order compromising a civil penalty action, may be appealed under any of those sections.

(i) *Compromise.* The FAA may compromise any civil penalty imposed under this section at any time before referring the action to the United States Attorney General, or the delegate of the Attorney General, for collection.

(1) When a civil penalty is compromised with a finding of violation, an agency attorney issues an order of assessment.

(2) When a civil penalty is compromised without a finding of violation, the agency attorney issues a compromise order of assessment that states the following:

(i) The individual has paid a civil penalty or has signed a promissory note providing for installment payments;

(ii) The FAA makes no finding of violation; and

(iii) The compromise order will not be used as evidence of a prior violation in any subsequent civil penalty proceeding or certificate action proceeding.

(j) *Payment.* (1) An individual must pay a civil penalty by:

(i) Sending a certified check or money order, payable to the Federal Aviation Administration, to the FAA office identified in the order of assessment; or

(ii) Making an electronic payment according to the directions specified in the order of assessment.

(2) The civil penalty must be paid within 30 days after service of the order of assessment, unless an appeal is filed with the National Transportation Safety Board. In cases where an appeal is filed with the National Transportation Safety Board, or a petition for review is filed with a U.S. court of appeals, the civil penalty must be paid within 30 days after all litigation in the matter is completed and the civil penalty is affirmed in whole or in part.

(k) *Collection of civil penalties.* If an individual does not pay a civil penalty imposed by an order of assessment or other final order, the Administrator may take action provided under the law to collect the penalty.

§ 13.19 **Certificate actions appealable to the National Transportation Safety Board.**

(a) This section applies to certificate actions by the Administrator that are appealable to the National Transportation Safety Board.

(1) Under 49 U.S.C. 44709(b) the Administrator may issue an order amending, modifying, suspending, or revoking all or part of any type certificate, production certificate, airworthiness certificate, airman certificate, air carrier operating certificate, air navigation facility certificate, or air agency certificate if as a result of a reinspection, reexamination, or other investigation, the Administrator determines that the public interest and safety in air commerce requires it, if a certificate holder has violated an aircraft noise or sonic boom standard or regulation prescribed under 49 U.S.C. 44715(a), or if the holder of the certificate is convicted of violating 16 U.S.C. 742j–1(a).

(2) The authority of the Administrator to issue orders under 49 U.S.C.

44709(b)(1)(A) and (b)(2) is delegated to the Chief Counsel, each Deputy Chief Counsel, and the Assistant Chief Counsel for Enforcement.

(b) The agency attorney will issue a notice before issuing a non-immediately effective order to amend, modify, suspend, or revoke a type certificate, production certificate, airworthiness certificate, airman certificate, air carrier operating certificate, air navigation facility certificate, air agency certificate, or to revoke an aircraft certificate of registration because the aircraft was used to carry out or facilitate an activity punishable under a law of the United States or a State related to a controlled substance (except a law related to simple possession of a controlled substance), by death or imprisonment for more than one year, and the owner of the aircraft permitted the use of the aircraft knowing that the aircraft was to be used for the activity.

(1) A notice of proposed certificate action will advise the certificate holder or aircraft owner of the charges or other reasons upon which the Administrator bases the proposed action, and allows the holder to answer any charges and to be heard as to why the certificate should not be amended, suspended, modified, or revoked.

(2) In response to a notice of proposed certificate action described in paragraph (b)(1) of this section, the certificate holder or aircraft owner, within 15 days of the date of receipt of the notice, may—

(i) Surrender the certificate and waive any right to contest or appeal the charged violations and sanction, in which case the Administrator will issue an order;

(ii) Answer the charges in writing by submitting information, including documents and witness statements, demonstrating that a violation of the regulations did not occur or that the proposed sanction is not warranted by the circumstances;

(iii) Submit a written request for an informal conference to discuss the matter with an agency attorney and submit relevant information or documents; or

(iv) Request that an order be issued in accordance with the notice of proposed certificate action so that the certificate holder or aircraft owner may appeal to the National Transportation Safety Board.

(c) In the case of an emergency order amending, modifying, suspending, or revoking a type certificate, production certificate, airworthiness certificate, airman certificate, air carrier operating certificate, air navigation facility certificate, or air agency certificate, a person affected by the immediate effectiveness of the Administrator's order may petition the National Transportation Safety Board for a review of the Administrator's determination that an emergency exists.

(d) A person may not petition the National Transportation Safety Board for a review of the Administrator's determination that safety in air transportation or air commerce requires the immediate effectiveness of an order where the action is based on the circumstances described in paragraph (d)(1), (2), or (3) of this section.

(1) The revocation of an individual's airman certificates for the reasons stated in paragraph (d)(1)(i) or (ii) of this section:

(i) A conviction under a law of the United States or a State related to a controlled substance (except a law related to simple possession of a controlled substance), of an offense punishable by death or imprisonment for more than one year if the Administrator finds that—

(A) An aircraft was used to commit, or facilitate the commission of the offense; and

(B) The individual served as an airman, or was on the aircraft, in connection with committing, or facilitating the commission of, the offense.

(ii) Knowingly carrying out an activity punishable, under a law of the United States or a State related to a controlled substance (except a law related to simple possession of a controlled substance), by death or imprisonment for more than one year; and—

(A) An aircraft was used to carry out or facilitate the activity; and

(B) The individual served as an airman, or was on the aircraft, in connec-

tion with carrying out, or facilitating the carrying out of, the activity.

(2) The revocation of a certificate of registration for an aircraft, and any other aircraft the owner of that aircraft holds, if the Administrator finds that—

(i) The aircraft was used to carry out or facilitate an activity punishable, under a law of the United States or a State related to a controlled substance (except a law related to simple possession of a controlled substance), by death or imprisonment for more than one year; and

(ii) The owner of the aircraft permitted the use of the aircraft knowing that the aircraft was to be used for the activity described in paragraph (d)(2)(i) of this section.

(3) The revocation of an airman certificate, design organization certificate, type certificate, production certificate, airworthiness certificate, air carrier operating certificate, airport operating certificate, air agency certificate, or air navigation facility certificate if the Administrator finds that the holder of the certificate or an individual who has a controlling or ownership interest in the holder—

(i) Was convicted in a court of law of a violation of a law of the United States relating to the installation, production, repair, or sale of a counterfeit or fraudulently-represented aviation part or material; or

(ii) Knowingly, and with the intent to defraud, carried out or facilitated an activity described in paragraph (d)(3)(i) of this section.

[Docket FAA–2018–1051, Amdt. 13–40, 86 FR 54527, Oct. 1, 2021, as amended by Docket FAA–2018–1051, Amdt. 13–40A, 87 FR 61233, Oct. 11, 2022]

§ 13.20 Orders of compliance, cease and desist orders, orders of denial, and other orders.

(a) *General.* This section applies to all of the following:

(1) Orders of compliance;

(2) Cease and desist orders;

(3) Orders of denial;

(4) Orders suspending or revoking a certificate of registration (but not revocation of a certificate of registration because the aircraft was used to carry

47

out or facilitate an activity punishable, under a law of the United States or a State related to a controlled substance (except a law related to simple possession of a controlled substance), by death or imprisonment for more than one year and the owner of the aircraft permitted the use of the aircraft knowing that the aircraft was to be used for the activity); and

(5) Other orders issued by the Administrator to carry out the provisions of the Federal aviation statute codified at 49 U.S.C. subtitle VII that apply this section by statute, rule, regulation, or order, or for which there is no specific administrative process provided by statute, rule, regulation, or order.

(b) *Applicability of procedures.* (1) Prior to the issuance of a non-immediately effective order covered by this section, the Administrator will provide the person who would be subject to the order with notice, advising the person of the charges or other reasons upon which the proposed action is based, and the provisions in paragraph (c) of this section apply.

(2) If the Administrator is of the opinion that an emergency exists related to safety in air commerce and requires immediate action and issues an order covered by this section that is immediately effective, the provisions of paragraph (d) of this section apply.

(c) *Non-emergency procedures.* (1) Within 30 days after service of the notice, the person subject to the notice may:

(i) Submit a written reply;

(ii) Agree to the issuance of the order as proposed in the notice of proposed action, waiving any right to contest or appeal the agreed-upon order issued under this option in any administrative or judicial forum;

(iii) Submit a written request for an informal conference to discuss the matter with an agency attorney; or

(iv) Request a hearing in accordance with the non-emergency procedures of subpart D of this part.

(2) After an informal conference is held or a reply is filed, if the agency attorney notifies the person that some or all of the proposed agency action will not be withdrawn, the person may, within 10 days after receiving the agency attorney's notification, request a hearing on the parts of the proposed agency action not withdrawn, in accordance with the non-emergency procedures of subpart D of this part.

(3) If a hearing is requested in accordance with paragraph (c)(1)(iv) or (c)(2) of this section, the non-emergency procedures of subpart D of this part apply.

(4) Failure to request a hearing within the periods provided in paragraph (c)(1)(iv) or (c)(2) of this section:

(i) Constitutes a waiver of the right to a hearing and appeal; and

(ii) Authorizes the agency to make appropriate findings of fact and to issue an appropriate order without further notice or proceedings.

(d) *Emergency procedures.* (1) If the Administrator is of the opinion that an emergency exists related to safety in air commerce and requires immediate action, the Administrator issues simultaneously:

(i) An immediately effective order that expires 80 days after the date of issuance and sets forth the charges or other reasons upon which the order is based; and

(ii) A notice of proposed action that:

(A) Sets forth the charges or other reasons upon which the notice of proposed action is based; and

(B) Advises that within 10 days after service of the notice, the person may appeal the notice by requesting an expedited hearing in accordance with the emergency procedures of subpart D of this part.

(2) The Administrator will serve the immediately effective order and the notice of proposed action together by personal or overnight delivery and by certified or registered mail to the person subject to the order and notice of proposed action.

(3) Failure to request a hearing challenging the notice of proposed action under the expedited procedures in subpart D of this part within 10 days after service of the notice:

(i) Constitutes a waiver of the right to a hearing and appeal under subpart D of this part; and

(ii) Authorizes the Administrator, without further notice or proceedings, to make appropriate findings of fact, issue an immediately effective order without expiration, and withdraw the 80-day immediately effective order.

48

(4) The filing of a request for hearing under subpart D of this part does not stay the effectiveness of the 80-day immediately effective order issued under this section.

(e) *Delegation of authority.* The authority of the Administrator under this section is delegated to the Chief Counsel, each Deputy Chief Counsel, and the Assistant Chief Counsel for Enforcement.

§§13.21–13.29 [Reserved]

Subpart D—Rules of Practice for FAA Hearings

SOURCE: Docket No. FAA-2018-1051; Amdt. No. 13-40, 86 FR 54532, Oct. 1, 2021, unless otherwise noted.

§13.31 Applicability.

This subpart applies to proceedings in which a hearing has been requested in accordance with §13.20 or §13.75. Hearings under this subpart are considered informal and are provided through the Office of Adjudication.

§13.33 Parties, representatives, and notice of appearance.

(a) *Parties.* Parties to proceedings under this subpart include the following: Complainant, respondent, and where applicable, intervenor.

(1) Complainant is the FAA Office that issued the notice of proposed action under the authorities listed in §13.31.

(2) Respondent is the party filing a request for hearing.

(3) Intervenor is a person permitted to participate as a party under §13.51.

(b) *Representatives.* Any party to a proceeding under this subpart may appear and be heard in person or by a representative. A representative is an attorney, or another representative designated by the party.

(c) *Notice of appearance*—(1) *Content.* The representative of a party must file a notice of appearance that includes the representative's name, address, telephone number, and, if available, fax number, and email address.

(2) *Filing.* A notice of appearance may be incorporated into an initial filing in a proceeding. A notice of appearance by additional representatives or substitutes after an initial filing in a proceeding must be filed independently.

§13.35 Request for hearing, complaint, and answer.

(a) *Initial filing and service.* A request for hearing must be filed with the FAA Hearing Docket, and a copy must be served on the official who issued the notice of proposed action, in accordance with the requirements in §13.43 for filing and service of documents. The request for hearing must be in writing and describe the action proposed by the FAA, and must contain a statement that a hearing is requested under this subpart.

(b) *Complaint.* Within 20 days after service of the copy of the request for hearing, the official who issued the notice of proposed action must forward a copy of that notice, which serves as the complaint, to the FAA Hearing Docket.

(c) *Answer.* Within 30 days after service of the copy of the complaint, the Respondent must file an answer to the complaint. All allegations in the complaint not specifically denied in the answer are deemed admitted.

§13.37 Hearing officer: Assignment and powers.

As soon as practicable after the filing of the complaint, the Director of the Office of Adjudication will assign a hearing officer to preside over the matter. The hearing officer may—

(a) Give notice concerning, and hold, prehearing conferences and hearings;

(b) Administer oaths and affirmations;

(c) Examine witnesses;

(d) Adopt procedures for the submission of evidence in written form;

(e) Issue subpoenas;

(f) Rule on offers of proof;

(g) Receive evidence;

(h) Regulate the course of proceedings, including but not limited to discovery, motions practice, imposition of sanctions, and the hearing;

(i) Hold conferences, before and during the hearing, to settle and simplify issues by consent of the parties;

(j) Dispose of procedural requests and similar matters;

(k) Issue protective orders governing the exchange and safekeeping of information otherwise protected by law, except that national security information may not be disclosed under such an order;

(l) Issue orders and decisions, and make findings of fact, as appropriate; and

(m) Take any other action authorized by this subpart.

§ 13.39 Disqualification of hearing officer.

(a) *Motion and supporting affidavit.* Any party may file a motion for disqualification under § 13.49(g). A party must state the grounds for disqualification, including, but not limited to, a financial or other personal interest that would be affected by the outcome of the enforcement action, personal animus against a party to the action or against a group to which a party belongs, prejudgment of the adjudicative facts at issue in the proceeding, or any other prohibited conflict of interest. A party must submit an affidavit with the motion for disqualification that sets forth, in detail, the matters alleged to constitute grounds for disqualification.

(b) *Timing.* A motion for disqualification must be filed prior to the issuance of the hearing officer's decision under § 13.63(b). Any party may file a response to a motion for disqualification, but must do so no later than 5 days after service of the motion for disqualification.

(c) *Decision on motion for disqualification.* The hearing officer must render a decision on the motion for disqualification no later than 15 days after the motion has been filed. If the hearing officer finds that the motion for disqualification and supporting affidavit show a basis for disqualification, the hearing officer must withdraw from the proceedings immediately. If the hearing officer finds that disqualification is not warranted, the hearing officer must deny the motion and state the grounds for the denial on the record. If the hearing officer fails to rule on a party's motion for disqualification within 15 days after the motion has been filed, the motion is deemed granted.

(d) *Self-disqualification.* A hearing officer may disqualify himself or herself at any time.

§ 13.41 Separation of functions and prohibition on ex parte communications.

(a) *Separation of powers.* The hearing officer independently exercises the powers under this subpart in a manner conducive to justice and the proper dispatch of business. The hearing officer must not participate in any appeal to the Administrator.

(b) *Ex parte communications.* (1) No substantive ex parte communications between the hearing officer and any party are permitted.

(2) A hearing, conference, or other event scheduled with prior notice will not constitute ex parte communication prohibited by this section. A hearing, conference, or other event scheduled with prior notice, may proceed in the hearing officer's sole discretion if a party fails to appear, respond, or otherwise participate, and will not constitute an ex parte communication prohibited by this section.

(3) For an appeal to the Administrator under this subpart, FAA attorneys representing the complainant must not advise the Administrator or engage in any ex parte communications with the Administrator or his advisors.

§ 13.43 Service and filing of pleadings, motions, and documents.

(a) *General rule.* A party must file all requests for hearing, pleadings, motions, and documents with the FAA Hearing Docket, and must serve a copy upon all parties to the proceedings.

(b) *Methods of filing.* Filing must be by email, personal delivery, expedited or overnight courier express service, mail, or fax.

(c) *Address for filing.* A person filing a document with the FAA Hearing Docket must use the address identified for the method of filing as follows:

(1) *If delivery is in person, or by expedited or overnight express courier service.* Federal Aviation Administration, 600 Independence Avenue SW, Wilbur Wright Building—Suite 2W100, Washington, DC 20597; Attention: FAA Hearing Docket, AGC–70.

(2) *If delivery is via U.S. mail, or U.S. certified or registered mail.* Federal Aviation Administration, 800 Independence Avenue SW, Washington, DC 20591; Attention: FAA Hearing Docket, AGC–70, Wilbur Wright Building—Suite 2W100.

(3) *Contact information.* The FAA Office of Adjudication will make available on its website an email address and fax number for the FAA Hearing Docket, as well as other contact information.

(d) *Requirement to file an original document and number of copies.* A party must file an original document and one copy when filing by personal delivery or by mail. Only one copy must be filed if filing is accomplished by email or fax.

(e) *Filing by email.* A document that is filed by email must be attached as a Portable Document Format (PDF) file to an email. The document must be signed in accordance with §13.207. The email message does not constitute a submission, but serves only to deliver the attached PDF file to the FAA Hearing Docket.

(f) *Methods of service*—(1) *General.* A person may serve any document by email, personal delivery, expedited or overnight courier express service, mail, or fax.

(2) *Service by email.* Service of documents by email is voluntary and requires the prior consent of the person to be served by email. A person may retract consent to be served by email by filing and serving a written retraction. A document that is served by email must be attached as a PDF file to an email message.

(g) *Certificate of service.* A certificate of service must accompany all documents filed with the FAA Hearing Docket. The certificate of service must be signed, describe the method of service, and state the date of service.

(h) *Date of filing and service.* If a document is sent by fax or email, the date of filing and service is the date the email or fax is sent. If a document is sent by personal delivery or by expedited or overnight express courier service, the date of filing and service is the date that delivery is accomplished. If a document is mailed, the date of filing and service is the date shown on the certificate of service, the date shown on the postmark if there is no certificate of service, or the mailing date shown by other evidence if there is no certificate of service or postmark.

§13.44 [Reserved]

§13.45 Computation of time and extension of time.

(a) In computing any period of time prescribed or allowed by this subpart, the date of the act, event, default, notice, or order is not to be included in the computation. The last day of the period so computed is to be included unless it is a Saturday, Sunday, or Federal holiday, in which event the period runs until the end of the next day that is not a Saturday, Sunday, or a Federal holiday.

(b) Whenever a party must respond within a prescribed period after service by mail, 5 days are added to the prescribed period.

(c) The parties may agree to extend the time for filing any document required by this subpart with the consent of—

(1) The Director of the Office of Adjudication prior to the designation of a hearing officer;

(2) The hearing officer prior to the filing of a notice of appeal; or

(3) The Director of the Office of Adjudication after the filing of a notice of appeal.

(d) If the parties do not agree, a party may make a written request to extend the time for filing to the appropriate official identified in paragraph (c) of this section. The appropriate official may grant the request for good cause shown.

§13.47 Withdrawal or amendment of the complaint, answer, or other filings.

(a) *Withdrawal.* At any time before the hearing, the complainant may withdraw the complaint, and the respondent may withdraw the request for hearing.

(b) *Amendments.* At any time more than 10 days before the date of hearing, any party may amend its complaint, answer, or other pleading, by filing the amendment with the FAA Hearing Docket and serving a copy of it on every other party. After that time, amendment requires approval of the hearing officer. If an initial pleading is amended, the hearing officer must allow the other parties a reasonable opportunity to respond.

§ 13.49 Motions.

(a) *Motions in lieu of an answer.* A respondent may file a motion to dismiss or a motion for a more definite statement in place of an answer. If the hearing officer denies the motion, the respondent must file an answer within 10 days.

(1) *Motion to dismiss.* The respondent may file a motion asserting that the allegations in the complaint fail to state a violation of Federal aviation statutes, a violation of regulations in this chapter, lack of qualification of the respondent, or other appropriate grounds.

(2) *Motion for more definite statement.* The respondent may file a motion that the allegations in the notice be made more definite and certain.

(b) *Motion to dismiss request for hearing.* The FAA may file a motion to dismiss a request for hearing based on jurisdiction, timeliness, or other appropriate grounds.

(c) *Motion for decision on the pleadings or for summary decision.* After the complaint and answer are filed, either party may move for a decision on the pleadings or for a summary decision, in the manner provided by Rules 12 and 56, respectively, of the Federal Rules of Civil Procedure.

(d) *Motion to strike.* Upon motion of either party, the hearing officer may order stricken, from any pleadings, any insufficient allegation or defense, or any redundant, immaterial, impertinent, or scandalous matter.

(e) *Motion to compel.* Any party may file a motion asking the hearing officer to order any other party to produce discovery requested in accordance with § 13.53 if—

(1) The other party has failed to timely produce the requested discovery; and

(2) The moving party certifies it has in good faith conferred with the other party in an attempt to obtain the requested discovery prior to filing the motion to compel.

(f) *Motion for protective order.* The hearing officer may order information contained in anything filed, or in any testimony given pursuant to this subpart withheld from public disclosure when, in the judgment of the hearing officer, disclosure would be detrimental to aviation safety; disclosure would not be in the public interest; or the information is not otherwise required to be made available to the public. Any person may make written objection to the public disclosure of any information, stating the ground for such objection.

(g) *Other motions.* Any application for an order or ruling not otherwise provided for in this subpart must be made by motion.

(h) *Responses to motions.* Any party may file a response to any motion under this subpart within 10 days after service of the motion.

§ 13.51 Intervention.

Any person may move for leave to intervene in a proceeding and may become a party thereto, if the hearing officer, after the case is sent to the hearing officer for hearing, finds that the person may be bound by the order to be issued in the proceedings or has a property or financial interest that may not be adequately represented by existing parties, and that the intervention will not unduly broaden the issues or delay the proceedings. Except for good cause shown, a motion for leave to intervene may not be considered if it is filed less than 10 days before the hearing.

§ 13.53 Discovery.

(a) *Filing.* Discovery requests and responses are not filed with the FAA Hearing Docket unless in support of a motion, offered for impeachment, or other permissible circumstances as approved by the hearing officer.

(b) *Scope of discovery.* Any party may discover any matter that is not privileged and is relevant to any party's claim or defense.

(c) *Time for response to written discovery requests.* (1) Written discovery includes interrogatories, requests for admission or stipulations, and requests for production of documents.

(2) Unless otherwise directed by the hearing officer, a party must serve its response to a discovery request no later than 30 days after service of the discovery request.

(d) *Depositions.* After the respondent has filed a request for hearing and an answer, either party may take testimony by deposition.

(e) *Limits on discovery.* The hearing officer may limit the frequency and extent of discovery upon a showing by a party that—

(1) The discovery requested is cumulative or repetitious;

(2) The discovery requested can be obtained from another less burdensome and more convenient source;

(3) The party requesting the information has had ample opportunity to obtain the information through other discovery methods permitted under this section; or

(4) The method or scope of discovery requested by the party is unduly burdensome or expensive.

§13.55 Notice of hearing.

The hearing officer must set a reasonable date, time, and location for the hearing, and must give the parties adequate notice thereof, and of the nature of the hearing. Due regard must be given to the convenience of the parties with respect to the location of the hearing.

§13.57 Subpoenas and witness fees.

(a) *Application.* The hearing officer, upon application by any party to the proceeding, may issue subpoenas requiring the attendance of witnesses or the production of documents or tangible things at a hearing or for the purpose of taking depositions, as permitted by law. The application for producing evidence must show its general relevance and reasonable scope. Absent good cause shown, a party must file a request for a subpoena at least:

(1) 15 days before a scheduled deposition under the subpoena; or

(2) 30 days before a scheduled hearing where attendance at the hearing is sought.

(b) *Procedure.* A party seeking the production of a document in the custody of an FAA employee must use the discovery procedure found in §13.53, and if necessary, a motion to compel under §13.49. A party that applies for the attendance of an FAA employee at a hearing must send the application, in writing, to the hearing officer. The application must set forth the need for that employee's attendance.

(c) *Fees.* Except for an employee of the agency who appears at the direction of the agency, a witness who appears at a deposition or hearing is entitled to the same fees and allowances as provided for under 28 U.S.C. 1821. The party who applies for a subpoena to compel the attendance of a witness at a deposition or hearing, or the party at whose request a witness appears at a deposition or hearing, must pay the witness fees and allowances described in this section.

(d) *Service of subpoenas.* Any person who is at least 18 years old and not a party may serve a subpoena. Serving a subpoena requires delivering a copy to the named person. Except for the complainant, the party that requested the subpoena must tender at the time of service the fees for 1 day's attendance and the allowances allowed by law if the subpoena requires that person's attendance. Proving service, if necessary, requires the filing with the FAA Hearing Docket of a statement showing the date and manner of service and the names of the persons served. The server must certify the statement.

(e) *Motion to quash or modify the subpoena.* A party, or any person served with a subpoena, may file a motion to quash or modify the subpoena with the hearing officer at or before the time specified in the subpoena for compliance. The movant must describe, in detail, the basis for the application to quash or modify the subpoena including, but not limited to, a statement that the testimony, document, or tangible thing is not relevant to the proceeding, that the subpoena is not reasonably tailored to the scope of the proceeding, or that the subpoena is unreasonable and oppressive. A motion to quash or modify the subpoena will stay the effect of the subpoena pending a decision by the hearing officer on the motion.

(f) *Enforcement of subpoena.* If a person disobeys a subpoena, a party may apply to a U.S. district court to seek judicial enforcement of the subpoena.

§13.59 Evidence.

(a) Each party to a hearing may present the party's case or defense by oral or documentary evidence, submit evidence in rebuttal, and conduct such cross-examination as may be needed for a full disclosure of the facts.

(b) Except with respect to affirmative defenses and notices of proposed denial, the burden of proof is upon the complainant.

§ 13.61 Argument and submittals.

The hearing officer must give the parties adequate opportunity to present arguments in support of motions, objections, and the final order. The hearing officer may determine whether arguments are to be oral or written. At the end of the hearing, the hearing officer may allow each party to submit written proposed findings and conclusions and supporting reasons for them.

§ 13.63 Record, decision, and aircraft registration proceedings.

(a) *The record.* (1) The testimony and exhibits admitted at a hearing, together with all papers, requests, and rulings filed in the proceedings, are the exclusive basis for the issuance of the hearing officer's decision.

(2) On appeal to the Administrator, the record shall include all of the information identified in paragraph (a)(1) of this section and evidence proffered but not admitted at the hearing.

(3) Any party may obtain a transcript of the hearing from the official reporter upon payment of the required fees.

(b) *Hearing officer's decision.* The decision by the hearing officer must include findings of fact based on the record, conclusions of law, and an appropriate order.

(c) *Certain aircraft registration proceedings.* If the hearing officer determines that an aircraft is ineligible for a certificate of aircraft registration in proceedings relating to aircraft registration orders suspending or revoking a certificate of registration under § 13.20, the hearing officer may suspend or revoke the aircraft registration certificate.

§ 13.65 Appeal to the Administrator, reconsideration, and judicial review.

(a) Any party to a hearing may appeal from the order of the hearing officer by filing with the FAA Hearing Docket a notice of appeal to the Administrator within 20 days after the date of issuance of the order. Filing and service of the notice of appeal, and any other papers, are accomplished according to the procedures in § 13.43.

(b) If a notice of appeal is not filed from the order issued by a hearing officer, such order is final with respect to the parties. Such order is not binding precedent and is not subject to judicial review.

(c) Any person filing an appeal authorized by paragraph (a) of this section must file an appeal brief with the Administrator within 40 days after the date of issuance of the order, and serve a copy on the other party. A reply brief must be filed within 40 days after service of the appeal brief and a copy served on the appellant.

(d) On appeal, the Administrator reviews the record of the proceeding and issues an order dismissing, reversing, modifying or affirming the order. The Administrator's order includes the reasons for the Administrator's action. The Administrator considers only whether:

(1) Each finding of fact is supported by a preponderance of the reliable, probative, and substantial evidence;

(2) Each conclusion is made in accordance with law, precedent, and policy; and

(3) The hearing officer committed any prejudicial error.

(e) The Director and legal personnel of the Office of Adjudication serve as the advisors to the Administrator for appeals under this section.

(1) The Director has the authority to:

(i) Manage all or portions of individual appeals; and to prepare written decisions and proposed final orders in such appeals;

(ii) Issue procedural and other interlocutory orders aimed at proper and efficient appeal management, including, without limitation, scheduling and sanctions orders;

(iii) Grant or deny motions to dismiss appeals;

(iv) Dismiss appeals upon request of the appellant or by agreement of the parties;

(v) Stay decisions and orders of the Administrator, pending judicial review or reconsideration by the Administrator;

(vi) Summarily dismiss repetitious or frivolous petitions to reconsider or modify orders;

(vii) Correct typographical, grammatical, and similar errors in the Administrator's decisions and orders, and to make non-substantive editorial changes; and

(viii) Take all other reasonable steps deemed necessary and proper for the management of the appeals process, in accordance with this part and applicable law.

(2) The Director's authority in paragraph (e)(1) of this section may be redelegated, as necessary, except to hearing officers and others materially involved in the hearing that is the subject of the appeal.

(f) Motions to reconsider the final order of the Administrator must be filed with the FAA Hearing Docket within thirty days of service of the Administrator's order.

(g) Judicial review of the Administrator's final order under this section is provided in accordance with 49 U.S.C. 5127 or 46110, as applicable.

§ 13.67 **Procedures for expedited proceedings.**

(a) When an expedited administrative hearing is requested in accordance with § 13.20(d), the procedures in this subpart will apply except as provided in paragraphs (a)(1) through (7) of this section.

(1) Service and filing of pleadings, motions, and documents must be by overnight delivery, and fax or email. Responses to motions must be filed within 7 days after service of the motion.

(2) Within 3 days after receipt of the request for hearing, the agency must file a copy of the notice of proposed action, which serves as the complaint, to the FAA Hearing Docket.

(3) Within 3 days after receipt of the complaint, the person that requested the hearing must file an answer to the complaint. All allegations in the complaint not specifically denied in the answer are deemed admitted. Failure to file a timely answer, absent a showing of good cause, constitutes withdrawal of the request for hearing.

(4) Within 3 days of the filing of the complaint, the Director of the Office of Adjudication will assign a hearing officer to preside over the matter.

(5) The parties must serve discovery as soon as possible and set time limits for compliance with discovery requests that accommodate the accelerated adjudication schedule set forth in this subpart. The hearing officer will resolve any failure of the parties to agree to a discovery schedule.

(6) The expedited hearing must commence within 40 days after the notice of proposed action was issued.

(7) The hearing officer must issue an oral decision and order dismissing, reversing, modifying, or affirming the notice of proposed action at the close of the hearing. If a notice of appeal is not filed, such order is final with respect to the parties and is not subject to judicial review.

(b) Any party to the expedited hearing may appeal from the initial decision of the hearing officer to the Administrator by filing a notice of appeal within 3 days after the date on which the decision was issued. The time limitations for the filing of documents for appeals under this section will not be extended by reason of the unavailability of the hearing transcript.

(1) Any appeal to the Administrator under this section must be perfected within 7 days after the date the notice of appeal was filed by filing a brief in support of the appeal. Any reply to the appeal brief must be filed within 7 days after the date the appeal brief was served on that party. The Administrator must issue an order deciding the appeal no later than 80 days after the date the notice of proposed action was issued.

(2) The Administrator's order is immediately effective and constitutes the final agency decision. The Administrator's order may be appealed pursuant to 49 U.S.C. 46110. The filing of an appeal under 49 U.S.C. 46110 does not stay the effectiveness of the Administrator's order.

(c) At any time after an immediately effective order is issued, the FAA may

55

request the United States Attorney General, or the delegate of the Attorney General, to bring an action for appropriate relief.

§ 13.69 Other matters: Alternative dispute resolution, standing orders, and forms.

(a) Parties may use mediation to achieve resolution of issues in controversy addressed by this subpart. Parties seeking alternative dispute resolution services may engage the services of a mutually acceptable mediator. The mediator must not participate in the adjudication under this subpart of any matter in which the mediator has provided mediation services. Mediation discussions and submissions will remain confidential consistent with the provisions of the Administrative Dispute Resolution Act, the principles of Federal Rule of Evidence 408, and other applicable Federal laws.

(b) The Director of the Office of Adjudication may issue standing orders and forms needed for the proper dispatch of business under this subpart.

Subpart E—Orders of Compliance Under the Hazardous Materials Transportation Act

SOURCE: Docket No. FAA-2018-1051; Amdt. No. 13-40, 86 FR 54536, Oct. 1, 2021, unless otherwise noted.

§ 13.70 Delegation of authority.

The authority of the Administrator under 49 U.S.C. 5121(a) and (d) is delegated to the Chief Counsel, each Deputy Chief Counsel, and the Assistant Chief Counsel for Enforcement.

[Docket FAA–2018–1051, Amdt. 13–40A, 87 FR 61233, Oct. 11, 2022]

§ 13.71 Applicability.

(a) An order of compliance may be issued after notice and an opportunity for a hearing in accordance with §§ 13.73 through 13.77 whenever the Chief Counsel, a Deputy Chief Counsel, or the Assistant Chief Counsel for Enforcement has reason to believe that a person is engaging in the transportation or shipment by air of hazardous materials in violation of the Hazardous Materials Transportation Act, as amended and

codified at 49 U.S.C. chapter 51, or any rule, regulation, or order issued under 49 U.S.C. chapter 51, for which the FAA exercises enforcement responsibility, and the circumstances do not require the issuance of an emergency order under 49 U.S.C. 5121(d).

(b) If circumstances require the issuance of an emergency order under 49 U.S.C. 5121(d), the Chief Counsel, a Deputy Chief Counsel, or the Assistant Chief Counsel for Enforcement will issue an emergency order of compliance as described in § 13.81.

§ 13.73 Notice of proposed order of compliance.

The Chief Counsel, a Deputy Chief Counsel, or the Assistant Chief Counsel for Enforcement may issue to an alleged violator a notice of proposed order of compliance advising the alleged violator of the charges and setting forth the remedial action sought in the form of a proposed order of compliance.

§ 13.75 Reply or request for hearing.

(a) Within 30 days after service upon the alleged violator of a notice of proposed order of compliance, the alleged violator may—

(1) Submit a written reply;

(2) Submit a written request for an informal conference to discuss the matter with an agency attorney; or

(3) Request a hearing in accordance with subpart D of this part.

(b) If, after an informal conference is held or a reply is filed, the agency attorney notifies the person named in the notice that some or all of the proposed agency action will not be withdrawn or will not be subject to a consent order of compliance, the alleged violator may, within 10 days after receiving the agency attorney's notification, request a hearing in accordance with subpart D of this part.

(c) Failure of the alleged violator to file a reply or request a hearing within the period provided in paragraph (a) or (b) of this section, as applicable—

(1) Constitutes a waiver of the right to a hearing under subpart D of this part and the right to petition for judicial review; and

(2) Authorizes the Administrator to make any appropriate findings of fact

and to issue an appropriate order of compliance, without further notice or proceedings.

§13.77 Consent order of compliance.

(a) At any time before the issuance of an order of compliance, an agency attorney and the alleged violator may agree to dispose of the case by the issuance of a consent order of compliance.

(b) The alleged violator may submit a proposed consent order to an agency attorney. The proposed consent order must include—

(1) An admission of all jurisdictional facts;

(2) An express waiver of the right to further procedural steps and of all rights to legal review in any forum;

(3) An express waiver of attorney's fees and costs;

(4) If a notice has been issued prior to the proposed consent order of compliance, an incorporation by reference of the notice and an acknowledgement that the notice may be used to construe the terms of the consent order of compliance; and

(5) If a request for hearing is pending in any forum, a provision that the alleged violator will withdraw the request for a hearing and request that the case be dismissed.

§13.79 [Reserved]

§13.81 Emergency orders.

(a) Notwithstanding §§13.73 through 13.77, the Chief Counsel, each Deputy Chief Counsel, or the Assistant Chief Counsel for Enforcement may issue an emergency order of compliance, which is effective upon issuance, in accordance with the procedures in subpart C of 49 CFR part 109, if the person who issues the order finds that there is an "imminent hazard" as defined in 49 CFR 109.1.

(b) The FAA official who issued the emergency order of compliance may rescind or suspend the order if the criteria set forth in paragraph (a) of this section are no longer satisfied, and, when appropriate, may issue a notice of proposed order of compliance under §13.73.

(c) If at any time in the course of a proceeding commenced in accordance with §13.73 the criteria set forth in paragraph (a) of this section are satisfied, the official who issued the notice may issue an emergency order of compliance, even if the period for filing a reply or requesting a hearing specified in §13.75 has not expired.

13.83–13.87 [Reserved]

Subpart F—Formal Fact-Finding Investigation Under an Order of Investigation

SOURCE: Docket No. FAA-2018-1051; Amdt. No. 13-40, 86 FR 54536, Oct. 1, 2021, unless otherwise noted.

§13.101 Applicability.

(a) This subpart applies to fact-finding investigations in which an investigation has been ordered under §13.3(c) or §13.5(f)(2).

(b) This subpart does not limit the authority of any person to issue subpoenas, administer oaths, examine witnesses, and receive evidence in any informal investigation as otherwise provided by law.

§13.103 Order of investigation.

The order of investigation—

(a) Defines the scope of the investigation by describing the information sought in terms of its subject matter or its relevancy to specified FAA functions;

(b) Sets forth the form of the investigation which may be either by individual deposition or investigative proceeding or both; and

(c) Names the official who is authorized to conduct the investigation and serve as the presiding officer.

§13.105 Notification.

Any person under investigation and any person required to testify and produce documentary or physical evidence during the investigation will be advised of the purpose of the investigation, and of the place where the investigative proceeding or deposition will be convened. This may be accomplished by a notice of investigation or by a subpoena. A copy of the order of investigation may be sent to such persons when appropriate.

§ 13.107 Designation of additional parties.

(a) The presiding officer may designate additional persons as parties to the investigation, if in the discretion of the presiding officer, it will aid in the conduct of the investigation.

(b) The presiding officer may designate any person as a party to the investigation if—

(1) The person petitions the presiding officer to participate as a party;

(2) The disposition of the investigation may as a practical matter impair the ability to protect the person's interest unless allowed to participate as a party; and

(3) The person's interest is not adequately represented by existing parties.

§ 13.109 Convening the investigation.

The presiding officer will conduct the investigation at a location convenient to the parties involved and as expeditious and efficient as handling of the investigation permits.

§ 13.111 Subpoenas.

(a) At the discretion of the presiding officer, or at the request of a party to the investigation, the presiding officer may issue a subpoena directing any person to appear at a designated time and place to testify or to produce documentary or physical evidence relating to any matter under investigation.

(b) Subpoenas must be served by personal service on the person or an agent designated in writing for the purpose, or by registered or certified mail addressed to the person or agent. Whenever service is made by registered or certified mail, the date of mailing will be considered the time when service is made.

(c) Subpoenas extend in jurisdiction throughout the United States and any territory or possession thereof.

§ 13.113 Noncompliance with the investigative process.

(a) If a person disobeys a subpoena, the Administrator or a party to the investigation may petition a court of the United States to enforce the subpoena in accordance with applicable statutes.

(b) If a party to the investigation fails to comply with the provisions of this subpart or an order issued by the presiding officer, the Administrator may bring a civil action to enforce the requirements of this subpart or any order issued under this subpart in a court of the United States in accordance with applicable statutes.

§ 13.115 Public proceedings.

(a) All investigative proceedings and depositions must be public unless the presiding officer determines that the public interest requires otherwise.

(b) The presiding officer may order information contained in any report or document filed or in any testimony given pursuant to this subpart withheld from public disclosure when, in the judgment of the presiding officer, disclosure would adversely affect the interests of any person and is not required in the public interest or is not otherwise required by statute to be made available to the public. Any person may make written objection to the public disclosure of information, stating the grounds for such objection.

§ 13.117 Conduct of investigative proceeding or deposition.

(a) The presiding officer may question witnesses.

(b) Any witness may be accompanied by counsel.

(c) Any party may be accompanied by counsel and either the party or counsel may—

(1) Question witnesses, provided the questions are relevant and material to the matters under investigation and would not unduly impede the progress of the investigation; and

(2) Make objections on the record and argue the basis for such objections.

(d) Copies of all notices or written communications sent to a party or witness must, upon request, be sent to that person's attorney of record.

§ 13.119 Immunity and orders requiring testimony or other information.

(a) Whenever a person refuses, on the basis of a privilege against self-incrimination, to testify or provide other information during the course of any investigation conducted under this subpart, the presiding officer may, with the approval of the United States Attorney General, or the delegate of the

Attorney General, issue an order requiring the person to give testimony or provide other information. However, no testimony or other information so compelled (or any information directly or indirectly derived from such testimony or other information) may be used against the person in any criminal case, except in a prosecution for perjury, giving a false statement, or otherwise failing to comply with the order.

(b) The presiding officer may issue an order under this section if—

(1) The testimony or other information from the witness may be necessary to the public interest; and

(2) The witness has refused or is likely to refuse to testify or provide other information on the basis of a privilege against self-incrimination.

(c) Immunity provided by this section will not become effective until the person has refused to testify or provide other information on the basis of a privilege against self-incrimination, and an order under this section has been issued. An order, however, may be issued prospectively to become effective in the event of a claim of the privilege.

§ 13.121 Witness fees.

All witnesses appearing, other than employees of the Federal Aviation Administration, are entitled to the same fees and allowances as provided for under 28 U.S.C. 1821.

§ 13.123 Submission by party to the investigation.

(a) During an investigation conducted under this subpart, a party may submit to the presiding officer—

(1) A list of witnesses to be called, specifying the subject matter of the expected testimony of each witness; and

(2) A list of exhibits to be considered for inclusion in the record.

(b) If the presiding officer determines that the testimony of a witness or the receipt of an exhibit in accordance with paragraph (a) of this section will be relevant, competent, and material to the investigation, the presiding officer may subpoena the witness or use the exhibit during the investigation.

§ 13.125 Depositions.

Depositions for investigative purposes may be taken at the discretion of the presiding officer with reasonable notice to the party under investigation. Depositions must be taken before the presiding officer or other person authorized to administer oaths and designated by the presiding officer. The testimony must be reduced to writing by the person taking the deposition, or under the direction of that person, and where possible must then be subscribed by the deponent. Any person may be compelled to appear and testify and to produce physical and documentary evidence.

§ 13.127 Reports, decisions, and orders.

The presiding officer must issue a written report based on the record developed during the formal investigation, including a summary of principal conclusions. A summary of principal conclusions must be prepared by the official who issued the order of investigation in every case that results in no action, or no action as to a particular party to the investigation. All such reports must be furnished to the parties to the investigation and made available to the public on request.

§ 13.129 Post-investigation action.

A decision on whether to initiate subsequent action must be made on the basis of the record developed during the formal investigation and any other information in the possession of the Administrator.

§ 13.131 Other procedures.

Any question concerning the scope or conduct of a formal investigation not covered in this subpart may be ruled on by the presiding officer on his or her own initiative, or on the motion of a party or a person testifying or producing evidence.

Subpart G—Rules of Practice In FAA Civil Penalty Actions

SOURCE: Docket No. FAA-2018-1051; Amdt. No. 13-40, 86 FR 54538, Oct. 1, 2021, unless otherwise noted.

§ 13.201　Applicability.

This subpart applies to all civil penalty actions initiated under § 13.16 in which a hearing has been requested.

§ 13.202　Definitions.

For this subpart only, the following definitions apply:

Administrative law judge means an administrative law judge appointed pursuant to the provisions of 5 U.S.C. 3105.

Agency attorney means the Deputy Chief Counsel or the Assistant Chief Counsel responsible for the prosecution of enforcement-related matters under this subpart, or attorneys who are supervised by those officials or are assigned to prosecute a particular enforcement-related matter under this subpart. Agency attorney does not include the Chief Counsel or anyone from the Office of Adjudication.

Complaint means a document issued by an agency attorney alleging a violation of a provision of the Federal aviation statute listed in the first sentence of 49 U.S.C. 46301(d)(2) or in 49 U.S.C. 47531, or of the Federal hazardous materials transportation statute, 49 U.S.C. 5121–5128, or a rule, regulation, or order issued under those statutes, that has been filed with the FAA Hearing Docket after a hearing has been requested under § 13.16(f)(3) or (g)(2)(ii).

Complainant means the FAA office that issued the notice of proposed civil penalty under § 13.16.

FAA decisionmaker means the Administrator of the Federal Aviation Administration, acting in the capacity of the decisionmaker on appeal, or any person to whom the Administrator has delegated the Administrator's decision-making authority in a civil penalty action. As used in this subpart, the FAA decisionmaker is the official authorized to issue a final decision and order of the Administrator in a civil penalty action.

Mail includes U.S. mail, U.S. certified mail, U.S. registered mail, or use of an expedited or overnight express courier service, but does not include email.

Office of Adjudication means the Federal Aviation Administration Office of Adjudication, including the FAA Hearing Docket, the Director of the Office of Adjudication and legal personnel, or any subsequently designated office (including its head and any legal personnel) that advises the FAA decisionmaker regarding appeals of initial decisions and orders to the FAA decisionmaker.

Order assessing civil penalty means a document that contains a finding of a violation of a provision of the Federal aviation statute listed in the first sentence of 49 U.S.C. 46301(d)(2) or in 49 U.S.C. 47531, or of the Federal hazardous materials transportation statute, 49 U.S.C. 5121–5128, or a rule, regulation, or order issued under those statutes, and may direct payment of a civil penalty. Unless an appeal is filed with the FAA decisionmaker in a timely manner, an initial decision or order of an administrative law judge is considered an order assessing civil penalty if an administrative law judge finds that an alleged violation occurred and determines that a civil penalty, in an amount found appropriate by the administrative law judge, is warranted. Unless a petition for review is filed with a U.S. Court of Appeals in a timely manner, a final decision and order of the Administrator is considered an order assessing civil penalty if the FAA decisionmaker finds that an alleged violation occurred and a civil penalty is warranted.

Party means the Respondent, the complainant and any intervenor.

Personal delivery includes hand-delivery or use of a contract or express messenger service. "Personal delivery" does not include the use of Federal Government interoffice mail service.

Pleading means a complaint, an answer, and any amendment of these documents permitted under this subpart.

Properly addressed means a document that shows an address contained in agency records; a residential, business, or other address submitted by a person on any document provided under this subpart; or any other address shown by other reasonable and available means.

Respondent means a person named in a complaint.

Writing or written includes paper or electronic documents that are filed or served by email, mail, personal delivery, or fax.

§ 13.203 Separation of functions.

(a) Civil penalty proceedings, including hearings, are prosecuted by an agency attorney.

(b) An agency employee who has engaged in the performance of investigative or prosecutorial functions in a civil penalty action must not participate in deciding or advising the administrative law judge or the FAA decisionmaker in that case, or a factually-related case, but may participate as counsel for the complainant or as a witness in the public proceedings.

(c) The Chief Counsel and the Director and legal personnel of the Office of Adjudication will advise the FAA decisionmaker regarding any appeal of an initial decision or order in a civil penalty action to the FAA decisionmaker.

§ 13.204 Appearances and rights of parties.

(a) Any party may appear and be heard in person.

(b) Any party may be accompanied, represented, or advised by an attorney or representative designated by the party, and may be examined by that attorney or representative in any proceeding governed by this subpart. An attorney or representative who represents a party must file a notice of appearance in the action, in the manner provided in § 13.210, and must serve a copy of the notice of appearance on each party, and on the administrative law judge, if assigned, in the manner provided in § 13.211, before participating in any proceeding governed by this subpart. The attorney or representative must include the name, address, and telephone number, and, if available, fax number and email address, of the attorney or representative in the notice of appearance.

(c) Any person may request a copy of a document in the record upon payment of reasonable costs. A person may keep an original document, data, or evidence, with the consent of the administrative law judge, by substituting a legible copy of the document for the record.

§ 13.205 Administrative law judges.

(a) *Powers of an administrative law judge.* In accordance with the rules of this subpart, an administrative law judge may:

(1) Give notice of, and hold, pre-hearing conferences and hearings;

(2) Administer oaths and affirmations;

(3) Issue subpoenas as authorized by law;

(4) Rule on offers of proof;

(5) Receive relevant and material evidence;

(6) Regulate the course of the hearing in accordance with the rules of this subpart;

(7) Hold conferences to settle or to simplify the issues by consent of the parties;

(8) Dispose of procedural motions and requests;

(9) Make findings of fact and conclusions of law, and issue an initial decision;

(10) Bar a person from a specific proceeding based on a finding of obstreperous or disruptive behavior in that specific proceeding; and

(11) Take any other action authorized by this subpart.

(b) *Limitations.* The administrative law judge must not issue an order of contempt, award costs to any party, or impose any sanction not specified in this subpart. If the administrative law judge imposes any sanction not specified in this subpart, a party may file an interlocutory appeal of right under § 13.219(c).

(c) *Disqualification.* The administrative law judge may disqualify himself or herself at any time. A party may file a motion for disqualification under § 13.218.

§ 13.206 Intervention.

(a) A person may submit a motion for leave to intervene as a party in a civil penalty action. Except for good cause shown, a motion for leave to intervene must be submitted not later than 10 days before the hearing.

(b) The administrative law judge may grant a motion for leave to intervene if the administrative law judge finds that intervention will not unduly broaden the issues or delay the proceedings and—

(1) The person seeking to intervene will be bound by any order or decision entered in the action; or

(2) The person seeking to intervene has a property, financial, or other legitimate interest that may not be addressed adequately by the parties.

(c) The administrative law judge may determine the extent to which an intervenor may participate in the proceedings.

§ 13.207 Certification of documents.

(a) *Signature required*. The attorney of record, the party, or the party's representative must sign, by hand, electronically, or by other method acceptable to the administrative law judge, or, if the matter is on appeal, to the FAA decisionmaker, each document tendered for filing with the FAA Hearing Docket or served on the administrative law judge and on each other party.

(b) *Effect of signing a document*. By signing a document, the attorney of record, the party, or the party's representative certifies that the attorney, the party, or the party's representative has read the document and, based on reasonable inquiry and to the best of that person's knowledge, information, and belief, the document is—

(1) Consistent with the rules in this subpart;

(2) Warranted by existing law or a good faith argument for extension, modification, or reversal of existing law; and

(3) Not unreasonable or unduly burdensome or expensive, not made to harass any person, not made to cause unnecessary delay, and not made to cause needless increase in the cost of the proceedings or for any other improper purpose.

(c) *Sanctions*. If the attorney of record, the party, or the party's representative signs a document in violation of this section, the administrative law judge or the FAA decisionmaker must:

(1) Strike the pleading signed in violation of this section;

(2) Strike the request for discovery or the discovery response signed in violation of this section and preclude further discovery by the party;

(3) Deny the motion or request signed in violation of this section;

(4) Exclude the document signed in violation of this section from the record;

(5) Dismiss the interlocutory appeal and preclude further appeal on that issue by the party who filed the appeal until an initial decision has been entered on the record; or

(6) Dismiss the appeal of the administrative law judge's initial decision to the FAA decisionmaker.

§ 13.208 Complaint.

(a) *Filing*. The agency attorney must file the complaint with the FAA Hearing Docket, or may file a written motion to dismiss a request for hearing under § 13.218 instead of filing a complaint, not later than 20 days after receipt by the agency attorney of a request for hearing. When filing the complaint, the agency attorney must follow the filing instructions in § 13.210. The agency attorney may suggest a location for the hearing when filing the complaint.

(b) *Service*. An agency attorney must serve a copy of the complaint on the respondent, the president of the corporation or company named as a respondent, or a person designated by the respondent to accept service of documents in the civil penalty action. When serving the complaint, the agency attorney must follow the service instructions in § 13.211.

(c) *Contents*. A complaint must set forth the facts alleged, any regulation allegedly violated by the respondent, and the proposed civil penalty in sufficient detail to provide notice of any factual or legal allegation and proposed civil penalty.

(d) *Motion to dismiss stale allegations or complaint*. Instead of filing an answer to the complaint, a respondent may move to dismiss the complaint, or that part of the complaint, alleging a violation that occurred more than 2 years before an agency attorney issued a notice of proposed civil penalty to the respondent.

(1) An administrative law judge may not grant the motion and dismiss the complaint or part of the complaint if the administrative law judge finds that the agency has shown good cause for any delay in issuing the notice of proposed civil penalty.

(2) If the agency fails to show good cause for any delay, an administrative law judge may dismiss the complaint, or that part of the complaint, alleging a violation that occurred more than 2 years before an agency attorney issued the notice of proposed civil penalty to the respondent.

(3) A party may appeal the administrative law judge's ruling on the motion to dismiss the complaint or any part of the complaint in accordance with §13.219(b).

§13.209 Answer.

(a) *Writing required.* A respondent must file in the FAA Hearing Docket a written answer to the complaint, or may file a written motion pursuant to §13.208 or §13.218 instead of filing an answer, not later than 30 days after service of the complaint. The answer must be dated and signed by the person responding to the complaint. An answer must be typewritten or legibly handwritten.

(b) *Filing.* A person filing an answer or motion under paragraph (a) of this section must follow the filing instructions in §13.210.

(c) *Service.* A person filing an answer or a motion under paragraph (a) of this section must serve a copy of the answer or motion in accordance with the service instructions in §13.211.

(d) *Contents.* An answer must specifically state any affirmative defense that the respondent intends to assert at the hearing. A person filing an answer may include a brief statement of any relief requested in the answer. The person filing an answer may recommend a location for the hearing when filing the answer.

(e) *Specific denial of allegations required.* A person filing an answer must admit, deny, or state that the person is without sufficient knowledge or information to admit or deny, each allegation in the complaint. All allegations in the complaint not specifically denied in the answer are deemed admitted. A general denial of the complaint is deemed a failure to file an answer.

(f) *Failure to file answer.* A person's failure to file an answer without good cause will be deemed an admission of the truth of each allegation contained in the complaint.

§13.210 Filing of documents.

(a) *General rule.* Unless provided otherwise in this subpart, all documents in proceedings under this subpart must be tendered for filing with the FAA Hearing Docket.

(b) *Methods of filing.* Filing must be by email, personal delivery, mail, or fax.

(c) *Address for filing.* A person filing a document with the FAA Hearing Docket must use the address identified for the method of filing as follows:

(1) *If delivery is in person, or by expedited or overnight express courier service.* Federal Aviation Administration, 600 Independence Avenue SW, Wilbur Wright Building—Suite 2W100, Washington, DC 20597; Attention: FAA Hearing Docket, AGC–70.

(2) *If delivery is via U.S. mail, or U.S. certified or registered mail.* Federal Aviation Administration, 800 Independence Avenue SW, Washington, DC 20591; Attention: FAA Hearing Docket, AGC–70, Wilbur Wright Building—Suite 2W100.

(3) *If delivery is via email or fax.* The email address and fax number for the FAA Hearing Docket, made available on the FAA Office of Adjudication website.

(d) *Date of filing.* If a document is filed by fax or email, the date of filing is the date the email or fax is sent. If a document is filed by personal delivery, the date of filing is the date that personal delivery is accomplished. If a document is filed by mail, the date of filing is the date shown on the certificate of service, the date shown on the postmark if there is no certificate of service, or the mailing date shown by other evidence if there is no certificate of service or postmark.

(e) *Form.* Each document must be typewritten or legibly handwritten.

(f) *Contents.* Unless otherwise specified in this subpart, each document must contain a short, plain statement of the facts on which the person's case rests and a brief statement of the action requested.

(g) *Requirement to file an original document and number of copies.* A party must file an original document and one copy when filing by personal delivery or by mail. Only one copy must be filed if filing is accomplished by email or fax.

(h) *Filing by email.* A document that is filed by email must be attached as a PDF file to an email. The document must be signed in accordance with § 13.207. The email message does not constitute a submission, but serves only to deliver the attached PDF file to the FAA Hearing Docket.

§ 13.211 Service of documents.

(a) *General.* A person must serve a copy of all documents on each party and the administrative law judge, if assigned, at the time of filing with the FAA Hearing Docket except as provided otherwise in this subpart.

(b) *Service by the FAA Hearing Docket, the administrative law judge, and the FAA decisionmaker.* The FAA Hearing Docket, the administrative law judge, and the FAA decisionmaker must send documents to a party by personal delivery, mail, fax, or email as provided in this section.

(c) *Methods of service*—(1) *General.* A person may serve any document by email, personal delivery, mail, or fax.

(2) *Service by email.* Service of documents by email is voluntary and requires the prior consent of the person to be served by email. A person may retract consent to be served by email by filing a written retraction with the FAA Hearing Docket and serving it on the other party and the administrative law judge. A document that is served by email must be attached as a PDF file to an email message.

(d) *Certificate of service.* A certificate of service must accompany all documents filed with the FAA Hearing Docket. The certificate of service must be signed, describe the method of service, and state the date of service.

(e) *Date of service.* If a document is served by fax or served by email, the date of service is the date the email or fax is sent. If a document is served by personal delivery, the date of service is the date that personal delivery is accomplished. If a document is mailed, the date of service is the date shown on the certificate of service, the date shown on the postmark if there is no certificate of service, or the mailing date shown by other evidence if there is no certificate of service or postmark.

(f) *Valid service.* A document served by mail or personal delivery that was properly addressed, was sent in accordance with this subpart, and that was returned as unclaimed, or that was refused or not accepted, is deemed to have been served in accordance with this subpart.

(g) *Additional time after service by mail.* Whenever a party must respond within a prescribed period after service by mail, 5 days are added to the prescribed period.

(h) *Presumption of service.* There is a presumption of service where a party or a person, who customarily receives mail, or receives it in the ordinary course of business, at either the person's residence or the person's principal place of business, acknowledges receipt of the document.

§ 13.212 Computation of time.

(a) This section applies to any period of time prescribed or allowed by this subpart, by notice or order of the administrative law judge, or by any applicable statute.

(b) The date of an act, event, or default is not included in a computation of time under this subpart.

(c) The last day of a time period is included unless it is a Saturday, Sunday, or a Federal holiday. If the last day is a Saturday, Sunday, or Federal holiday, the time period runs until the end of the next day that is not a Saturday, Sunday, or Federal holiday.

§ 13.213 Extension of time.

(a) The parties may agree to extend for a reasonable period the time for filing a document under this subpart. The party seeking the extension of time must submit a draft order to the administrative law judge to be signed by the administrative law judge and filed with the FAA Hearing Docket. The administrative law judge must sign and issue the order if the extension agreed to by the parties is reasonable.

(b) A party may file a written motion for an extension of time. A written motion for an extension of time must be filed with the FAA Hearing Docket in accordance with § 13.210. The motion must be filed no later than seven days before the document is due unless good cause for the late filing is shown. The party filing the motion must serve a copy of the motion in accordance with

§13.211. The administrative law judge may grant the extension of time if good cause for the extension is shown.

(c) If the administrative law judge fails to rule on a motion for an extension of time by the date the document was due, the motion for an extension of time is deemed granted for no more than 20 days after the original date the document was to be filed.

§13.214 Amendment of pleadings.

(a) *Filing and service.* A party must file the amendment with the FAA Hearing Docket and must serve a copy of the amendment on the administrative law judge, if assigned, and on all parties to the proceeding.

(b) *Time.* (1) Not later than 15 days before the scheduled date of a hearing, a party may amend a complaint or an answer without the consent of the administrative law judge.

(2) Less than 15 days before the scheduled date of a hearing, the administrative law judge may allow amendment of a complaint or an answer only for good cause shown in a motion to amend.

(c) *Responses.* The administrative law judge must allow a reasonable time, but not more than 20 days from the date of filing, for other parties to respond if an amendment to a complaint, answer, or other pleading has been filed with the FAA Hearing Docket and served on the administrative law judge and other parties.

§13.215 Withdrawal of complaint or request for hearing.

At any time before or during a hearing, an agency attorney may withdraw a complaint or a party may withdraw a request for a hearing without the consent of the administrative law judge. If an agency attorney withdraws the complaint or a party withdraws the request for a hearing and the answer, the administrative law judge must dismiss the proceedings under this subpart with prejudice.

§13.216 Waivers.

Waivers of any rights provided by statute or regulation must be in writing or by stipulation made at a hearing and entered into the record. The parties must set forth the precise terms of the waiver and any conditions.

§13.217 Joint procedural or discovery schedule.

(a) *General.* The parties may agree to submit a schedule for filing all prehearing motions, conducting discovery in the proceedings, or both.

(b) *Form and content of schedule.* If the parties agree to a joint procedural or discovery schedule, one of the parties must file the joint schedule setting forth the dates to which the parties have agreed, in accordance with §13.210, and must also serve a copy of the joint schedule in accordance with §13.211. The filing of the joint schedule must include a draft order establishing a joint schedule to be signed by the administrative law judge.

(1) The joint schedule may include, but need not be limited to, requests for discovery, objections to discovery requests, responses to discovery requests to which there are no objections, submission of prehearing motions, responses to prehearing motions, exchange of exhibits to be introduced at the hearing, and a list of witnesses that may be called at the hearing.

(2) Each party must sign the joint schedule.

(c) *Time.* The parties may agree to submit all prehearing motions and responses and may agree to close discovery in the proceedings under the joint schedule within a reasonable time before the date of the hearing, but not later than 15 days before the hearing.

(d) *Joint scheduling order.* The joint schedule filed by the parties is a proposed schedule that requires approval of the administrative law judge to become the joint scheduling order.

(e) *Disputes.* The administrative law judge must resolve disputes regarding discovery or disputes regarding compliance with the joint scheduling order as soon as possible so that the parties may continue to comply with the joint scheduling order.

(f) *Sanctions for failure to comply with joint schedule.* If a party fails to comply with a joint scheduling order, the administrative law judge may impose any of the following sanctions, proportional to the party's failure to comply with the order:

(1) Strike the relevant portion of a party's pleadings;

(2) Preclude prehearing or discovery motions by that party;

(3) Preclude admission of the relevant portion of a party's evidence at the hearing; or

(4) Preclude the relevant portion of the testimony of that party's witnesses at the hearing.

§ 13.218   Motions.

(a) *General.* A party applying for an order or ruling not specifically provided in this subpart must do so by filing a motion in accordance with § 13.210. A party must serve a copy of each motion in accordance with § 13.211.

(b) *Form and contents.* A party must state the relief sought by the motion and the particular grounds supporting that relief. If a party has evidence in support of a motion, the party must attach any supporting evidence, including affidavits, to the motion.

(c) *Filing of motions.* A motion made prior to the hearing must be in writing. Unless otherwise agreed by the parties or for good cause shown, a party must file any prehearing motion not later than 30 days before the hearing in the FAA Hearing Docket in accordance with § 13.210, and must serve a copy on the administrative law judge, if assigned, and on each party in accordance with § 13.211. Motions introduced during a hearing may be made orally on the record unless the administrative law judge directs otherwise.

(d) *Responses to motions.* Any party may file a response, with affidavits or other evidence in support of the response, not later than 10 days after service of a written motion on that party. When a motion is made during a hearing, the response may be made at the hearing on the record, orally or in writing, within a reasonable time determined by the administrative law judge.

(e) *Rulings on motions.* The administrative law judge must rule on all motions as follows:

(1) *Discovery motions.* The administrative law judge must resolve all pending discovery motions not later than 10 days before the hearing.

(2) *Prehearing motions.* The administrative law judge must resolve all pending prehearing motions not later than 7 days before the hearing. If the administrative law judge issues a ruling or order orally, the administrative law judge must serve a written copy of the ruling or order, within 3 days, on each party. In all other cases, the administrative law judge must issue rulings and orders in writing and must serve a copy of the ruling or order on each party.

(3) *Motions made during the hearing.* The administrative law judge must issue rulings and orders on oral motions. Oral rulings or orders on motions must be made on the record.

(f) *Specific motions.* The motions that a party may file include but are not limited to the following:

(1) *Motion to dismiss for insufficiency.* A respondent may file a motion to dismiss the complaint for insufficiency instead of filing an answer. If the administrative law judge denies the motion to dismiss the complaint for insufficiency, the respondent must file an answer not later than 10 days after service of the administrative law judge's denial of the motion. A motion to dismiss the complaint for insufficiency must show that the complaint fails to state a violation of a provision of the Federal aviation statute listed in the first sentence in 49 U.S.C. 46301(d)(2) or in 49 U.S.C. 47531, or any implementing rule, regulation, or order, or a violation of the Federal hazardous materials transportation statute, 49 U.S.C. 5121–5128, or any implementing rule, regulation, or order.

(2) *Motion to dismiss.* A party may file a motion to dismiss, specifying the grounds for dismissal. If an administrative law judge grants a motion to dismiss in part, a party may appeal the administrative law judge's ruling on the motion to dismiss under § 13.219(b).

(i) *Motion to dismiss a request for a hearing.* An agency attorney may file a motion to dismiss a request for a hearing instead of filing a complaint. If the motion to dismiss is not granted, the agency attorney must file the complaint in the FAA Hearing Docket and must serve a copy of the complaint on the administrative law judge and on each party not later than 10 days after service of the administrative law judge's ruling or order on the motion to dismiss. If the motion to dismiss is granted and the proceedings are terminated without a hearing, the respondent may appeal to the FAA decisionmaker under § 13.233. If required by the decision on appeal, the agency attorney must file a complaint in the FAA Hearing Docket and must serve a copy of the complaint on the administrative law judge and each party not later than 10 days after service of the FAA decisionmaker's decision on appeal.

(ii) *Motion to dismiss a complaint.* A respondent may file a motion to dismiss a complaint instead of filing an answer, including a motion to dismiss a stale complaint or allegations as provided in § 13.208. If the motion to dismiss is not granted, the respondent must file an answer in the FAA Hearing Docket and must serve a copy of the answer on the administrative law judge and on each party not later than 10 days after service of the administrative law judge's ruling or order on the motion to dismiss. If the motion to dismiss is granted and the proceedings are terminated without a hearing, the agency attorney may file an appeal in the FAA Hearing Docket under § 13.233 and must serve each other party. If required by the FAA decisionmaker's decision on appeal, the respondent must file an answer in the FAA Hearing Docket, and must serve a copy of the answer on the administrative law judge and on each party not later than 10 days after service of the decision on appeal.

(3) *Motion for a more definite statement.* A party may file a motion for a more definite statement of any pleading which requires a response under this subpart. A party must set forth, in detail, the indefinite or uncertain allegations contained in a complaint or response to any pleading and must submit the details that the party believes would make the allegation or response definite and certain.

(i) *Complaint.* A respondent may file a motion requesting a more definite statement of the allegations contained in the complaint instead of filing an answer. If the administrative law judge grants the motion, the agency attorney must supply a more definite statement not later than 15 days after service of the ruling granting the motion. If the agency attorney fails to supply a more

definite statement, the administrative law judge may strike the allegations in the complaint to which the motion is directed. If the administrative law judge denies the motion, the respondent must file an answer in the FAA Hearing Docket and must serve a copy of the answer on the administrative law judge and on each party not later than 10 days after service of the order of denial.

(ii) *Answer.* An agency attorney may file a motion requesting a more definite statement if an answer fails to respond clearly to the allegations in the complaint. If the administrative law judge grants the motion, the respondent must supply a more definite statement not later than 15 days after service of the ruling on the motion. If the respondent fails to supply a more definite statement, the administrative law judge may strike those statements in the answer to which the motion is directed. The respondent's failure to supply a more definite statement may be deemed an admission of unanswered allegations in the complaint.

(4) *Motion to strike.* Any party may make a motion to strike any insufficient allegation or defense, or any redundant, immaterial, impertinent, or scandalous matter in a pleading. A party must file a motion to strike before a response is required under this subpart or, if a response is not required, not later than 10 days after service of the pleading. A motion to strike must be filed in the FAA Hearing Docket and served on the administrative law judge, if assigned, and on each other party.

(5) *Motion for decision.* A party may make a motion for decision, regarding all or any part of the proceedings, at any time before the administrative law judge has issued an initial decision in the proceedings. The administrative law judge must grant a party's motion for decision if the pleadings, depositions, answers to interrogatories, admissions, matters that the administrative law judge has officially noticed, or evidence introduced during the hearing shows that there is no genuine issue of material fact and that the party making the motion is entitled to a decision as a matter of law. The party making the motion for decision has the burden

67

of showing that there is no genuine issue of material fact disputed by the parties.

(6) *Motion for disqualification.* A party may file a motion for disqualification in the FAA Hearing Docket and must serve a copy on the administrative law judge and on each party. A party may file the motion at any time after the administrative law judge has been assigned to the proceedings but must make the motion before the administrative law judge files an initial decision in the proceedings.

(i) *Motion and supporting affidavit.* A party must state the grounds for disqualification in a motion for disqualification, including, but not limited to, a financial or other personal interest that would be affected by the outcome of the enforcement action, personal animus against a party to the action or against a group to which a party belongs, prejudgment of the adjudicative facts at issue in the proceeding, or any other prohibited conflict of interest. A party must submit an affidavit with the motion for disqualification that sets forth, in detail, the matters alleged to constitute grounds for disqualification.

(ii) *Response.* A party must respond to the motion for disqualification not later than 5 days after service of the motion for disqualification.

(iii) *Decision on motion for disqualification.* The administrative law judge must render a decision on the motion for disqualification not later than 15 days after the motion has been filed. If the administrative law judge finds that the motion for disqualification and supporting affidavit show a basis for disqualification, the administrative law judge must withdraw from the proceedings immediately. If the administrative law judge finds that disqualification is not warranted, the administrative law judge must deny the motion and state the grounds for the denial on the record. If the administrative law judge fails to rule on a party's motion for disqualification within 15 days after the motion has been filed, the motion is deemed granted.

(iv) *Appeal.* A party may appeal the administrative law judge's denial of the motion for disqualification in accordance with § 13.219(b).

(7) *Motions for reconsideration of an initial decision, order dismissing a complaint, order dismissing a request for hearing or order dismissing a request for hearing and answer.* The FAA decisionmaker may treat motions for reconsideration of an initial decision, order dismissing a complaint, order dismissing a request for hearing, or order dismissing a request for hearing and answer as a notice of appeal under § 13.233, and if the motion was filed within the time allowed for the filing of a notice of appeal, the FAA decisionmaker will issue a briefing schedule.

§ 13.219   Interlocutory appeals.

(a) *General.* Unless otherwise provided in this subpart, a party may not appeal a ruling or decision of the administrative law judge to the FAA decisionmaker until the initial decision has been entered on the record. A decision or order of the FAA decisionmaker on the interlocutory appeal does not constitute a final order of the Administrator for the purposes of judicial appellate review as provided in § 13.235.

(b) *Interlocutory appeal for cause.* If a party orally requests or files a written request for an interlocutory appeal for cause, the proceedings are stayed until the administrative law judge issues a decision on the request. Any written request for interlocutory appeal for cause must be filed in the FAA Hearing Docket and served on each party and on the administrative law judge. If the administrative law judge grants the request, the proceedings are stayed until the FAA decisionmaker issues a decision on the interlocutory appeal. The administrative law judge must grant the request if a party shows that delay of the appeal would be detrimental to the public interest or would result in undue prejudice to any party.

(c) *Interlocutory appeals of right.* If a party notifies the administrative law judge of an interlocutory appeal of right, the proceedings are stayed until the FAA decisionmaker issues a decision on the interlocutory appeal. A party may file an interlocutory appeal of right, without the consent of the administrative law judge, before an initial decision has been entered in the case of:

(1) A ruling or order by the administrative law judge barring a person from the proceedings;

(2) Failure of the administrative law judge to dismiss the proceedings in accordance with §13.215; or

(3) A ruling or order by the administrative law judge in violation of §13.205(b).

(d) *Procedure.* A party must file a notice of interlocutory appeal, with supporting documents, with the FAA Hearing Docket, and must serve a copy of the notice and supporting documents on each party and the administrative law judge not later than 10 days after the administrative law judge's decision forming the basis of an interlocutory appeal of right, or not later than 10 days after the administrative law judge's decision granting an interlocutory appeal for cause, as appropriate. A party must file a reply, if any, with the FAA Hearing Docket, and serve a copy on each party and the administrative law judge not later than 10 days after service of the appeal. The FAA decisionmaker must render a decision on the interlocutory appeal on the record and as a part of the decision in the proceedings, within a reasonable time after receipt of the interlocutory appeal.

(e) *Summary rejection.* The FAA decisionmaker may reject frivolous, repetitive, or dilatory appeals, and may issue an order precluding one or more parties from making further interlocutory appeals in a proceeding in which there have been frivolous, repetitive, or dilatory interlocutory appeals.

§ 13.220 Discovery.

(a) *Initiation of discovery.* Any party may initiate discovery described in this section without the consent or approval of the administrative law judge at any time after a complaint has been filed in the proceedings.

(b) *Methods of discovery.* The following methods of discovery are permitted under this section: Depositions on oral examination or written questions of any person; written interrogatories directed to a party; requests for production of documents or tangible items to any person; and requests for admission by a party. A party must not file written interrogatories and re-sponses, requests for production of documents or tangible items and responses, and requests for admission and response with the FAA Hearing Docket or serve them on the administrative law judge. In the event of a discovery dispute, a party must attach a copy of the relevant documents in support of a motion made under this section.

(c) *Service on the agency.* A party must serve each discovery request directed to the agency or any agency employee on the agency attorney of record.

(d) *Time for response to discovery requests.* Unless otherwise directed by this subpart or agreed by the parties, a party must respond to a request for discovery, including filing objections to a request for discovery, not later than 30 days after service of the request.

(e) *Scope of discovery.* Subject to the limits on discovery set forth in paragraph (f) of this section, a party may discover any matter that is not privileged and that is relevant to any party's claim or defense, including the existence, description, nature, custody, condition, and location of any document or other tangible item and the identity and location of any person having knowledge of discoverable matter. A party may discover facts known, or opinions held, by an expert who any other party expects to call to testify at the hearing. A party has no ground to object to a discovery request on the basis that the information sought would not be admissible at the hearing.

(f) *Limiting discovery.* The administrative law judge must limit the frequency and extent of discovery permitted by this section if a party shows that—

(1) The information requested is cumulative or repetitious;

(2) The information requested can be obtained from another less burdensome and more convenient source;

(3) The party requesting the information has had ample opportunity to obtain the information through other discovery methods permitted under this section; or

(4) The method or scope of discovery requested by the party is unduly burdensome or expensive.

(g) *Confidential orders.* A party or person who has received a discovery request for information that is related to a trade secret, confidential or sensitive material, competitive or commercial information, proprietary data, or information on research and development, may file a motion for a confidential order in the FAA Hearing Docket in accordance with § 13.210, and must serve a copy of the motion for a confidential order on each party and on the administrative law judge in accordance with § 13.211.

(1) The party or person making the motion must show that the confidential order is necessary to protect the information from disclosure to the public.

(2) If the administrative law judge determines that the requested material is not necessary to decide the case, the administrative law judge must preclude any inquiry into the matter by any party.

(3) If the administrative law judge determines that the requested material may be disclosed during discovery, the administrative law judge may order that the material may be discovered and disclosed under limited conditions or may be used only under certain terms and conditions.

(4) If the administrative law judge determines that the requested material is necessary to decide the case and that a confidential order is warranted, the administrative law judge must provide:

(i) An opportunity for review of the document by the parties off the record;

(ii) Procedures for excluding the information from the record; and

(iii) Order that the parties must not disclose the information in any manner and the parties must not use the information in any other proceeding.

(h) *Protective orders.* A party or a person who has received a request for discovery may file a motion for protective order in the FAA Hearing Docket and must serve a copy of the motion for protective order on the administrative law judge and each other party. The party or person making the motion must show that the protective order is necessary to protect the party or the person from annoyance, embarrassment, oppression, or undue burden or expense. As part of the protective order, the administrative law judge may:

(1) Deny the discovery request;

(2) Order that discovery be conducted only on specified terms and conditions, including a designation of the time or place for discovery or a determination of the method of discovery; or

(3) Limit the scope of discovery or preclude any inquiry into certain matters during discovery.

(i) *Duty to supplement or amend responses.* A party who has responded to a discovery request has a duty to supplement or amend the response, as soon as the information is known, as follows:

(1) A party must supplement or amend any response to a question requesting the identity and location of any person having knowledge of discoverable matters.

(2) A party must supplement or amend any response to a question requesting the identity of each person who will be called to testify at the hearing as an expert witness and the subject matter and substance of that witness's testimony.

(3) A party must supplement or amend any response that was incorrect when made or any response that was correct when made but is no longer correct, accurate, or complete.

(j) *Depositions*—(1) *Form.* A deposition must be taken on the record and reduced to writing. The person being deposed must sign the deposition unless the parties agree to waive the requirement of a signature.

(2) *Administration of oaths.* Within the United States, or a territory or possession subject to the jurisdiction of the United States, a party must take a deposition before a person authorized to administer oaths by the laws of the United States or authorized by the law of the place where the examination is held. In foreign countries, a party must take a deposition in any manner allowed by the Federal Rules of Civil Procedure.

(3) *Notice of deposition.* A party must serve a notice of deposition, stating the time and place of the deposition and the name and address of each person to be examined, on the person to be deposed, the administrative law judge, and each party not later than 7 days before the deposition. The notice must

be filed in the FAA Hearing Docket simultaneously. A party may serve a notice of deposition less than 7 days before the deposition only with consent of the administrative law judge. The party noticing a deposition must attach a copy of any subpoena *duces tecum* requesting that materials be produced at the deposition to the notice of deposition.

(4) *Use of depositions.* A party may use any part or all of a deposition at a hearing authorized under this subpart only upon a showing of good cause. The deposition may be used against any party who was present or represented at the deposition or who had reasonable notice of the deposition.

(k) *Interrogatories.* A party, the party's attorney, or the party's representative may sign the party's responses to interrogatories. A party must answer each interrogatory separately and completely in writing. If a party objects to an interrogatory, the party must state the objection and the reasons for the objection. An opposing party may use any part or all of a party's responses to interrogatories at a hearing authorized under this subpart to the extent that the response is relevant, material, and not repetitious.

(1) A party must not serve more than 30 interrogatories to each other party. Each subpart of an interrogatory must be counted as a separate interrogatory.

(2) A party must file a motion for leave to serve additional interrogatories on a party with the administrative law judge before serving additional interrogatories on a party. The administrative law judge may grant the motion only if the party shows good cause.

(l) *Requests for admission.* A party may serve a written request for admission of the truth of any matter within the scope of discovery under this section or the authenticity of any document described in the request. A party must set forth each request for admission separately. A party must serve copies of documents referenced in the request for admission unless the documents have been provided or are reasonably available for inspection and copying.

(1) *Time.* A party's failure to respond to a request for admission, in writing and signed by the attorney or the party, not later than 30 days after service of the request, is deemed an admission of the truth of the statement or statements contained in the request for admission. The administrative law judge may determine that a failure to respond to a request for admission is not deemed an admission of the truth if a party shows that the failure was due to circumstances beyond the control of the party or the party's attorney.

(2) *Response.* A party may object to a request for admission and must state the reasons for objection. A party may specifically deny the truth of the matter or describe the reasons why the party is unable to truthfully deny or admit the matter. If a party is unable to deny or admit the truth of the matter, the party must show that the party has made reasonable inquiry into the matter or that the information known to, or readily obtainable by, the party is insufficient to enable the party to admit or deny the matter. A party may admit or deny any part of the request for admission. If the administrative law judge determines that a response does not comply with the requirements of this paragraph (l)(2) or that the response is insufficient, the matter is deemed admitted.

(3) *Effect of admission.* Any matter admitted or deemed admitted under this section is conclusively established for the purpose of the hearing and appeal.

(m) *Motion to compel discovery.* A party may make a motion to compel discovery if a person refuses to answer a question during a deposition, a party fails or refuses to answer an interrogatory, if a person gives an evasive or incomplete answer during a deposition or when responding to an interrogatory, or a party fails or refuses to produce documents or tangible items. During a deposition, the proponent of a question may complete the deposition or may adjourn the examination before making a motion to compel if a person refuses to answer. Any motion to compel must be filed with the FAA Hearing Docket and served on the administrative law judge and other parties in accordance with §§13.210 and 13.211, respectively.

(n) *Failure to comply with a discovery order.* If a party fails to comply with a

71

discovery order, the administrative law judge may impose any of the following sanctions proportional to the party's failure to comply with the order:

(1) Strike the relevant portion of a party's pleadings;

(2) Preclude prehearing or discovery motions by that party;

(3) Preclude admission of the relevant portion of a party's evidence at the hearing; or

(4) Preclude the relevant portion of the testimony of that party's witnesses at the hearing.

§ 13.221 Notice of hearing.

(a) *Notice.* The administrative law judge must provide each party with notice of the date, time, and location of the hearing at least 60 days before the hearing date.

(b) *Date, time, and location of the hearing.* The administrative law judge to whom the proceedings have been assigned must set a reasonable date, time, and location for the hearing. The administrative law judge must consider the need for discovery and any joint procedural or discovery schedule submitted by the parties when determining the hearing date. The administrative law judge must give due regard to the convenience of the parties, the location where the majority of the witnesses reside or work, and whether the location is served by a scheduled air carrier.

(c) *Earlier hearing.* With the consent of the administrative law judge, the parties may agree to hold the hearing on an earlier date than the date specified in the notice of hearing.

§ 13.222 Evidence.

(a) *General.* A party is entitled to present the party's case or defense by oral, documentary, or demonstrative evidence, to submit rebuttal evidence, and to conduct any cross-examination that may be required for a full and true disclosure of the facts.

(b) *Admissibility.* A party may introduce any oral, documentary, or demonstrative evidence in support of the party's case or defense. The administrative law judge must admit any relevant oral, documentary, or demonstrative evidence introduced by a party, but must exclude irrelevant, immaterial, or unduly repetitious evidence.

(c) *Hearsay evidence.* Hearsay evidence is admissible in proceedings governed by this subpart. The fact that evidence submitted by a party is hearsay goes only to the weight of the evidence and does not affect its admissibility.

§ 13.223 Standard of proof.

The administrative law judge must issue an initial decision or must rule in a party's favor only if the decision or ruling is supported by, and in accordance with, the reliable, probative, and substantial evidence contained in the record. In order to prevail, the party with the burden of proof must prove the party's case or defense by a preponderance of reliable, probative, and substantial evidence.

§ 13.224 Burden of proof.

(a) Except in the case of an affirmative defense, the burden of proof is on the agency.

(b) Except as otherwise provided by statute or rule, the proponent of a motion, request, or order has the burden of proof.

(c) A party who has asserted an affirmative defense has the burden of proving the affirmative defense.

§ 13.225 Offer of proof.

A party whose evidence has been excluded by a ruling of the administrative law judge may offer the evidence for the record on appeal.

§ 13.226 Public disclosure of information.

(a) The administrative law judge may order that any information contained in the record be withheld from public disclosure. Any party or interested person may object to disclosure of information in the record by filing and serving a written motion to withhold specific information in accordance with §§ 13.210 and 13.211 respectively. A party may file a motion seeking to protect from public disclosure information contained in a document that the party is filing at the same time it files the document. The person or party must state the specific grounds for nondisclosure in the motion.

(b) The administrative law judge must grant the motion to withhold if, based on the motion and any response to the motion, the administrative law judge determines that: Disclosure would be detrimental to aviation safety; disclosure would not be in the public interest; or the information is not otherwise required to be made available to the public.

§13.227 Expert or opinion witnesses.

An employee of the agency may not be called as an expert or opinion witness for any party other than the FAA in any proceeding governed by this subpart. An employee of a respondent may not be called by an agency attorney as an expert or opinion witness for the FAA in any proceeding governed by this subpart to which the respondent is a party.

§13.228 Subpoenas.

(a) *Request for subpoena.* The administrative law judge, upon application by any party to the proceeding, may issue subpoenas requiring the attendance of witnesses or the production of documents or tangible things at a hearing or for the purpose of taking depositions, as permitted by law. A request for a subpoena must show its general relevance and reasonable scope. The party must serve the subpoena on the witness or the holder of the documents or tangible items as permitted by applicable statute. A request for a subpoena must be filed and served in accordance with §§13.210 and 13.211, respectively. Absent good cause shown, the filing and service must be completed as follows:

(1) Not later than 15 days before a scheduled deposition under the subpoena; or

(2) Not later than 30 days before a scheduled hearing where attendance at the hearing is sought.

(b) *Motion to quash or modify the subpoena.* A party, or any person upon whom a subpoena has been served, may file in the FAA Hearing Docket a motion to quash or modify the subpoena and must serve a copy on the administrative law judge and each party at or before the time specified in the subpoena for compliance. The movant must describe, in detail, the basis for the motion to quash or modify the subpoena including, but not limited to, a statement that the testimony, document, or tangible evidence is not relevant to the proceeding, that the subpoena is not reasonably tailored to the scope of the proceeding, or that the subpoena is unreasonable and oppressive. A motion to quash or modify the subpoena will stay the effect of the subpoena pending a decision by the administrative law judge on the motion.

(c) *Enforcement of subpoena.* Upon a showing that a person has failed or refused to comply with a subpoena, a party may apply to the appropriate U.S. district court to seek judicial enforcement of the subpoena.

§13.229 Witness fees.

(a) *General.* The party who applies for a subpoena to compel the attendance of a witness at a deposition or hearing, or the party at whose request a witness appears at a deposition or hearing, must pay the witness fees described in this section.

(b) *Amount.* Except for an employee of the agency who appears at the direction of the agency, a witness who appears at a deposition or hearing is entitled to the same fees and allowances provided for under 28 U.S.C. 1821.

§13.230 Record.

(a) *Exclusive record.* The pleadings, transcripts of the hearing and pre-hearing conferences, exhibits admitted into evidence, rulings, motions, applications, requests, briefs, and responses thereto, constitute the exclusive record for decision of the proceedings and the basis for the issuance of any orders in the proceeding. Any proceedings regarding the disqualification of an administrative law judge must be included in the record. Though only exhibits admitted into evidence are part of the record before an administrative law judge, evidence proffered but not admitted is also part of the record on appeal, as provided by §13.225.

(b) *Examination and copying of record.* The parties may examine the record at the FAA Hearing Docket and may obtain copies of the record upon payment of applicable fees. Any other person may obtain copies of the releasable portions of the record in accordance with applicable law.

§ 13.231 Argument before the administrative law judge.

(a) *Arguments during the hearing.* During the hearing, the administrative law judge must give the parties a reasonable opportunity to present arguments on the record supporting or opposing motions, objections, and rulings if the parties request an opportunity for argument. The administrative law judge may request written arguments during the hearing if the administrative law judge finds that submission of written arguments would be reasonable.

(b) *Final oral argument.* At the conclusion of the hearing and before the administrative law judge issues an initial decision in the proceedings, the administrative law judge must allow the parties to submit oral proposed findings of fact and conclusions of law, exceptions to rulings of the administrative law judge, and supporting arguments for the findings, conclusions, or exceptions. At the conclusion of the hearing, a party may waive final oral argument.

(c) *Post-hearing briefs.* The administrative law judge may request written post-hearing briefs before the administrative law judge issues an initial decision in the proceedings if the administrative law judge finds that submission of written arguments would be reasonable. If a party files a written post-hearing brief, the party must include proposed findings of fact and conclusions of law, exceptions to rulings of the administrative law judge, and supporting arguments for the findings, conclusions, or exceptions. The administrative law judge must give the parties a reasonable opportunity, but not more than 30 days after receipt of the transcript, to prepare and submit the briefs. A party must file and serve any post-hearing brief in in accordance with §§ 13.210 and 13.211, respectively.

§ 13.232 Initial decision.

(a) *Contents.* The administrative law judge must issue an initial decision at the conclusion of the hearing. In each oral or written decision, the administrative law judge must include findings of fact and conclusions of law, as well as the grounds supporting those find-ings and conclusions, for all material issues of fact, the credibility of witnesses, the applicable law, any exercise of the administrative law judge's discretion, and the amount of any civil penalty found appropriate by the administrative law judge. The administrative law judge must also include a discussion of the basis for any order issued in the proceedings. The administrative law judge is not required to provide a written explanation for rulings on objections, procedural motions, and other matters not directly relevant to the substance of the initial decision. If the administrative law judge refers to any previous unreported or unpublished initial decision, the administrative law judge must make copies of that initial decision available to all parties and the FAA decisionmaker.

(b) *Oral decision.* Except as provided in paragraph (c) of this section, at the conclusion of the hearing, the administrative law judge's oral initial decision and order must be on the record.

(c) *Written decision.* The administrative law judge may issue a written initial decision not later than 30 days after the conclusion of the hearing or submission of the last post-hearing brief if the administrative law judge finds that issuing a written initial decision is reasonable. The administrative law judge must serve a copy of any written initial decision on each party.

(d) *Reconsideration of an initial decision.* The FAA decisionmaker may treat a motion for reconsideration of an initial decision as a notice of appeal under § 13.233, and if the motion was filed within the time allowed for the filing of a notice of appeal, the FAA decisionmaker will issue a briefing schedule, as provided in § 13.218.

(e) *Order assessing civil penalty.* Unless appealed pursuant to § 13.233, the initial decision issued by the administrative law judge is considered an order assessing civil penalty if the administrative law judge finds that an alleged violation occurred and determines that a civil penalty, in an amount found appropriate by the administrative law judge, is warranted. The administrative law judge may not assess a civil penalty exceeding the amount sought in the complaint.

§ 13.233 Appeal from initial decision.

(a) *Notice of appeal.* A party may appeal the administrative law judge's initial decision, and any decision not previously appealed to the FAA decisionmaker on interlocutory appeal pursuant to § 13.219, by filing a notice of appeal in accordance with § 13.210 no later than 10 days after entry of the oral initial decision on the record or service of the written initial decision on the parties. The party must serve a copy of the notice of appeal on each party in accordance with § 13.211. A party is not required to serve any documents under § 13.233 on the administrative law judge.

(b) *Issues on appeal.* In any appeal from a decision of an administrative law judge, the FAA decisionmaker considers only the following issues:

(1) Whether each finding of fact is supported by a preponderance of reliable, probative, and substantial evidence;

(2) Whether each conclusion of law is made in accordance with applicable law, precedent, and public policy; and

(3) Whether the administrative law judge committed any prejudicial errors.

(c) *Perfecting an appeal.* Except as follows in paragraphs (c)(1) and (2) of this section, a party must perfect an appeal to the FAA decisionmaker no later than 50 days after entry of the oral initial decision on the record or service of the written initial decision on the parties by filing an appeal brief in accordance with § 13.210 and serving a copy on every other party in accordance with § 13.211.

(1) *Extension of time by agreement of the parties.* The parties may agree to extend the time for perfecting the appeal with the consent of the FAA decisionmaker. If the FAA decisionmaker grants an extension of time to perfect the appeal, the FAA decisionmaker must serve a letter confirming the extension of time on each party.

(2) *Written motion for extension.* If the parties do not agree to an extension of time for perfecting an appeal, a party desiring an extension of time may file a written motion for an extension in accordance with § 13.210 and must serve a copy of the motion on each party under § 13.211. Any party may file a

written response to the motion for extension no later than 10 days after service of the motion. The FAA decisionmaker may grant an extension if good cause for the extension is shown in the motion.

(d) *Appeal briefs.* A party must file the appeal brief in accordance with § 13.210 and must serve a copy of the appeal brief on each party in accordance with § 13.211.

(1) A party must set forth, in detail, the party's specific objections to the initial decision or rulings in the appeal brief. A party also must set forth, in detail, the basis for the appeal, the reasons supporting the appeal, and the relief requested in the appeal. If the party relies on evidence contained in the record for the appeal, the party must specifically refer to the pertinent evidence contained in the transcript in the appeal brief.

(2) The FAA decisionmaker may dismiss an appeal, on the FAA decisionmaker's own initiative or upon motion of any other party, where a party has filed a notice of appeal but fails to perfect the appeal by timely filing an appeal brief with the FAA decisionmaker.

(e) *Reply brief.* Except as follows in paragraphs (e)(1) and (2) of this section, any party may file a reply brief in accordance with § 13.210 not later than 35 days after the appeal brief has been served on that party. The party filing the reply brief must serve a copy of the reply brief on each party in accordance with § 13.211. If the party relies on evidence contained in the record for the reply, the party must specifically refer to the pertinent evidence contained in the transcript in the reply brief.

(1) *Extension of time by agreement of the parties.* The parties may agree to extend the time for filing a reply brief with the consent of the FAA decisionmaker. If the FAA decisionmaker grants an extension of time to file the reply brief, the FAA decisionmaker must serve a letter confirming the extension of time on each party.

(2) *Written motion for extension.* If the parties do not agree to an extension of time for filing a reply brief, a party desiring an extension of time may file a written motion for an extension in accordance with § 13.210 and must serve a

copy of the motion on each party in accordance with §13.211. Any party choosing to respond to the motion must file and serve a written response to the motion no later than 10 days after service of the motion The FAA decisionmaker may grant an extension if good cause for the extension is shown in the motion.

(f) *Other briefs.* The FAA decisionmaker may allow any person to submit an *amicus curiae* brief in an appeal of an initial decision. A party may not file more than one brief unless permitted by the FAA decisionmaker. A party may petition the FAA decisionmaker, in writing, for leave to file an additional brief and must serve a copy of the petition on each party. The party may not file the additional brief with the petition. The FAA decisionmaker may grant leave to file an additional brief if the party demonstrates good cause for allowing additional argument on the appeal. The FAA decisionmaker will allow a reasonable time for the party to file the additional brief.

(g) *Number of copies.* A party must file the original plus one copy of the appeal brief or reply brief, but only one copy if filing by email or fax, as provided in §13.210.

(h) *Oral argument.* The FAA decisionmaker may permit oral argument on the appeal. On the FAA decisionmaker's own initiative, or upon written motion by any party, the FAA decisionmaker may find that oral argument will contribute substantially to the development of the issues on appeal and may grant the parties an opportunity for oral argument.

(i) *Waiver of objections on appeal.* If a party fails to object to any alleged error regarding the proceedings in an appeal or a reply brief, the party waives any objection to the alleged error. The FAA decisionmaker is not required to consider any objection in an appeal brief, or any argument in the reply brief, if a party's objection or argument is based on evidence contained on the record and the party does not specifically refer to the pertinent evidence from the record in the brief.

(j) *FAA decisionmaker's decision on appeal.* The FAA decisionmaker will review the record, the briefs on appeal, and the oral argument, if any, when considering the issues on appeal. The FAA decisionmaker may affirm, modify, or reverse the initial decision, make any necessary findings, or remand the case for any proceedings that the FAA decisionmaker determines may be necessary. The FAA decisionmaker may assess a civil penalty but must not assess a civil penalty in an amount greater than that sought in the complaint.

(1) The FAA decisionmaker may raise any issue, on the FAA decisionmaker's own initiative, that is required for proper disposition of the proceedings. The FAA decisionmaker will give the parties a reasonable opportunity to submit arguments on the new issues before making a decision on appeal. If an issue raised by the FAA decisionmaker requires the consideration of additional testimony or evidence, the FAA decisionmaker will remand the case to the administrative law judge for further proceedings and an initial decision related to that issue. If an issue raised by the FAA decisionmaker is solely an issue of law, or the issue was addressed at the hearing but was not raised by a party in the briefs on appeal, a remand of the case to the administrative law judge for further proceedings is not required but may be provided in the discretion of the FAA decisionmaker.

(2) The FAA decisionmaker will issue the final decision and order of the Administrator on appeal in writing and will serve a copy of the decision and order on each party. Unless a petition for review is filed pursuant to §13.235, a final decision and order of the Administrator will be considered an order assessing civil penalty if the FAA decisionmaker finds that an alleged violation occurred and a civil penalty is warranted.

(3) A final decision and order of the Administrator after appeal is precedent in any other civil penalty action. Any issue, finding or conclusion, order, ruling, or initial decision of an administrative law judge that has not been appealed to the FAA decisionmaker is not precedent in any other civil penalty action.

§ 13.234   Petition to reconsider or modify a final decision and order of the FAA decisionmaker on appeal.

(a) *General.* Any party may petition the FAA decisionmaker to reconsider or modify a final decision and order issued by the FAA decisionmaker on appeal from an initial decision. A party must file a petition to reconsider or modify in accordance with § 13.210 not later than 30 days after service of the FAA decisionmaker's final decision and order on appeal and must serve a copy of the petition on each party in accordance with § 13.211. A party is not required to serve any documents under this section on the administrative law judge. The FAA decisionmaker will not reconsider or modify an initial decision and order issued by an administrative law judge that has not been appealed by any party to the FAA decisionmaker.

(b) *Number of copies.* The parties must file the original plus one copy of the petition or the reply to the petition, but only one copy if filing by email or fax, as provided in § 13.210.

(c) *Contents.* A party must state briefly and specifically the alleged errors in the final decision and order on appeal, the relief sought by the party, and the grounds that support the petition to reconsider or modify.

(1) If the petition is based, in whole or in part, on allegations regarding the consequences of the FAA decisionmaker's decision, the party must describe these allegations and must describe, and support, the basis for the allegations.

(2) If the petition is based, in whole or in part, on new material not previously raised in the proceedings, the party must set forth the new material and include affidavits of prospective witnesses and authenticated documents that would be introduced in support of the new material. The party must explain, in detail, why the new material was not discovered through due diligence prior to the hearing.

(d) *Repetitious and frivolous petitions.* The FAA decisionmaker will not consider repetitious or frivolous petitions. The FAA decisionmaker may summarily dismiss repetitious or frivolous petitions to reconsider or modify.

(e) *Reply petitions.* Any party replying to a petition to reconsider or modify must file the reply in accordance with § 13.210 no later than 10 days after service of the petition on that party, and must also serve a copy of the reply on each party in accordance with § 13.211.

(f) *Effect of filing petition.* The filing of a timely petition under this section will stay the effective date of the FAA decisionmaker's decision and order on appeal until final disposition of the petition by the FAA decisionmaker.

(g) *FAA decisionmaker's decision on petition.* The FAA decisionmaker has discretion to grant or deny a petition to reconsider. The FAA decisionmaker will grant or deny a petition to reconsider within a reasonable time after receipt of the petition or receipt of the reply petition, if any. The FAA decisionmaker may affirm, modify, or reverse the final decision and order on appeal, or may remand the case for any proceedings that the FAA decisionmaker determines may be necessary.

§ 13.235   Judicial review of a final decision and order.

(a) In cases under the Federal aviation statute, a party may seek judicial review of a final decision and order of the Administrator, as provided in 49 U.S.C. 46110(a), and, as applicable, in 49 U.S.C. 46301(d)(7)(D)(iii), 46301(g), or 47532.

(b) In cases under the Federal hazardous materials transportation statute, a party may seek judicial review of a final decision and order of the Administrator, as provided in 49 U.S.C. 5127.

(c) A party seeking judicial review of a final order issued by the Administrator may file a petition for review in the United States Court of Appeals for the District of Columbia Circuit or in the United States Court of Appeals for the circuit in which the party resides or has its principal place of business.

(d) The party must file the petition for review no later than 60 days after service of the Administrator's final decision and order.

§ 13.236 Alternative dispute resolution.

Parties may use mediation to achieve resolution of issues in controversy addressed by this subpart. Parties seeking alternative dispute resolution services may engage the services of a mutually acceptable mediator. The mediator must not participate in the adjudication under this subpart of any matter in which the mediator has provided mediation services. Mediation discussions and submissions will remain confidential consistent with the provisions of the Administrative Dispute Resolution Act and other applicable Federal laws.

Subpart H—Civil Monetary Penalty Inflation Adjustment

SOURCE: Docket No. 28762, 61 FR 67445, Dec. 20, 1996, unless otherwise noted.

§ 13.301 Inflation adjustments of civil monetary penalties.

(a) This subpart provides the maximum civil monetary penalties or range of minimum and maximum civil monetary penalties for each statutory civil penalty subject to FAA jurisdiction, as adjusted for inflation.

(b) Each adjustment to a maximum civil monetary penalty or to minimum and maximum civil monetary penalties that establish a civil monetary penalty range applies to actions initiated under this part for violations occurring on or after December 28, 2023, notwithstanding references to specific civil penalty amounts elsewhere in this part.

(c) Minimum and maximum civil monetary penalties are as follows:

TABLE 1 TO § 13.301(c)—MINIMUM AND MAXIMUM CIVIL MONETARY PENALTY AMOUNTS FOR CERTAIN VIOLATIONS

United States Code citation	Civil monetary penalty description	2023 minimum penalty amount	New adjusted minimum penalty amount for violations occurring on or after December 28, 2023	2023 maximum penalty amount	New adjusted maximum penalty amount for violations occurring on or after December 28, 2023
49 U.S.C. 5123(a)(1)	Violation of hazardous materials transportation law.	N/A	N/A	$96,624	$99,756.
49 U.S.C. 5123(a)(2)	Violation of hazardous materials transportation law resulting in death, serious illness, severe injury, or substantial property destruction.	N/A	N/A	$225,455	$232,762.
49 U.S.C. 5123(a)(3)	Violation of hazardous materials transportation law relating to training.	$582	$601	$96,624	$99,756.
49 U.S.C. 44704(d)(3) ..	Knowing presentation of a nonconforming aircraft for issuance of an initial airworthiness certificate by a production certificate holder.	N/A	N/A	$1,144,488	$1,181,581.
49 U.S.C. 44704(e)(4) ..	Knowing failure by an applicant for or holder of a type certificate to submit safety critical information or include certain such information in an airplane flight manual or flight crew operating manual.	N/A	N/A	$1,144,488	$1,181,581.

TABLE 1 TO § 13.301(c)—MINIMUM AND MAXIMUM CIVIL MONETARY PENALTY AMOUNTS FOR CERTAIN VIOLATIONS—Continued

United States Code citation	Civil monetary penalty description	2023 minimum penalty amount	New adjusted minimum penalty amount for violations occurring on or after December 28, 2023	2023 maximum penalty amount	New adjusted maximum penalty amount for violations occurring on or after December 28, 2023
49 U.S.C. 44704(e)(5) ..	Knowing false statement by an airline transport pilot (ATP) certificate holder with respect to the submission of certain safety critical information.	N/A	N/A	See entries for 49 U.S.C. 46301(a)(1) and (a)(5).	See entries for 49 U.S.C. 46301(a)(1) and (a)(5).
49 U.S.C. 44742	Interference by a supervisory employee of an organization designation authorization (ODA) holder that manufactures a transport category airplane with an ODA unit member's performance of authorized functions.	N/A	N/A	See entries for 49 U.S.C. 46301(a)(1).	See entries for 49 U.S.C. 46301(a)(1).
49 U.S.C. 44802 note ...	Operation of an unmanned aircraft or unmanned aircraft system equipped or armed with a dangerous weapon.	N/A	N/A	$29,462	$30,417.
49 U.S.C. 46301(a)(1) ..	Violation by a person other than an individual or small business concern under 49 U.S.C. 46301(a)(1)(A) or (B).	N/A	N/A	$40,272	$41,577.
49 U.S.C. 46301(a)(1) ..	Violation by an airman serving as an airman under 49 U.S.C. 46301(a)(1)(A) or (B) (but not covered by 46301(a)(5)(A) or (B)).	N/A	N/A	$1,771	$1,828.
49 U.S.C. 46301(a)(1) ..	Violation by an individual or small business concern under 49 U.S.C. 46301(a)(1)(A) or (B) (but not covered in 49 U.S.C. 46301(a)(5)).	N/A	N/A	$1,771	$1,828.
49 U.S.C. 46301(a)(3) ..	Violation of 49 U.S.C. 47107(b) (or any assurance made under such section) or 49 U.S.C. 47133.	N/A	N/A	Increase above otherwise applicable maximum amount not to exceed 3 times the amount of revenues used in violation of such section.	No change.
49 U.S.C. 46301(a)(5)(A).	Violation by an individual or small business concern (except an airman serving as an airman) under 49 U.S.C. 46301(a)(5)(A)(i) or (ii).	N/A	N/A	$16,108	$16,630.
49 U.S.C. 46301(a)(5)(B)(i).	Violation by an individual or small business concern related to the transportation of hazardous materials.	N/A	N/A	$16,108	$16,630.

TABLE 1 TO § 13.301(c)—MINIMUM AND MAXIMUM CIVIL MONETARY PENALTY AMOUNTS FOR CERTAIN VIOLATIONS—Continued

United States Code citation	Civil monetary penalty description	2023 minimum penalty amount	New adjusted minimum penalty amount for violations occurring on or after December 28, 2023	2023 maximum penalty amount	New adjusted maximum penalty amount for violations occurring on or after December 28, 2023
49 U.S.C. 46301(a)(5)(B)(ii).	Violation by an individual or small business concern related to the registration or recordation under 49 U.S.C. chapter 441, of an aircraft not used to provide air transportation.	N/A	N/A	$16,108	$16,630.
49 U.S.C. 46301(a)(5)(B)(iii).	Violation by an individual or small business concern of 49 U.S.C. 44718(d), relating to limitation on construction or establishment of landfills.	N/A	N/A	$16,108	$16,630.
49 U.S.C. 46301(a)(5)(B)(iv).	Violation by an individual or small business concern of 49 U.S.C. 44725, relating to the safe disposal of life-limited aircraft parts.	N/A	N/A	$16,108	$16,630.
49 U.S.C. 46301 note ...	Individual who aims the beam of a laser pointer at an aircraft in the airspace jurisdiction of the United States, or at the flight path of such an aircraft.	N/A	N/A	$30,820	$31,819.
49 U.S.C. 46301(b)	Tampering with a smoke alarm device.	N/A	N/A	$5,171	$5,339.
49 U.S.C. 46302	Knowingly providing false information about alleged violation involving the special aircraft jurisdiction of the United States.	N/A	N/A	$28,085	$28,995.
49 U.S.C. 46318	Physical or sexual assault or threat to physically or sexually assault crewmember or other individual on an aircraft, or action that poses an imminent threat to the safety of the aircraft or individuals on board.	N/A	N/A	$42,287	$43,658.
49 U.S.C. 46319	Permanent closure of an airport without providing sufficient notice.	N/A	N/A	$16,108	$16,630.
49 U.S.C. 46320	Operating an unmanned aircraft and in so doing knowingly or recklessly interfering with a wildfire suppression, law enforcement, or emergency response effort.	N/A	N/A	$24,656	$25,455.
49 U.S.C. 47531	Violation of 49 U.S.C. 47528–47530 or 47534, relating to the prohibition of operating certain aircraft not complying with stage 3 noise levels.	N/A	N/A	See entries for 49 U.S.C. 46301(a)(1) and (a)(5).	See entries for 49 U.S.C. 46301(a)(1) and (a)(5).

[84 FR 37068, July 31, 2019, as amended at 86 FR 1753, Jan. 11, 2021; 86 FR 23249, May 3, 2021; 87 FR 15863, Mar. 21, 2022; 88 FR 1122, Jan. 6, 2023; 88 FR 6971, Feb. 2, 2023; 88 FR 89557, Dec. 28, 2023]

Subpart I—Flight Operational Quality Assurance Programs

§ 13.401 Flight Operational Quality Assurance Program: Prohibition against use of data for enforcement purposes.

(a) *Applicability.* This section applies to any operator of an aircraft who operates such aircraft under an approved Flight Operational Quality Assurance (FOQA) program.

(b) *Definitions.* For the purpose of this section, the terms—

(1) *Flight Operational Quality Assurance (FOQA) program* means an FAA-approved program for the routine collection and analysis of digital flight data gathered during aircraft operations, including data currently collected pursuant to existing regulatory provisions, when such data is included in an approved FOQA program.

(2) *FOQA data* means any digital flight data that has been collected from an individual aircraft pursuant to an FAA-approved FOQA program, regardless of the electronic format of that data.

(3) *Aggregate FOQA data* means the summary statistical indices that are associated with FOQA event categories, based on an analysis of FOQA data from multiple aircraft operations.

(c) *Requirements.* In order for paragraph (e) of this section to apply, the operator must submit, maintain, and adhere to a FOQA Implementation and Operation Plan that is approved by the Administrator and which contains the following elements:

(1) A description of the operator's plan for collecting and analyzing flight recorded data from line operations on a routine basis, including identification of the data to be collected;

(2) Procedures for taking corrective action that analysis of the data indicates is necessary in the interest of safety;

(3) Procedures for providing the FAA with aggregate FOQA data;

(4) Procedures for informing the FAA as to any corrective action being undertaken pursuant to paragraph (c)(2) of this section.

(d) *Submission of aggregate data.* The operator will provide the FAA with aggregate FOQA data in a form and manner acceptable to the Administrator.

(e) *Enforcement.* Except for criminal or deliberate acts, the Administrator will not use an operator's FOQA data or aggregate FOQA data in an enforcement action against that operator or its employees when such FOQA data or aggregate FOQA data is obtained from a FOQA program that is approved by the Administrator.

(f) *Disclosure.* FOQA data and aggregate FOQA data, if submitted in accordance with an order designating the information as protected under part 193 of this chapter, will be afforded the nondisclosure protections of part 193 of this chapter.

(g) *Withdrawal of program approval.* The Administrator may withdraw approval of a previously approved FOQA program for failure to comply with the requirements of this chapter. Grounds for withdrawal of approval may include, but are not limited to—

(1) Failure to implement corrective action that analysis of available FOQA data indicates is necessary in the interest of safety; or

(2) Failure to correct a continuing pattern of violations following notice by the agency; or also

(3) Willful misconduct or willful violation of the FAA regulations in this chapter.

[Doc. No. FAA–2000–7554, 66 FR 55048, Oct. 31, 2001; Amdt. 13–30, 67 FR 31401, May 9, 2002]

PART 14—RULES IMPLEMENTING THE EQUAL ACCESS TO JUSTICE ACT OF 1980

Subpart A—General Provisions

Subpart B—Information Required From Applicants

AUTHORITY: 5 U.S.C. 504; 49 U.S.C. 106(f), 40113, 46104 and 47122.

SOURCE: Docket No. 25958, 54 FR 46199, Nov. 1, 1989, unless otherwise noted.

Subpart A—General Provisions

§ 14.01   Purpose of these rules.

The Equal Access to Justice Act, 5 U.S.C. 504 (the Act), provides for the award of attorney fees and other expenses to eligible individuals and entities who are parties to certain administrative proceedings (adversary adjudications) before the Federal Aviation Administration (FAA). An eligible party may receive an award when it prevails over the FAA, unless the agency's position in the proceeding was substantially justified or special circumstances make an award unjust. The rules in this part describe the parties eligible for awards and the proceedings that are covered. They also explain how to apply for awards, and the procedures and standards that the FAA Decisionmaker will use to make them. As used hereinafter, the term "agency" applies to the FAA.

§ 14.02   Proceedings covered.

(a) The Act applies to certain adversary adjudications conducted by the FAA under 49 CFR part 17 and the Acquisition Management System (AMS). These are adjudications under 5 U.S.C. 554, in which the position of the FAA is represented by an attorney or other representative who enters an appearance and participates in the proceeding. This subpart applies to proceedings under 49 U.S.C. 46301, 46302, and 46303 and to the Default Adjudicative Process under part 17 of this chapter and the AMS.

(b) If a proceeding includes both matters covered by the Act and matters specifically excluded from coverage, any award made will include only fees and expenses related to covered issues.

(c) Fees and other expenses may not be awarded to a party for any portion of the adversary adjudication in which such party has unreasonably protracted the proceedings.

[54 FR 46199, Nov. 1, 1989, as amended by Amdt. 14–03, 64 FR 32935, June 18, 1999]

§ 14.03   Eligibility of applicants.

(a) To be eligible for an award of attorney fees and other expenses under the Act, the applicant must be a party to the adversary adjudication for which it seeks an award. The term "party" is defined in 5 U.S.C. 504(b)(1)(B) and 5 U.S.C. 551(3). The applicant must show that it meets all conditions or eligibility set out in this subpart.

(b) The types of eligible applicants are as follows:

(1) An individual with a net worth of not more than $2 million at the time the adversary adjudication was initiated;

(2) The sole owner of an unincorporated business who has a net worth of not more than $7 million, including both personal and business interests, and not more than 500 employees at the time the adversary adjudication was initiated;

(3) A charitable or other tax-exempt organization described in section 501(c)(3) of the Internal Revenue Code (26 U.S.C. 501(c)(3)) with not more than 500 employees at the time the adversary adjudication was initiated; and

(4) A cooperative association as defined in section 15(a) of the Agricultural Marketing Act (12 U.S.C. 1141j(a)) with not more than 500 employees at the time the adversary adjudication was initiated; and

(5) Any other partnership, corporation, association, or public or private organization with a net worth of not more than $7 million and not more than 500 employees at the time the adversary adjudication was initiated.

(c) For the purpose of eligibility, the net worth and number of employees of an applicant shall be determined as of the date the proceeding was initiated.

(d) An applicant who owns an unincorporated business will be considered an "individual" rather than a "sole owner of an unincorporated business" if the issues on which the applicant prevails are related primarily to personal interests rather than to business interest.

(e) The employees of an applicant include all persons who regularly perform services for remuneration for the applicant, under the applicant's direction and control. Part-time employees shall be included on a proportional basis.

(f) The net worth and number of employees of the applicant and all of its affiliates shall be aggregated to determine eligibility. Any individual, corporation, or other entity that directly or indirectly controls or owns a majority of the voting shares or other interest of the applicant, or any corporation or other entity of which the applicant directly or indirectly owns or controls a majority of the voting shares or other interest, will be considered an affiliate for purposes of this part, unless the ALJ or adjudicative officer determines that such treatment would be unjust and contrary to the purposes of the Act in light of the actual relationship between the affiliated entities. In addition, the ALJ or adjudicative officer may determine that financial relationships of the applicant, other than those described in this paragraph, constitute special circumstances that would make an award unjust.

(g) An applicant that participates in a proceeding primarily on behalf of one or more other persons or entities that would be ineligible if not itself eligible for an award.

[54 FR 46199, Nov. 1, 1989, as amended by Amdt. 14–03, 64 FR 32935, June 18, 1999]

§ 14.04 Standards for awards.

(a) A prevailing applicant may receive an award for attorney fees and other expenses incurred in connection with a proceeding, or in a significant and discrete substantive portion of the proceeding, unless the position of the agency over which the applicant has prevailed was substantially justified. Whether or not the position of the FAA

was substantially justified shall be determined on the basis of the record (including the record with respect to the action or failure to act by the agency upon which the civil action is based) which was made in the civil action for which fees and other expenses are sought. The burden of proof that an award should not be made to an eligible prevailing applicant is on the agency counsel, who may avoid an award by showing that the agency's position was reasonable in law and fact.

(b) An award will be reduced or denied if the applicant has unduly or unreasonably protracted the proceeding or if special circumstances make the award sought unjust.

§ 14.05 Allowance fees and expenses.

(a) Awards will be based on rates customarily charged by persons engaged in the business of acting as attorneys, agents, and expert witnesses, even if the services were made available without charge or at a reduced rate to the applicant.

(b) No award for the fee of an attorney or agent under this part may exceed $125 per hour, or such rate as prescribed by 5 U.S.C. 504. No award to compensate an expert witness may exceed the highest rate at which the agency pays expert witnesses. However, an award may also include the reasonable expenses of the attorney, agent, or witness as a separate item, if the attorney, agent, or witness ordinarily charges clients separately for such expenses.

(c) In determining the reasonableness of the fee sought for an attorney, agent, or expert witness, the ALJ or adjudicative officer shall consider the following:

(1) If the attorney, agent, or witness is in private practice, his or her customary fee for similar services, or if an employee of the applicant, the fully allocated cost of the services;

(2) The prevailing rate for similar services in the community in which the attorney, agent, or witness ordinarily performs services;

(3) The time actually spent in the representation of the applicant;

(4) The time reasonably spent in light of the difficulty or complexity of the issues in the proceeding; and

(5) Such other factors as may bear on the value of the services provided.

(d) The reasonable cost of any study, analysis, engineering report, test, project, or similar matter prepared on behalf of a party may be awarded, to the extent that the charge for the service does not exceed the prevailing rate for similar services, and the study or other matter was necessary for preparation of the applicant's case.

(e) Fees may be awarded only for work performed after the issuance of a complaint, or in the Default Adjudicative Process for a protest or contract dispute under part 17 of this chapter and the AMS.

[Amdt. 13–18, 53 FR 34655, Sept. 7, 1988, as amended by Amdt. 14–1, 55 FR 15131, Apr. 20, 1990; Amdt. 14–03, 64 FR 32935, June 18, 1999]

Subpart B—Information Required From Applicants

§ 14.10 Contents of application.

(a) An application for an award of fees and expenses under the Act shall identify the applicant and the proceeding for which an award is sought. The application shall show that the applicant has prevailed and identify the position of the agency in the proceeding that the applicant alleges was not substantially justified. Unless the applicant is an individual, the application shall also state the number of employees of the applicant and describe briefly the type and purpose of its organization or business.

(b) The application shall also include a statement that the applicant's net worth does not exceed $2 million (if an individual) or $7 million (for all other applicants, including their affiliates) at the time the adversary adjudication was initiated. However, an applicant may omit this statement if:

(1) It attaches a copy of a ruling by the Internal Revenue Service that it qualifies as an organization described in section 501(c)(3) of the Internal Revenue Code (26 U.S.C. 501(c)(3)), or in the case of a tax-exempt organization not required to obtain a ruling from the Internal Revenue Service on its exempt status, a statement that describes the basis for the applicant's belief that it qualifies under such section; or

(2) It states that it is a cooperative association as defined in section 15(a) of the Agricultural Marketing Act (12 U.S.C. 1141j(a)).

(c) The application shall state the amount of fees and expenses for which an award is sought.

(d) The application may also include any other matters that the applicant wishes this agency to consider in determining whether and in what amount an award should be made.

(e) The application shall be signed by the applicant or an authorized officer or attorney for the applicant. It shall also contain or be accompanied by a written verification under oath or under penalty of perjury that the information provided in the application is true and correct.

(f) If the applicant is a partnership, corporation, association, organization, or sole owner of an unincorporated business, the application shall state that the applicant did not have more than 500 employees at the time the adversary adjudication was initiated, giving the number of its employees and describing briefly the type and purpose of its organization or business.

§ 14.11 Net worth exhibit.

(a) Each applicant except a qualified tax-exempt organization or cooperative association must provide with its application a detailed exhibit showing the net worth of the applicant and any affiliates when the proceeding was initiated. If any individual, corporation, or other entity directly or indirectly controls or owns a majority of the voting shares or other interest of the applicant, or if the applicant directly or indirectly owns or controls a majority of the voting shares or other interest of any corporation or other entity, the exhibit must include a showing of the net worth of all such affiliates or of the applicant including the affiliates. The exhibit may be in any form convenient to the applicant that provides full disclosure of the applicant's and its affiliates' assets and liabilities and is sufficient to determine whether the applicant qualifies under the standards in this part. The administrative law judge may require an applicant to file additional information to determine the eligibility for an award.

(b) The net worth exhibit shall describe any transfers of assets from, or obligations incurred by, the applicant or any affiliate, occurring in the one-year period prior to the date on which the proceeding was initiated, that reduced the net worth of the applicant and its affiliates below the applicable net worth ceiling. If there were no such transactions, the applicant shall so state.

(c) Ordinarily, the net worth exhibit will be included in the public record of the proceeding. However, an applicant that objects to public disclosure of the net worth exhibit, or any part of it, may submit that portion of the exhibit directly to the ALJ or adjudicative officer in a sealed envelope labeled "Confidential Financial Information," accompanied by a motion to withhold the information.

(1) The motion shall describe the information sought to be withheld and explain, in detail, why it should be exempt under applicable law or regulation, why public disclosure would adversely affect the applicant, and why disclosure is not required in the public interest.

(2) The net worth exhibit shall be served on the FAA counsel, but need not be served on any other party to the proceeding.

(3) If the ALJ or adjudicative officer finds that the net worth exhibit, or any part of it, should not be withheld from disclosure, it shall be placed in the public record of the proceeding. Otherwise, any request to inspect or copy the exhibit shall be disposed of in accordance with the FAA's established procedures.

[54 FR 46199, Nov. 1, 1989, as amended by Amdt. 14–03, 64 FR 32935, June 18, 1999]

§14.12 Documentation of fees and expenses.

The application shall be accompanied by full documentation of the fees and expenses, including the cost of any study, analysis, engineering report, test, project or similar matter, for which an award is sought. A separate itemized statement shall be submitted for each professional firm or individual whose services are covered by the application, showing the hours spent in connection with the proceedings by each individual, a description of the specific services performed, the rate at which each fee has been computed, any expenses for which reimbursement is sought, the total amount claimed, and the total amount paid or payable by the applicant or by any other person or entity for the services provided. The administrative law judge may require the applicant to provide vouchers, receipts, or other substantiation for any expenses claimed.

Subpart C—Procedures for Considering Applications

§14.20 When an application may be filed.

(a) An application may be filed whenever the applicant has prevailed in the proceeding, but in no case later than 30 days after the FAA Decisionmaker's final disposition of the proceeding, or service of the order of the Administrator in a proceeding under the AMS.

(b) If review or reconsideration is sought or taken of a decision to which an applicant believes it has prevailed, proceedings for the award of fees shall be stayed pending final disposition of the underlying controversy.

(c) For purposes of this part, final disposition means the later of:

(1) Under part 17 of this chapter and the AMS, the date on which the order of the Administrator is served;

(2) The date on which an unappealed initial decision becomes administratively final;

(3) Issuance of an order disposing of any petitions for reconsideration of the FAA Decisionmaker's final order in the proceeding;

(4) If no petition for reconsideration is filed, the last date on which such a petition could have been filed; or

(5) Issuance of a final order or any other final resolution of a proceeding, such as a settlement or voluntary dismissal, which is not subject to a petition for reconsideration.

[54 FR 46199, Nov. 1, 1989, as amended by Amdt. 14–03, 64 FR 32936, June 18, 1999]

§14.21 Filing and service of documents.

Any application for an award or other pleading or document related to an application shall be filed and served

85

on all parties to the proceeding in the same manner as other pleadings in the proceeding, except as provided in § 14.11(b) for confidential financial information. Where the proceeding was held under part 17 of this chapter and the AMS, the application shall be filed with the FAA's attorney and with the Office of Dispute Resolution for Acquisition.

[Doc. No. FAA–1998–4379, 64 FR 32936, June 18, 1999]

§ 14.22 Answer to application.

(a) Within 30 days after service of an application, counsel representing the agency against which an award is sought may file an answer to the application. Unless agency counsel requests an extension of time for filing or files a statement of intent to negotiate under paragraph (b) of the section, failure to file an answer within the 30-day period may be treated as a consent to the award requested.

(b) If the FAA's counsel and the applicant believe that the issues in the fee application can be settled, they may jointly file a statement of their intent to negotiate a settlement. The filing of this statement shall extend the time for filing an answer for an additional 30 days, and further extensions may be granted by the ALJ or adjudicative officer upon request by the FAA's counsel and the applicant.

(c) The answer shall explain in detail any objections to the award requested and identify the facts relied on in support of agency counsel's position. If the answer is based on any alleged facts not already in the record of the proceeding, agency counsel shall include with the answer either supporting affidavits or a request for further proceedings under § 14.26.

[54 FR 46199, Nov. 1, 1989, as amended by Amdt. 14–03, 64 FR 32936, June 18, 1999]

§ 14.23 Reply.

Within 15 days after service of an answer, the applicant may file a reply. If the reply is based on any alleged facts not already in the record of the proceeding, the applicant shall include with the reply either supporting affidavits or a request for further proceedings under § 14.26.

§ 14.24 Comments by other parties.

Any party to a proceeding other than the applicant and the FAA's counsel may file comments on an application within 30 days after it is served, or on an answer within 15 days after it is served. A commenting party may not participate further in proceedings on the application unless the ALJ or adjudicative officer determines that the public interest requires such participation in order to permit full exploration of matters raised in the comments.

[Doc. No. FAA–1998–4379, 64 FR 32936, June 18, 1999]

§ 14.25 Settlement.

The applicant and agency counsel may agree on a proposed settlement of the award before final action on the application, either in connection with a settlement of the underlying proceeding, or after the underlying proceeding has been concluded. If a prevailing party and agency counsel agree on a proposed settlement of an award before an application has been filed, the application shall be filed with the proposed settlement.

§ 14.26 Further proceedings.

(a) Ordinarily the determination of an award will be made on the basis of the written record; however, on request of either the applicant or agency counsel, or on his or her own initiative, the ALJ or adjudicative officer assigned to the matter may order further proceedings, such as an informal conference, oral argument, additional written submissions, or an evidentiary hearing. Such further proceedings shall be held only when necessary for full and fair resolution of the issues arising from the application and shall be conducted as promptly as possible.

(b) A request that the administrative law judge order further proceedings under this section shall specifically identify the information sought or the disputed issues and shall explain why the additional proceedings are necessary to resolve the issues.

[54 FR 46199, Nov. 1, 1989, as amended by Amdt. 14–03, 64 FR 32936, June 18, 1999]

§ 14.27 Decision.

(a) The ALJ shall issue an initial decision on the application within 60 days after completion of proceedings on the application.

(b) An adjudicative officer in a proceeding under part 17 of this chapter and the AMS shall prepare a findings and recommendations for the Office of Dispute Resolution for Acquisition.

(c) A decision under paragraph (a) or (b) of this section shall include written findings and conclusions on the applicant's eligibility and status as prevailing party and an explanation of the reasons for any difference between the amount requested and the amount awarded. The decision shall also include, if at issue, findings on whether the FAA's position was substantially justified, or whether special circumstances make an award unjust.

[Doc. No. FAA–1998–4379, 64 FR 32936, June 18, 1999]

§ 14.28 Review by FAA decisionmaker.

(a) In proceedings other than those under part 17 of this chapter and the AMS, either the applicant or the FAA counsel may seek review of the initial decision on the fee application in accordance with subpart G of part 13 of this chapter, specifically § 13.233. Additionally, the FAA Decisionmaker may decide to review the decision on his/her own initiative. If neither the applicant nor the FAA's counsel seeks review within 30 days after the decision is issued, it shall become final. Whether to review a decision is a matter within the discretion of the FAA Decisionmaker. If review is taken, the FAA Decisionmaker will issue a final decision on the application or remand the application to the ALJ who issue the initial fee award determination for further proceedings.

(b) In proceedings under part 17 of this chapter and the AMS, the adjudicative officer shall prepare findings and recommendations for the Office of Dispute Resolution for Acquisition with recommendations as to whether or not an award should be made, the amount of the award, and the reasons therefor. The Office of Dispute Resolution for Acquisition shall submit a recommended order to the Administrator after the completion of all submissions related to the EAJA application. Upon the Administrator's action, the order shall become final, and may be reviewed under 49 U.S.C. 46110.

[Doc. No. FAA–1998–4379, 64 FR 32936, June 18, 1999, as amended at 70 FR 8238, Feb. 18, 2005]

§ 14.29 Judicial review.

If an applicant is dissatisfied with the determination of fees and other expenses made under this subsection, pursuant 5 U.S.C. 504(c)(2), that applicant may, within thirty (30) days after the determination is made, appeal the determination to the court of the United States having jurisdiction to review the merits of the underlying decision of the FAA adversary adjudication. The court's determination on any appeal heard under this paragraph shall be based solely on the factual record made before the FAA. The court may modify the determination of fees and other expenses only if the court finds that the failure to make an award of fees and other expenses, or the calculation of the amount of the award, was unsupported by substantial evidence.

§ 14.30 Payment of award.

An applicant seeking payment of an award shall submit to the disbursing official of the FAA a copy of the FAA Decisionmaker's final decision granting the award, accompanied by a statement that the applicant will not seek review of the decision in the United States courts. Applications for award grants in cases involving the FAA shall be sent to: The Office of Accounting and Audit, AAA–1, Federal Aviation Administration, 800 Independence Avenue, SW., Washington, DC 20591. The agency will pay the amount awarded to the applicant within 60 days, unless judicial review of the award or of the underlying decision of the adversary adjudication has been sought by the applicant or any other party to the proceeding.

PART 15—ADMINISTRATIVE CLAIMS UNDER FEDERAL TORT CLAIMS ACT

Subpart A—General Procedures

Subpart B—Indemnification Under Section 1118 of the Federal Aviation Act of 1958

AUTHORITY: 5 U.S.C. 301; 28 U.S.C. 2672, 2675; 49 U.S.C. 106(g), 40113, 44721.

Subpart A—General Procedures

SOURCE: Docket No. 25264, 52 FR 18171, May 13, 1987, unless otherwise noted.

§ 15.1　Scope of regulations.

(a) These regulations apply to claims asserted under the Federal Tort Claims Act, as amended, for money damages against the United States for injury to, or loss of property, or for personal injury or death, caused by the negligent or wrongful act or omission of an employee of the FAA acting within the scope of office or employment. The regulations in this part supplement the Attorney General's regulations in 28 CFR part 14, as amended. The regulations in 28 CFR part 14, as amended, and the regulations in this part apply to consideration by the FAA of administrative claims under the Federal Tort Claims Act.

§ 15.3　Administrative claim, when presented; appropriate office.

(a) A claim is deemed to have been presented when the FAA receives, at a place designated in paragraph (b) of this section, an executed Standard Form 95 or other written notification of an incident, accompanied by a claim for money damages in a sum certain for injury to, or loss of, property or for personal injury or death, alleged to have occurred by reason of the incident. A claim which should have been presented to the FAA but which was mistakenly filed with another Federal agency, is deemed presented to the FAA on the date the claim is received by the FAA at a place designated in paragraph (b) of this section. A claim addressed to, or filed with, the FAA by mistake will be transferred to the appropriate Federal agency, if that agency can be determined, or returned to the claimant.

(b) Claims shall be delivered or mailed to the Assistant Chief Counsel, Litigation Division, AGC–400, Federal Aviation Administration, 800 Independence Avenue, SW., Washington, DC 20591, or alternatively, may be mailed or delivered to the Regional Counsel in any of the FAA Regional Offices or the Assistant Chief Counsel, Europe, Africa, and Middle East Area Office.

(c) Claim forms are available at each location listed in paragraph (b) of this section.

(d) A claim presented in accordance with this section may be amended by the claimant at any time prior to final FAA action or prior to the exercise of the claimant's option, under 28 U.S.C. 2675(a), to deem the agency's failure to make a final disposition of his or her claim within 6 months after it was filed as a final denial. Each amendment to a claim shall be submitted in writing and signed by the claimant or the claimant's duly authorized agent or legal representative. Upon the timely filing of an amendment to a pending claim, the FAA has 6 months thereafter in which to make a final disposition of the claim as amended, and the claimant's option under 28 U.S.C. 2675(a) does not accrue until 6 months after the filing of the amendment.

[Doc. No. 18884, 44 FR 63723, Nov. 5, 1979, as amended by Amdt. 15–1, 54 FR 39290, Sept. 25, 1989; Amdt. 15–4, 62 FR 46866, Sept. 4, 1997]

§ 15.5　Administrative claim, who may file.

(a) A claim for injury to, or loss of, property may be presented by the owner of the property interest which is the subject of the claim or by the owner's duly authorized agent or legal representative.

(b) A claim for personal injury may be presented by the injured person or that person's duly authorized agent or legal representative.

(c) A claim based on death may be presented by the executor or administrator of the decedent's estate or by any other person legally entitled to assert such a claim under applicable State law.

(d) A claim for loss wholly compensated by an insurer with the rights of a subrogee may be presented by the insurer. A claim for loss partially compensated by an insurer with the rights of a subrogee may be presented by the insurer or the insured individually, as their respective interest appear, or jointly. Whenever an insurer presents a claim asserting the rights of a subrogee, it shall present with its claim appropriate evidence that it has the rights of a subrogee.

(e) A claim presented by an agent or legal representative shall be presented in the name of the claimant, be signed by the agent or legal representative, show the title or legal capacity of the person signing, and be accompanied by evidence of authority to present a claim on behalf of the claimant as agent, executor, administrator, parent, guardian, or other representative.

§ 15.7 Administrative claims; evidence and information to be submitted.

(a) *Death.* In support of a claim based on death, the claimant may be required to submit the following evidence or information:

(1) An authenticated death certificate or other competent evidence showing cause of death, date of death, and age of the decedent.

(2) The decedent's employment or occupation at time of death, including monthly or yearly salary or earnings (if any), and the duration of last employment or occupation.

(3) Full names, addresses, birth dates, kinship, and marital status of the decedent's survivors, including identification of those survivors who were dependent for support upon the decedent at the time of death.

(4) Degree of support afforded by the decedent to each survivor dependent upon decedent for support at the time of death.

(5) Decedent's general, physical, and mental conditions before death.

(6) Itemized bills for medical and burial expenses incurred by reason of the incident causing death or itemized receipts of payment for such expenses.

(7) If damages for pain and suffering prior to death are claimed, a physician's detailed statement specifying the injuries suffered, duration of pain and suffering, any drugs administered for pain, and the decedent's physical condition in the interval between injury and death.

(8) Any other evidence or information which may have a bearing on either the responsibility of the United States for the death or the amount of damages claimed.

(b) *Personal injury.* In support of a claim for personal injury, including pain and suffering, the claimant may be required to submit the following evidence or information:

(1) A written report by the attending physician or dentist setting forth the nature and extent of the injuries, nature and extent of treatment, any degree of temporary or permanent disability, the prognosis, period of hospitalization, and any diminished earning capacity.

(2) In addition to the report required by paragraph (b)(1) of this section, the claimant may be required to submit to a physical or mental examination by a physician employed by the FAA or another Federal agency. A copy of the report of the examining physician is made available to the claimant upon the claimant's written request if the claimant has, upon request, furnished the report required by paragraph (b)(1), and has made or agrees to make available to the FAA any other physician's reports previously or thereafter made on the physical or mental condition which is the subject matter of the claim.

(3) Itemized bills for medical, dental, and hospital expenses incurred or itemized receipts of payment for such expenses.

(4) If the prognosis reveals the necessity for future treatment, a statement

89

of expected expenses for such treatment.

(5) If a claim is made for loss of time from employment, a written statement from the claimant's employer showing actual time lost from employment, whether the claimant is a full or part-time employee, and wages or salary actually lost.

(6) If a claim is made for loss of income and the claimant is self-employed, documentary evidence showing the amount of earnings actually lost.

(7) Any other evidence or information which may have a bearing on the responsibility of the United States for the personal injury or the damages claimed.

(c) *Property damage.* In support of a claim for injury to or loss of property, real or personal, the claimant may be required to submit the following evidence or information:

(1) Proof of ownership of the property interest which is the subject of the claim.

(2) A detailed statement of the amount claimed with respect to each item of property.

(3) An itemized receipt of payment for necessary repairs or itemized written estimates of the cost of such repairs.

(4) A statement listing date of purchase, purchase price, and salvage value, where repair is not economical.

(5) Any other evidence or information which may have a bearing on either the responsibility of the United States for the injury to or loss of property or the damages claimed.

§ 15.9 Investigation and examination.

The FAA may investigate a claim or conduct a physical examination of a claimant. The FAA may request any other Federal agency to investigate a claim or conduct a physical examination of a claimant and provide a report of the investigation or examination to the FAA.

Subpart B—Indemnification Under Section 1118 of the Federal Aviation Act of 1958

SOURCE: Amdt. 15–2, 55 FR 18710, May 3, 1990, unless otherwise noted.

§ 15.101 Applicability.

This subpart prescribes procedural requirements for the indemnification of a publisher of aeronautical charts or maps under section 1118 of the Federal Aviation Act of 1958, as amended, when the publisher incurs liability as a result of publishing—

(a) A chart or map accurately depicting a defective or deficient flight procedure or airway that was promulgated by the FAA; or

(b) Aeronautical data that—

(1) Is visually displayed in the cockpit of an aircraft; and

(2) When visually displayed, accurately depicts a defective or deficient flight procedure or airway promulgated by the FAA.

§ 15.103 Exclusions.

A publisher that requests indemnification under this part will not be indemnified if—

(a) The complaint filed against the publisher, or demand for payment against the publisher, first occurred before December 19, 1985;

(b) The publisher does not negotiate a good faith settlement;

(c) The publisher does not conduct a good faith defense;

(d) The defective or deficient flight procedure or airway—

(1) Was not promulgated by the FAA;

(2) Was not accurately depicted on the publisher's chart or map;

(3) Was not accurately displayed on a visual display in the cockpit, or

(4) Was obviously defective or deficient;

(e) The publisher does not give notice as required by § 15.107 of this part and that failure is prejudicial to the Government; or

(f) The publisher does not appeal a lower court's decision pursuant to a request by the Administrator under § 15.111(d)(2) of this part.

§ 15.105 Filing of requests for indemnification.

A request for indemnification under this part—

(a) May be filed by—

(1) A publisher described in § 15.101 of this part; or

(2) The publisher's duly authorized agent or legal representative;

(b) Shall be filed with the Chief Counsel, Federal Aviation Administration, 800 Independence Avenue SW., Washington, DC 20591; and

(c) Shall state the basis for the publisher's assertion that indemnification under this part is required.

§15.107 Notification requirements.

A request for indemnification will not be considered by the FAA unless the following conditions are met:

(a) The publisher must notify the Chief Counsel of the FAA, within the time limits prescribed in paragraph (b) or (c) of this section, of the publisher's first receipt of a demand for payment, or service of a complaint in any proceeding, federal or state, in which it appears that indemnification under this part may be required.

(b) For each complaint filed, or demand for payment made, on or after December 19, 1985, and before June 4, 1990, the notice required by paragraph (a) of this section must be received by the FAA on or before July 2, 1990.

(c) For each complaint filed, or demand for payment made, on or after June 4, 1990, the notice required by paragraph (a) of this section must be received by the FAA within 60 days after the day the publisher first receives the demand for payment or service of the complaint.

(d) Within 5 days after the day a judgment is rendered against the publisher in any proceeding, or within 30 days of the denial of an appeal, whichever is later, the publisher must notify the FAA Chief Counsel that—

(1) There is an adverse judgment against the publisher; and

(2) The publisher has a claim for indemnification against the FAA arising out of that judgment.

§15.109 Settlements.

(a) A publisher may not settle a claim with another party, for which the publisher has sought, or intends to seek, indemnification under this part, unless—

(1) The publisher submits a copy of the proposed settlement, and a statement justifying the settlement, to the Chief Counsel of the FAA; and

(2) The Administrator and where necessary, the appropriate official of the Department of Justice, approves the proposed settlement.

(3) The publisher submits a signed release that clearly releases the United States from any further liability to the publisher and the claimant.

(b) If the Administrator does not approve the proposed settlement, the Administrator will—

(1) So notify the publisher by registered mail within 60 days of receipt of the proposed settlement; and

(2) Explain why the request for indemnification was not approved.

(c) If the Administrator approves the proposed settlement, the Administrator will so notify the publisher by registered mail within 60 days after the FAA's receipt of the proposed settlement.

(d) If the Administrator does not have sufficient information to approve or disapprove the proposed settlement, the Administrator will request, within 60 days after receipt of the proposed settlement, the additional information needed to make a determination.

§15.111 Conduct of litigation.

(a) If a lawsuit is filed against the publisher and the publisher has sought, or intends to seek, indemnification under this part, the publisher shall—

(1) Give notice as required by §15.107 of this part;

(2) If requested by the United States—

(i) Implead the United States as a third-party defendant in the action; and

(ii) Arrange for the removal of the action to Federal Court;

(3) Promptly provide any additional information requested by the United States; and

(4) Cooperate with the United States in the defense of the lawsuit.

(b) If the lawsuit filed against the publisher results in a proposed settlement, the publisher shall submit that proposed settlement to the FAA for approval in accordance with §15.109 of this part.

(c) If the lawsuit filed against the publisher results in a judgment against the publisher and the publisher has sought, or intends to seek, indemnification under this part as a result of the adverse judgment, the publisher shall—

(1) Give notice to the FAA as required by § 15.107(d) of this part;

(2) Submit a copy of the trial court's decision to the FAA Chief Counsel not more than 5 business days after the adverse judgment is rendered; and

(3) If an appeal is taken from the adverse judgment, submit a copy of the appellate decision to the FAA Chief Counsel not more than 30 days after that decision is rendered.

(d) Within 60 days after receipt of the trial court's decision, the Administrator by registered mail will—

(1) Notify the publisher that indemnification is required under this part;

(2) Request that the publisher appeal the trial court's adverse decision; or

(3) Notify the publisher that it is not entitled to indemnification under this part and briefly state the basis for the denial.

§ 15.113 Indemnification agreements.

(a) Upon a finding of the Administrator that indemnification is required under this part, and after obtaining the concurrence of the United States Department of Justice, the FAA will promptly enter into an indemnification agreement providing for the payment of the costs specified in paragraph (c) of this section.

(b) The indemnification agreement will be signed by the Chief Counsel and the publisher.

(c) The FAA will indemnify the publisher for—

(1) Compensatory damages awarded by the court against the publisher;

(2) Reasonable costs and fees, including reasonable attorney fees at a rate not to exceed that permitted under the Equal Access to Justice Act (5 U.S.C. 504), and any postjudgment interest, if the publisher conducts a good faith defense, or pursues a good faith appeal, at the request, or with the concurrence, of the FAA.

(d) Except as otherwise provided in this section, the FAA will not indemnify the publisher for—

(1) Punitive or exemplary damages;

(2) Civil or criminal fines or any other litigation sanctions;

(3) Postjudgment interest;

(4) Costs;

(5) Attorney fees; or

(6) Other incidental expenses.

(e) The indemnification agreement must provide that the Government will be subrogated to all claims or rights of the publisher, including third-party claims, cross-claims, and counterclaims.

§ 15.115 Payment.

After execution of the indemnification agreement, the FAA will submit the agreement to the United States Department of Justice and request payment, in accordance with the agreement, from the Judgment Fund.

PART 16—RULES OF PRACTICE FOR FEDERALLY-ASSISTED AIRPORT ENFORCEMENT PROCEEDINGS

Subpart A—General Provisions

Subpart B—General Rules Applicable to Complaints, Proceedings Initiated by the FAA, and Appeals

Subpart C—Special Rules Applicable to Complaints

Subpart D—Special Rules Applicable to Proceedings Initiated by the FAA

AUTHORITY: 49 U.S.C. 106(g), 322, 1110, 1111, 1115, 1116, 1718(a) and (b), 1719, 1723, 1726, 1727, 40103(e), 40113, 40116, 44502(b), 46101, 46104, 46110, 47104, 47106(e), 47107, 47108, 47111(d), 47122, 47123–47125, 47133, 47151–47153, 48103.

SOURCE: Docket No. 27783, 61 FR 54004, Oct. 16, 1996, unless otherwise noted.

Subpart A—General Provisions

§ 16.1 Applicability and description of part.

(a) *General.* The provisions of this part govern all Federal Aviation Administration (FAA) proceedings involving Federally-assisted airports, except for complaints or requests for determination filed with the Secretary under 14 CFR part 302, whether the proceedings are instituted by order of the FAA or by filing a complaint with the FAA under the following authorities:

(1) 49 U.S.C. 40103(e), prohibiting the grant of exclusive rights for the use of any landing area or air navigation facility on which Federal funds have been expended (formerly section 308 of the Federal Aviation Act of 1958, as amended).

(2) Requirements of the Anti-Head Tax Act, 49 U.S.C. 40116.

(3) The assurances and other Federal obligations contained in grant-in-aid agreements issued under the Federal Airport Act of 1946, 49 U.S.C. 1101 *et seq.* (repealed 1970).

(4) The assurances and other Federal obligations contained in grant-in-aid agreements issued under the Airport and Airway Development Act of 1970, as amended, 49 U.S.C. 1701 *et seq.*

(5) The assurances and other Federal obligations contained in grant-in-aid agreements issued under the Airport and Airway Improvement Act of 1982 (AAIA), as amended and recodified, 49 U.S.C. 47101 *et seq.*, specifically section 511(a), 49 U.S.C. 47107, and 49 U.S.C. 47133.

(6) Section 505(d) of the Airport and Airway Improvement Act of 1982, and the requirements concerning civil rights and/or Disadvantaged Business Enterprise (DBE) issues contained in 49 U.S.C. 47107(e) and 49 U.S.C. 47113; 49 U.S.C. 47123; 49 U.S.C. 322, as amended; 49 CFR parts 23 and/or 26; and/or grant assurance 30 and/or grant assurance 37.

(7) Obligations contained in property deeds for property transferred pursuant to section 16 of the Federal Airport Act (49 U.S.C. 1115), section 23 of the Airport and Airway Development Act (49 U.S.C. 1723), or section 516 of the Airport and Airway Improvement Act (49 U.S.C. 47125).

(8) Obligations contained in property deeds for property transferred under the Surplus Property Act (49 U.S.C. 47151–47153).

(b) *Other agencies.* Where a grant assurance concerns a statute, executive order, regulation, or other authority that provides an administrative process for the investigation or adjudication of complaints by a Federal agency other than the FAA, persons shall use the administrative process established

93

by those authorities. Where a grant assurance concerns a statute, executive order, regulation, or other authority that enables a Federal agency other than the FAA to investigate, adjudicate, and enforce compliance under those authorities on its own initiative, the FAA may defer to that Federal agency.

(c) *Other enforcement.* If a complaint or action initiated by the FAA involves a violation of the 49 U.S.C. subtitle VII or FAA regulations, except as specified in paragraphs (a)(1) and (a)(2) of this section, the FAA may take investigative and enforcement action under 14 CFR part 13, "Investigative and Enforcement Procedures."

(d) *Effective date.* This part applies to a complaint filed with the FAA and to an investigation initiated by the FAA on or after December 16, 1996.

[Doc. No. 27783, 61 FR 54004, Oct. 16, 1996, as amended at Amdt. 16–1, 78 FR 56141, Sept. 12, 2013]

§ 16.3 Definitions.

Terms defined in the Acts are used as so defined. As used in this part:

Act means a statute listed in § 16.1 and any regulation, agreement, or document of conveyance issued or made under that statute.

Administrator means the Administrator of the FAA.

Agency means the FAA.

Agency attorney means the Deputy Chief Counsel; the Assistant Chief Counsel and attorneys in the Airports/Environmental Law Division of the Office of the Chief Counsel; the Assistant Chief Counsel and attorneys in an FAA region or center who represent the FAA during the investigation of a complaint or at a hearing on a complaint, and who prosecute on behalf of the FAA, as appropriate. An agency attorney shall not include the Chief Counsel; the Assistant Chief Counsel for Litigation, or any attorney on the staff of the Assistant Chief Counsel for Litigation, who advises the Associate Administrator regarding an initial decision of the hearing officer or any appeal to the Associate Administrator or who is supervised in that action by a person who provides such advice in an action covered by this part.

Agency employee means any employee of the FAA.

Associate Administrator means the FAA Associate Administrator for Airports or a designee. For the purposes of this part only, Associate Administrator also means the Assistant Administrator for Civil Rights or a designee for complaints that the FAA Associate Administrator for Airports transfers to the Assistant Administrator for Civil Rights.

Complainant means the person submitting a complaint.

Complaint means a written document meeting the requirements of this part and filed under this part:

(1) By a person directly and substantially affected by anything allegedly done or omitted to be done by any person in contravention of any provision of any Act, as defined in this section, as to matters within the jurisdiction of the Administrator, or

(2) By a person under 49 CFR 26.105(c) against a recipient of FAA funds alleged to have violated a provision of 49 CFR parts 23 and/or 26.

Decisional employee means the Administrator, Deputy Administrator, Associate Administrator, Director, hearing officer, or other FAA employee who is or who may reasonably be expected to be involved in the decisional process of the proceeding.

Director means the Director of the FAA Office of Airport Compliance and Management Analysis, or a designee. For the purposes of this part only, Director also means the Deputy Assistant Administrator for Civil Rights for complaints that the Director of the FAA Office of Airport Compliance and Management Analysis transfers to the Deputy Assistant Administrator for Civil Rights or designee.

Electronic filing means the process of sending electronic mail (email) to the FAA Part 16 Docket Clerk, with scanned documents attached, as a Portable Document Format (PDF) file.

Ex parte communication means an oral or written communication not on the public record with respect to which reasonable prior notice to all parties is not given, but it shall not include requests for status reports on any matter or proceeding covered by this part, or communications between FAA employees who participate as parties to a hearing pursuant to 16.203(b) of this part and other parties to a hearing.

Hearing officer means an attorney designated by the Deputy Chief Counsel in a hearing order to serve as a hearing officer in a hearing under this part. The following are not designated as hearing officers: the Chief Counsel and Deputy Chief Counsel; the Regional or Center Counsel and attorneys in the FAA region or center in which the noncompliance has allegedly occurred or is occurring; the Assistant Chief Counsel and attorneys in the Airports and Environmental Law Division of the FAA Office of the Chief Counsel; and the Assistant Chief Counsel and attorneys in the Litigation Division of the FAA Office of Chief Counsel.

Initial decision means a decision made by the hearing officer in a hearing under subpart F of this part.

Mail means U.S. first class mail; U.S. certified mail; and U.S. express mail. Unless otherwise noted, mail also means electronic mail containing PDF copies of pleadings or documents required herein.

Noncompliance means anything done or omitted to be done by any person in contravention of any provision of any Act, as defined in this section, as to matters within the jurisdiction of the Administrator.

Party means the complainant(s) and the respondent(s) named in the complaint and, after an initial determination providing an opportunity for hearing is issued under § 16.31 and subpart E of this part, the agency.

Person in addition to its meaning under 49 U.S.C. 40102(a)(33), includes a public agency as defined in 49 U.S.C. 47102(a)(15).

Personal delivery means same-day hand delivery or overnight express delivery service.

Respondent means any person named in a complaint as a person responsible for noncompliance.

Sponsor means:

(1) Any public agency which, either individually or jointly with one or more other public agencies, has received Federal financial assistance for airport development or planning under the Federal Airport Act, Airport and Airway Development Act or Airport and Airway Improvement Act;

(2) Any private owner of a public-use airport that has received financial assistance from the FAA for such airport; and

(3) Any person to whom the Federal Government has conveyed property for airport purposes under section 13(g) of the Surplus Property Act of 1944, as amended.

Writing or written includes paper documents that are filed and/or served by mail, personal delivery, facsimile, or email (as attached PDF files).

[Doc. No. 27783, 61 FR 54004, Oct. 16, 1996, as amended at Amdt. 16–1, 78 FR 56141, Sept. 12, 2013]

§ 16.5 Separation of functions.

(a) Proceedings under this part, including hearings under subpart F of this part, will be prosecuted by an agency attorney.

(b) After issuance of an initial determination in which the FAA provides the opportunity for a hearing, an agency employee engaged in the performance of investigative or prosecutorial functions in a proceeding under this part will not, in that case or a factually related case, participate or give advice in an initial decision by the hearing officer, or a final decision by the Associate Administrator or designee on written appeal, and will not, except as counsel or as witness in the public proceedings, engage in any substantive communication regarding that case or a related case with the hearing officer, the Associate Administrator on written appeal, or agency employees advising those officials in that capacity.

(c) The Chief Counsel, the Assistant Chief Counsel for Litigation, or an attorney on the staff of the Assistant Chief Counsel for Litigation advises the Associate Administrator regarding an initial decision, an appeal, or a final decision regarding any case brought under this part.

Subpart B—General Rules Applicable to Complaints, Proceedings Initiated by the FAA, and Appeals

§ 16.11 General processes.

(a) Under the authority of 49 U.S.C. 40113 and 47121, the Director may conduct investigations, issue orders, and take such other actions as are necessary to fulfill the purposes of this part. This includes the extension of any time period prescribed, where necessary or appropriate for a fair and complete consideration of matters before the agency, prior to issuance of the Director's Determination.

(b) Notwithstanding any other provision of this part, upon finding that circumstances require expedited handling of a particular case or controversy, the Director may issue an order directing any of the following prior to the issuance of the Director's Determination:

(1) Shortening the time period for any action under this part consistent with due process;

(2) If other adequate opportunity to respond to pleadings is available, eliminating the reply, rebuttal, or other actions prescribed by this part;

(3) Designating alternative methods of service; or

(4) Directing such other measures as may be required.

(c) Other than those matters concerning a Corrective Action Plan, the jurisdiction of the Director terminates upon the issuance of the Director's Determination. All matters arising during the appeal period, such as requests for extension of time to make an appeal, will be addressed by the Associate Administrator.

(d) The Director may transfer to the FAA Deputy Assistant Administrator for Civil Rights or Office of Civil Rights designee the authority to prepare and issue Director's Determinations pursuant to § 16.31 for complaints alleging violations of section 505(d) of the Airport and Airway Improvement Act of 1982, and the requirements concerning civil rights and/or Disadvantaged Business Enterprise (DBE) issues contained in 49 U.S.C. 47107(e) and 49 U.S.C. 47113; 49 U.S.C. 47123; 49 U.S.C. 322, as amended; 49 CFR parts 23 and/or 26; and/or grant assurance 30 and/or grant assurance 37.

[Doc. No. 27783, 61 FR 54004, Oct. 16, 1996, as amended at Amdt. 16–1, 78 FR 56142, Sept. 12, 2013]

§ 16.13 Filing of documents.

Except as otherwise provided in this part, documents shall be filed with the FAA during a proceeding under this part as follows:

(a) *Filing address.* Documents filed under this Part shall be filed with the Office of the Chief Counsel, Attention: FAA Part 16 Docket Clerk, AGC–600, Federal Aviation Administration, 800 Independence Avenue SW., Washington, DC 20591. Documents to be filed with a hearing officer shall be filed at the address and in the manner stated in the hearing order.

(b) *Date and method of filing.* Filing of any document shall be by personal delivery or mail as defined in this part, by facsimile (when confirmed by filing on the same date by one of the foregoing methods), or electronically as set forth in paragraph (h) of this section. Unless the date is shown to be inaccurate, documents filed with the FAA shall be deemed to be filed on the date of personal delivery, on the mailing date shown on the certificate of service, on the date shown on the postmark if there is no certificate of service, on the send date shown on the facsimile (provided filing has been confirmed through one of the foregoing methods), or on the mailing date shown by other evidence if there is no certificate of service and no postmark. Unless the date is shown to be inaccurate, documents filed electronically shall be deemed to be filed on the date shown on the certificate of service or, if none, the date of electronic transmission to the last party required to be served.

(c) *Number of copies.* With the exception of electronic filing or unless otherwise specified, an executed original and three copies of each document shall be filed with the FAA Part 16 Docket Clerk. One of the three copies shall not be stapled, bound or hole-punched. Copies need not be signed, but the name of the person signing the original shall be shown. If a hearing order has been issued in the case, one of the three copies shall be filed with the hearing officer unless otherwise prescribed by the hearing officer.

(d) *Form.* Documents filed under this part shall:

(1) Be typewritten or legibly printed;

(2) Include, in the case of docketed proceedings, the docket number of the proceeding on the front page; and

(3) Be marked to identify personal, privileged or proprietary information. Decisions for the publication and release of these documents will be made in accordance with 5 U.S.C. 552 and 49 CFR part 7.

(e) *Signing of documents and other papers.* The original of every document filed shall be signed by the person filing it or the person's duly authorized representative. The signature shall serve as a certification that the signer has read the document and, based on reasonable inquiry and to the best of the signer's knowledge, information, and belief, the document is—

(1) Consistent with this part;

(2) Warranted by existing law or that a good faith argument exists for extension, modification, or reversal of existing law; and

(3) Not interposed for any improper purpose, such as to harass or to cause unnecessary delay or needless increase in the cost of the administrative process.

(f) *Designation of person to receive service.* The initial document filed by any person shall state on the first page the name, physical address, telephone number, facsimile number, if any, and email address, if filing electronically, of the person(s) to be served with documents in the proceeding. If any of these items change during the proceeding, the person shall promptly file notice of the change with the FAA Part 16 Docket Clerk and the hearing officer and shall serve the notice on all parties.

(g) *Docket numbers.* Each submission identified as a complaint under this part by the submitting person will be assigned a docket number.

(h) *Electronic filing.* (1) The initial complaint may be served electronically upon the respondent only if the respondent has previously agreed with the complainant in writing to participate in electronic filing. Documents may be filed under this Part electronically by sending an email containing

(an) attachment(s) of (a) PDF file(s) of the required pleading to the FAA Docket Clerk, and the person designated in paragraph (h)(3) of this section.

(2) The subject line of the email must contain the names of the complainant and respondent, and must contain the FAA docket number (if assigned). The size of each email must be less than 10 MB. Email attachments containing executable files (e.g., .exe and .vbs files) will not be accepted.

(3) The email address at which the parties may file the documents described in this section is *9-AWA-AGC-Part-16@faa.gov.* No acknowledgement or receipt will be provided by the FAA to parties using this method. A party filing electronically as described in this section must provide to the FAA Part 16 Docket Clerk and the opposing party an email address of the person designated by the party to receive pleadings.

(4) By filing a pleading or document electronically as described in this section, a party waives the rights under this part for service by the opposing party and the FAA by methods other than email. If a party subsequently decides to "opt-out" of electronic filing, that party must so notify the FAA Part 16 Docket Clerk and the other party in writing, from which time the FAA and the parties will begin serving the opting-out party in accordance with §§16.13 and 16.15. This subsection only exempts the parties from the filing and service requirements in §16.13(a) (with the exception that "Documents to be filed with a hearing officer shall be filed at the address and in the manner stated in the hearing order."), the method of filing requirements in §16.13(b), and the number of documents requirements in §16.13(c).

(i) *Internet accessibility of documents filed in the Hearing Docket.* (1) Unless protected from public disclosure, all documents filed in the Hearing Docket are accessible through the Federal Docket Management System (FDMS): *http://www.regulations.gov.* To access a particular case file, use the FDMS number assigned to the case.

(2) Determinations issued by the Director and Associate Administrator in

Part 16 cases, indexes of decisions, contact information for the FAA Hearing Docket, the rules of practice, and other information are available on the FAA Office of Airports' Web site at: *http:// part16.airports.faa.gov/index.cfm.*

[Doc. No. 27783, 61 FR 54004, Oct. 16, 1996, as amended at Amdt. 16–1, 78 FR 56142, Sept. 12, 2013]

§ 16.15 Service of documents on the parties and the agency.

Except as otherwise provided in this part, documents shall be served as follows:

(a) *Whom must be served.* Copies of all documents filed with the FAA Part 16 Docket Clerk shall be served by the persons filing them on all parties to the proceeding. A certificate of service shall accompany all documents when they are tendered for filing and shall certify concurrent service on the FAA and all parties. Certificates of service shall be in substantially the following form:

I hereby certify that I have this day served the foregoing [name of document] on the following persons at the following addresses, facsimile numbers (if also served by facsimile), or email address (if served electronically in accordance with § 16.13(h)), by [specify method of service]:

[list persons, addresses, facsimile numbers, email addresses (as applicable)]

Dated this _ day of _, 20 _.

[signature], for [party]

(b) Method of service. Except as otherwise agreed by the parties and, if applicable, the hearing officer, the method of service is the same as set forth in § 16.13(b) for filing documents.

(c) *Where service shall be made.* Service shall be made to the persons identified in accordance with § 16.13(f). If no such person has been designated, service shall be made on the party.

(d) *Presumption of service.* There shall be a presumption of lawful service—

(1) When acknowledgment of receipt is by a person who customarily or in the ordinary course of business receives mail at the address of the party or of the person designated under § 16.13(f);

(2) When a properly addressed envelope, sent to the most current address submitted under § 16.13(f), has been re-turned as undeliverable, unclaimed, or refused; or

(3) When the party serving the document electronically has a confirmation statement demonstrating that the email was properly sent to a party correctly addressed.

(e) *Date of service.* The date of service shall be determined in the same manner as the filing date under § 16.13(b).

[Doc. No. 27783, 61 FR 54004, Oct. 16, 1996, as amended at Amdt. 16–1, 78 FR 56143, Sept. 12, 2013]

§ 16.17 Computation of time.

This section applies to any period of time prescribed or allowed by this part, by notice or order of the hearing officer, or by an applicable statute.

(a) The date of an act, event, or default, after which a designated time period begins to run, is not included in a computation of time under this part.

(b) The last day of a time period is included in a computation of time unless it is a Saturday, Sunday, or legal holiday for the FAA, in which case, the time period runs until the end of the next day that is not a Saturday, Sunday, or legal holiday.

(c) Whenever a party has the right or is required to do some act within a prescribed period after service of a document upon the party, and the document is served on the party by first class mail or certified mail, 5 days shall be added to the prescribed period.

[Doc. No. 27783, 61 FR 54004, Oct. 16, 1996, as amended at Amdt. 16–1, 78 FR 56143, Sept. 12, 2013]

§ 16.19 Motions.

(a) *General.* An application for an order or ruling not otherwise specifically provided for in this part shall be by motion. Unless otherwise ordered by the agency, the filing of a motion will not stay the date that any action is permitted or required by this part.

(b) *Form and contents.* Unless made during a hearing, motions shall be made in writing, shall state with particularity the relief sought and the grounds for the relief sought, and shall be accompanied by affidavits or other evidence relied upon. Motions introduced during hearings may be made orally on the record, unless the hearing officer directs otherwise.

(c) *Answers to motions.* Except as otherwise provided in this part, or except when a motion is made during a hearing, any party may file an answer in support of or in opposition to a motion, accompanied by affidavits or other evidence relied upon, provided that the answer to the motion is filed within 10 days after the motion has been served upon the person answering, or any other period set by the hearing officer. Where a motion is made during a hearing, the answer and the ruling thereon may be made at the hearing, or orally or in writing within the time set by the hearing officer.

(d) *Deferred actions on motions.* A ruling on a motion made before the time set for the issuance of the Director's Determination may be deferred to and included with the Director's Determination.

(e) *Extension by motion.* A party shall file a written motion for an extension of time not later than 3 business days before the document is due unless good cause for the late filing is shown. A party filing a motion for extension should attempt to obtain the concurrence of the opposing party. A party filing a written motion for an extension of time shall file the motion as required under § 16.13, and serve a copy of the motion on all parties and the docket clerk as required under § 16.15.

[Doc. No. 27783, 61 FR 54004, Oct. 16, 1996, as amended at Amdt. 16–1, 78 FR 56143, Sept. 12, 2013]

Subpart C—Special Rules Applicable to Complaints

§ 16.21   Pre-complaint resolution.

(a) Except for those persons filing under 49 CFR 26.105(c), prior to filing a complaint under this part, a person directly and substantially affected by the alleged noncompliance shall initiate and engage in good faith efforts to resolve the disputed matter informally with those individuals or entities believed responsible for the noncompliance. These efforts at informal resolution may include, without limitation, at the parties' expense, mediation, arbitration, or the use of a dispute resolution board, or other form of third party assistance. The FAA Airports District Office, FAA Airports Field Office, FAA Regional Airports Division responsible for administering financial assistance to the sponsor, or the FAA Office of Civil Rights will be available upon request to assist the parties with informal resolution.

(b) Except for complaints filed under 49 CFR 26.105(c), a complaint will be dismissed under § 16.27 unless the person or authorized representative filing the complaint certifies that:

(1) The complainant has made substantial and reasonable good faith efforts to resolve the disputed matter informally prior to filing the complaint; and

(2) There is no reasonable prospect for practical and timely resolution of the dispute.

(c) The certification required under paragraph (b) of this section, shall include a brief description of the party's efforts to obtain informal resolution but shall not include information on monetary or other settlement offers made but not agreed upon in writing by all parties. Such efforts to resolve informally should be relatively recent and be demonstrated by pertinent documentation. There is no required form or process for informal resolution, but in each case the requirements to resolve the matter informally must meet the requirements of this paragraph.

[Doc. No. 27783, 61 FR 54004, Oct. 16, 1996, as amended at Amdt. 16–1, 78 FR 56143, Sept. 12, 2013]

§ 16.23   Pleadings.

(a) A person directly and substantially affected by any alleged noncompliance or a person qualified under 49 CFR 26.105(c) may file a complaint under this part. A person doing business with an airport and paying fees or rentals to the airport shall be considered directly and substantially affected by alleged revenue diversion as defined in 49 U.S.C. 47107(b).

(b) Complaints filed under this part shall—

(1) State the name and address of each person who is the subject of the complaint and, with respect to each person, the specific provisions of each Act that the complainant believes were violated;

(2) Include all documents then available in the exercise of reasonable diligence, to be offered in support of the complaint, and to be served upon all persons named in the complaint as persons responsible for the alleged action(s) or omission(s) upon which the complaint is based;

(3) Provide a concise but complete statement of the facts relied upon to substantiate each allegation; and

(4) Except for complaints filed under 49 CFR 26.105(c), describe how the complainant was directly and substantially affected by the things done or omitted to be done by the respondents.

(c) Unless the complaint is dismissed pursuant to § 16.25 or § 16.27, the FAA notifies the complainant and respondent in writing within 20 days after the date the FAA receives the complaint that the complaint has been docketed.

(d) The respondent shall file an answer within 20 days of the date of service of the FAA notification or, if a motion is filed under § 16.26, within 20 days of the date of service of an FAA order denying all or part of that motion.

(e) The complainant may file a reply within 10 days of the date of service of the answer.

(f) The respondent may file a rebuttal within 10 days of the date of service of the complainant's reply.

(g) The answer, reply, and rebuttal shall, like the complaint, be accompanied by supporting documentation upon which the parties rely.

(h) The answer shall deny or admit the allegations made in the complaint or state that the person filing the document is without sufficient knowledge or information to admit or deny an allegation, and shall assert any affirmative defense.

(i) The answer, reply, and rebuttal shall each contain a concise but complete statement of the facts relied upon to substantiate the answers, admissions, denials, or averments made.

(j) Amendments or supplements to the pleadings described in this section will not be allowed without showing good cause through a motion and supporting documents.

(k) *Burden of proof.* Except as used in subpart F of this part,

(1) The burden of proof is on the complainant to show noncompliance with an Act or any regulation, order, agreement or document of conveyance issued under the authority of an Act.

(2) Except as otherwise provided by statute or rule, the proponent of a motion, request, or order has the burden of proof.

(3) A party who has asserted an affirmative defense has the burden of proving the affirmative defense.

(l) Except for good cause shown through motion and supporting documents, discovery is not permitted except as provided in §§ 16.213 and 16.215.

[Doc. No. 27783, 61 FR 54004, Oct. 16, 1996, as amended at Amdt. 16-1, 78 FR 56143, Sept. 12, 2013]

§ 16.25 Dismissals.

(a) Within 20 days after the receipt of the complaint, unless a motion has been filed under § 16.26, the Director will dismiss a complaint, or any claim made in a complaint, with prejudice if:

(1) It appears on its face to be outside the jurisdiction of the Administrator under the Acts listed in § 16.1;

(2) On its face it does not state a claim that warrants an investigation or further action by the FAA; or

(3) The complainant lacks standing to file a complaint under §§ 16.3 and 16.23.

(b) A dismissal under this section will include the reasons for the dismissal.

[Amdt. 16-1, 78 FR 56144, Sept. 12, 2013]

§ 16.26 Motions to dismiss and motions for summary judgment.

(a) In lieu of an answer, the respondent may file a motion to dismiss the complaint or a motion for summary judgment on the complaint. The respondent may move for dismissal of the entire complaint or move for dismissal of particular issues from adjudication. The motion must be filed within 20 days after the date of service of the FAA notification of docketing.

(b) *Motions to dismiss.* (1) A motion to dismiss shall be accompanied by a concise statement of the reasons for seeking dismissal. The respondent must show that the complaint should be dismissed, with prejudice, if:

(i) It appears on its face to be outside the jurisdiction of the Administrator under the Acts listed in § 16.1;

(ii) On its face it does not state a claim that warrants an investigation or further action by the FAA; or

(iii) The complainant lacks standing to file a complaint under §§ 16.3 and 16.23.

(2) A motion to dismiss may seek dismissal of the entire complaint or the dismissal of specified claims in the complaint. A motion to dismiss shall be accompanied by a supporting memorandum of points and authorities.

(3) A complainant may file an answer to the motion to dismiss within 10 days of the date the motion is served on the complainant, or within any other period set by the Director. The answer shall be accompanied by a concise statement of reasons for opposing dismissal, and may be accompanied by affidavits and other documentary evidence in support of that contention.

(4) Within 30 days of the date an answer to a motion to dismiss is due under this section, the Director may issue an order disposing of the motion. If the Director denies the motion to dismiss in whole or in part, or grants the motion in part, then within 20 days of when the order is served on the respondent, the respondent shall file an answer to the complaint.

(5) If the Director does not act on the motion to dismiss within 30 days of the date an answer to a motion is due under this section, the respondent shall file an answer to the complaint within the next 20 days.

(c) *Motions for summary judgment.* (1) A motion for summary judgment may be based upon the ground that there is no genuine issue of material fact for adjudication and that the complaint, when viewed in the light most favorable to the complainant, should be summarily adjudicated in favor of the respondent as a matter of law. A motion for summary judgment may seek dismissal of the entire complaint or dismissal of specified claims or issues in the complaint.

(2) The motion for summary judgment shall be accompanied by a concise statement of the material facts as to which the respondent contends there is no genuine issue of material fact. The motion may include affidavits and documentary evidence in support of the

contention that there is no genuine issue of material fact in dispute.

(3) A complainant may file an answer to the motion for summary judgment within 10 days of the date the motion is served on the complainant, or within any other period set by the Director. The answer shall be accompanied by a concise statement of the material facts as to which the complainant contends there is a genuine issue, and may be accompanied by affidavits and other documentary evidence in support of that contention.

(4) Within 30 days of the date an answer to a motion for summary judgment is due under this section, the Director may issue an order disposing of the motion. If the Director denies the motion in whole or in part, or grants the motion in part, then within 20 days of when the order is served on the respondent, the respondent shall file an answer to the complaint.

(5) If the Director does not act on the motion for summary judgment within 30 days of the date an answer to a motion is due under this section, the respondent shall file an answer to the complaint within the next 20 days.

[Amdt. 16–1, 78 FR 56144, Sept. 12, 2013]

§ 16.27 Incomplete complaints.

(a) If a complaint is not dismissed pursuant to § 16.25, but is deficient as to one or more of the requirements set forth in § 16.21 or § 16.23(b), the Director will dismiss the complaint within 20 days after receiving it. Dismissal will be without prejudice to the refiling of the complaint after amendment to correct the deficiencies. The Director's dismissal will include the reasons for the dismissal.

(b) Dismissals under this section are not initial determinations, and appeals from decisions under this section will not be permitted.

[Amdt. 16–1, 78 FR 56144, Sept. 12, 2013]

§ 16.29 Investigations.

(a) If, based on the pleadings, there appears to be a reasonable basis for further investigation, the FAA investigates the subject matter of the complaint.

(b) The investigation may include one or more of the following, at the sole discretion of the FAA:

(1) A review of the written submissions or pleadings of the parties, as supplemented by any informal investigation the FAA considers necessary and by additional information furnished by the parties at FAA request. In rendering its initial determination, the FAA may rely entirely on the complaint and the responsive pleadings provided under this subpart. Each party shall file documents that it considers sufficient to present all relevant facts and argument necessary for the FAA to determine whether the sponsor is in compliance.

(2) Obtaining additional oral and documentary evidence by use of the agency's authority to compel production of such evidence under 49 U.S.C. 40113 and 46104, and 49 U.S.C. 47122. The Administrator's statutory authority to issue compulsory process has been delegated to the Chief Counsel, the Deputy Chief Counsel, the Assistant Chief Counsel for Airports and Environmental Law, and each Assistant Chief Counsel for a region or center.

(3) Conducting or requiring that a sponsor conduct an audit of airport financial records and transactions as provided in 49 U.S.C. 47107 and 47121.

[Doc. No. 27783, 61 FR 54004, Oct. 16, 1996, as amended at Amdt. 16–1 78 FR 56145, Sept. 12, 2013]

§ 16.31 Director's Determinations after investigations.

(a) After consideration of the pleadings and other information obtained by the FAA after investigation, the Director will render an initial determination and serve it upon each party within 120 days of the date the last pleading specified in § 16.23 was due.

(b)(1) The Director's Determination shall include findings of fact and conclusions of law, accompanied by explanations and based upon all material issues of fact, credibility of the evidence, law and discretion presented on the record, together with a statement of the reasons therefor.

(2) The Director shall issue a determination or rule in a party's favor only if the determination or ruling is in accordance with law and supported by a preponderance of the reliable, probative, and substantial evidence contained in the record.

(c) A party adversely affected by the Director's Determination may appeal the initial determination as provided in § 16.33. However, if the Director's Determination that is appealed contains a Corrective Action Plan, the Director has the discretion to suspend the Corrective Action Plan until the appeal is resolved.

(d) If the Director's Determination finds the respondent in noncompliance and proposes the issuance of a compliance order, the initial determination will include notice of opportunity for a hearing under subpart F of this part if a hearing is required by statute or otherwise provided by the FAA. A hearing may be required by statute if the FAA determination would terminate eligibility for grants under 49 U.S.C. 47114(c) or (e), or terminate payments on a grant agreement under 49 U.S.C. subchapter 471. The respondent may elect or waive a hearing, as provided in subpart E of this part.

(e) The Director will not consider requests for rehearing, reargument, reconsideration, or modification of a Director's Determination without a finding of good cause.

[Amdt. 16–1, 78 FR 56145, Sept. 12, 2013]

§ 16.33 Final decisions without hearing.

(a) The Associate Administrator may transfer to the FAA Assistant Administrator for Civil Rights the responsibility to prepare and issue Final Agency Decisions pursuant to this section for appeals with issues concerning civil rights.

(b) The Associate Administrator will issue a final decision on appeal from the Director's Determination, without a hearing, where—

(1) The complaint is dismissed after investigation;

(2) A hearing is not required by statute and is not otherwise made available by the FAA; or

(3) The FAA provides opportunity for a hearing to the respondent and the respondent waives the opportunity for a hearing as provided in subpart E of this part.

(c) In the cases described in paragraph (b) of this section, within 30 days after the date of service of the initial determination, a party adversely affected by the Director's Determination may file in accordance with § 16.13 and serve in accordance with § 16.15 a simultaneous Notice of Appeal and Brief.

(d) A reply to an appeal brief may be filed within 20 days after the date of service of the appeal.

(e) On appeal, the Associate Administrator will consider the issues addressed in any order on a motion to dismiss or motion for summary judgment and any issues accepted in the Director's Determination using the following analysis:

(1) Are the findings of fact each supported by a preponderance of reliable, probative, and substantial evidence contained in the record?

(2) Are conclusions made in accordance with law, precedent and policy?

(3) Are the questions on appeal substantial?

(4) Have any prejudicial errors occurred?

(f) Any new issues or evidence presented in an appeal or reply will not be considered unless accompanied by a petition and good cause found as to why the new issue or evidence was not presented to the Director. Such a petition must:

(1) Set forth the new matter;

(2) Contain affidavits of prospective witnesses, authenticated documents, or both, or an explanation of why such substantiation is unavailable; and

(3) Contain a statement explaining why such new issue or evidence could not have been discovered in the exercise of due diligence prior to the date on which the evidentiary record closed.

(g) The Associate Administrator will issue a final decision and order within 60 days after the due date of the reply.

(h) If no appeal is filed within the time period specified in paragraph (c) of this section, the Director's Determination becomes the final decision and order of the FAA without further action. A Director's Determination that becomes final, because there is no administrative appeal, is not judicially reviewable.

(i) No requests for rehearing, reargument, reconsideration, or modification of a final order will be considered without a finding of good cause.

[Amdt. 16–1, 78 FR 56145, Sept. 12, 2013]

§ 16.34 Consent orders.

(a) The parties may agree at any time before the issuance of a final agency decision to dispose of the case by proposing a consent order. Good faith efforts to resolve a complaint through issuance of a consent order may continue throughout the administrative process. However, except as provided in § 16.11(a), such efforts may not serve as the basis for extensions of the times set forth in this part.

(b) A proposal for a consent order, specified in paragraph (a) of this section, shall include:

(1) A proposed consent order;

(2) An admission of all jurisdictional facts; and

(3) An express waiver of the right to further procedural steps and of all rights of judicial review.

(c) If the parties agree to dispose of a case by issuance of a consent order before the FAA issues a Director's Determination, the proposal for a consent order is submitted jointly by the parties to the Director, together with a request to adopt the consent order and dismiss the case. The Director may issue the consent order as an order of the FAA and terminate the proceeding.

[Amdt. 16–1, 78 FR 56145, Sept. 12, 2013]

Subpart D—Special Rules Applicable to Proceedings Initiated by the FAA

§ 16.101 Basis for the initiation of agency action.

The FAA may initiate its own investigation of any matter within the applicability of this part without having received a complaint. The investigation may include, without limitation, any of the actions described in § 16.29(b).

§ 16.103 Notice of investigation.

Following the initiation of an investigation under § 16.101, the FAA sends a notice to the person(s) subject to investigation. The notice will set forth the areas of the agency's concern and the reasons therefor; request a response to the notice within 30 days of the date of service; and inform the respondent that the FAA will, in its discretion, invite good faith efforts to resolve the matter.

§ 16.105 Failure to resolve informally.

If the matters addressed in the FAA notices are not resolved informally, the FAA may issue a Director's Determination under § 16.31.

[Doc. No. 27783, 61 FR 54004, Oct. 16, 1996, as amended at Amdt. 16–1, 78 FR 56146, Sept. 12, 2013]

Subpart E—Proposed Orders of Compliance

§ 16.109 Orders terminating eligibility for grants, cease and desist orders, and other compliance orders.

(a) The agency will provide the opportunity for a hearing if, in the Director's determination, the agency issues or proposes to issue an order terminating eligibility for grants pursuant to 49 U.S.C. 47106(d), an order suspending the payment of grant funds pursuant to 49 U.S.C. 47111(d); an order withholding approval of any new application to impose a passenger facility charge pursuant to 49 U.S.C. 47111(e); a cease and desist order; an order directing the refund of fees unlawfully collected; or any other compliance order issued by the Administrator to carry out the provisions of the Acts, and required to be issued after notice and opportunity for a hearing. In cases in which a hearing is not required by statute, the FAA may provide opportunity for a hearing at its discretion.

(b) In a case in which the agency provides the opportunity for a hearing, the Director's Determination issued under § 16.31 will include a statement of the availability of a hearing under subpart F of this part.

(1) Within 20 days after service of a Director's Determination under § 16.31 that provides an opportunity for a hearing a person subject to the proposed compliance order may—

(i) Request a hearing under subpart F of this part;

(ii) Waive hearing and appeal the Director's Determination in writing, as provided in § 16.33;

(iii) File, jointly with a complainant, a motion to withdraw the complaint and to dismiss the proposed compliance action; or

(iv) Submit, jointly with the agency, a proposed consent order under § 16.34(c).

(2) If the respondent fails to file an appeal in writing within the time periods provided in paragraph (c) of this section, the Director's Determination becomes final.

(c) The Director may either direct the respondent to submit a Corrective Action Plan or initiate proceedings to revoke and/or deny the respondent's application for Airport Improvement Program discretionary grants under 49 U.S.C. 47115 and general aviation airport grants under 49 U.S.C. 47114(d) when a Director's Determination finds a respondent in noncompliance and does not provide for a hearing.

(d) In the event that the respondent fails to submit, in accordance with a Director's Determination, a Corrective Action Plan acceptable to the FAA within the time provided, unless extended by the FAA for good cause, and/or if the respondent fails to complete the Corrective Action Plan as specified therein, the Director may initiate action to revoke and/or deny applications for Airport Improvement Program discretionary grants under 49 U.S.C. 47115 and general aviation airport grants under 49 U.S.C. 47114(d).

(e) For those violations that cannot be remedied through corrective action, the Director may initiate action to revoke and/or deny the respondent's applications for Airport Improvement Program discretionary grants under 49 U.S.C. 47115 and general aviation airport grants under 49 U.S.C. 47114(d).

(f) When the Director concludes that the respondent has fully complied with the Corrective Action Plan and/or when the Director determines that the respondent has corrected the areas of noncompliance, the Director will terminate the proceeding.

(g) A complainant's standing terminates upon the issuance of a Director's Determination that finds a respondent in noncompliance on all identified issues. The complainant may not appeal the Director's Determination if the Director finds noncompliance on all identified issues.

[Amdt. 16–1, 78 FR 56146, Sept. 12, 2013]

Subpart F—Hearings

§ 16.201 Notice and order of hearing.

(a) If a respondent is provided the opportunity for hearing in an initial determination and does not waive hearing, the Deputy Chief Counsel within 10 days after the respondent elects a hearing will issue and serve on the respondent and complainant a hearing order. The hearing order will set forth:

(1) The allegations in the complaint, or notice of investigation, and the chronology and results of the investigation preliminary to the hearing;

(2) The relevant statutory, judicial, regulatory, and other authorities;

(3) The issues to be decided;

(4) Such rules of procedure as may be necessary to supplement the provisions of this part;

(5) The name and address of the person designated as hearing officer, and the assignment of authority to the hearing officer to conduct the hearing in accordance with the procedures set forth in this part; and

(6) The date by which the hearing officer is directed to issue an initial decision.

(b) Where there are no genuine issues of material fact requiring oral examination of witnesses, the hearing order may contain a direction to the hearing officer to conduct a hearing by submission of briefs and oral argument without the presentation of testimony or other evidence.

[Doc. No. 27783, 61 FR 54004, Oct. 16, 1996, as amended at Amdt. 16–1, 78 FR 56146, Sept. 12, 2013]

§ 16.202 Powers of a hearing officer.

In accordance with the rules of this subpart, a hearing officer may:

(a) Give notice of, and hold, prehearing conferences and hearings;

(b) Administer oaths and affirmations;

(c) Issue subpoenas authorized by law and issue notices of deposition requested by the parties;

(d) Limit the frequency and extent of discovery;

(e) Rule on offers of proof;

(f) Receive relevant and material evidence;

(g) Regulate the course of the hearing in accordance with the rules of this part to avoid unnecessary and duplicative proceedings in the interest of prompt and fair resolution of the matters at issue;

(h) Hold conferences to settle or to simplify the issues by consent of the parties;

(i) Dispose of procedural motions and requests;

(j) Examine witnesses; and

(k) Make findings of fact and conclusions of law, and issue an initial decision.

§ 16.203 Appearances, parties, and rights of parties.

(a) *Appearances.* Any party may appear and be heard in person.

(1) Any party may be accompanied, represented, or advised by an attorney licensed by a State, the District of Columbia, or a territory of the United States to practice law or appear before the courts of that State or territory, or by another person authorized by the hearing officer to be the party's representative.

(2) An attorney, or other duly authorized representative, who represents a party shall file a notice of appearance in accordance with § 16.13.

(b) *Parties and agency participation.* (1) The parties to the hearing are the complainant(s) and respondent(s) named in the hearing order, and the agency. The style of any pleadings filed under this Subpart shall name the respondent as the Appellant, and the Federal Aviation Administration as the Agency.

(2) Unless otherwise specified in the hearing order, the agency attorney will serve as prosecutor for the agency from the date of issuance of the Director's Determination providing an opportunity for hearing.

[Doc. No. 27783, 61 FR 54004, Oct. 16, 1996, as amended at Amdt. 16–1, 78 FR 56146, Sept. 12, 2013]

§ 16.207 Intervention and other participation.

(a) Intervention and participation by other persons are permitted only at the hearing stage of the complaint process

105

and with the written approval of the hearing officer.

(b) A person may submit a written motion for leave to intervene as a party. Except for good cause shown, a motion for leave to intervene shall be submitted not later than 10 days after the notice of hearing and hearing order.

(c) If the hearing officer finds that intervention will not unduly broaden the issues or delay the proceedings and, if the person has an interest that will benefit the proceedings, the hearing officer may grant a motion for leave to intervene. The hearing officer may determine the extent to which an intervenor may participate in the proceedings.

(d) Other persons may petition the hearing officer for leave to participate in the hearing. Participation is limited to the filing of a posthearing brief and reply to the hearing officer and the Associate Administrator. Such a brief shall be filed and served on all parties in the same manner as the parties' posthearing briefs are filed.

(e) Participation under this section is at the discretion of the hearing officer, and no decision permitting participation shall be deemed to constitute an expression that the participant has such a substantial interest in the proceeding as would entitle it to judicial review of such decision.

[Amdt. 16–1, 78 FR 56146, Sept. 12, 2013]

§ 16.209   Extension of time.

(a) *Extension by oral agreement.* The parties may agree to extend for a reasonable period of time for filing a document under this part. If the parties agree, the hearing officer shall grant one extension of time to each party. The party seeking the extension of time shall submit a draft order to the hearing officer to be signed by the hearing officer and filed with the hearing docket. The hearing officer may grant additional oral requests for an extension of time where the parties agree to the extension.

(b) *Extension by motion.* A party shall file a written motion for an extension of time with the hearing officer not later than 7 days before the document is due unless good cause for the late filing is shown. A party filing a written

motion for an extension of time shall serve a copy of the motion on each party.

(c) *Failure to rule.* If the hearing officer fails to rule on a written motion for an extension of time by the date the document was due, the motion for an extension of time is deemed denied.

(d) *Effect on time limits.* In a hearing required by section 519(b) of the Airport and Airways Improvement Act, as amended in 1987, 49 U.S.C. 47106(e) and 47111(d), the due date for the hearing officer's initial decision and for the final agency decision are extended by the length of the extension granted by the hearing officer only if the hearing officer grants an extension of time as a result of an agreement by the parties as specified in paragraph (a) of this section or, if the hearing officer grants an extension of time as a result of the sponsor's failure to adhere to the hearing schedule. In any other hearing, an extension of time granted by the hearing officer for any reason extends the due date for the hearing officer's initial decision and for the final agency decision by the length of time of the hearing officer's decision.

§ 16.211   Prehearing conference.

(a) *Prehearing conference notice.* The hearing officer schedules a prehearing conference and serves a prehearing conference notice on the parties promptly after being designated as a hearing officer.

(1) The prehearing conference notice specifies the date, time, place, and manner (in person or by telephone) of the prehearing conference.

(2) The prehearing conference notice may direct the parties to exchange proposed witness lists, requests for evidence and the production of documents in the possession of another party, responses to interrogatories, admissions, proposed procedural schedules, and proposed stipulations before the date of the prehearing conference.

(b) *The prehearing conference.* The prehearing conference is conducted by telephone or in person, at the hearing officer's discretion. The prehearing conference addresses matters raised in the prehearing conference notice and such other matters as the hearing officer determines will assist in a prompt, full and fair hearing of the issues.

(c) *Prehearing conference report.* At the close of the prehearing conference, the hearing officer rules on any requests for evidence and the production of documents in the possession of other parties, responses to interrogatories, and admissions; on any requests for depositions; on any proposed stipulations; and on any pending applications for subpoenas as permitted by § 16.219. In addition, the hearing officer establishes the schedule, which shall provide for the issuance of an initial decision not later than 110 days after issuance of the Director's Determination order unless otherwise provided in the hearing order.

[Doc. No. 27783, 61 FR 54004, Oct. 16, 1996, as amended at Amdt. 16–1, 78 FR 56147, Sept. 12, 2013]

§ 16.213 Discovery.

(a) Discovery is limited to requests for admissions, requests for production of documents, interrogatories, and depositions as authorized by § 16.215.

(b) The hearing officer shall limit the frequency and extent of discovery permitted by this section if a party shows that—

(1) The information requested is cumulative or repetitious;

(2) The information requested may be obtained from another less burdensome and more convenient source;

(3) The party requesting the information has had ample opportunity to obtain the information through other discovery methods permitted under this section; or

(4) The method or scope of discovery requested by the party is unduly burdensome or expensive.

§ 16.215 Depositions.

(a) *General.* For good cause shown, the hearing officer may order that the testimony of a witness may be taken by deposition and that the witness produce documentary evidence in connection with such testimony. Generally, an order to take the deposition of a witness is entered only if:

(1) The person whose deposition is to be taken would be unavailable at the hearing;

(2) The deposition is deemed necessary to perpetuate the testimony of the witness; or

(3) The taking of the deposition is necessary to prevent undue and excessive expense to a party and will not result in undue burden to other parties or in undue delay.

(b) *Application for deposition.* Any party desiring to take the deposition of a witness shall make application therefor to the hearing officer in writing, with a copy of the application served on each party. The application shall include:

(1) The name and residence of the witness;

(2) The time and place for the taking of the proposed deposition;

(3) The reasons why such deposition should be taken; and

(4) A general description of the matters concerning which the witness will be asked to testify.

(c) *Order authorizing deposition.* If good cause is shown, the hearing officer, in his or her discretion, issues an order authorizing the deposition and specifying the name of the witness to be deposed, the location and time of the deposition and the general scope and subject matter of the testimony to be taken.

(d) *Procedures for deposition.* (1) Witnesses whose testimony is taken by deposition shall be sworn or shall affirm before any questions are put to them. Each question propounded shall be recorded and the answers of the witness transcribed verbatim.

(2) Objections to questions or evidence shall be recorded in the transcript of the deposition. The interposing of an objection shall not relieve the witness of the obligation to answer questions, except where the answer would violate a privilege.

(3) The written transcript shall be subscribed by the witness, unless the parties by stipulation waive the signing, or the witness is ill, cannot be found, or refuses to sign. The reporter shall note the reason for failure to sign.

(e) *Depositions of agency employees.* (1) Depositions of Agency Employees will not be allowed except under the provisions of 49 CFR part 9.

(2) Such depositions will be allowed only with the specific written permission of the Chief Counsel or his or her designee.

[Doc. No. 27783, 61 FR 54004, Oct. 16, 1996, as amended at Amdt. 16–1, 78 FR 56147, Sept. 12, 2013]

§ 16.217 Witnesses.

(a) Each party may designate as a witness any person who is able and willing to give testimony that is relevant and material to the issues in the hearing case, subject to the limitation set forth in paragraph (b) of this section.

(b) The hearing officer may exclude testimony of witnesses that would be irrelevant, immaterial, or unduly repetitious.

(c) Any witness may be accompanied by counsel. Counsel representing a nonparty witness has no right to examine the witness or otherwise participate in the development of testimony.

§ 16.219 Subpoenas.

(a) Request for subpoena. A party may apply to the hearing officer, within the time specified for such applications in the prehearing conference report, for a subpoena to compel testimony at a hearing or to require the production of documents only from the following persons:

(1) Another party;

(2) An officer, employee, or agent of another party;

(3) Any other person named in the complaint as participating in or benefiting from the actions of the respondent alleged to have violated any Act;

(4) An officer, employee, or agent of any other person named in the complaint as participating in or benefiting from the actions of the respondent alleged to have violated any Act.

(b) Issuance and service of subpoena. (1) The hearing officer issues the subpoena if the hearing officer determines that the evidence to be obtained by the subpoena is relevant and material to the resolution of the issues in the case.

(2) Subpoenas shall be served by personal service, or upon an agent designated in writing for the purpose, or by certified mail, return receipt addressed to such person or agent. Whenever service is made by registered or certified mail, the date of mailing shall be considered as the time when service is made.

(3) A subpoena issued under this part is effective throughout the United States or any territory or possession thereof.

(c) Motions to quash or modify subpoena. (1) A party or any person upon whom a subpoena has been served may file a motion to quash or modify the subpoena with the hearing officer at or before the time specified in the subpoena for the filing of such motions. The applicant shall describe in detail the basis for the application to quash or modify the subpoena including, but not limited to, a statement that the testimony, document, or tangible evidence is not relevant to the proceeding, that the subpoena is not reasonably tailored to the scope of the proceeding, or that the subpoena is unreasonable and oppressive.

(2) A motion to quash or modify the subpoena stays the effect of the subpoena pending a decision by the hearing officer on the motion.

§ 16.221 Witness fees.

(a) The party on whose behalf a witness appears is responsible for paying any witness fees and mileage expenses.

(b) Except for employees of the United States summoned to testify as to matters related to their public employment, witnesses summoned by subpoena shall be paid the same fees and mileage expenses as are paid to a witness in a court of the United States in comparable circumstances.

§ 16.223 Evidence.

(a) General. A party may submit direct and rebuttal evidence in accordance with this section.

(b) Requirement for written testimony and evidence. Except in the case of evidence obtained by subpoena, or in the case of a special ruling by the hearing officer to admit oral testimony, a party's direct and rebuttal evidence shall be submitted in written form in advance of the oral hearing pursuant to the schedule established in the hearing officer's prehearing conference report. Written direct and rebuttal fact testimony shall be certified by the witness as true and correct. Subject to the same exception (for evidence obtained by subpoena or subject to a special ruling by the hearing officer), oral examination of a party's own witness is limited to certification of the accuracy of written evidence, including correction and updating, if necessary, and reexamination following cross-examination by other parties.

(c) *Subpoenaed testimony.* Testimony of witnesses appearing under subpoena may be obtained orally.

(d) *Cross-examination.* A party may conduct cross-examination that may be required for disclosure of the facts, subject to control by the hearing officer for fairness, expedition and exclusion of extraneous matters.

(e) *Hearsay evidence.* Hearsay evidence is admissible in proceedings governed by this part. The fact that evidence is hearsay goes to the weight of evidence and does not affect its admissibility.

(f) *Admission of evidence.* The hearing officer admits evidence introduced by a party in support of its case in accordance with this section, but may exclude irrelevant, immaterial, or unduly repetitious evidence.

(g) *Expert or opinion witnesses.* An employee of the FAA or DOT may not be called as an expert or opinion witness for any party other than the agency except as provided in Department of Transportation regulations at 49 CFR part 9.

§16.225 Public disclosure of evidence.

(a) Except as provided in this section, the hearing shall be open to the public.

(b) The hearing officer may order that any information contained in the record be withheld from public disclosure. Any person may object to disclosure of information in the record by filing a written motion to withhold specific information with the hearing officer. The person shall state specific grounds for nondisclosure in the motion.

(c) The hearing officer shall grant the motion to withhold information from public disclosure if the hearing officer determines that disclosure would be in violation of the Privacy Act, would reveal trade secrets or privileged or confidential commercial or financial information, or is otherwise prohibited by law.

§16.227 Standard of proof.

The hearing officer shall issue an initial decision or rule in a party's favor only if the decision or ruling is in accordance with law and supported by a preponderance of the reliable, probative, and substantial evidence contained in the record.

[Amdt. 16–1, as amended at 78 FR 56147, Sept. 12, 2013]

§16.229 Burden of proof.

As used in this subpart, the burden of proof is as follows:

(a) The burden of proof of noncompliance with an Act or any regulation, order, agreement or document of conveyance issued under the authority of an Act is on the agency.

(b) Except as otherwise provided by statute or rule, the proponent of a motion, request, or order has the burden of proof.

(c) A party who has asserted an affirmative defense has the burden of proving the affirmative defense.

[Doc. No. 27783, 61 FR 54004, Oct. 16, 1996, as amended at Amdt. 16–1, 78 FR 56147, Sept. 12, 2013]

§16.231 Offer of proof.

A party whose evidence has been excluded by a ruling of the hearing officer may offer the evidence on the record when filing an appeal.

§16.233 Record.

(a) *Exclusive record.* The transcript of all testimony in the hearing, all exhibits received into evidence, all motions, applications requests and rulings, all documents included in the hearing record and the Director's Determination shall constitute the exclusive record for decision in the proceedings and the basis for the issuance of any orders.

(b) *Examination and copy of record.* A copy of the record will be filed by the FAA Part 16 Docket Clerk in the Federal Docket Management System (FDMS). Any person desiring to review the record may then do so at *http://www.regulations.gov.*

[Amdt. 16–1, 78 FR 56147, Sept. 12, 2013]

§ 16.235 Argument before the hearing officer.

(a) *Argument during the hearing.* During the hearing, the hearing officer shall give the parties reasonable opportunity to present oral argument on the record supporting or opposing motions, objections, and rulings if the parties request an opportunity for argument. The hearing officer may direct written argument during the hearing if the hearing officer finds that submission of written arguments would not delay the hearing.

(b) *Posthearing briefs.* The hearing officer may request or permit the parties to submit posthearing briefs. The hearing officer may provide for the filing of simultaneous reply briefs as well, if such filing will not unduly delay the issuance of the hearing officer's initial decision. Posthearing briefs shall include proposed findings of fact and conclusions of law; exceptions to rulings of the hearing officer; references to the record in support of the findings of fact; and supporting arguments for the proposed findings, proposed conclusions, and exceptions.

[Doc. No. 27783, 61 FR 54004, Oct. 16, 1996, as amended at Amdt. 16–1, 78 FR 56147, Sept. 12, 2013]

§ 16.237 Waiver of procedures.

(a) The hearing officer shall waive such procedural steps as all parties to the hearing agree to waive before issuance of an initial decision.

(b) Consent to a waiver of any procedural step bars the raising of this issue on appeal.

(c) The parties may not by consent waive the obligation of the hearing officer to enter an initial decision on the record.

§ 16.241 Initial decisions, order, and appeals.

(a) The hearing officer shall issue an initial decision based on the record developed during the proceeding and shall send the initial decision to the parties not later than 110 days after the Director's Determination unless otherwise provided in the hearing order.

(b) Each party adversely affected by the hearing officer's initial decision may file an appeal with the Associate Administrator within 15 days of the date the initial decision is issued. Each party may file a reply to an appeal within 10 days after it is served on the party. Filing and service of appeals and replies shall be by personal delivery.

(c) If an appeal is filed, the Associate Administrator reviews the entire record and issues a final agency decision and order within 60 days of the due date of the reply. If no appeal is filed, the Associate Administrator may take review of the case on his or her own motion. If the Associate Administrator finds that the respondent is not in compliance with any Act or any regulation, agreement, or document of conveyance issued or made under such Act, the final agency order includes, in accordance with § 16.245(d), a statement of corrective action, if appropriate, and identifies sanctions for continued noncompliance.

(d) If no appeal is filed, and the Associate Administrator does not take review of the initial decision on the Associate Administrator's own motion, the initial decision shall take effect as the final agency decision and order on the sixteenth day after the actual date the initial decision is issued.

(e) The failure to file an appeal is deemed a waiver of any rights to seek judicial review of an initial decision that becomes a final agency decision by operation of paragraph (d) of this section.

[Doc. No. 27783, 61 FR 54004, Oct. 16, 1996, as amended at Amdt. 16–1, 78 FR 56147, Sept. 12, 2013]

§ 16.243 Consent orders.

(a) The agency attorney and the respondents may agree at any time before the issuance of a final decision and order to dispose of the case by issuance of a consent order. Good faith efforts to resolve a complaint through issuance of a consent order may continue throughout the administrative process. Except as provided in § 16.209, such efforts may not serve as the basis for extensions of the times set forth in this part.

(b) A proposal for a consent order, specified in paragraph (a) of this section, shall include:

(1) A proposed consent order;

(2) An admission of all jurisdictional facts;

(3) An express waiver of the right to further procedural steps and of all rights of judicial review; and

(4) The hearing order, if issued, and an acknowledgment that the hearing order may be used to construe the terms of the consent order.

(c) If the issuance of a consent order has been agreed upon by all parties to the hearing, the proposed consent order shall be filed with the hearing officer, along with a draft order adopting the consent decree and dismissing the case, for the hearing officer's adoption.

(d) The deadline for the hearing officer's initial decision and the final agency decision is extended by the amount of days elapsed between the filing of the proposed consent order with the hearing officer and the issuance of the hearing officer's order continuing the hearing.

(e) If the agency attorney and sponsor agree to dispose of a case by issuance of a consent order before the FAA issues a hearing order, the proposal for a consent order is submitted jointly to the official authorized to issue a hearing order, together with a request to adopt the consent order and dismiss the case. The official authorized to issue the hearing order issues the consent order as an order of the FAA and terminates the proceeding.

[Doc. No. 27783, 61 FR 54004, Oct. 16, 1996, as amended at Amdt. 16–1, 78 FR 56147, Sept. 12, 2013]

§ 16.245 Associate Administrator review after a hearing.

(a) The Associate Administrator may transfer to the FAA Assistant Administrator for Civil Rights the authority to prepare and issue Final Agency Decisions pursuant to § 16.241 for appeals from a hearing concerning civil rights issues.

(b) After a hearing is held, and, after considering the issues as set forth in § 16.245(e), if the Associate Administrator determines that the hearing officer's initial decision or order should be changed, the Associate Administrator may:

(1) Make any necessary findings and issue an order in lieu of the hearing officer's initial decision or order, or

(2) Remand the proceeding for any such purpose as the Associate Administrator may deem necessary.

(c) If the Associate Administrator takes review of the hearing officer's initial decision on the Associate Administrator's own motion, the Associate Administrator will issue a notice of review within 20 days of the actual date the initial decision is issued.

(1) The notice sets forth the specific findings of fact and conclusions of law in the initial decision that are subject to review by the Associate Administrator.

(2) Parties may file one brief on review to the Associate Administrator or rely on their posthearing brief to the hearing officer. A brief on review shall be filed not later than 10 days after service of the notice of review. Filing and service of a brief on review shall be by personal delivery.

(3) The Associate Administrator issues a final agency decision and order within 30 days of the due date of the brief. If the Associate Administrator finds that the respondent is not in compliance with any Act or any regulation, agreement or document of conveyance issued under such Act, the final agency order includes a statement of corrective action, if appropriate.

(d) When the final agency decision finds a respondent in noncompliance, and where a respondent fails to properly seek judicial review of the final agency decision as set forth in subpart G of this part, the Associate Administrator will issue an order remanding the case to the Director for the following action:

(1) In the event that the respondent fails to submit, in accordance with the final agency decision, a Corrective Action Plan acceptable to the FAA within the time provided, unless extended by the FAA for good cause, and/or if the respondent fails to complete the Corrective Action Plan as specified therein, the Director may initiate action to revoke and/or deny applications for Airport Improvement Program grants issued under 49 U.S.C. 47114(c)–(e) and 47115. When the Director concludes that the respondent has fully complied with the Corrective Action Plan, the Director will issue an Order terminating the proceeding.

(2) For those violations that cannot be remedied through corrective action, the Director may initiate action to revoke and/or deny the respondent's applications for Airport Improvement Program grants issued under 49 U.S.C. 47114(c)–(e) and 47115.

(e) On appeal from a hearing officer's initial decision, the Associate Administrator will consider the following questions:

(1) Are the findings of fact each supported by a preponderance of reliable, probative and substantial evidence?

(2) Are conclusions made in accordance with law, precedent and policy?

(3) Are the questions on appeal substantial?

(4) Have any prejudicial errors occurred?

(f) Any new issues or evidence presented in an appeal or reply will not be allowed unless accompanied by a certified petition and good cause found as to why the new matter was not presented to the Director. Such a petition must:

(1) Set forth the new issues or evidence;

(2) Contain affidavits of prospective witnesses, authenticated documents, or both, or an explanation of why such substantiation is unavailable; and

(3) Contain a statement explaining why such new matter could not have been discovered in the exercise of due diligence prior to the date on which the evidentiary record closed.

(g) A Final Agency Decision may be appealed in accordance with subpart G of this part.

[Amdt. 16–1, 78 FR 56147, Sept. 12, 2013]

Subpart G—Judicial Review

SOURCE: Docket No. 27783, 61 FR 54004, Oct. 16, 1996, unless otherwise noted. Redesignated by Amdt. 16–1, 78 FR 56148, Sept. 12, 2013.

§ 16.247 Judicial review of a final decision and order.

(a) A person may seek judicial review, in a United States Court of Appeals, of a final decision and order of the Associate Administrator, and of an order of dismissal with prejudice issued by the Director, as provided in 49 U.S.C. 46110 or 49 U.S.C. 47106(d) and 47111(d). A party seeking judicial review shall file a petition for review with the Court not later than 60 days after the order has been served on the party or within 60 days after the entry of an order under 49 U.S.C. 46110.

(b) The following do not constitute final decisions and orders subject to judicial review:

(1) An FAA decision to dismiss a complaint without prejudice, as set forth in § 16.27;

(2) A Director's Determination;

(3) An initial decision issued by a hearing officer at the conclusion of a hearing;

(4) A Director's Determination or an initial decision of a hearing officer becomes the final decision of the Associate Administrator because it was not appealed within the applicable time periods provided under §§ 16.33(c) and 16.241(b).

[Doc. No. 27783, 61 FR 54004, Oct. 16, 1996. Redesignated and amended by Amdt. 16–1, 78 FR 56148, Sept. 12, 2013]

Subpart H—Ex Parte Communications

SOURCE: Docket No. 27783, 61 FR 54004, Oct. 16, 1996, unless otherwise noted. Redesignated at 78 FR 56148, Sept. 12, 2013.

§ 16.301 Prohibited ex parte communications.

(a) The prohibitions of this section shall apply from the time a proceeding is noticed for hearing unless the person responsible for the communication has knowledge that it will be noticed, in which case the prohibitions shall apply at the time of the acquisition of such knowledge.

(b) Except to the extent required for the disposition of ex parte matters as authorized by law:

(1) No interested person outside the FAA and no FAA employee participating as a party shall make or knowingly cause to be made to any decisional employee an ex parte communication relevant to the merits of the proceeding;

(2) No FAA employee shall make or knowingly cause to be made to any interested person outside the FAA an ex parte communication relevant to the merits of the proceeding; or

(3) Ex parte communications regarding solely matters of agency procedure or practice are not prohibited by this section.

[Doc. No. 27783, 61 FR 54004, Oct. 16, 1996. Redesignated at Amdt. 16–1, 78 FR 56148, Sept. 12, 2013]

§ 16.303 Procedures for handling ex parte communications.

A decisional employee who receives or who makes or knowingly causes to be made a communication prohibited by § 16.303 shall place in the public record of the proceeding:

(a) All such written communications;

(b) Memoranda stating the substance of all such oral communications; and

(c) All written responses, and memoranda stating the substance of all oral responses, to the materials described in paragraphs (a) and (b) of this section.

[Doc. No. 27783, 61 FR 54004, Oct. 16, 1996. Redesignated at Amdt. 16–1, 78 FR 56148, Sept. 12, 2013]

§ 16.305 Requirement to show cause and imposition of sanction.

(a) Upon receipt of a communication knowingly made or knowingly caused to be made by a party in violation of § 16.303, the Associate Administrator or his designee or the hearing officer may, to the extent consistent with the interests of justice and the policy of the underlying statutes, require the party to show cause why his or her claim or interest in the proceeding should not be dismissed, denied, disregarded, or otherwise adversely affected on account of such violation.

(b) The Associate Administrator may, to the extent consistent with the interests of justice and the policy of the underlying statutes administered by the FAA, consider a violation of this subpart sufficient grounds for a decision adverse to a party who has knowingly committed such violation or knowingly caused such violation to occur.

[Doc. No. 27783, 61 FR 54004, Oct. 16, 1996. Redesignated at Amdt. 16–1, 78 FR 56148, Sept. 12, 2013]

PART 17—PROCEDURES FOR PROTESTS AND CONTRACT DISPUTES

AUTHORITY: 5 U.S.C. 570–581, 49 U.S.C. 106(f)(2), 40110, 40111, 40112, 46102, 46014, 46105, 46109, and 46110.

SOURCE: 76 FR 55221, Sept. 7, 2011, unless otherwise noted.

Subpart A—General

§ 17.1 Applicability.

This part applies to all Acquisition Management System (AMS) bid protests and contract disputes involving the FAA that are filed at the Office of Dispute Resolution for Acquisition (ODRA) on or after October 7, 2011, with the exception of those contract disputes arising under or related to FAA contracts entered into prior to April 1, 1996, where such contracts have not been modified to be made subject to the FAA AMS. This part also applies to pre-disputes as described in subpart G of this part.

§ 17.3 Definitions.

(a) *Accrual* means to come into existence as a legally enforceable claim.

(b) *Accrual of a contract claim* means that all events relating to a claim have occurred, which fix liability of either the government or the contractor and permit assertion of the claim, regardless of when the claimant actually discovered those events. For liability to be fixed, some injury must have occurred. Monetary damages need not have been incurred, but if the claim is for money, such damages must be capable of reasonable estimation. The accrual of a claim or the running of the limitations period may be tolled on equitable grounds, including but not limited to active concealment, fraud, or if the facts were inherently unknowable.

(c) *Acquisition Management System* (AMS) establishes the policies, guiding principles, and internal procedures for the FAA's acquisition system.

(d) *Adjudicative Process* is an administrative adjudicatory process used to decide protests and contract disputes where the parties have not achieved resolution through informal communication or the use of ADR. The Adjudicative Process is conducted by a Dispute Resolution Officer (DRO) or Special Master selected by the ODRA Director to preside over the case in accordance with Public Law 108–176, Section 224, Codified at 49 U.S.C. 40110(d)(4).

(e) *Administrator* means the Administrator of the Federal Aviation Administration.

(f) *Alternative Dispute Resolution* (ADR) is the primary means of voluntary dispute resolution that is employed by the ODRA. See Appendix A of this part.

(g) *Compensated Neutral* refers to an impartial third party chosen by the parties to act as a facilitator, mediator, or arbitrator functioning to resolve the protest or contract dispute under the auspices of the ODRA. The parties pay equally for the services of a compensated neutral, unless otherwise agreed to by the parties. An ODRA DRO or neutral cannot be a compensated neutral.

(h) *Contract Dispute*, as used in this part, means a written request to the ODRA seeking, as a matter of right under an FAA contract subject to the AMS, the payment of money in a sum certain, the adjustment or interpretation of contract terms, or for other relief arising under, relating to, or involving an alleged breach of that contract. A contract dispute does not require, as a prerequisite, the issuance of a Contracting Officer final decision. Contract disputes, for purposes of ADR only, may also involve contracts not subject to the AMS.

(i) *Counsel* refers to a Legal Representative who is an attorney licensed by a State, the District of Columbia, or a territory of the United States to practice law or appear before the courts of that State or territory.

(j) *Contractor* is a party in contractual privity with the FAA and responsible for performance of a contract's requirements.

(k) *Discovery* is the procedure whereby opposing parties in a protest or contract dispute may, either voluntarily or to the extent ordered by the ODRA, obtain testimony from, or documents and information held by, other parties or non-parties.

(l) *Dispute Resolution Officer* (DRO) is an attorney and member of the ODRA staff. The term DRO can include the Director of the ODRA.

(m) *Interested party*, in the context of a bid protest, is one whose direct economic interest has been or would be affected by the award or failure to award an FAA contract. Proposed subcontractors are not "interested parties" within this definition and are not eligible to submit protests to the ODRA. Subcontractors not in privity with the FAA are not interested parties in the context of a contract dispute.

(n) *Intervenor* is an interested party other than the protester whose participation in a protest is allowed by the ODRA. For a post-award protest, the awardee of the contract that is the subject of the protest will be allowed, upon timely request, to participate as an intervenor in the protest. In such a protest, no other interested parties will be allowed to participate as intervenors.

(o) *Legal Representative* is an individual(s) designated to act on behalf of a party in matters before the ODRA. Unless otherwise provided under §§ 17.15(c)(2), 17.27(a)(1), or 17.59(a)(6), a Notice of Appearance must be filed with the ODRA containing the name, address, telephone and facsimile (Fax) numbers of a party's legal representative.

(p) *Neutral* refers to an impartial third party in the ADR process chosen by the parties to act as a facilitator, mediator, arbitrator, or otherwise to aid the parties in resolving a protest or contract dispute. A neutral can be a DRO or a person not an employee of the ODRA.

(q) *ODRA* is the FAA's exclusive forum acting on behalf of the Administrator, pursuant to the statutory authority granted by Public Law 108–176, Section 224, to provide dispute resolution services and to adjudicate matters within its jurisdiction. The ODRA may also provide non-binding dispute resolution services in matters outside of its jurisdiction where mutually requested to do so by the parties involved.

(r) *Parties* include the protester(s) or the contractor, the FAA, and any intervenor(s).

(s) *Pre-Disputes* mean an issue(s) in controversy concerning an FAA contract or solicitation that, by mutual agreement of the parties, is filed with the ODRA. See subpart G of this part.

(t) *Product Team*, as used in these rules, refers to the FAA organization(s) responsible for the procurement or contracting activity, without regard to funding source, and includes the Contracting Officer (CO). The Product Team, acting through assigned FAA counsel, is responsible for all communications with and submissions to the ODRA in pending matters.

(u) *Screening Information Request* (SIR or Solicitation) means a request by the FAA for documentation, information, presentations, proposals, or binding offers concerning an approach to meeting potential acquisition requirements established by the FAA.

(v) A *Special Master* is a non-FAA attorney or judge who has been assigned by the ODRA to act as its finder of fact, and to make findings and recommendations based upon AMS policy and applicable law and authorities in the Adjudicative Process.

§ 17.5 Delegation of authority.

(a) The authority of the Administrator to conduct dispute resolution and adjudicative proceedings concerning acquisition matters is delegated to the Director of the ODRA.

(b) The Director of the ODRA may redelegate to Special Masters and DROs such delegated authority in paragraph (a) of this section as deemed necessary by the Director for efficient resolution of an assigned protest or contract dispute, including the imposition of sanctions for the filing of frivolous pleadings, making false statements, or other disciplinary actions. See subpart F of this part.

§ 17.7 Filing and computation of time.

(a) Filing of a protest or contract dispute may be accomplished by overnight delivery, by hand delivery, by Fax, or, if permitted by Order of the ODRA, by electronic filing. A protest or contract dispute is considered to be filed on the date it is received by the ODRA during normal business hours. The ODRA's normal business hours are from 8:30 a.m. to 5 p.m. Eastern Time. A protest or contract dispute received after the time period prescribed for filing shall not be considered timely filed. Service shall also be made on the Contracting

Officer (CO) pursuant to §§ 17.15(e) and 17.27(d).

(b) Submissions to the ODRA after the initial filing of a protest or contract dispute may be accomplished by any means available in paragraph (a) of this section. Copies of all such submissions shall be served on the opposing party or parties.

(c) The time limits stated in this part are calculated in business days, which exclude weekends, Federal holidays and other days on which Federal Government offices in Washington, DC are not open. In computing time, the day of the event beginning a period of time shall not be included. If the last day of a period falls on a weekend or a Federal holiday, the first business day following the weekend or holiday shall be considered the last day of the period.

(d) Electronic Filing—Procedures for electronic filing may be utilized where permitted by Order of the ODRA on a case-by-case basis or pursuant to a Standing Order of the ODRA permitting electronic filing.

§ 17.9 Protective orders.

(a) The ODRA may issue protective orders addressing the treatment of protected information, including protected information in electronic form, either at the request of a party or upon its own initiative. Such information may include proprietary, confidential, or source-selection-sensitive material, or other information the release of which could result in a competitive advantage to one or more firms.

(b) The terms of the ODRA's standard protective order may be altered to suit particular circumstances, by negotiation of the parties, subject to the approval of the ODRA. The protective order establishes procedures for application for access to protected information, identification and safeguarding of that information, and submission of redacted copies of documents omitting protected information.

(c) After a protective order has been issued, counsel or consultants retained by counsel appearing on behalf of a party may apply for access to the material under the order by submitting an application to the ODRA, with copies furnished simultaneously to all parties. The application shall establish that the applicant is not involved in competitive decision-making for any firm that could gain a competitive advantage from access to the protected information and that the applicant will diligently protect any protected information received from inadvertent disclosure. Objections to an applicant's admission shall be raised within two (2) days of the application, although the ODRA may consider objections raised after that time for good cause.

(d) Any violation of the terms of a protective order may result in the imposition of sanctions, including but not limited to removal of the violator from the protective order and reporting of the violator to his or her bar association(s), and the taking of other actions as the ODRA deems appropriate. Additional civil or criminal penalties may apply.

Subpart B—Protests

§ 17.11 Matters not subject to protest.

The following matters may not be protested before the ODRA, except for review of compliance with the AMS:

(a) FAA purchases from or through, State, local, and tribal governments and public authorities;

(b) FAA purchases from or through other Federal agencies;

(c) Grants;

(d) Cooperative agreements;

(e) Other transactions.

§ 17.13 Dispute resolution process for protests.

(a) Protests concerning FAA SIRs, solicitations, or contract awards shall be resolved pursuant to this part.

(b) Potential protestors should, where possible, attempt to resolve any issues concerning potential protests with the CO. Such attempts are not a prerequisite to filing a protest with the ODRA.

(c) Offerors or prospective offerors shall file a protest with the ODRA in accordance with § 17.15. The protest time limitations set forth in § 17.15 will not be extended by attempts to resolve a potential protest with the CO. Other than the time limitations specified in § 17.15 for the filing of protests, the ODRA retains the discretion to modify any timeframes established herein in connection with protests.

(d) In accordance with §17.17(b), the ODRA shall convene an initial status conference for the purpose of scheduling proceedings in the protest and to encourage the parties to consider using the ODRA's ADR process to attempt to resolve the protest, pursuant to subpart D of this part. It is the Agency's policy to use voluntary ADR to the maximum extent practicable. If the parties elect not to attempt ADR, or if ADR efforts do not completely resolve the protest, the protest will proceed under the ODRA Adjudicative Process set forth in subpart E of this part. Informal ADR techniques may be utilized simultaneously with ongoing adjudication.

(e) The ODRA Director shall designate DROs, outside neutrals or Special Masters as potential neutrals for the resolution of protests through ADR. The ultimate choice of an ADR neutral is made by the parties participating in the ADR. The ODRA Director also shall, at his or her sole discretion, designate an adjudicating DRO or Special Master for each matter. A person serving as a neutral in an ADR effort in a matter, shall not serve as an adjudicating DRO or Special Master for that matter.

(f) Multiple protests concerning the same SIR, solicitation, or contract award may be consolidated at the discretion of the ODRA Director, and assigned to a single DRO or Special Master for adjudication.

(g) Procurement activities, and, where applicable, contractor performance pending resolution of a protest, shall continue during the pendency of a protest, unless there is a compelling reason to suspend all or part of the procurement activities or contractor performance. Pursuant to §§17.15(d) and 17.17(a), the ODRA may impose a temporary suspension and recommend suspension of award or contract performance, in whole or in part, for a compelling reason. A decision to suspend procurement activities or contractor performance is made in writing by the Administrator or the Administrator's delegee upon recommendation of the ODRA.

§ 17.15 **Filing a protest.**

(a) An interested party may initiate a protest by filing with the ODRA in accordance with §17.7(a) within the timeframes set forth in this Section. Protests that are not timely filed shall be dismissed. The timeframes applicable to the filing of protests are as follows:

(1) Protests based upon alleged SIR or solicitation improprieties that are apparent prior to bid opening or the time set for receipt of initial proposals shall be filed prior to bid opening or the time set for the receipt of initial proposals.

(2) In procurements where proposals are requested, alleged improprieties that do not exist in the initial solicitation, but which are subsequently incorporated into the solicitation, must be protested not later than the next closing time for receipt of proposals following the incorporation.

(3) For protests other than those related to alleged solicitation improprieties, the protest must be filed on the later of the following two dates:

(i) Not later than seven (7) business days after the date the protester knew or should have known of the grounds for the protest; or

(ii) If the protester has requested a post-award debriefing from the FAA Product Team, not later than five (5) business days after the date on which the Product Team holds that debriefing.

(b) Protests shall be filed with the ODRA, AGC–70, Federal Aviation Administration, telephone (202) 267–3290 as follows:

(1) 600 Independence Avenue SW., Room 2W100, Washington, DC 20591 for filing by hand delivery, courier or other form of in-person delivery;

(2) 800 Independence Avenue SW., Washington, DC 20591 [Attention: AGC–70, Wilbur Wright Bldg., Room 2W100] for filing by U.S. Mail; or

(3) Numbers (202) 267–3720 or alternate (202) 267–1293 for filing by facsimile.

(c) A protest shall be in writing, and set forth:

(1) The protester's name, address, telephone number, and FAX number;

(2) The name, address, telephone number, and FAX number of the protester's legal representative, and who

117

shall be duly authorized to represent the protester, to be the point of contact;

(3) The SIR number or, if available, the contract number and the name of the CO;

(4) The basis for the protester's status as an interested party;

(5) The facts supporting the timeliness of the protest;

(6) Whether the protester requests a protective order, the material to be protected, and attach a redacted copy of that material;

(7) A detailed statement of both the legal and factual grounds of the protest, and one (1) copy of each relevant document;

(8) The remedy or remedies sought by the protester, as set forth in § 17.23;

(9) The signature of the legal representative, or another person duly authorized to represent the protester.

(d) If the protester wishes to request a suspension of the procurement or contract performance, in whole or in part, and believes that a compelling reason(s) exists to suspend the procurement or contract performance because of the protested action, the protester shall, in its initial filing:

(1) Set forth such compelling reason(s), supply all facts and documents supporting the protester's position; and

(2) Demonstrate—

(i) The protester has alleged a substantial case;

(ii) The lack of a suspension would be likely to cause irreparable injury;

(iii) The relative hardships on the parties favor a suspension; and

(iv) That a suspension is in the public interest.

(3) Failure of a protester to provide information or documents in support of a requested suspension or failure to address the elements of paragraph (d)(2) of this section may result in the summary rejection of the request for suspension, or a requirement that the protester supplement its request prior to the scheduling of a Product Team response to the request under § 17.17(a).

(e) Concurrent with the filing of a protest with the ODRA, the protester shall serve a copy of the protest on the CO and any other official designated in the SIR for receipt of protests, by means reasonably calculated to be received by the CO on the same day as it is to be received by the ODRA. The protest shall include a signed statement from the protester, certifying to the ODRA the manner of service, date, and time when a copy of the protest was served on the CO and other designated official(s).

(f) Upon receipt of the protest, the CO shall notify the awardee of a challenged contract award in writing of the existence of the protest. The awardee and/or interested parties shall notify the ODRA in writing, of their interest in participating in the protest as intervenors within two (2) business days of receipt of the CO's notification, and shall, in such notice, designate a person as the point of contact for the ODRA.

(g) The ODRA has discretion to designate the parties who shall participate in the protest as intervenors. In protests of awarded contracts, only the awardee may participate as an intervenor as a matter of right.

[76 FR 55221, Sept. 7, 2011, as amended by Doc. No. FAA–2017–0075, 82 FR 14429, Mar. 21, 2017]

§ 17.17 Initial protest procedures.

(a) If, as part of its initial protest filing, the protester requests a suspension of procurement activities or contractor performance in whole or in part, in accordance with § 17.15(d), the Product Team shall submit a response to the request to the ODRA by no later than the close of business on the date of the initial scheduling conference or on such other date as is established by the ODRA. Copies of the response shall be furnished to the protester and any intervenor(s) so as to be received within the same timeframe. The protester and any intervenor(s) shall have the opportunity of providing additional comments on the response within two (2) business days of receiving it. Based on its review of such submissions, the ODRA, in its discretion, may—

(1) Decline the suspension request; or

(2) Recommend such suspension to the Administrator or the Administrator's designee. The ODRA also may impose a temporary suspension of no more than ten (10) business days, where it is recommending that the Administrator impose a suspension.

(b) Within five (5) business days of the filing of a protest, or as soon thereafter as practicable, the ODRA shall convene an initial status conference for purposes of:

(1) Reviewing the ODRA's ADR and adjudication procedures and establishing a preliminary schedule;

(2) Identifying legal or other preliminary or potentially dispositive issues and answering the parties' questions regarding the ODRA process;

(3) Dealing with issues related to protected information and the issuance of any needed protective order;

(4) Encouraging the parties to consider using ADR;

(5) Appointing a DRO as a potential ADR neutral to assist the parties in considering ADR options and developing an ADR agreement; and

(6) For any other reason deemed appropriate by the DRO or by the ODRA.

(c) The Product Team and protester will have five (5) business days from the date of the initial status conference to decide whether they will attempt to use an ADR process in the case. With the agreement of the ODRA, ADR may be used concurrently with the adjudication of a protest. See § 17.37(e).

(d) If the Product Team and protester elect to use ADR proceedings to resolve the protest, they will agree upon the neutral to conduct the ADR proceedings (either an ODRA DRO or a compensated neutral of their own choosing) pursuant to § 17.37, and shall execute and file with the ODRA a written ADR agreement. Agreement of any intervenor(s) to the use of ADR or the resolution of a dispute through ADR shall not be required.

(e) If the Product Team or protester indicate that ADR proceedings will not be used, or if ADR is not successful in resolving the entire protest, the ODRA Director upon being informed of the situation, will schedule an adjudication of the protest.

§ 17.19 Motions practice and dismissal or summary decision of protests.

(a) Separate motions generally are discouraged in ODRA bid protests. Counsel and parties are encouraged to incorporate any such motions in their respective agency responses or comments. Parties and counsel are encouraged to attempt to resolve typical motions issues through the ODRA ADR process. The ODRA may rule on any non-dispositive motion, where appropriate and necessary, after providing an opportunity for briefing on the motion by all affected parties. Unjustifiable, inappropriate use of motions may result in the imposition of sanctions. Where appropriate, a party may request by dispositive motion to the ODRA, or the ODRA may recommend or order, that:

(1) The protest, or any count or portion of a protest, be dismissed for lack of jurisdiction, timeliness, or standing to pursue the protest;

(2) The protest, or any count or portion of a protest, be dismissed, if frivolous or without basis in fact or law, or for failure to state a claim upon which relief may be had;

(3) A summary decision be issued with respect to the protest, or any count or portion of a protest, if there are no material facts in dispute and a party is entitled to summary decision as a matter of law.

(b) In connection with consideration of possible dismissal or summary decision, the ODRA shall consider any material facts in dispute, in a light most favorable to the party against whom the dismissal or summary decision would operate and draw all factual inferences in favor of the non-moving party.

(c) Either upon motion by a party or on its own initiative, the ODRA may, at any time, exercise its discretion to:

(1) Recommend to the Administrator dismissal or the issuance of a summary decision with respect to the entire protest;

(2) Dismiss the entire protest or issue a summary decision with respect to the entire protest, if delegated that authority by the Administrator; or

(3) Dismiss or issue a summary decision with respect to any count or portion of a protest.

(d) A dismissal or summary decision regarding the entire protest by either the Administrator, or the ODRA by delegation, shall be construed as a final agency order. A dismissal or summary decision that does not resolve all counts or portions of a protest shall

119

not constitute a final agency order, unless and until such dismissal or decision is incorporated or otherwise adopted in a decision by the Administrator (or the ODRA, by delegation) regarding the entire protest.

(e) Prior to recommending or entering either a dismissal or a summary decision, either in whole or in part, the ODRA shall afford all parties against whom the dismissal or summary decision is to be entered the opportunity to respond to the proposed dismissal or summary decision.

§ 17.21 Adjudicative Process for protests.

(a) Other than for the resolution of preliminary or dispositive matters, the Adjudicative Process for protests will be commenced by the ODRA Director pursuant to § 17.17(e).

(b) The Director of the ODRA shall appoint a DRO or a Special Master to conduct the adjudication proceedings, develop the administrative record, and prepare findings and recommendations for review of the ODRA Director.

(c) The DRO or Special Master may conduct such proceedings and prepare procedural orders for the proceedings as deemed appropriate; and may require additional submissions from the parties.

(d) The Product Team response to the protest will be due to be filed and served ten (10) business days from the commencement of the ODRA Adjudication process. The Product Team response shall consist of a written chronological, supported statement of proposed facts, and a written presentation of applicable legal or other defenses. The Product Team response shall cite to and be accompanied by all relevant documents, which shall be chronologically indexed, individually tabbed, and certified as authentic and complete. A copy of the response shall be furnished so as to be received by the protester and any intervenor(s) on the same date it is filed with the ODRA. In all cases, the Product Team shall indicate the method of service used.

(e) Comments of the protester and the intervenor on the Product Team response will be due to be filed and served five (5) business days after their receipt of the response. Copies of such comments shall be provided to the other participating parties by the same means and on the same date as they are furnished to the ODRA. Comments may include any supplemental relevant documents.

(f) The ODRA may alter the schedule for filing of the Product Team response and the comments for good cause or to accommodate the circumstances of a particular protest.

(g) The DRO or Special Master may convene the parties and/or their representatives, as needed for the Adjudicative Process.

(h) If, in the sole judgment of the DRO or Special Master, the parties have presented written material sufficient to allow the protest to be decided on the record presented, the DRO or Special Master shall have the discretion to decide the protest on that basis.

(i) The parties may engage in limited, focused discovery with one another and, if justified, with non-parties, so as to obtain information relevant to the allegations of the protest.

(1) The DRO or Special Master shall manage the discovery process, including limiting its length and availability, and shall establish schedules and deadlines for discovery, which are consistent with timeframes established in this part and with the FAA policy of providing fair and expeditious dispute resolution.

(2) The DRO or Special Master may also direct the parties to exchange, in an expedited manner, relevant, non-privileged documents.

(3) Where justified, the DRO or Special Master may direct the taking of deposition testimony, however, the FAA dispute resolution process does not contemplate extensive discovery.

(4) The use of interrogatories and requests for admission is not permitted in ODRA bid protests.

(5) Where parties cannot voluntarily reach agreement on a discovery-related issue, they may timely seek assistance from an ODRA ADR neutral or may file an appropriate motion with the ODRA. Parties may request a subpoena.

(6) Discovery requests and responses are not part of the record and will not be filed with the ODRA, except in connection with a motion or other permissible filing.

(7) Unless timely objection is made, documents properly filed with the ODRA will be deemed admitted into the administrative record.

(j) Hearings are not typically held in bid protests. The DRO or Special Master may conduct hearings, and may limit the hearings to the testimony of specific witnesses and/or presentations regarding specific issues. The DRO or Special Master shall control the nature and conduct of all hearings, including the sequence and extent of any testimony. Hearings will be conducted:

(1) Where the DRO or Special Master determines that there are complex factual issues in dispute that cannot adequately or efficiently be developed solely by means of written presentations and/or that resolution of the controversy will be dependent on his/her assessment of the credibility of statements provided by individuals with first-hand knowledge of the facts; or

(2) Upon request of any party to the protest, unless the DRO or Special Master finds specifically that a hearing is unnecessary and that no party will be prejudiced by limiting the record in the adjudication to the parties' written submissions. All witnesses at any such hearing shall be subject to cross-examination by the opposing party and to questioning by the DRO or Special Master.

(k) The Director of the ODRA may review the status of any protest in the Adjudicative Process with the DRO or Special Master.

(l) After the closing of the administrative record, the DRO or Special Master will prepare and submit findings and recommendations to the ODRA that shall contain the following:

(1) Findings of fact;

(2) Application of the principles of the AMS, and any applicable law or authority to the findings of fact;

(3) A recommendation for a final FAA order; and

(4) If appropriate, suggestions for future FAA action.

(m) In preparing findings and recommendations in protests, the DRO or Special Master, using the preponderance of the evidence standard, shall consider whether the Product Team actions in question were consistent with the requirements of the AMS, had a rational basis, and whether the Product Team decision was arbitrary, capricious or an abuse of discretion. Notwithstanding the above, allegations that government officials acted with bias or in bad faith must be established by clear and convincing evidence.

(n) The DRO or Special Master has broad discretion to recommend a remedy that is consistent with § 17.23.

(o) A DRO or Special Master shall submit findings and recommendations only to the Director of the ODRA or the Director's designee. The findings and recommendations will be released to the parties and to the public upon issuance of the final FAA order in the case. If an ODRA protective order was issued in connection with the protest, or if a protest involves proprietary or competition-sensitive information, a redacted version of the findings and recommendations, omitting any protected information, shall be prepared wherever possible and released to the public, as soon as is practicable, along with a copy of the final FAA order. Only persons admitted by the ODRA under the protective order and Government personnel shall be provided copies of the unredacted findings and recommendations that contain proprietary or competition-sensitive information.

(p) Other than communications regarding purely procedural matters or ADR, there shall be no substantive *ex parte* communication between ODRA personnel and any principal or representative of a party concerning a pending or potentially pending matter. A potential or serving ADR neutral may communicate on an *ex parte* basis to establish or conduct the ADR.

§ 17.23 Protest remedies.

(a) The ODRA has broad discretion to recommend and impose protest remedies that are consistent with the AMS and applicable law. Such remedies may include, but are not limited to one or a combination of, the following:

(1) Amend the SIR;

(2) Refrain from exercising options under the contract;

(3) Issue a new SIR;

(4) Require a recompetition or revaluation;

(5) Terminate an existing contract for the FAA's convenience;

(6) Direct an award to the protester;

(7) Award bid and proposal costs; or

(8) Any other remedy consistent with the AMS that is appropriate under the circumstances.

(b) In determining the appropriate recommendation, the ODRA may consider the circumstances surrounding the procurement or proposed procurement including, but not limited to: the nature of the procurement deficiency; the degree of prejudice to other parties or to the integrity of the acquisition system; the good faith of the parties; the extent of performance completed; the feasibility of any proposed remedy; the urgency of the procurement; the cost and impact of the recommended remedy; and the impact on the Agency's mission.

(c) Attorney's fees of a prevailing protester are allowable to the extent permitted by the Equal Access to Justice Act, 5 U.S.C. 504(a)(1) (EAJA) and 14 CFR part 14.

Subpart C—Contract Disputes

§ 17.25 Dispute resolution process for contract disputes.

(a) All contract disputes arising under contracts subject to the AMS shall be resolved under this subpart.

(b) Contract disputes shall be filed with the ODRA pursuant to § 17.27.

(c) The ODRA has broad discretion to recommend remedies for a contract dispute that are consistent with the AMS and applicable law, including such equitable remedies or other remedies as it deems appropriate.

§ 17.27 Filing a contract dispute.

(a) Contract disputes must be in writing and should contain:

(1) The contractor's name, address, telephone and Fax numbers and the name, address, telephone and Fax numbers of the contractor's legal representative(s) (if any) for the contract dispute;

(2) The contract number and the name of the Contracting Officer;

(3) A detailed chronological statement of the facts and of the legal grounds underlying the contract dispute, broken down by individual claim item, citing to relevant contract provisions and attaching copies of the contract and other relevant documents;

(4) Information establishing the ODRA's jurisdiction and the timeliness of the contract dispute;

(5) A request for a specific remedy, and the amount, if known, of any monetary remedy requested, together with pertinent cost information and documentation (e.g., invoices and cancelled checks). Supporting documentation should be broken down by individual claim item and summarized; and

(6) The signature of a duly authorized representative of the initiating party.

(b) Contract Disputes shall be filed with the ODRA, AGC–70, Federal Aviation Administration, telephone (202) 267–3290 as follows:

(1) 600 Independence Avenue SW., Room 2W100, Washington, DC 20591 for filing by hand delivery, courier or other form of in-person delivery;

(2) 800 Independence Avenue SW., Washington, DC 20591 [Attention: AGC–70, Wilbur Wright Bldg., Room 2W100] for filing by U.S. Mail; or

(3) Numbers (202) 267–3720 or alternate (202) 267–1293 for filing by facsimile.

(c) A contract dispute against the FAA shall be filed with the ODRA within two (2) years of the accrual of the contract claim involved. A contract dispute by the FAA against a contractor (excluding contract disputes alleging warranty issues, fraud or latent defects) likewise shall be filed within two (2) years of the accrual of the contract claim. If an underlying contract entered into prior to the effective date of this part provides for time limitations for filing of contract disputes with the ODRA, which differ from the aforesaid two (2) year period, the limitation periods in the contract shall control over the limitation period of this section. In no event will either party be permitted to file with the ODRA a contract dispute seeking an equitable adjustment or other damages after the contractor has accepted final contract payment, with the exception of FAA contract disputes related to warranty issues, gross mistakes amounting to fraud or latent defects. FAA contract disputes against the contractor based on warranty issues must be filed within the time specified under applicable contract warranty provisions. Any FAA contract disputes against the contractor based on gross mistakes amounting to fraud or latent defects shall be filed with the ODRA within two (2) years of the date on which the FAA knew or should have known of the presence of the fraud or latent defect.

(d) A party shall serve a copy of the contract dispute upon the other party, by means reasonably calculated to be received on the same day as the filing is received by the ODRA.

(e) With the exception of the time limitations established herein for the filing of contract disputes, the ODRA retains the discretion to modify any timeframe established herein in connection with contract disputes.

[76 FR 55221, Sept. 7, 2011, as amended by Doc. No. FAA–2017–0075, 82 FR 14429, Mar. 21, 2017]

§17.29 Informal resolution period.

(a) The ODRA process for contract disputes includes an informal resolution period of twenty (20) business days from the date of filing in order for the parties to attempt to informally resolve the contract dispute either through direct negotiation or with the assistance of the ODRA. The CO, with the advice of FAA legal counsel, has full discretion to settle contract disputes, except where the matter involves fraud.

(b) During the informal resolution period, if the parties request it, the ODRA will appoint a DRO for ADR who will discuss ADR options with the parties, offer his or her services as a potential neutral, and assist the parties to enter into an agreement for a formal ADR process. A person serving as a neutral in an ADR effort in a matter shall not serve as an adjudicating DRO or Special Master for that matter.

(c) The informal resolution period may be extended at the request of the parties for good cause.

(d) If the matter has not been resolved informally, the parties shall file joint or separate statements with the ODRA no later than twenty (20) business days after the filing of the contract dispute. The ODRA may extend this time, pursuant to §17.27(e). The statement(s) shall include either:

(1) A joint request for ADR, or an executed ADR agreement, pursuant to §17.37(d), specifying which ADR techniques will be employed; or

(2) Written explanation(s) as to why ADR proceedings will not be used and why the Adjudicative Process will be needed.

(e) If the contract dispute is not completely resolved during the informal resolution period, the ODRA's Adjudicative Process will commence unless the parties have reached an agreement to attempt a formal ADR effort. As part of such an ADR agreement the parties, with the concurrence of the ODRA, may agree to defer commencement of the adjudication process pending completion of the ADR or that the ADR and adjudication process will run concurrently. If a formal ADR is attempted but does not completely resolve the contract dispute, the Adjudicative Process will commence.

(f) The ODRA shall hold a status conference with the parties within ten (10) business days, or as soon thereafter as is practicable, of the ODRA's receipt of a written notification that ADR proceedings will not be used, or have not fully resolved the Contract Dispute. The purpose of the status conference will be to commence the Adjudicative Process and establish the schedule for adjudication.

(g) The submission of a statement which indicates that ADR will not be utilized will not in any way preclude the parties from engaging in non-binding ADR techniques during the Adjudicative Process, pursuant to subpart D of this part.

§17.31 Dismissal or summary decision of contract disputes.

(a) Any party may request by motion, or the ODRA on its own initiative may recommend or direct, that a contract dispute be dismissed, or that a count or portion thereof be stricken, if:

(1) It was not timely filed;

(2) It was filed by a subcontractor or other person or entity lacking standing;

(3) It fails to state a matter upon which relief may be had; or

(4) It involves a matter not subject to the jurisdiction of the ODRA.

(b) Any party may request by motion, or the ODRA on its own initiative may recommend or direct, that a summary decision be issued with respect to a contract dispute, or any count or portion thereof if there are no material facts in dispute and a party is entitled to a summary decision as a matter of law.

(c) In connection with any potential dismissal of a contract dispute, or summary decision, the ODRA will consider any material facts in dispute in a light most favorable to the party against whom the dismissal or summary decision would be entered, and draw all factual inferences in favor of that party.

(d) At any time, whether pursuant to a motion or on its own initiative and at its discretion, the ODRA may:

(1) Dismiss or strike a count or portion of a contract dispute or enter a partial summary decision;

(2) Recommend to the Administrator that the entire contract dispute be dismissed or that a summary decision be entered; or

(3) With a delegation from the Administrator, dismiss the entire contract dispute or enter a summary decision with respect to the entire contract dispute.

(e) An order of dismissal of the entire contract dispute or summary decision with respect to the entire contract dispute, issued either by the Administrator or by the ODRA, on the grounds set forth in this section, shall constitute a final agency order. An ODRA order dismissing or striking a count or portion of a contract dispute or entering a partial summary judgment shall not constitute a final agency order, unless and until such ODRA order is incorporated or otherwise adopted in a final agency decision of the Administrator or the Administrator's delegee regarding the remainder of the dispute.

(f) Prior to recommending or entering either a dismissal or a summary decision, either in whole or in part, the ODRA shall afford all parties against whom the dismissal or summary decision would be entered the opportunity to respond to a proposed dismissal or summary decision.

§ 17.33 Adjudicative Process for contract disputes.

(a) The Adjudicative Process for contract disputes will be commenced by the ODRA Director upon being notified by the ADR neutral or by any party that either—

(1) The parties will not be attempting ADR; or

(2) The parties have not settled all of the dispute issues via ADR, and it is unlikely that they can do so within the time period allotted and/or any reasonable extension.

(b) In cases initiated by a contractor against the FAA, within twenty (20) business days of the commencement of the Adjudicative Process or as scheduled by the ODRA, the Product Team shall prepare and submit to the ODRA, with a copy to the contractor, a chronologically arranged and indexed substantive response, containing a legal and factual position regarding the dispute and all documents relevant to the facts and issues in dispute. The contractor will be entitled, at a specified time, to supplement the record with additional documents.

(c) In cases initiated by the FAA against a contractor, within twenty (20) business days of the commencement of the Adjudicative Process or as scheduled by the ODRA, the contractor shall prepare and submit to the ODRA, with a copy to the Product Team counsel, a chronologically arranged and indexed substantive response, containing a legal and factual position regarding the dispute and all documents relevant to the facts and issues in dispute. The Product Team will be entitled, at a specified time, to supplement the record with additional documents.

(d) Unless timely objection is made, documents properly filed with the ODRA will be deemed admitted into the administrative record. Discovery requests and responses are not part of the record and will not be filed with the ODRA, except in connection with a motion or other permissible filing. Designated, relevant portions of such documents may be filed, with the permission of the ODRA.

(e) The Director of the ODRA shall assign a DRO or a Special Master to conduct adjudicatory proceedings, develop the administrative adjudication record and prepare findings and recommendations for the review of the ODRA Director or the Director's designee.

(f) The DRO or Special Master may conduct a status conference(s) as necessary and issue such orders or decisions as are necessary to promote the efficient resolution of the contract dispute.

(g) At any such status conference, or as necessary during the Adjudicative Process, the DRO or Special Master will:

(1) Determine the appropriate amount of discovery required;

(2) Review the need for a protective order, and if one is needed, prepare a protective order pursuant to §17.9;

(3) Determine whether any issue can be stricken; and

(4) Prepare necessary procedural orders for the proceedings.

(h) Unless otherwise provided by the DRO or Special Master, or by agreement of the parties with the concurrence of the DRO or Special Master, responses to written discovery shall be due within thirty (30) business days from the date received.

(i) At a time or at times determined by the DRO or Special Master, and in advance of the decision of the case, the parties shall make individual final submissions to the ODRA and to the DRO or Special Master, which submissions shall include the following:

(1) A statement of the issues;

(2) A proposed statement of undisputed facts related to each issue together with citations to the administrative record or other supporting materials;

(3) Separate statements of disputed facts related to each issue, with appropriate citations to documents in the Dispute File, to pages of transcripts of any hearing or deposition, or to any affidavit or exhibit which a party may wish to submit with its statement;

(4) Separate legal analyses in support of the parties' respective positions on disputed issues.

(j) Each party shall serve a copy of its final submission on the other party by means reasonably calculated so that the other party receives such submissions on the same day it is received by the ODRA.

(k) The DRO or Special Master may decide the contract dispute on the basis of the administrative record and the submissions referenced in this section, or may, in the DRO or Special Master's discretion, direct the parties to make additional presentations in writing. The DRO or Special Master may conduct hearings, and may limit the hearings to the testimony of specific witnesses and/or presentations regarding specific issues. The DRO or Special Master shall control the nature and conduct of all hearings, including the sequence and extent of any testimony. Evidentiary hearings on the record shall be conducted by the ODRA:

(1) Where the DRO or Special Master determines that there are complex factual issues in dispute that cannot adequately or efficiently be developed solely by means of written presentations and/or that resolution of the controversy will be dependent on his/her assessment of the credibility of statements provided by individuals with first-hand knowledge of the facts; or

(2) Upon request of any party to the contract dispute, unless the DRO or Special Master finds specifically that a hearing is unnecessary and that no party will be prejudiced by limiting the record in the adjudication to the parties' written submissions. All witnesses at any such hearing shall be subject to cross-examination by the opposing party and to questioning by the DRO or Special Master.

(l) The DRO or Special Master shall prepare findings and recommendations, which will contain findings of fact, application of the principles of the AMS and other law or authority applicable to the findings of fact, and a recommendation for a final FAA order.

(m) The DRO or Special Master shall conduct a de novo review using the preponderance of the evidence standard, unless a different standard is prescribed for a particular issue. Notwithstanding the above, allegations that government officials acted with bias or in bad faith must be established by clear and convincing evidence.

(n) The Director of the ODRA may review the status of any contract dispute in the Adjudicative Process with the DRO or Special Master.

(o) A DRO or Special Master shall submit findings and recommendations to the Director of the ODRA or the Director's designee. The findings and recommendations will be released to the parties and to the public, upon issuance of the final FAA order in the case. Should an ODRA protective order be issued in connection with the contract dispute, or should the matter involve proprietary or competition-sensitive information, a redacted version of the findings and recommendations omitting any protected information, shall be prepared wherever possible and released to the public, as soon as is practicable, along with a copy of the final FAA order. Only persons admitted by the ODRA under the protective order and Government personnel shall be provided copies of the unredacted findings and recommendations.

(p) Attorneys' fees of a qualified prevailing contractor are allowable to the extent permitted by the EAJA, 5 U.S.C. 504(a)(1). *See* 14 CFR part 14.

(q) Other than communications regarding purely procedural matters or ADR, there shall be no substantive *ex parte* communication between ODRA personnel and any principal or representative of a party concerning a pending or potentially pending matter. A potential or serving ADR neutral may communicate on an ex parte basis to establish or conduct the ADR.

Subpart D—Alternative Dispute Resolution

§ 17.35 Use of alternative dispute resolution.

(a) By statutory mandate, it is the policy of the FAA to use voluntary ADR to the maximum extent practicable to resolve matters pending at the ODRA. The ODRA therefore uses voluntary ADR as its primary means of resolving all factual, legal, and procedural controversies.

(b) The parties are encouraged to make a good faith effort to explore ADR possibilities in all cases and to employ ADR in every appropriate case. The ODRA uses ADR techniques such as mediation, neutral evaluation, binding arbitration or variations of these techniques as agreed by the parties and approved by the ODRA. At the beginning of each case, the ODRA assigns a DRO as a potential neutral to explore ADR options with the parties and to convene an ADR process. See § 17.35(b).

(c) The ODRA Adjudicative Process will be used where the parties cannot achieve agreement on the use of ADR; where ADR has been employed but has not resolved all pending issues in dispute; or where the ODRA concludes that ADR will not provide an expeditious means of resolving a particular dispute. Even where the Adjudicative Process is to be used, the ODRA, with the parties' consent, may employ informal ADR techniques concurrently with the adjudication.

§ 17.37 Election of alternative dispute resolution process.

(a) The ODRA will make its personnel available to serve as Neutrals in ADR proceedings and, upon request by the parties, will attempt to make qualified non-FAA personnel available to serve as Neutrals through neutral-sharing programs and other similar arrangements. The parties may elect to employ a mutually acceptable compensated neutral at their expense.

(b) The parties using an ADR process to resolve a protest shall submit an executed ADR agreement containing the information outlined in paragraph (d) of this section to the ODRA pursuant to § 17.17(c). The ODRA may extend this time for good cause.

(c) The parties using an ADR process to resolve a contract dispute shall submit an executed ADR agreement containing the information outlined in paragraph (d) of this section to the ODRA pursuant to § 17.29.

(d) The parties to a protest or contract dispute who elect to use ADR must submit to the ODRA an ADR agreement setting forth:

(1) The agreed ADR procedures to be used; and

(2) The name of the neutral. If a compensated neutral is to be used, the agreement must address how the cost of the neutral's services will be reimbursed.

(e) Non-binding ADR techniques are not mutually exclusive, and may be used in combination if the parties agree that a combination is most appropriate to the dispute. The techniques to be employed must be determined in advance by the parties and shall be expressly described in their ADR agreement. The agreement may provide for the use of any fair and reasonable ADR technique that is designed to achieve a prompt resolution of the matter. An ADR agreement for non-binding ADR shall provide for a termination of ADR proceedings and the commencement of adjudication under the Adjudicative Process, upon the election of any party. Notwithstanding such termination, the parties may still engage with the ODRA in ADR techniques (neutral evaluation and/or informal mediation) concurrently with adjudication.

(f) Binding arbitration is available through the ODRA, subject to the provisions of applicable law and the ODRA Binding Arbitration Guidance dated October 2001 as developed in consultation with the Department of Justice.

(g) The parties may, where appropriate in a given case, submit to the ODRA a negotiated protective order for use in ADR in accordance with the requirements of §17.9.

§17.39 Confidentiality of ADR.

(a) The provisions of the Administrative Dispute Resolution Act of 1996, 5 U.S.C. 571, *et seq.*, shall apply to ODRA ADR proceedings.

(b) The ODRA looks to the principles of Rule 408 of the Federal Rules of Evidence in deciding admissibility issues related to ADR communications.

(c) ADR communications are not part of the administrative record unless otherwise agreed by the parties.

Subpart E—Finality and Review

§17.41 Final orders.

All final FAA orders regarding protests or contract disputes under this part are to be issued by the FAA Administrator or by a delegee of the Administrator.

§17.43 Judicial review.

(a) A protester or contractor may seek review of a final FAA order, pursuant to 49 U.S.C. 46110, only after the administrative remedies of this part have been exhausted.

(b) A copy of the petition for review shall be filed with the ODRA and the FAA Chief Counsel on the date that the petition for review is filed with the appropriate circuit court of appeals.

§17.45 Conforming amendments.

The FAA shall amend pertinent provisions of the AMS, standard contract forms and clauses, and any guidance to contracting officials, so as to conform to the provisions of this part.

§17.47 Reconsideration.

The ODRA will not entertain requests for reconsideration as a routine matter, or where such requests evidence mere disagreement with a decision or restatements of previous arguments. A party seeking reconsideration must demonstrate either clear errors of fact or law in the underlying decision or previously unavailable evidence that warrants reversal or modification of the decision. In order to be considered, requests for reconsideration must be filed within ten (10) business days of the date of issuance of the public version of the subject decision or order.

Subpart F—Other Matters

§17.49 Sanctions.

If any party or its representative fails to comply with an Order or Directive of the ODRA, the ODRA may enter such orders and take such other actions as it deems necessary and in the interest of justice.

§17.51 Decorum and professional conduct.

Legal representatives are expected to conduct themselves at all times in a civil and respectful manner appropriate to an administrative forum. Additionally, counsel are expected to conduct themselves at all times in a professional manner and in accordance with all applicable rules of professional conduct.

§ 17.53 Orders and subpoenas for testimony and document production.

(a) Parties are encouraged to seek cooperative and voluntary production of documents and witnesses prior to requesting a subpoena or an order under this section.

(b) Upon request by a party, or on his or her own initiative, a DRO or Special Master may, for good cause shown, order a person to give testimony by deposition and to produce records. Section 46104(c) of Title 49 of the United States Code governs the conduct of depositions or document production.

(c) Upon request by a party, or on his or her own initiative, a DRO or Special Master may, for good cause shown, subpoena witnesses or records related to a hearing from any place in the United States to the designated place of a hearing.

(d) A subpoena or order under this section may be served by a United States marshal or deputy marshal, or by any other person who is not a party and not less than 18 years of age. Service upon a person named therein shall be made by personally delivering a copy to that person and tendering the fees for one day's attendance and the mileage provided by 28 U.S.C. 1821 or other applicable law; however, where the subpoena is issued on behalf of the Product Team, money payments need not be tendered in advance of attendance. The person serving the subpoena or order shall file a declaration of service with the ODRA, executed in the form required by 28 U.S.C. 1746. The declaration of service shall be filed promptly with the ODRA, and before the date on which the person served must respond to the subpoena or order.

(e) Upon written motion by the person subpoenaed or ordered under this section, or by a party, made within ten (10) business days after service, but in any event not later than the time specified in the subpoena or order for compliance, the DRO may—

(1) Rescind or modify the subpoena or order if it is unreasonable and oppressive or for other good cause shown, or

(2) Require the party on whose behalf the subpoena or order was issued to advance the reasonable cost of producing documentary evidence. Where circumstances require, the DRO may act upon such a motion at any time after a copy has been served upon all parties.

(f) The party that requests the DRO to issue a subpoena or order under this section shall be responsible for the payment of fees and mileage, as required by 49 U.S.C. 46104(d), for witnesses, officers who serve the order, and the officer before whom a deposition is taken.

(g) Subpoenas and orders issued under this section may be enforced in a judicial proceeding under 49 U.S.C. 46104(b).

§ 17.55 Standing orders of the ODRA Director.

The Director may issue such Standing Orders as necessary for the orderly conduct of business before the ODRA.

Subpart G—Pre-Disputes

§ 17.57 Dispute resolution process for Pre-disputes.

(a) All potential disputes arising under contracts or solicitations with the FAA may be resolved with the consent of the parties to the dispute under this subpart.

(b) Pre-disputes shall be filed with the ODRA pursuant to § 17.59.

(c) The time limitations for the filing of Protests and Contract Disputes established in §§ 17.15(a) and 17.27(c) will not be extended by efforts to resolve the dispute under this subpart.

§ 17.59 Filing a Pre-dispute.

(a) A Pre-dispute must be in writing, affirmatively state that it is a Pre-dispute pursuant to this subpart, and shall contain:

(1) The party's name, address, telephone and Fax numbers and the name, address, telephone and Fax numbers of the contractor's legal representative(s) (if any);

(2) The contract or solicitation number and the name of the Contracting Officer;

(3) A chronological statement of the facts and of the legal grounds for the party's positions regarding the dispute citing to relevant contract or solicitation provisions and documents and attaching copies of those provisions and documents; and

(4) The signature of a duly authorized legal representative of the initiating party.

(b) Pre-disputes shall be filed with the ODRA, AGC–70, Federal Aviation Administration, telephone (202) 267–3290 as follows:

(1) 600 Independence Avenue SW., Room 2W100, Washington, DC 20591 for filing by hand delivery, courier or other form of in-person delivery;

(2) 800 Independence Avenue SW., Washington, DC 20591 [Attention: AGC–70, Wilbur Wright Bldg., Room 2W100] for filing by U.S. Mail; or

(3) Numbers (202) 267–3720 or alternate (202) 267–1293 for filing by facsimile.

(c) Upon the filing of a Pre-dispute with the ODRA, the ODRA will contact the opposing party to offer its services pursuant to §17.57. If the opposing party agrees, the ODRA will provide Pre-dispute services. If the opposing party does not agree, the ODRA Pre-dispute file will be closed and no service will be provided.

[76 FR 55221, Sept. 7, 2011, as amended by Doc. No. FAA–2017–0075, 82 FR 14429, Mar. 21, 2017]

§ 17.61 Use of alternative dispute resolution.

(a) Only non-binding, voluntary ADR will be used to attempt to resolve a Pre-dispute pursuant to §17.37.

(b) ADR conducted under this subpart is subject to the confidentiality requirements of §17.39.

APPENDIX A TO PART 17—ALTERNATIVE DISPUTE RESOLUTION (ADR)

A. The FAA dispute resolution procedures encourage the parties to protests and contract disputes to use ADR as the primary means to resolve protests and contract disputes, pursuant to the Administrative Dispute Resolution Act of 1996, Public Law 104–320, 5 U.S.C. 570–579, and Department of Transportation and FAA policies to utilize ADR to the maximum extent practicable. Under the procedures presented in this part, the ODRA encourages parties to consider ADR techniques such as case evaluation, mediation, or arbitration.

B. ADR encompasses a number of processes and techniques for resolving protests or contract disputes. The most commonly used types include:

(1) *Mediation.* The neutral or compensated neutral ascertains the needs and interests of both parties and facilitates discussions between or among the parties and an amicable resolution of their differences, seeking approaches to bridge the gaps between the parties" respective positions. The neutral or compensated neutral can meet with the parties separately, conduct joint meetings with the parties" representatives, or employ both methods in appropriate cases.

(2) *Neutral Evaluation.* At any stage during the ADR process, as the parties may agree, the neutral or compensated neutral will provide a candid assessment and opinion of the strengths and weaknesses of the parties" positions as to the facts and law, so as to facilitate further discussion and resolution.

(3) *Binding Arbitration.* The ODRA, after consultation with the United States Department of Justice in accordance with the provisions of the Administrative Disputes Resolution Act offers true binding arbitration in cases within its jurisdiction. The ODRA's Guidance for the Use of Binding Arbitration may be found on its website at: *http:// www.faa.gov/go/odra.*

SUBCHAPTER C—AIRCRAFT

PART 21—CERTIFICATION PROCEDURES FOR PRODUCTS AND ARTICLES

Special Federal Aviation Regulation No. 88

131

Subpart P—Special Federal Aviation Regulations

21.700　SFAR No. 111—Lavatory oxygen systems.

AUTHORITY: 42 U.S.C. 7572; 49 U.S.C. 106(f), 106(g), 40105, 40113, 44701–44702, 44704, 44707, 44709, 44711, 44713, 44715, 45303.

EDITORIAL NOTES: 1. For miscellaneous amendments to cross references in this 21 see Amdt. 21–10, 31 FR 9211, July 6, 1966.

2. Nomenclature changes to part 21 appear at 74 FR 53384, Oct. 16, 2009.

SPECIAL FEDERAL AVIATION REGULATION NO. 88—FUEL TANK SYSTEM FAULT TOLERANCE EVALUATION REQUIREMENTS

1. *Applicability.* This SFAR applies to the holders of type certificates, and supplemental type certificates that may affect the airplane fuel tank system, for turbine-powered transport category airplanes, provided the type certificate was issued after January 1, 1958, and the airplane has either a maximum type certificated passenger capacity of 30 or more, or a maximum type certificated payload capacity of 7,500 pounds or more. This SFAR also applies to applicants for type certificates, amendments to a type certificate, and supplemental type certificates affecting the fuel tank systems for those airplanes identified above, if the application was filed before June 6, 2001, the effective date of this SFAR, and the certificate was not issued before June 6, 2001.

2. *Compliance:* Each type certificate holder, and each supplemental type certificate holder of a modification affecting the airplane fuel tank system, must accomplish the following within the compliance times specified in paragraph (e) of this section:

(a) Conduct a safety review of the airplane fuel tank system to determine that the design meets the requirements of §§ 25.901 and 25.981(a) and (b) of this chapter. If the current design does not meet these requirements, develop all design changes to the fuel tank system that are necessary to meet these requirements. The responsible Aircraft Certification Service office for the affected airplane may grant an extension of the 18-month compliance time for development of design changes if:

(1) The safety review is completed within the compliance time;

(2) Necessary design changes are identified within the compliance time; and

(3) Additional time can be justified, based on the holder's demonstrated aggressiveness in performing the safety review, the complexity of the necessary design changes, the availability of interim actions to provide an acceptable level of safety, and the resulting level of safety.

(b) Develop all maintenance and inspection instructions necessary to maintain the design features required to preclude the existence or development of an ignition source within the fuel tank system of the airplane.

(c) Submit a report for approval to the responsible Aircraft Certification Service office for the affected airplane, that:

(1) Provides substantiation that the airplane fuel tank system design, including all necessary design changes, meets the requirements of §§ 25.901 and 25.981(a) and (b) of this chapter; and

(2) Contains all maintenance and inspection instructions necessary to maintain the design features required to preclude the existence or development of an ignition source within the fuel tank system throughout the operational life of the airplane.

(d) The responsible Aircraft Certification Service office for the affected airplane, may approve a report submitted in accordance with paragraph 2(c) if it determines that any provisions of this SFAR not complied with are compensated for by factors that provide an equivalent level of safety.

(e) Each type certificate holder must comply no later than December 6, 2002, or within 18 months after the issuance of a type certificate for which application was filed before June 6, 2001, whichever is later; and each supplemental type certificate holder of a modification affecting the airplane fuel tank system must comply no later than June 6, 2003, or within 18 months after the issuance of a supplemental type certificate for which application was filed before June 6, 2001, whichever is later.

[Doc. No. 1999–6411, 66 FR 23129, May 7, 2001, as amended by Amdt. 21–82, 67 FR 57493, Sept. 10, 2002; 67 FR 70809, Nov. 26, 2002; Amdt. 21–83, 67 FR 72833, Dec. 9, 2002; Doc. No. FAA–2018–0119, Amdt. 21–101, 83 FR 9169, Mar. 5, 2018]

Subpart A—General

§ 21.1　Applicability and definitions.

(a) This part prescribes—

(1) Procedural requirements for issuing and changing—

(i) Design approvals;

(ii) Production approvals;

(iii) Airworthiness certificates; and

(iv) Airworthiness approvals;

(2) Rules governing applicants for, and holders of, any approval or certificate specified in paragraph (a)(1) of this section; and

(3) Procedural requirements for the approval of articles.

(b) For the purposes of this part—

(1) *Airworthiness approval* means a document, issued by the FAA for an aircraft, aircraft engine, propeller, or article, which certifies that the aircraft, aircraft engine, propeller, or article conforms to its approved design and is in a condition for safe operation, unless otherwise specified;

(2) *Article* means a material, part, component, process, or appliance;

(3) *Commercial part* means an article that is listed on an FAA-approved Commercial Parts List included in a design approval holder's Instructions for Continued Airworthiness required by § 21.50;

(4) *Design approval* means a type certificate (including amended and supplemental type certificates) or the approved design under a PMA, TSO authorization, letter of TSO design approval, or other approved design;

(5) *Interface component* means an article that serves as a functional interface between an aircraft and an aircraft engine, an aircraft engine and a propeller, or an aircraft and a propeller. An interface component is designated by the holder of the type certificate or the supplemental type certificate who controls the approved design data for that article;

(6) *Product* means an aircraft, aircraft engine, or propeller;

(7) *Production approval* means a document issued by the FAA to a person that allows the production of a product or article in accordance with its approved design and approved quality system, and can take the form of a production certificate, a PMA, or a TSO authorization;

(8) *State of Design* means the country or jurisdiction having regulatory authority over the organization responsible for the design and continued airworthiness of a civil aeronautical product or article;

(9) *State of Manufacture* means the country or jurisdiction having regulatory authority over the organization responsible for the production and airworthiness of a civil aeronautical product or article.

(10) *Supplier* means a person at any tier in the supply chain who provides a product, article, or service that is used or consumed in the design or manufacture of, or installed on, a product or article.

[Doc. No. FAA–2006–25877, Amdt. 21–92, 74 FR 53384, Oct. 16, 2009; Doc. No. FAA–2013–0933, Amdt. 21–98, 80 FR 59031, Oct. 1, 2015; Amdt. 21–98A, 80 FR 59031, Dec. 17, 2015; Docket FAA–2015–0150, Amdt. 21–99, 81 FR 42207, June 28, 2016; Docket FAA–2018–1087, Amdt. 21–105, 86 FR 4381, Jan. 15, 2021]

§ 21.2 Falsification of applications, reports, or records.

(a) A person may not make or cause to be made—

(1) Any fraudulent, intentionally false, or misleading statement on any application for a certificate or approval under this part;

(2) Any fraudulent, intentionally false, or misleading statement in any record or report that is kept, made, or used to show compliance with any requirement of this part;

(3) Any reproduction for a fraudulent purpose of any certificate or approval issued under this part.

(4) Any alteration of any certificate or approval issued under this part.

(b) The commission by any person of an act prohibited under paragraph (a) of this section is a basis for—

(1) Denying issuance of any certificate or approval under this part; and

(2) Suspending or revoking any certificate or approval issued under this part and held by that person.

[Doc. No. 23345, 57 FR 41367, Sept. 9, 1992, as amended by Amdt. 21–92, 74 FR 53384, Oct. 16, 2009; Amdt. 21–92A, 75 FR 9095, Mar. 1, 2010]

§ 21.3 Reporting of failures, malfunctions, and defects.

(a) The holder of a type certificate (including amended or supplemental type certificates), a PMA, or a TSO authorization, or the licensee of a type certificate must report any failure, malfunction, or defect in any product or article manufactured by it that it determines has resulted in any of the occurrences listed in paragraph (c) of this section.

(b) The holder of a type certificate (including amended or supplemental type certificates), a PMA, or a TSO authorization, or the licensee of a type certificate must report any defect in any product or article manufactured by it that has left its quality system and that it determines could result in any of the occurrences listed in paragraph (c) of this section.

(c) The following occurrences must be reported as provided in paragraphs (a) and (b) of this section:

(1) Fires caused by a system or equipment failure, malfunction, or defect.

(2) An engine exhaust system failure, malfunction, or defect which causes damage to the engine, adjacent aircraft structure, equipment, or components.

(3) The accumulation or circulation of toxic or noxious gases in the crew compartment or passenger cabin.

(4) A malfunction, failure, or defect of a propeller control system.

(5) A propeller or rotorcraft hub or blade structural failure.

(6) Flammable fluid leakage in areas where an ignition source normally exists.

(7) A brake system failure caused by structural or material failure during operation.

(8) A significant aircraft primary structural defect or failure caused by any autogenous condition (fatigue, understrength, corrosion, etc.).

(9) Any abnormal vibration or buffeting caused by a structural or system malfunction, defect, or failure.

(10) An engine failure.

(11) Any structural or flight control system malfunction, defect, or failure which causes an interference with normal control of the aircraft for which derogates the flying qualities.

(12) A complete loss of more than one electrical power generating system or hydraulic power system during a given operation of the aircraft.

(13) A failure or malfunction of more than one attitude, airspeed, or altitude instrument during a given operation of the aircraft.

(d) The requirements of paragraph (a) of this section do not apply to—

(1) Failures, malfunctions, or defects that the holder of a type certificate (including amended or supplemental type certificates), PMA, TSO authorization, or the licensee of a type certificate determines—

(i) Were caused by improper maintenance or use;

(ii) Were reported to the FAA by another person under this chapter; or

(iii) Were reported under the accident reporting provisions of 49 CFR part 830 of the regulations of the National Transportation Safety Board.

(2) Failures, malfunctions, or defects in products or articles—

(i) Manufactured by a foreign manufacturer under a U.S. type certificate issued under § 21.29 or under an approval issued under § 21.621; or

(ii) Exported to the United States under § 21.502.

(e) Each report required by this section—

(1) Must be made to the FAA within 24 hours after it has determined that the failure, malfunction, or defect required to be reported has occurred. However, a report that is due on a Saturday or a Sunday may be delivered on the following Monday and one that is due on a holiday may be delivered on the next workday;

(2) Must be transmitted in a manner and form acceptable to the FAA and by the most expeditious method available; and

(3) Must include as much of the following information as is available and applicable:

(i) The applicable product and article identification information required by part 45 of this chapter;

(ii) Identification of the system involved; and

(iii) Nature of the failure, malfunction, or defect.

(f) If an accident investigation or service difficulty report shows that a product or article manufactured under this part is unsafe because of a manufacturing or design data defect, the holder of the production approval for that product or article must, upon request of the FAA, report to the FAA the results of its investigation and any action taken or proposed by the holder of that production approval to correct that defect. If action is required to correct the defect in an existing product or article, the holder of that production approval must send the data necessary for issuing an appropriate airworthiness directive to the FAA.

[Amdt. 21–36, 35 FR 18187, Nov. 28, 1970, as amended by Amdt. 21–37, 35 FR 18450, Dec. 4, 1970; Amdt. 21–50, 45 FR 38346, June 9, 1980; Amdt. 21–67, 54 FR 39291, Sept. 25, 1989; Amdt. 21–92, 74 FR 53385, Oct. 16, 2009; Doc. No. FAA–2018–0119, Amdt. 21–101, 83 FR 9169, Mar. 5, 2018]

§ 21.4 ETOPS reporting requirements.

(a) *Early ETOPS: reporting, tracking, and resolving problems.* The holder of a type certificate for an airplane-engine combination approved using the Early ETOPS method specified in part 25, Appendix K, of this chapter must use a system for reporting, tracking, and resolving each problem resulting in one of the occurrences specified in paragraph (a)(6) of this section.

(1) The system must identify how the type certificate holder will promptly identify problems, report them to the responsible Aircraft Certification Service office, and propose a solution to the FAA to resolve each problem. A proposed solution must consist of—

(i) A change in the airplane or engine type design;

(ii) A change in a manufacturing process;

(iii) A change in an operating or maintenance procedure; or

(iv) Any other solution acceptable to the FAA.

(2) For an airplane with more than two engines, the system must be in place for the first 250,000 world fleet engine-hours for the approved airplane-engine combination.

(3) For two-engine airplanes, the system must be in place for the first 250,000 world fleet engine-hours for the approved airplane-engine combination and after that until—

(i) The world fleet 12-month rolling average IFSD rate is at or below the rate required by paragraph (b)(2) of this section; and

(ii) The FAA determines that the rate is stable.

(4) For an airplane-engine combination that is a derivative of an airplane-engine combination previously approved for ETOPS, the system need only address those problems specified in the following table, provided the type certificate holder obtains prior authorization from the FAA:

If the change does not require a new airplane type certificate and . . .	Then the Problem Tracking and Resolution System must address . . .
(i) Requires a new engine type certificate	All problems applicable to the new engine installation, and for the remainder of the airplane, problems in changed systems only.
(ii) Does not require a new engine type certificate	Problems in changed systems only.

(5) The type certificate holder must identify the sources and content of data that it will use for its system. The data must be adequate to evaluate the specific cause of any in-service problem reportable under this section or § 21.3(c) that could affect the safety of ETOPS.

(6) In implementing this system, the type certificate holder must report the following occurrences:

(i) IFSDs, except planned IFSDs performed for flight training.

(ii) For two-engine airplanes, IFSD rates.

(iii) Inability to control an engine or obtain desired thrust or power.

(iv) Precautionary thrust or power reductions.

(v) Degraded ability to start an engine in flight.

(vi) Inadvertent fuel loss or unavailability, or uncorrectable fuel imbalance in flight.

(vii) Turn backs or diversions for failures, malfunctions, or defects associated with an ETOPS group 1 significant system.

(viii) Loss of any power source for an ETOPS group 1 significant system, including any power source designed to provide backup power for that system.

(ix) Any event that would jeopardize the safe flight and landing of the airplane on an ETOPS flight.

(x) Any unscheduled engine removal for a condition that could result in one of the reportable occurrences listed in this paragraph.

(b) *Reliability of two-engine airplanes—* (1) *Reporting of two-engine airplane in-service reliability.* The holder of a type certificate for an airplane approved for ETOPS and the holder of a type certificate for an engine installed on an airplane approved for ETOPS must report monthly to their respective Aircraft Certification Service office on the reliability of the world fleet of those airplanes and engines. The report provided by both the airplane and engine type certificate holders must address each airplane-engine combination approved for ETOPS. The FAA may approve quarterly reporting if the airplane-engine combination demonstrates an IFSD rate at or below those specified in paragraph (b)(2) of this section for a period acceptable to the FAA. This reporting may be combined with the reporting required by § 21.3. The responsible type certificate holder must investigate any cause of an IFSD resulting from an occurrence attributable to the design of its product and report the results of that investigation to its responsible Aircraft Certification Service office. Reporting must include:

(i) Engine IFSDs, except planned IFSDs performed for flight training.

(ii) The world fleet 12-month rolling average IFSD rates for all causes, except planned IFSDs performed for flight training.

(iii) ETOPS fleet utilization, including a list of operators, their ETOPS diversion time authority, flight hours, and cycles.

(2) *World fleet IFSD rate for two-engine airplanes.* The holder of a type certificate for an airplane approved for ETOPS and the holder of a type certificate for an engine installed on an airplane approved for ETOPS must issue service information to the operators of those airplanes and engines, as appropriate, to maintain the world fleet 12-month rolling average IFSD rate at or below the following levels:

(i) A rate of 0.05 per 1,000 world-fleet engine-hours for an airplane-engine combination approved for up to and including 120-minute ETOPS. When all ETOPS operators have complied with the corrective actions required in the configuration, maintenance and procedures (CMP) document as a condition for ETOPS approval, the rate to be maintained is at or below 0.02 per 1,000 world-fleet engine-hours.

(ii) A rate of 0.02 per 1,000 world-fleet engine-hours for an airplane-engine combination approved for up to and including 180-minute ETOPS, including airplane-engine combinations approved for 207-minute ETOPS in the North Pacific operating area under appendix P, section I, paragraph (h), of part 121 of this chapter.

(iii) A rate of 0.01 per 1,000 world-fleet engine-hours for an airplane-engine combination approved for ETOPS beyond 180 minutes, excluding airplane-engine combinations approved for 207-minute ETOPS in the North Pacific operating area under appendix P, section I, paragraph (h), of part 121 of this chapter.

[Doc. No. FAA-2002-6717, 72 FR 1872, Jan. 16, 2007, as amended by Doc. No. FAA-2018-0119, Amdt. 21-101, 83 FR 9169, Mar. 5, 2018]

§ **21.5 Airplane or Rotorcraft Flight Manual.**

(a) With each airplane or rotorcraft not type certificated with an Airplane or Rotorcraft Flight Manual and having no flight time before March 1, 1979, the holder of a type certificate (including amended or supplemental type certificates) or the licensee of a type certificate must make available to the owner at the time of delivery of the aircraft a current approved Airplane or Rotorcraft Flight Manual.

(b) The Airplane or Rotorcraft Flight Manual required by paragraph (a) of this section must contain the following information:

(1) The operating limitations and information required to be furnished in an Airplane or Rotorcraft Flight Manual or in manual material, markings, and placards, by the applicable regulations under which the airplane or rotorcraft was type certificated.

(2) The maximum ambient atmospheric temperature for which engine cooling was demonstrated must be stated in the performance information section of the Flight Manual, if the applicable regulations under which the aircraft was type certificated do not require ambient temperature on engine

cooling operating limitations in the Flight Manual.

[Amdt. 21–46, 43 FR 2316, Jan. 16, 1978, as amended by Amdt. 21–92, 74 FR 53385, Oct. 16, 2009]

§ 21.6 Manufacture of new aircraft, aircraft engines, and propellers.

(a) Except as specified in paragraphs (b) and (c) of this section, no person may manufacture a new aircraft, aircraft engine, or propeller based on a type certificate unless the person—

(1) Is the holder of the type certificate or has a licensing agreement from the holder of the type certificate to manufacture the product; and

(2) Meets the requirements of subpart F or G of this part.

(b) A person may manufacture one new aircraft based on a type certificate without meeting the requirements of paragraph (a) of this section if that person can provide evidence acceptable to the FAA that the manufacture of the aircraft by that person began before August 5, 2004.

(c) The requirements of this section do not apply to—

(1) New aircraft imported under the provisions of §§ 21.183(c), 21.184(b), or 21.185(c); and

(2) New aircraft engines or propellers imported under the provisions of § 21.500.

[Doc. No. FAA–2003–14825, 71 FR 52258, Sept. 1, 2006]

§ 21.7 Continued airworthiness and safety improvements for transport category airplanes.

(a) On or after December 10, 2007, the holder of a design approval and an applicant for a design approval must comply with the applicable continued airworthiness and safety improvement requirements of part 26 of this subchapter.

(b) For new transport category airplanes manufactured under the authority of the FAA, the holder or licensee of a type certificate must meet the applicable continued airworthiness and safety improvement requirements specified in part 26 of this subchapter for new production airplanes. Those requirements only apply if the FAA has jurisdiction over the organization responsible for final assembly of the airplane.

[Doc. No. FAA–2004–18379, Amdt. 21–90, 72 FR 63404, Nov. 8, 2007]

§ 21.8 Approval of articles.

If an article is required to be approved under this chapter, it may be approved—

(a) Under a PMA;

(b) Under a TSO;

(c) In conjunction with type certification procedures for a product; or

(d) In any other manner approved by the FAA.

[Doc. No. FAA–2006–5877, Amdt. 21–92, 74 FR 53385, Oct. 16, 2009]

§ 21.9 Replacement and modification articles.

(a) If a person knows, or should know, that a replacement or modification article is reasonably likely to be installed on a type-certificated product, the person may not produce that article unless it is—

(1) Produced under a type certificate;

(2) Produced under an FAA production approval;

(3) A standard part (such as a nut or bolt) manufactured in compliance with a government or established industry specification;

(4) A commercial part as defined in § 21.1 of this part;

(5) Produced by an owner or operator for maintaining or altering that owner or operator's product;

(6) Fabricated by an appropriately rated certificate holder with a quality system, and consumed in the repair or alteration of a product or article in accordance with part 43 of this chapter; or

(7) Produced in any other manner approved by the FAA.

(b) Except as provided in paragraphs (a)(1) through (a)(2) of this section, a person who produces a replacement or modification article for sale may not represent that part as suitable for installation on a type-certificated product.

(c) Except as provided in paragraphs (a)(1) through (a)(2) of this section, a person may not sell or represent an article as suitable for installation on an aircraft type-certificated under

137

§§ 21.25(a)(2) or 21.27 unless that article—

(1) Was declared surplus by the U.S. Armed Forces, and

(2) Was intended for use on that aircraft model by the U.S. Armed Forces.

[Doc. No. FAA–2006–25877, Amdt. 21–92, 74 FR 53385, Oct. 16, 2009; Amdt. 21–92A, 75 FR 9095, Mar. 1, 2010; Doc. No. FAA–2015–1621, Amdt. 21–100, 81 FR 96688, Dec. 30, 2016]

Subpart B—Type Certificates

SOURCE: Docket No. 5085, 29 FR 14564, Oct. 24, 1964, unless otherwise noted.

§ 21.11 Applicability.

This subpart prescribes—

(a) Procedural requirements for the issue of type certificates for aircraft, aircraft engines, and propellers; and

(b) Rules governing the holders of those certificates.

§ 21.13 Eligibility.

Any interested person may apply for a type certificate.

[Amdt. 21–25, 34 FR 14068, Sept. 5, 1969]

§ 21.15 Application for type certificate.

(a) An application for a type certificate is made on a form and in a manner prescribed by the FAA.

(b) An application for an aircraft type certificate must be accompanied by a three-view drawing of that aircraft and available preliminary basic data.

(c) An application for an aircraft engine type certificate must be accompanied by a description of the engine design features, the engine operating characteristics, and the proposed engine operating limitations.

[Doc. No. 5085, 29 FR 14564, Oct. 24, 1964, as amended by Amdt. 21–40, 39 FR 35459, Oct. 1, 1974; Amdt. 21–67, 54 FR 39291, Sept. 25, 1989; Amdt. 21–92, 74 FR 53385, Oct. 16, 2009; Doc. No. FAA–2018–0119, Amdt. 21–101, 83 FR 9169, Mar. 5, 2018]

§ 21.16 Special conditions.

If the FAA finds that the airworthiness regulations of this subchapter do not contain adequate or appropriate safety standards for an aircraft, aircraft engine, or propeller because of a novel or unusual design feature of the aircraft, aircraft engine or propeller, he prescribes special conditions and amendments thereto for the product. The special conditions are issued in accordance with Part 11 of this chapter and contain such safety standards for the aircraft, aircraft engine or propeller as the FAA finds necessary to establish a level of safety equivalent to that established in the regulations.

[Amdt. 21–19, 32 FR 17851, Dec. 13, 1967, as amended by Amdt. 21–51, 45 FR 60170, Sept. 11, 1980]

§ 21.17 Designation of applicable regulations.

(a) Except as provided in §§ 25.2, 27.2, 29.2, and in parts 26, 34, and 36 of this subchapter, an applicant for a type certificate must show that the aircraft, aircraft engine, or propeller concerned meets—

(1) The applicable requirements of this subchapter that are effective on the date of application for that certificate unless—

(i) Otherwise specified by the FAA; or

(ii) Compliance with later effective amendments is elected or required under this section; and

(2) Any special conditions prescribed by the FAA.

(b) For special classes of aircraft, including the engines and propellers installed thereon (e.g., gliders, airships, and other nonconventional aircraft), for which airworthiness standards have not been issued under this subchapter, the applicable requirements will be the portions of those other airworthiness requirements contained in Parts 23, 25, 27, 29, 31, 33, and 35 found by the FAA to be appropriate for the aircraft and applicable to a specific type design, or such airworthiness criteria as the FAA may find provide an equivalent level of safety to those parts.

(c) An application for type certification of a transport category aircraft is effective for 5 years and an application for any other type certificate is effective for 3 years, unless an applicant shows at the time of application that his product requires a longer period of time for design, development, and testing, and the FAA approves a longer period.

(d) In a case where a type certificate has not been issued, or it is clear that a type certificate will not be issued, within the time limit established under paragraph (c) of this section, the applicant may—

(1) File a new application for a type certificate and comply with all the provisions of paragraph (a) of this section applicable to an original application; or

(2) File for an extension of the original application and comply with the applicable airworthiness requirements of this subchapter that were effective on a date, to be selected by the applicant, not earlier than the date which precedes the date of issue of the type certificate by the time limit established under paragraph (c) of this section for the original application.

(e) If an applicant elects to comply with an amendment to this subchapter that is effective after the filing of the application for a type certificate, he must also comply with any other amendment that the FAA finds is directly related.

(f) For primary category aircraft, the requirements are:

(1) The applicable airworthiness requirements contained in parts 23, 27, 31, 33, and 35 of this subchapter, or such other airworthiness criteria as the FAA may find appropriate and applicable to the specific design and intended use and provide a level of safety acceptable to the FAA.

(2) The noise standards of part 36 applicable to primary category aircraft.

[Doc. No. 5085, 29 FR 14564, Oct. 24, 1964, as amended by Amdt. 21–19, 32 FR 17851, Dec. 13, 1967; Amdt. 21–24, 34 FR 364, Jan. 10, 1969; Amdt. 21–42, 40 FR 1033, Jan. 6, 1975; Amdt. 21–58, 50 FR 46877, Nov. 13, 1985; Amdt. 21–60, 52 FR 8042, Mar. 13, 1987; Amdt. 21–68, 55 FR 32860, Aug. 10, 1990; Amdt. 21–69, 56 FR 41051, Aug. 16, 1991; Amdt. 21–70, 57 FR 41367, Sept. 9, 1992; Amdt. 21–90, 72 FR 63404, Nov. 8, 2007; Doc. No. FAA–2015–1621, Amdt. 21–100, 81 FR 96688, Dec. 30, 2016]

§21.19 Changes requiring a new type certificate.

Each person who proposes to change a product must apply for a new type certificate if the FAA finds that the proposed change in design, power, thrust, or weight is so extensive that a substantially complete investigation of compliance with the applicable regulations is required.

[Doc. No. 28903, 65 FR 36265, June 7, 2000]

§21.20 Compliance with applicable requirements.

The applicant for a type certificate, including an amended or supplemental type certificate, must—

(a) Show compliance with all applicable requirements and must provide the FAA the means by which such compliance has been shown; and

(b) Provide a statement certifying that the applicant has complied with the applicable requirements.

[Doc. No. FAA–2006–25877, Amdt. 21–92, 74 FR 53385, Oct. 16, 2009]

§21.21 Issue of type certificate: normal, utility, acrobatic, commuter, and transport category aircraft; manned free balloons; special classes of aircraft; aircraft engines; propellers.

An applicant is entitled to a type certificate for an aircraft in the normal, utility, acrobatic, commuter, or transport category, or for a manned free balloon, special class of aircraft, or an aircraft engine or propeller, if—

(a) The product qualifies under §21.27; or

(b) The applicant submits the type design, test reports, and computations necessary to show that the product to be certificated meets the applicable airworthiness, aircraft noise, fuel venting, and exhaust emission requirements of this subchapter and any special conditions prescribed by the FAA, and the FAA finds—

(1) Upon examination of the type design, and after completing all tests and inspections, that the type design and the product meet the applicable noise, fuel venting, and emissions requirements of this subchapter, and further finds that they meet the applicable airworthiness requirements of this subchapter or that any airworthiness provisions not complied with are compensated for by factors that provide an equivalent level of safety; and

(2) For an aircraft, that no feature or characteristic makes it unsafe for the

category in which certification is requested.

[Doc. No. 5085, 29 FR 14564, Oct. 24, 1964, as amended by Amdt. 21–15, 32 FR 3735, Mar. 4, 1967; Amdt. 21–27, 34 FR 18368, Nov. 18, 1969; Amdt. 21–60, 52 FR 8042, Mar. 13, 1987; Amdt. 21–68, 55 FR 32860, Aug. 10, 1990; Amdt. 21–92, 74 FR 53385, Oct. 16, 2009]

§ 21.23 [Reserved]

§ 21.24 Issuance of type certificate: primary category aircraft.

(a) The applicant is entitled to a type certificate for an aircraft in the primary category if—

(1) The aircraft—

(i) Is unpowered; is an airplane powered by a single, naturally aspirated engine with a 61-knot or less V_{so} stall speed as determined under part 23 of this chapter; or is a rotorcraft with a 6-pound per square foot main rotor disc loading limitation, under sea level standard day conditions;

(ii) Weighs not more than 2,700 pounds; or, for seaplanes, not more than 3,375 pounds;

(iii) Has a maximum seating capacity of not more than four persons, including the pilot; and

(iv) Has an unpressurized cabin.

(2) The applicant has submitted—

(i) Except as provided by paragraph (c) of this section, a statement, in a form and manner acceptable to the FAA, certifying that: the applicant has completed the engineering analysis necessary to demonstrate compliance with the applicable airworthiness requirements; the applicant has conducted appropriate flight, structural, propulsion, and systems tests necessary to show that the aircraft, its components, and its equipment are reliable and function properly; the type design complies with the airworthiness standards and noise requirements established for the aircraft under § 21.17(f); and no feature or characteristic makes it unsafe for its intended use;

(ii) The flight manual required by § 21.5(b), including any information required to be furnished by the applicable airworthiness standards;

(iii) Instructions for continued airworthiness in accordance with § 21.50(b); and

(iv) A report that: summarizes how compliance with each provision of the type certification basis was determined; lists the specific documents in which the type certification data information is provided; lists all necessary drawings and documents used to define the type design; and lists all the engineering reports on tests and computations that the applicant must retain and make available under § 21.49 to substantiate compliance with the applicable airworthiness standards.

(3) The FAA finds that—

(i) The aircraft complies with those applicable airworthiness requirements approved under § 21.17(f) of this part; and

(ii) The aircraft has no feature or characteristic that makes it unsafe for its intended use.

(b) An applicant may include a special inspection and preventive maintenance program as part of the aircraft's type design or supplemental type design.

(c) For aircraft manufactured outside of the United States in a country with which the United States has a bilateral airworthiness agreement for the acceptance of these aircraft, and from which the aircraft is to be imported into the United States—

(1) The statement required by paragraph (a)(2)(i) of this section must be made by the civil airworthiness authority of the exporting country; and

(2) The required manuals, placards, listings, instrument markings, and documents required by paragraphs (a) and (b) of this section must be submitted in English.

[Doc. No. 23345, 57 FR 41367, Sept. 9, 1992, as amended by Amdt. 21–75, 62 FR 62808, Nov. 25, 1997; Doc. No. FAA–2015–1621, Amdt. 21–100, 81 FR 96689, Dec. 30, 2016]

§ 21.25 Issue of type certificate: Restricted category aircraft.

(a) An applicant is entitled to a type certificate for an aircraft in the restricted category for special purpose operations if he shows compliance with the applicable noise requirements of Part 36 of this chapter, and if he shows that no feature or characteristic of the aircraft makes it unsafe when it is operated under the limitations prescribed for its intended use, and that the aircraft—

(1) Meets the airworthiness requirements of an aircraft category except those requirements that the FAA finds inappropriate for the special purpose for which the aircraft is to be used; or

(2) Is of a type that has been manufactured in accordance with the requirements of and accepted for use by, an Armed Force of the United States and has been later modified for a special purpose.

(b) For the purposes of this section, "special purpose operations" includes—

(1) Agricultural (spraying, dusting, and seeding, and livestock and predatory animal control);

(2) Forest and wildlife conservation;

(3) Aerial surveying (photography, mapping, and oil and mineral exploration);

(4) Patrolling (pipelines, power lines, and canals);

(5) Weather control (cloud seeding);

(6) Aerial advertising (skywriting, banner towing, airborne signs and public address systems); and

(7) Any other operation specified by the FAA.

[Doc. No. 5085, 29 FR 14564, Oct. 24, 1964, as amended by Amdt. 21–42, 40 FR 1033, Jan. 6, 1975]

§21.27 Issue of type certificate: surplus aircraft of the Armed Forces.

(a) Except as provided in paragraph (b) of this section an applicant is entitled to a type certificate for an aircraft in the normal, utility, acrobatic, commuter, or transport category that was designed and constructed in the United States, accepted for operational use, and declared surplus by, an Armed Force of the United States, and that is shown to comply with the applicable certification requirements in paragraph (f) of this section.

(b) An applicant is entitled to a type certificate for a surplus aircraft of the Armed Forces of the United States that is a counterpart of a previously type certificated civil aircraft, if he shows compliance with the regulations governing the original civil aircraft type certificate.

(c) Aircraft engines, propellers, and their related accessories installed in surplus Armed Forces aircraft, for which a type certificate is sought under this section, will be approved for use on those aircraft if the applicant shows that on the basis of the previous military qualifications, acceptance, and service record, the product provides substantially the same level of airworthiness as would be provided if the engines or propellers were type certificated under Part 33 or 35 of this subchapter.

(d) The FAA may relieve an applicant from strict compliance with a specific provision of the applicable requirements in paragraph (f) of this section, if the FAA finds that the method of compliance proposed by the applicant provides substantially the same level of airworthiness and that strict compliance with those regulations would impose a severe burden on the applicant. The FAA may use experience that was satisfactory to an Armed Force of the United States in making such a determination.

(e) The FAA may require an applicant to comply with special conditions and later requirements than those in paragraphs (c) and (f) of this section, if the FAA finds that compliance with the listed regulations would not ensure an adequate level of airworthiness for the aircraft.

(f) Except as provided in paragraphs (b) through (e) of this section, an applicant for a type certificate under this section must comply with the appropriate regulations listed in the following table:

Type of aircraft	Date accepted for operational use by the Armed Forces of the United States	Regulations that apply [1]
Small reciprocating-engine powered airplanes	Before May 16, 1956	CAR Part 3, as effective May 15, 1956.
	After May 15, 1956	CAR Part 3, or 14 CFR Part 23.
Small turbine engine-powered airplanes	Before Oct. 2, 1959	CAR Part 3, as effective Oct. 1, 1959.
	After Oct. 1, 1959	CAR Part 3 or 14 CFR Part 23.
Commuter category airplanes	After (Feb. 17, 1987)	FAR Part 23 as of (Feb. 17, 1987)..

141

Type of aircraft	Date accepted for operational use by the Armed Forces of the United States	Regulations that apply [1]
Large reciprocating-engine powered airplanes	Before Aug. 26, 1955	CAR Part 4b, as effective Aug. 25, 1955.
	After Aug. 25, 1955	CAR Part 4b or 14 CFR Part 25.
Large turbine engine-powered airplanes	Before Oct. 2, 1959	CAR Part 4b, as effective Oct. 1, 1959.
	After Oct. 1, 1959	CAR Part 4b or 14 CFR Part 25.
Rotorcraft with maximum certificated takeoff weight of:		
6,000 pounds or less	Before Oct. 2, 1959	CAR Part 6, as effective Oct. 1, 1959.
	After Oct. 1, 1959	CAR Part 6, or 14 CFR Part 27.
Over 6,000 pounds	Before Oct. 2, 1959	CAR Part 7, as effective Oct. 1, 1959.
	After Oct. 1, 1959	CAR Part 7, or 14 CFR Part 29.

[1] Where no specific date is listed, the applicable regulations are those in effect on the date that the first aircraft of the particular model was accepted for operational use by the Armed Forces.

[Doc. No. 5085, 29 FR 14564, Oct. 24, 1964, as amended by Amdt. 21–59, 52 FR 1835, Jan. 15, 1987; 52 FR 7262, Mar. 9, 1987; 70 FR 2325, Jan. 13, 2005; Amdt. 21–92, 74 FR 53386, Oct. 16, 2009]

§ 21.29 Issue of type certificate: import products.

(a) The FAA may issue a type certificate for a product that is manufactured in a foreign country or jurisdiction with which the United States has an agreement for the acceptance of these products for export and import and that is to be imported into the United States if—

(1) The applicable State of Design certifies that the product has been examined, tested, and found to meet—

(i) The applicable aircraft noise, fuel venting, and exhaust emissions requirements of this subchapter as designated in § 21.17, or the applicable aircraft noise, fuel venting, and exhaust emissions requirements of the State of Design, and any other requirements the FAA may prescribe to provide noise, fuel venting, and exhaust emission levels no greater than those provided by the applicable aircraft noise, fuel venting, and exhaust emission requirements of this subchapter as designated in § 21.17; and

(ii) The applicable airworthiness requirements of this subchapter as designated in § 21.17, or the applicable airworthiness requirements of the State of Design and any other requirements the FAA may prescribe to provide a level of safety equivalent to that provided by the applicable airworthiness requirements of this subchapter as designated in § 21.17;

(2) The applicant has provided technical data to show the product meets the requirements of paragraph (a)(1) of this section; and

(3) The manuals, placards, listings, and instrument markings required by the applicable airworthiness (and noise, where applicable) requirements are presented in the English language.

(b) A product type certificated under this section is considered to be type certificated under the noise standards of part 36 of this subchapter and the fuel venting and exhaust emission standards of part 34 of this subchapter. Compliance with parts 36 and 34 of this subchapter is certified under paragraph (a)(1)(i) of this section, and the applicable airworthiness standards of this subchapter, or an equivalent level of safety, with which compliance is certified under paragraph (a)(1)(ii) of this section.

[Amdt. 21–92, 74 FR 53386, Oct. 16, 2009]

§ 21.31 Type design.

The type design consists of—

(a) The drawings and specifications, and a listing of those drawings and specifications, necessary to define the configuration and the design features of the product shown to comply with the requirements of that part of this subchapter applicable to the product;

(b) Information on dimensions, materials, and processes necessary to define the structural strength of the product;

(c) The Airworthiness Limitations section of the Instructions for Continued Airworthiness as required by parts 23, 25, 26, 27, 29, 31, 33 and 35 of this subchapter, or as otherwise required by the FAA; and as specified in the applicable airworthiness criteria for special classes of aircraft defined in §21.17(b); and

(d) For primary category aircraft, if desired, a special inspection and preventive maintenance program designed to be accomplished by an appropriately rated and trained pilot-owner.

(e) Any other data necessary to allow, by comparison, the determination of the airworthiness, noise characteristics, fuel venting, and exhaust emissions (where applicable) of later products of the same type.

[Doc. No. 5085, 29 FR 14564, Oct. 24, 1964, as amended by Amdt. 21–27, 34 FR 18363, Nov. 18, 1969; Amdt. 21–51, 45 FR 60170, Sept. 11, 1980; Amdt. 21–60, 52 FR 8042, Mar. 13, 1987; Amdt. 21–68, 55 FR 32860, Aug. 10, 1990; Amdt. 21–70, 57 FR 41368, Sept. 9, 1992; Amdt. 21–90, 72 FR 63404, Nov. 8, 2007]

§21.33 Inspection and tests.

(a) Each applicant must allow the FAA to make any inspection and any flight and ground test necessary to determine compliance with the applicable requirements of this subchapter. However, unless otherwise authorized by the FAA—

(1) No aircraft, aircraft engine, propeller, or part thereof may be presented to the FAA for test unless compliance with paragraphs (b)(2) through (b)(4) of this section has been shown for that aircraft, aircraft engine, propeller, or part thereof; and

(2) No change may be made to an aircraft, aircraft engine, propeller, or part thereof between the time that compliance with paragraphs (b)(2) through (b)(4) of this section is shown for that aircraft, aircraft engine, propeller, or part thereof and the time that it is presented to the FAA for test.

(b) Each applicant must make all inspections and tests necessary to determine—

(1) Compliance with the applicable airworthiness, aircraft noise, fuel venting, and exhaust emission requirements;

(2) That materials and products conform to the specifications in the type design;

(3) That parts of the products conform to the drawings in the type design; and

(4) That the manufacturing processes, construction and assembly conform to those specified in the type design.

[Doc. No. 5085, 29 FR 14564, Oct. 24, 1964, as amended by Amdt. 21–17, 32 FR 14926, Oct. 28, 1967; Amdt. 21–27, 34 FR 18363, Nov. 18, 1969; Amdt. 21–44, 41 FR 55463, Dec. 20, 1976; Amdt. 21–68, 55 FR 32860, Aug. 10, 1990; Amdt. 21–68, 55 FR 32860, Aug. 10, 1990; Amdt. 21–92, 74 FR 53386, Oct. 16, 2009]

§21.35 Flight tests.

(a) Each applicant for an aircraft type certificate (other than under §§21.24 through 21.29) must make the tests listed in paragraph (b) of this section. Before making the tests the applicant must show—

(1) Compliance with the applicable structural requirements of this subchapter;

(2) Completion of necessary ground inspections and tests;

(3) That the aircraft conforms with the type design; and

(4) That the FAA received a flight test report from the applicant (signed, in the case of aircraft to be certificated under Part 25 [New] of this chapter, by the applicant's test pilot) containing the results of his tests.

(b) Upon showing compliance with paragraph (a) of this section, the applicant must make all flight tests that the FAA finds necessary—

(1) To determine compliance with the applicable requirements of this subchapter; and

(2) For aircraft to be certificated under this subchapter, except gliders and low-speed, certification level 1 or 2 airplanes, as defined in part 23 of this chapter, to determine whether there is reasonable assurance that the aircraft, its components, and its equipment are reliable and function properly.

(c) Each applicant must, if practicable, make the tests prescribed in paragraph (b)(2) of this section upon the aircraft that was used to show compliance with—

(1) Paragraph (b)(1) of this section; and

(2) For rotorcraft, the rotor drive endurance tests prescribed in §27.923 or §29.923 of this chapter, as applicable.

(d) Each applicant must show for each flight test (except in a glider or a manned free balloon) that adequate provision is made for the flight test crew for emergency egress and the use of parachutes.

(e) Except in gliders and manned free balloons, an applicant must discontinue flight tests under this section until he shows that corrective action has been taken, whenever—

(1) The applicant's test pilot is unable or unwilling to make any of the required flight tests; or

(2) Items of noncompliance with requirements are found that may make additional test data meaningless or that would make further testing unduly hazardous.

(f) The flight tests prescribed in paragraph (b)(2) of this section must include—

(1) For aircraft incorporating turbine engines of a type not previously used in a type certificated aircraft, at least 300 hours of operation with a full complement of engines that conform to a type certificate; and

(2) For all other aircraft, at least 150 hours of operation.

[Doc. No. 5085, 29 FR 14564, Oct. 24, 1964, as amended by Amdt. 21–40, 39 FR 35459, Oct. 1, 1974; Amdt. 21–51, 45 FR 60170, Sept. 11, 1980; Amdt. 21–70, 57 FR 41368, Sept. 9, 1992; Amdt. 21–95, 76 FR 64233, Oct. 18, 2011; Doc. No. FAA–2015–1621, Amdt. 21–100, 81 FR 96689, Dec. 30, 2016]

§ 21.37   Flight test pilot.

Each applicant for a normal, utility, acrobatic, commuter, or transport category aircraft type certificate must provide a person holding an appropriate pilot certificate to make the flight tests required by this part.

[Doc. No. 5085, 29 FR 14564, Oct. 24, 1964, as amended by Amdt. 21–59, 52 FR 1835, Jan. 15, 1987]

§ 21.39   Flight test instrument calibration and correction report.

(a) Each applicant for a normal, utility, acrobatic, commuter, or transport category aircraft type certificate must submit a report to the FAA showing the computations and tests required in connection with the calibration of instruments used for test purposes and in the correction of test results to standard atmospheric conditions.

(b) Each applicant must allow the FAA to conduct any flight tests that he finds necessary to check the accu-

racy of the report submitted under paragraph (a) of this section.

[Doc. No. 5085, 29 FR 14564, Oct. 24, 1964, as amended by Amdt. 21–59, 52 FR 1835, Jan. 15, 1987]

§ 21.41   Type certificate.

Each type certificate is considered to include the type design, the operating limitations, the certificate data sheet, the applicable regulations of this subchapter with which the FAA records compliance, and any other conditions or limitations prescribed for the product in this subchapter.

§ 21.43   Location of manufacturing facilities.

Except as provided in § 21.29, the FAA does not issue a type certificate if the manufacturing facilities for the product are located outside of the United States, unless the FAA finds that the location of the manufacturer's facilities places no undue burden on the FAA in administering applicable airworthiness requirements.

§ 21.45   Privileges.

The holder or licensee of a type certificate for a product may—

(a) In the case of aircraft, upon compliance with §§ 21.173 through 21.189, obtain airworthiness certificates;

(b) In the case of aircraft engines or propellers, obtain approval for installation on certificated aircraft;

(c) In the case of any product, upon compliance with subpart G of this part, obtain a production certificate for the type certificated product;

(d) Obtain approval of replacement parts for that product.

[Doc. No. 5085, 29 FR 14564, Oct. 24, 1964, as amended by Amdt. 21–92, 74 FR 53386, Oct. 16, 2009]

§ 21.47   Transferability.

(a) A holder of a type certificate may transfer it or make it available to other persons by licensing agreements.

(b) For a type certificate transfer in which the State of Design will remain the same, each transferor must, before such a transfer, notify the FAA in writing. This notification must include the applicable type certificate number, the name and address of the transferee, and the anticipated date of the transfer.

(c) For a type certificate transfer in which the State of Design is changing, a type certificate may only be transferred to or from a person subject to the authority of another State of Design if the United States has an agreement with that State of Design for the acceptance of the affected product for export and import. Each transferor must notify the FAA before such a transfer in a form and manner acceptable to the FAA. This notification must include the applicable type certificate number; the name, address, and country of residence of the transferee; and the anticipated date of the transfer.

(d) Before executing or terminating a licensing agreement that makes a type certificate available to another person, the type certificate holder must notify the FAA in writing. This notification must include the type certificate number addressed by the licensing agreement, the name and address of the licensee, the extent of authority granted the licensee, and the anticipated date of the agreement.

[Doc. No. FAA–2006–25877, Amdt. 21–92, 74 FR 53386, Oct. 16, 2009; Doc. No. FAA–2018–0119, Amdt. 21–101, 83 FR 9169, Mar. 5, 2018]

§21.49 Availability.

The holder of a type certificate must make the certificate available for examination upon the request of the FAA or the National Transportation Safety Board.

[Doc. No. 5085, 29 FR 14564, Oct. 24, 1964, as amended by Doc. No. 8084, 32 FR 5769, Apr. 11, 1967]

§21.50 Instructions for continued airworthiness and manufacturer's maintenance manuals having airworthiness limitations sections.

(a) The holder of a type certificate for a rotorcraft for which a Rotorcraft Maintenance Manual containing an "Airworthiness Limitations" section has been issued under §27.1529 (a)(2) or §29.1529 (a)(2) of this chapter, and who obtains approval of changes to any replacement time, inspection interval, or related procedure in that section of the manual, must make those changes available upon request to any operator of the same type of rotorcraft.

(b) The holder of a design approval, including either a type certificate or supplemental type certificate for an aircraft, aircraft engine, or propeller for which application was made after January 28, 1981, must furnish at least one set of complete Instructions for Continued Airworthiness to the owner of each type aircraft, aircraft engine, or propeller upon its delivery, or upon issuance of the first standard airworthiness certificate for the affected aircraft, whichever occurs later. The Instructions for Continued Airworthiness must be prepared in accordance with §§23.1529, 25.1529, 25.1729, 27.1529, 29.1529, 31.82, 33.4, 35.4, or part 26 of this subchapter, or as specified in the applicable airworthiness criteria for special classes of aircraft defined in §21.17(b), as applicable. If the holder of a design approval chooses to designate parts as commercial, it must include in the Instructions for Continued Airworthiness a list of commercial parts submitted in accordance with the provisions of paragraph (c) of this section. Thereafter, the holder of a design approval must make those instructions available to any other person required by this chapter to comply with any of the terms of those instructions. In addition, changes to the Instructions for Continued Airworthiness shall be made available to any person required by this chapter to comply with any of those instructions.

(c) To designate commercial parts, the holder of a design approval, in a manner acceptable to the FAA, must submit:

(1) A Commercial Parts List;

(2) Data for each part on the List showing that:

(i) The failure of the commercial part, as installed in the product, would not degrade the level of safety of the product; and

(ii) The part is produced only under the commercial part manufacturer's specification and marked only with the commercial part manufacturer's markings; and

(3) Any other data necessary for the FAA to approve the List.

[Amdt. 21–23, 33 FR 14105, Sept. 18, 1968, as amended by Amdt. 21–51, 45 FR 60170, Sept. 11, 1980; Amdt. 21–60, 52 FR 8042, Mar. 13, 1987; Amdt. 21–90, 72 FR 63404, Nov. 8, 2007; Amdt. 21–92, 74 FR 53386, Oct. 16, 2009; Doc. No. FAA–2015–1621, Amdt. 21–100, 81 FR 96689, Dec. 30, 2016]

§ 21.51 Duration.

A type certificate is effective until surrendered, suspended, revoked, or a termination date is otherwise established by the FAA.

§ 21.53 Statement of conformity.

(a) Each applicant must provide, in a form and manner acceptable to the FAA, a statement that each aircraft engine or propeller presented for type certification conforms to its type design.

(b) Each applicant must submit a statement of conformity to the FAA for each aircraft or part thereof presented to the FAA for tests. This statement of conformity must include a statement that the applicant has complied with § 21.33(a) (unless otherwise authorized under that paragraph).

[Amdt. 21–17, 32 FR 14926, Oct. 28, 1967, as amended by Amdt. 21–92, 74 FR 53386, Oct. 16, 2009]

§ 21.55 Responsibility of type certificate holders to provide written licensing agreements.

A type certificate holder who allows a person to use the type certificate to manufacture a new aircraft, aircraft engine, or propeller must provide that person with a written licensing agreement acceptable to the FAA.

[Doc. No. FAA–2003–14825, 71 FR 52258, Sept. 1, 2006]

Subpart C—Provisional Type Certificates

SOURCE: Docket No. 5085, 29 FR 14566, Oct. 24, 1964, unless otherwise noted.

§ 21.71 Applicability.

This subpart prescribes—
(a) Procedural requirements for the issue of provisional type certificates, amendments to provisional type certificates, and provisional amendments to type certificates; and
(b) Rules governing the holders of those certificates.

§ 21.73 Eligibility.

(a) Any manufacturer of aircraft manufactured within the United States who is a United States citizen may apply for Class I or Class II provisional type certificates, for amendments to provisional type certificates held by him, and for provisional amendments to type certificates held by him.

(b) Any manufacturer of aircraft in a State of Manufacture subject to the provisions of an agreement with the United States for the acceptance of those aircraft for export and import may apply for a Class II provisional type certificate, for amendments to provisional type certificates held by him, and for provisional amendments to type certificates held by him.

(c) An aircraft engine manufacturer who is a United States citizen and who has altered a type certificated aircraft by installing different type certificated aircraft engines manufactured by him within the United States may apply for a Class I provisional type certificate for the aircraft, and for amendments to Class I provisional type certificate held by him, if the basic aircraft, before alteration, was type certificated in the normal, utility, acrobatic, commuter, or transport category.

[Doc. No. 5085, 29 FR 14566, Oct. 24, 1964, as amended by Amdt. 21–12, 31 FR 13380, Oct. 15, 1966; Amdt. 21–59, 52 FR 1836, Jan. 15, 1987; Amdt. 21–92, 74 FR 53387, Oct. 16, 2009]

§ 21.75 Application.

Each applicant for a provisional type certificate, for an amendment thereto, or for a provisional amendment to a type certificate must apply to the FAA and provide the information required by this subpart.

[Doc. No. FAA–2006–25877, Amdt. 21–92, 74 FR 53387, Oct. 16, 2009; Doc. No. FAA–2018–0119, Amdt. 21–101, 83 FR 9169, Mar. 5, 2018]

§ 21.77 Duration.

(a) Unless sooner surrendered, superseded, revoked, or otherwise terminated, provisional type certificates and amendments thereto are effective for the periods specified in this section.

(b) A Class I provisional type certificate is effective for 24 months after the date of issue.

(c) A Class II provisional type certificate is effective for twelve months after the date of issue.

(d) An amendment to a Class I or Class II provisional type certificate is effective for the duration of the amended certificate.

(e) A provisional amendment to a type certificate is effective for six months after its approval or until the amendment of the type certificate is approved, whichever is first.

[Doc. No. 5085, 29 FR 14566, Oct. 24, 1964, as amended by Amdt. 21–7, 30 FR 14311, Nov. 16, 1965]

§21.79 Transferability.

Provisional type certificates are not transferable.

§21.81 Requirements for issue and amendment of Class I provisional type certificates.

(a) An applicant is entitled to the issue or amendment of a Class I provisional type certificate if he shows compliance with this section and the FAA finds that there is no feature, characteristic, or condition that would make the aircraft unsafe when operated in accordance with the limitations established in paragraph (e) of this section and in §91.317 of this chapter.

(b) The applicant must apply for the issue of a type or supplemental type certificate for the aircraft.

(c) The applicant must certify that—

(1) The aircraft has been designed and constructed in accordance with the airworthiness requirements applicable to the issue of the type or supplemental type certificate applied for;

(2) The aircraft substantially meets the applicable flight characteristic requirements for the type or supplemental type certificate applied for; and

(3) The aircraft can be operated safely under the appropriate operating limitations specified in paragraph (a) of this section.

(d) The applicant must submit a report showing that the aircraft had been flown in all maneuvers necessary to show compliance with the flight requirements for the issue of the type or supplemental type certificate applied

for, and to establish that the aircraft can be operated safely in accordance with the limitations contained in this subchapter.

(e) The applicant must establish all limitations required for the issue of the type or supplemental type certificate applied for, including limitations on weights, speeds, flight maneuvers, loading, and operation of controls and equipment unless, for each limitation not so established, appropriate operating restrictions are established for the aircraft.

(f) The applicant must establish an inspection and maintenance program for the continued airworthiness of the aircraft.

(g) The applicant must show that a prototype aircraft has been flown for at least 50 hours under an experimental certificate issued under §§21.191 through 21.195, or under the auspices of an Armed Force of the United States. However, in the case of an amendment to a provisional type certificate, the FAA may reduce the number of required flight hours.

[Doc. No. 5085, 29 FR 14566, Oct. 24, 1964, as amended by Amdt. 21–66, 54 FR 34329, Aug. 18, 1989]

§21.83 Requirements for issue and amendment of Class II provisional type certificates.

(a) An applicant who manufactures aircraft within the United States is entitled to the issue or amendment of a Class II provisional type certificate if he shows compliance with this section and the FAA finds that there is no feature, characteristic, or condition that would make the aircraft unsafe when operated in accordance with the limitations in paragraph (h) of this section, and §§91.317 and 121.207 of this chapter.

(b) An applicant who manufactures aircraft in a country with which the United States has an agreement for the acceptance of those aircraft for export and import is entitled to the issue or amendment of a Class II provisional type certificate if the country in which the aircraft was manufactured certifies that the applicant has shown compliance with this section, that the aircraft meets the requirements of paragraph (f) of this section and that there

147

is no feature, characteristic, or condition that would make the aircraft unsafe when operated in accordance with the limitations in paragraph (h) of this section and §§ 91.317 and 121.207 of this chapter.

(c) The applicant must apply for a type certificate, in the transport category, for the aircraft.

(d) The applicant must hold a U.S. type certificate for at least one other aircraft in the same transport category as the subject aircraft.

(e) The FAA's official flight test program or the flight test program conducted by the authorities of the country in which the aircraft was manufactured, with respect to the issue of a type certificate for that aircraft, must be in progress.

(f) The applicant or, in the case of a foreign manufactured aircraft, the country in which the aircraft was manufactured, must certify that—

(1) The aircraft has been designed and constructed in accordance with the airworthiness requirements applicable to the issue of the type certificate applied for;

(2) The aircraft substantially complies with the applicable flight characteristic requirements for the type certificate applied for; and

(3) The aircraft can be operated safely under the appropriate operating limitations in this subchapter.

(g) The applicant must submit a report showing that the aircraft has been flown in all maneuvers necessary to show compliance with the flight requirements for the issue of the type certificate and to establish that the aircraft can be operated safely in accordance with the limitations in this subchapter.

(h) The applicant must prepare a provisional aircraft flight manual containing all limitations required for the issue of the type certificate applied for, including limitations on weights, speeds, flight maneuvers, loading, and operation of controls and equipment unless, for each limitation not so established, appropriate operating restrictions are established for the aircraft.

(i) The applicant must establish an inspection and maintenance program for the continued airworthiness of the aircraft.

(j) The applicant must show that a prototype aircraft has been flown for at least 100 hours. In the case of an amendment to a provisional type certificate, the FAA may reduce the number of required flight hours.

[Amdt. 21–12, 31 FR 13386, Oct. 15, 1966, as amended by Amdt. 21–66, 54 FR 34329, Aug. 18, 1989]

§ 21.85 Provisional amendments to type certificates.

(a) An applicant who manufactures aircraft within the United States is entitled to a provisional amendment to a type certificate if he shows compliance with this section and the FAA finds that there is no feature, characteristic, or condition that would make the aircraft unsafe when operated under the appropriate limitations contained in this subchapter.

(b) An applicant who manufactures aircraft in a foreign country with which the United States has an agreement for the acceptance of those aircraft for export and import is entitled to a provisional amendment to a type certificate if the country in which the aircraft was manufactured certifies that the applicant has shown compliance with this section, that the aircraft meets the requirements of paragraph (e) of this section and that there is no feature, characteristic, or condition that would make the aircraft unsafe when operated under the appropriate limitations contained in this subchapter.

(c) The applicant must apply for an amendment to the type certificate.

(d) The FAA's official flight test program or the flight test program conducted by the authorities of the country in which the aircraft was manufactured, with respect to the amendment of the type certificate, must be in progress.

(e) The applicant or, in the case of foreign manufactured aircraft, the country in which the aircraft was manufactured, must certify that—

(1) The modification involved in the amendment to the type certificate has

been designed and constructed in accordance with the airworthiness requirements applicable to the issue of the type certificate for the aircraft;

(2) The aircraft substantially complies with the applicable flight characteristic requirements for the type certificate; and

(3) The aircraft can be operated safely under the appropriate operating limitations in this subchapter.

(f) The applicant must submit a report showing that the aircraft incorporating the modifications involved has been flown in all maneuvers necessary to show compliance with the flight requirements applicable to those modifications and to establish that the aircraft can be operated safely in accordance with the limitations specified in §§91.317 and 121.207 of this chapter.

(g) The applicant must establish and publish, in a provisional aircraft flight manual or other document and on appropriate placards, all limitations required for the issue of the type certificate applied for, including weight, speed, flight maneuvers, loading, and operation of controls and equipment, unless, for each limitation not so established, appropriate operating restrictions are established for the aircraft.

(h) The applicant must establish an inspection and maintenance program for the continued airworthiness of the aircraft.

(i) The applicant must operate a prototype aircraft modified in accordance with the corresponding amendment to the type certificate for the number of hours found necessary by the FAA.

[Amdt. 21–12, 31 FR 13388, Oct. 15, 1966, as amended by Amdt. 21–66, 54 FR 34329, Aug. 18, 1989]

Subpart D—Changes to Type Certificates

SOURCE: Docket No. 5085, 29 FR 14567, Oct. 24, 1964, unless otherwise noted.

§21.91 Applicability.

This subpart prescribes procedural requirements for the approval of changes to type certificates.

§21.93 Classification of changes in type design.

(a) In addition to changes in type design specified in paragraph (b) of this section, changes in type design are classified as minor and major. A "minor change" is one that has no appreciable effect on the weight, balance, structural strength, reliability, operational characteristics, or other characteristics affecting the airworthiness of the product. All other changes are "major changes" (except as provided in paragraph (b) of this section).

(b) For the purpose of complying with Part 36 of this chapter, and except as provided in paragraphs (b)(2), (b)(3), and (b)(4) of this section, any voluntary change in the type design of an aircraft that may increase the noise levels of that aircraft is an "acoustical change" (in addition to being a minor or major change as classified in paragraph (a) of this section) for the following aircraft:

(1) Transport category large airplanes.

(2) Jet (Turbojet powered) airplanes (regardless of category). For airplanes to which this paragraph applies, "acoustical changes" do not include changes in type design that are limited to one of the following—

(i) Gear down flight with one or more retractable landing gear down during the entire flight, or

(ii) Spare engine and nacelle carriage external to the skin of the airplane (and return of the pylon or other external mount), or

(iii) Time-limited engine and/or nacelle changes, where the change in type design specifies that the airplane may not be operated for a period of more than 90 days unless compliance with the applicable acoustical change provisions of Part 36 of this chapter is shown for that change in type design.

(3) Propeller driven commuter category and small airplanes in the primary, normal, utility, acrobatic, transport, and restricted categories, except for airplanes that are:

(i) Designated for "agricultural aircraft operations" (as defined in §137.3 of this chapter, effective January 1, 1966) to which §36.1583 of this chapter does not apply, or

(ii) Designated for dispensing fire fighting materials to which § 36.1583 of this chapter does not apply, or

(iii) U.S. registered, and that had flight time prior to January 1, 1955 or

(iv) Land configured aircraft reconfigured with floats or skis. This reconfiguration does not permit further exception from the requirements of this section upon any acoustical change not enumerated in § 21.93(b).

(4) Helicopters except:

(i) Those helicopters that are designated exclusively:

(A) For "agricultural aircraft operations", as defined in § 137.3 of this chapter, as effective on January 1, 1966;

(B) For dispensing fire fighting materials; or

(C) For carrying external loads, as defined in § 133.1(b) of this chapter, as effective on December 20, 1976.

(ii) Those helicopters modified by installation or removal of external equipment. For purposes of this paragraph, "external equipment" means any instrument, mechanism, part, apparatus, appurtenance, or accessory that is attached to, or extends from, the helicopter exterior but is not used nor is intended to be used in operating or controlling a helicopter in flight and is not part of an airframe or engine. An "acoustical change" does not include:

(A) Addition or removal of external equipment;

(B) Changes in the airframe made to accommodate the addition or removal of external equipment, to provide for an external load attaching means, to facilitate the use of external equipment or external loads, or to facilitate the safe operation of the helicopter with external equipment mounted to, or external loads carried by, the helicopter;

(C) Reconfiguration of the helicopter by the addition or removal of floats and skis;

(D) Flight with one or more doors and/or windows removed or in an open position; or

(E) Any changes in the operational limitations placed on the helicopter as a consequence of the addition or removal of external equipment, floats, and skis, or flight operations with doors and/or windows removed or in an open position.

(5) Tiltrotors.

(c) For purposes of complying with part 34 of this chapter, any voluntary change in the type design of the airplane or engine which may increase fuel venting or exhaust emissions is an "emissions change."

[Amdt. 21–27, 34 FR 18363, Nov. 18, 1969]

EDITORIAL NOTE: For FEDERAL REGISTER citations affecting § 21.93, see the List of CFR Sections Affected, which appears in the Finding Aids section of the printed volume and at *www.govinfo.gov.*

§ 21.95 Approval of minor changes in type design.

Minor changes in a type design may be approved under a method acceptable to the FAA before submitting to the FAA any substantiating or descriptive data.

§ 21.97 Approval of major changes in type design.

(a) An applicant for approval of a major change in type design must—

(1) Provide substantiating data and necessary descriptive data for inclusion in the type design;

(2) Show that the change and areas affected by the change comply with the applicable requirements of this subchapter, and provide the FAA the means by which such compliance has been shown; and

(3) Provide a statement certifying that the applicant has complied with the applicable requirements.

(b) Approval of a major change in the type design of an aircraft engine is limited to the specific engine configuration upon which the change is made unless the applicant identifies in the necessary descriptive data for inclusion in the type design the other configurations of the same engine type for which approval is requested and shows that the change is compatible with the other configurations.

[Amdt. 21–40, 39 FR 35459, Oct. 1, 1974, as amended by Amdt. 21–92, 74 FR 53387, Oct. 16, 2009; Amdt. 21–96, 77 FR 71695, Dec. 4, 2012]

§ 21.99 Required design changes.

(a) When an Airworthiness Directive is issued under Part 39 the holder of the type certificate for the product concerned must—

(1) If the FAA finds that design changes are necessary to correct the unsafe condition of the product, and upon his request, submit appropriate design changes for approval; and

(2) Upon approval of the design changes, make available the descriptive data covering the changes to all operators of products previously certificated under the type certificate.

(b) In a case where there are no current unsafe conditions, but the FAA or the holder of the type certificate finds through service experience that changes in type design will contribute to the safety of the product, the holder of the type certificate may submit appropriate design changes for approval. Upon approval of the changes, the manufacturer must make information on the design changes available to all operators of the same type of product.

[Doc. No. 5085, 29 FR 14567, Oct. 24, 1964, as amended by Amdt. 21–3, 30 FR 8826, July 24, 1965]

§ 21.101 Designation of applicable regulations.

(a) An applicant for a change to a type certificate must show that the change and areas affected by the change comply with the airworthiness requirements applicable to the category of the product in effect on the date of the application for the change and with parts 34 and 36 of this chapter. Exceptions are detailed in paragraphs (b) and (c) of this section.

(b) Except as provided in paragraph (g) of this section, if paragraphs (b)(1), (2), or (3) of this section apply, an applicant may show that the change and areas affected by the change comply with an earlier amendment of a regulation required by paragraph (a) of this section, and of any other regulation the FAA finds is directly related. However, the earlier amended regulation may not precede either the corresponding regulation included by reference in the type certificate, or any regulation in §§ 25.2, 27.2, or 29.2 of this chapter that is related to the change. The applicant may show compliance with an earlier amendment of a regulation for any of the following:

(1) A change that the FAA finds not to be significant. In determining whether a specific change is signifi-
cant, the FAA considers the change in context with all previous relevant design changes and all related revisions to the applicable regulations incorporated in the type certificate for the product. Changes that meet one of the following criteria are automatically considered significant:

(i) The general configuration or the principles of construction are not retained.

(ii) The assumptions used for certification of the product to be changed do not remain valid.

(2) Each area, system, component, equipment, or appliance that the FAA finds is not affected by the change.

(3) Each area, system, component, equipment, or appliance that is affected by the change, for which the FAA finds that compliance with a regulation described in paragraph (a) of this section would not contribute materially to the level of safety of the product or would be impractical.

(c) An applicant for a change to an aircraft (other than a rotorcraft) of 6,000 pounds or less maximum weight, to a non-turbine rotorcraft of 3,000 pounds or less maximum weight, to a level 1 low-speed airplane, or to a level 2 low-speed airplane may show that the change and areas affected by the change comply with the regulations included in the type certificate. However, if the FAA finds that the change is significant in an area, the FAA may designate compliance with an amendment to the regulation incorporated by reference in the type certificate that applies to the change and any regulation that the FAA finds is directly related, unless the FAA also finds that compliance with that amendment or regulation would not contribute materially to the level of safety of the product or would be impractical.

(d) If the FAA finds that the regulations in effect on the date of the application for the change do not provide adequate standards with respect to the proposed change because of a novel or unusual design feature, the applicant must also comply with special conditions, and amendments to those special conditions, prescribed under the provisions of § 21.16, to provide a level of safety equal to that established by the

151

regulations in effect on the date of the application for the change.

(e) An application for a change to a type certificate for a transport category aircraft is effective for 5 years, and an application for a change to any other type certificate is effective for 3 years. If the change has not been approved, or if it is clear that it will not be approved under the time limit established under this paragraph, the applicant may do either of the following:

(1) File a new application for a change to the type certificate and comply with all the provisions of paragraph (a) of this section applicable to an original application for a change.

(2) File for an extension of the original application and comply with the provisions of paragraph (a) of this section. The applicant must then select a new application date. The new application date may not precede the date the change is approved by more than the time period established under this paragraph (e).

(f) For aircraft certificated under §§ 21.17(b), 21.24, 21.25, and 21.27 the airworthiness requirements applicable to the category of the product in effect on the date of the application for the change include each airworthiness requirement that the FAA finds to be appropriate for the type certification of the aircraft in accordance with those sections.

(g) Notwithstanding paragraph (b) of this section, for transport category airplanes, the applicant must show compliance with each applicable provision of part 26 of this chapter, unless the applicant has elected or was required to comply with a corresponding amendment to part 25 of this chapter that was issued on or after the date of the applicable part 26 provision.

[Doc. No. 28903, 65 FR 36266, June 7, 2000, as amended by Amdt. 21–90, 72 FR 63404, Nov. 8, 2007; Amdt. 21–96, 77 FR 71695, Dec. 4, 2012; Doc. No. FAA–2015–1621, Amdt. 21–100, 81 FR 96689, Dec. 30, 2016]

Subpart E—Supplemental Type Certificates

Source: Docket No. 5085, 29 FR 14568, Oct. 24, 1964, unless otherwise noted.

§ 21.111 Applicability.

This subpart prescribes procedural requirements for the issue of supplemental type certificates.

§ 21.113 Requirement for supplemental type certificate.

(a) If a person holds the TC for a product and alters that product by introducing a major change in type design that does not require an application for a new TC under § 21.19, that person must apply to the FAA either for an STC, or to amend the original type certificate under subpart D of this part.

(b) If a person does not hold the TC for a product and alters that product by introducing a major change in type design that does not require an application for a new TC under § 21.19, that person must apply to the FAA for an STC.

(c) The application for an STC must be made in the form and manner prescribed by the FAA.

[Doc. No. FAA–2006–25877, Amdt. 21–92, 74 FR 53387, Oct. 16, 2009; Doc. No. FAA–2018–0119, Amdt. 21–101, 83 FR 9169, Mar. 5, 2018]

§ 21.115 Applicable requirements.

(a) Each applicant for a supplemental type certificate must show that the altered product meets applicable requirements specified in § 21.101 and, in the case of an acoustical change described in § 21.93(b), show compliance with the applicable noise requirements of part 36 of this chapter and, in the case of an emissions change described in § 21.93(c), show compliance with the applicable fuel venting and exhaust emissions requirements of part 34 of this chapter.

(b) Each applicant for a supplemental type certificate must meet §§ 21.33 and 21.53 with respect to each change in the type design.

[Amdt. 21–17, 32 FR 14927, Oct. 28, 1967, as amended by Amdt. 21–42, 40 FR 1033, Jan. 6, 1975; Amdt. 21–52A, 45 FR 79009, Nov. 28, 1980; Amdt. 21–61, 53 FR 3540, Feb. 5, 1988; Amdt. 21–68, 55 FR 32860, Aug. 10, 1990; Amdt. 21–71, 57 FR 42854, Sept. 16, 1992; Amdt. 21–77, 65 FR 36266, June 7, 2000]

§ 21.117 Issue of supplemental type certificates.

(a) An applicant is entitled to a supplemental type certificate if the FAA finds that the applicant meets the requirements of §§ 21.113 and 21.115.

(b) A supplemental type certificate consists of—

(1) The approval by the FAA of a change in the type design of the product; and

(2) The type certificate previously issued for the product.

[Doc. No. 5085, 29 FR 14568, Oct. 24, 1964, as amended by Amdt. 21–92, 74 FR 53387, Oct. 16, 2009]

§ 21.119 Privileges.

The holder of a supplemental type certificate may—

(a) In the case of aircraft, obtain airworthiness certificates;

(b) In the case of other products, obtain approval for installation on certificated aircraft; and

(c) Obtain a production certificate in accordance with the requirements of subpart G of this part for the change in the type design approved by the supplemental type certificate.

[Doc. No. 5085, 29 FR 14568, Oct. 24, 1964, as amended by Amdt. 21–92, 74 FR 53387, Oct. 16, 2009]

§ 21.120 Responsibility of supplemental type certificate holders to provide written permission for alterations.

A supplemental type certificate holder who allows a person to use the supplemental type certificate to alter an aircraft, aircraft engine, or propeller must provide that person with written permission acceptable to the FAA.

[Doc. No. FAA–2003–14825, 71 FR 52258, Sept. 1, 2006]

Subpart F—Production Under Type Certificate

SOURCE: Docket No. 5085, 29 FR 14568, Oct. 24, 1964, unless otherwise noted.

§ 21.121 Applicability.

This subpart prescribes rules for production under a type certificate.

§ 21.122 Location of or change to manufacturing facilities.

(a) A type certificate holder may utilize manufacturing facilities located outside of the United States if the FAA finds no undue burden in administering the applicable requirements of Title 49 U.S.C. and this subchapter.

(b) The type certificate holder must obtain FAA approval before making any changes to the location of any of its manufacturing facilities.

(c) The type certificate holder must immediately notify the FAA, in writing, of any change to the manufacturing facilities that may affect the inspection, conformity, or airworthiness of its product or article.

[Doc. No. FAA–2006–25877, Amdt. 21–92, 74 FR 53387, Oct. 16, 2009; Amdt. 21–92A, 75 FR 9095, Mar. 1, 2010]

§ 21.123 Production under type certificate.

Each manufacturer of a product being manufactured under a type certificate must—

(a) Maintain at the place of manufacture all information and data specified in §§ 21.31 and 21.41;

(b) Make each product and article thereof available for inspection by the FAA;

(c) Maintain records of the completion of all inspections and tests required by §§ 21.127, 21.128, and 21.129 for at least 5 years for the products and articles thereof manufactured under the approval and at least 10 years for critical components identified under § 45.15(c) of this chapter;

(d) Allow the FAA to make any inspection or test, including any inspection or test at a supplier facility, necessary to determine compliance with this subchapter;

(e) Mark the product in accordance with part 45 of this chapter, including any critical parts;

(f) Identify any portion of that product (*e.g.*, sub-assemblies, component parts, or replacement articles) that leave the manufacturer's facility as FAA approved with the manufacturer's part number and name, trademark, symbol, or other FAA-approved manufacturer's identification; and

(g) Except as otherwise authorized by the FAA, obtain a production certificate for that product in accordance with subpart G of this part within 6 months after the date of issuance of the type certificate.

[Doc. No. FAA–2006–25877, Amdt. 21–92, 74 FR 53387, Oct. 16, 2009]

§ 21.125 [Reserved]

§ 21.127 Tests: aircraft.

(a) Each person manufacturing aircraft under a type certificate must establish an approved production flight test procedure and flight check-off form, and in accordance with that form, flight test each aircraft produced.

(b) Each production flight test procedure must include the following:

(1) An operational check of the trim, controllability, or other flight characteristics to establish that the production aircraft has the same range and degree of control as the prototype aircraft.

(2) An operational check of each part or system operated by the crew while in flight to establish that, during flight, instrument readings are within normal range.

(3) A determination that all instruments are properly marked, and that all placards and required flight manuals are installed after flight test.

(4) A check of the operational characteristics of the aircraft on the ground.

(5) A check on any other items peculiar to the aircraft being tested that can best be done during the ground or flight operation of the aircraft.

§ 21.128 Tests: aircraft engines.

(a) Each person manufacturing aircraft engines under a type certificate must subject each engine (except rocket engines for which the manufacturer must establish a sampling technique) to an acceptable test run that includes the following:

(1) Break-in runs that include a determination of fuel and oil consumption and a determination of power characteristics at rated maximum continuous power or thrust and, if applicable, at rated takeoff power or thrust.

(2) At least five hours of operation at rated maximum continuous power or thrust. For engines having a rated takeoff power or thrust higher than rated maximum continuous power or thrust, the five-hour run must include

30 minutes at rated takeoff power or thrust.

(b) The test runs required by paragraph (a) of this section may be made with the engine appropriately mounted and using current types of power and thrust measuring equipment.

[Doc. No. 5085, 29 FR 14568, Oct. 24, 1964, as amended by Amdt. 21–5, 32 FR 3735, Mar. 4, 1967]

§ 21.129 Tests: propellers.

Each person manufacturing propellers under a type certificate must give each variable pitch propeller an acceptable functional test to determine if it operates properly throughout the normal range of operation.

§ 21.130 Statement of conformity.

Each holder or licensee of a type certificate who manufactures a product under this subpart must provide, in a form and manner acceptable to the FAA, a statement that the product for which the type certificate has been issued conforms to its type certificate and is in a condition for safe operation.

[Doc. No. FAA–2006–25877, Amdt. 21–92, 74 FR 53387, Oct. 16, 2009]

Subpart G—Production Certificates

Source: Docket No. FAA–2006–25877, Amdt. 21–92, 74 FR 53387, Oct. 16, 2009, unless otherwise noted.

§ 21.131 Applicability.

This subpart prescribes—

(a) Procedural requirements for issuing production certificates; and

(b) Rules governing holders of those certificates.

§ 21.132 Eligibility.

Any person may apply for a production certificate if that person holds, for the product concerned—

(a) A current type certificate,

(b) A supplemental type certificate, or

(c) Rights to the benefits of that type certificate or supplemental type certificate under a licensing agreement.

§21.133 Application.

Each applicant must apply for a production certificate in a form and manner prescribed by the FAA.

§21.135 Organization.

(a) Each applicant for or holder of a production certificate must provide the FAA with a document—

(1) Describing how its organization will ensure compliance with the provisions of this subpart;

(2) Describing assigned responsibilities, delegated authorities, and the functional relationship of those responsible for quality to management and other organizational components; and

(3) Identifying an accountable manager.

(b) The accountable manager specified in paragraph (a) of this section must be responsible within the applicant's or production approval holder's organization for, and have authority over, all production operations conducted under this part. The accountable manager must confirm that the procedures described in the quality manual required by §21.138 are in place and that the production approval holder satisfies the requirements of the applicable regulations of subchapter C, Aircraft. The accountable manager must serve as the primary contact with the FAA.

[Doc. No. FAA-2013-0933, Amdt. 21-98, 80 FR 59031, Oct. 1, 2015]

§21.137 Quality system.

Each applicant for or holder of a production certificate must establish and describe in writing a quality system that ensures that each product and article conforms to its approved design and is in a condition for safe operation. This quality system must include:

(a) *Design data control.* Procedures for controlling design data and subsequent changes to ensure that only current, correct, and approved data is used.

(b) *Document control.* Procedures for controlling quality system documents and data and subsequent changes to ensure that only current, correct, and approved documents and data are used.

(c) *Supplier control.* Procedures that—

(1) Ensure that each supplier-provided product, article, or service conforms to the production approval holder's requirements; and

(2) Establish a supplier-reporting process for products, articles, or services that have been released from or provided by the supplier and subsequently found not to conform to the production approval holder's requirements.

(d) *Manufacturing process control.* Procedures for controlling manufacturing processes to ensure that each product and article conforms to its approved design.

(e) *Inspecting and testing.* Procedures for inspections and tests used to ensure that each product and article conforms to its approved design. These procedures must include the following, as applicable:

(1) A flight test of each aircraft produced unless that aircraft will be exported as an unassembled aircraft.

(2) A functional test of each aircraft engine and each propeller produced.

(f) *Inspection, measuring, and test equipment control.* Procedures to ensure calibration and control of all inspection, measuring, and test equipment used in determining conformity of each product and article to its approved design. Each calibration standard must be traceable to a standard acceptable to the FAA.

(g) *Inspection and test status.* Procedures for documenting the inspection and test status of products and articles supplied or manufactured to the approved design.

(h) *Nonconforming product and article control.* (1) Procedures to ensure that only products or articles that conform to their approved design are installed on a type-certificated product. These procedures must provide for the identification, documentation, evaluation, segregation, and disposition of nonconforming products and articles. Only authorized individuals may make disposition determinations.

(2) Procedures to ensure that discarded articles are rendered unusable.

(i) *Corrective and preventive actions.* Procedures for implementing corrective and preventive actions to eliminate the causes of an actual or potential nonconformity to the approved design or noncompliance with the approved quality system.

155

(j) *Handling and storage.* Procedures to prevent damage and deterioration of each product and article during handling, storage, preservation, and packaging.

(k) *Control of quality records.* Procedures for identifying, storing, protecting, retrieving, and retaining quality records. A production approval holder must retain these records for at least 5 years for the products and articles manufactured under the approval and at least 10 years for critical components identified under § 45.15(c) of this chapter.

(l) *Internal audits.* Procedures for planning, conducting, and documenting internal audits to ensure compliance with the approved quality system. The procedures must include reporting results of internal audits to the manager responsible for implementing corrective and preventive actions.

(m) *In-service feedback.* Procedures for receiving and processing feedback on in-service failures, malfunctions, and defects. These procedures must include a process for assisting the design approval holder to—

(1) Address any in-service problem involving design changes; and

(2) Determine if any changes to the Instructions for Continued Airworthiness are necessary.

(n) *Quality escapes.* Procedures for identifying, analyzing, and initiating appropriate corrective action for products or articles that have been released from the quality system and that do not conform to the applicable design data or quality system requirements.

(o) *Issuing authorized release documents.* Procedures for issuing authorized release documents for aircraft engines, propellers, and articles if the production approval holder intends to issue those documents. These procedures must provide for the selection, appointment, training, management, and removal of individuals authorized by the production approval holder to issue authorized release documents. Authorized release documents may be issued for new aircraft engines, propellers, and articles manufactured by the production approval holder; and for used aircraft engines, propellers, and articles when rebuilt, or altered, in accordance with § 43.3(j) of this chapter.

When a production approval holder issues an authorized release document for the purpose of export, the production approval holder must comply with the procedures applicable to the export of new and used aircraft engines, propellers, and articles specified in § 21.331 and the responsibilities of exporters specified in § 21.335.

[Docket No. FAA–2006–25877, Amdt. 21–92, 74 FR 53387, Oct. 16, 2009, as amended by Doc. No. FAA–2013–0933, Amdt. 21–98, 80 FR 59031, Oct. 1, 2015; Amdt. 21–98A, 80 FR 59031, Dec. 17, 2015]

§ 21.138 Quality manual.

Each applicant for or holder of a production certificate must provide a manual describing its quality system to the FAA for approval. The manual must be in the English language and retrievable in a form acceptable to the FAA.

§ 21.139 Location of or change to manufacturing facilities.

(a) An applicant may obtain a production certificate for manufacturing facilities located outside of the United States if the FAA finds no undue burden in administering the applicable requirements of Title 49 U.S.C. and this subchapter.

(b) The production certificate holder must obtain FAA approval before making any changes to the location of any of its manufacturing facilities.

(c) The production certificate holder must immediately notify the FAA, in writing, of any change to the manufacturing facilities that may affect the inspection, conformity, or airworthiness of its product or article.

§ 21.140 Inspections and tests.

Each applicant for or holder of a production certificate must allow the FAA to inspect its quality system, facilities, technical data, and any manufactured products or articles and witness any tests, including any inspections or tests at a supplier facility, necessary to determine compliance with this subchapter.

§ 21.141 Issuance.

The FAA issues a production certificate after finding that the applicant complies with the requirements of this subpart.

§ 21.142 Production limitation record.

The FAA issues a production limitation record as part of a production certificate. The record lists the type certificate number and model of every product that the production certificate holder is authorized to manufacture, and identifies every interface component that the production certificate holder is authorized to manufacture and install under this part.

[Doc. No. FAA–2013–0933, Amdt. 21–98, 80 FR 59031, Oct. 1, 2015, as amended by Amdt. 21–98A, 80 FR 59031, Dec. 17, 2015]

§ 21.143 Duration.

A production certificate is effective until surrendered, suspended, revoked, or the FAA otherwise establishes a termination date.

§ 21.144 Transferability.

The holder of a production certificate may not transfer the production certificate.

§ 21.145 Privileges.

(a) The holder of a production certificate may—

(1) Obtain an aircraft airworthiness certificate without further showing, except that the FAA may inspect the aircraft for conformity with the type design; or

(2) In the case of other products, obtain approval from the FAA for installation on type-certificated aircraft.

(b) Notwithstanding the provisions of § 147.3 of this chapter, the holder of a production certificate for a primary category aircraft, or for a normal, utility, or acrobatic category aircraft of a type design that is eligible for a special airworthiness certificate in the primary category under § 21.184(c), may—

(1) Conduct training for persons in the performance of a special inspection and preventive maintenance program approved as a part of the aircraft's type design under § 21.24(b), provided a person holding a mechanic certificate with appropriate airframe and powerplant ratings issued under part 65 of this chapter gives the training; and

(2) Issue a certificate of competency to persons successfully completing the approved training program, provided the certificate specifies the aircraft make and model to which the certificate applies.

§ 21.146 Responsibility of holder.

The holder of a production certificate must—

(a) Amend the document required by § 21.135 as necessary to reflect changes in the organization and provide these amendments to the FAA.

(b) Maintain the quality system in compliance with the data and procedures approved for the production certificate;

(c) Ensure that each completed product or article for which a production certificate has been issued, including primary category aircraft assembled under a production certificate by another person from a kit provided by the holder of the production certificate, presented for airworthiness certification or approval conforms to its approved design and is in a condition for safe operation;

(d) Mark the product or article for which a certificate or approval has been issued. Marking must be in accordance with part 45 of this chapter, including any critical parts;

(e) Identify any portion of the product or article (e.g., sub-assemblies, component parts, or replacement articles) that leave the manufacturer's facility as FAA approved with the manufacturer's part number and name, trademark, symbol, or other FAA approved manufacturer's identification;

(f) Have access to type design data necessary to determine conformity and airworthiness for each product and article produced under the production certificate;

(g) Retain its production certificate and make it available to the FAA upon request; and

(h) Make available to the FAA information regarding all delegation of authority to suppliers.

§ 21.147 Amendment of production certificates.

(a) A holder of a production certificate must apply for an amendment to a production certificate in a form and manner prescribed by the FAA.

(b) An applicant for an amendment to a production certificate to add a type certificate or model, or both, must comply with §§ 21.137, 21.138, and 21.150.

(c) An applicant may apply to amend its production limitation record to allow the manufacture and installation of an interface component, provided—

(1) The applicant owns or has a license to use the design and installation data for the interface component and makes that data available to the FAA upon request;

(2) The applicant manufactures the interface component;

(3) The applicant's product conforms to its approved type design and the interface component conforms to its approved type design;

(4) The assembled product with the installed interface component is in a condition for safe operation; and

(5) The applicant complies with any other conditions and limitations the FAA considers necessary.

[Doc. No. FAA–2013–0933, Amdt. 21–98, 80 FR 59031, Oct. 1, 2015, as amended by Amdt. 21–98A, 80 FR 59031, Dec. 17, 2015]

§ 21.150   Changes in quality system.

After the issuance of a production certificate—

(a) Each change to the quality system is subject to review by the FAA; and

(b) The holder of a production certificate must immediately notify the FAA, in writing, of any change that may affect the inspection, conformity, or airworthiness of its product or article.

Subpart H—Airworthiness Certificates

Source: Docket No. 5085, 29 FR 14569, Oct. 24, 1964, unless otherwise noted.

§ 21.171   Applicability.

This subpart prescribes procedural requirements for the issue of airworthiness certificates.

§ 21.173   Eligibility.

Any registered owner of a U.S.-registered aircraft (or the agent of the owner) may apply for an airworthiness certificate for that aircraft. An application for an airworthiness certificate must be made in a form and manner acceptable to the FAA, and may be submitted to any FAA office.

[Amdt. 21–26, 34 FR 15244, Sept. 30, 1969]

§ 21.175   Airworthiness certificates: classification.

(a) Standard airworthiness certificates are airworthiness certificates issued for aircraft type certificated in the normal, utility, acrobatic, commuter, or transport category, and for manned free balloons, and for aircraft designated by the FAA as special classes of aircraft.

(b) Special airworthiness certificates are primary, restricted, limited, light-sport, and provisional airworthiness certificates, special flight permits, and experimental certificates.

[Amdt. 21–21, 33 FR 6858, May 7, 1968, as amended by Amdt. 21–60, 52 FR 8043, Mar. 13, 1987; Amdt. 21–70, 57 FR 41368, Sept. 9, 1992; Amdt. 21–85, 69 FR 44861, July 27, 2004]

§ 21.177   Amendment or modification.

An airworthiness certificate may be amended or modified only upon application to the FAA.

§ 21.179   Transferability.

An airworthiness certificate is transferred with the aircraft.

§ 21.181   Duration.

(a) Unless sooner surrendered, suspended, revoked, or a termination date is otherwise established by the FAA, airworthiness certificates are effective as follows:

(1) Standard airworthiness certificates, special airworthiness certificates—primary category, and airworthiness certificates issued for restricted or limited category aircraft are effective as long as the maintenance, preventive maintenance, and alterations are performed in accordance with Parts 43 and 91 of this chapter and the aircraft are registered in the United States.

(2) A special flight permit is effective for the period of time specified in the permit.

(3) A special airworthiness certificate in the light-sport category is effective as long as—

(i) The aircraft meets the definition of a light-sport aircraft;

(ii) The aircraft conforms to its original configuration, except for those alterations performed in accordance with an applicable consensus standard and authorized by the aircraft's manufacturer or a person acceptable to the FAA;

(iii) The aircraft has no unsafe condition and is not likely to develop an unsafe condition; and

(iv) The aircraft is registered in the United States.

(4) An experimental certificate for research and development, showing compliance with regulations, crew training, or market surveys is effective for 1 year after the date of issue or renewal unless the FAA prescribes a shorter period. The duration of an experimental certificate issued for operating amateur-built aircraft, exhibition, air-racing, operating primary kit-built aircraft, or operating light-sport aircraft is unlimited, unless the FAA establishes a specific period for good cause.

(b) The owner, operator, or bailee of the aircraft must, upon request, make it available for inspection by the FAA.

(c) Upon suspension, revocation, or termination by order of the FAA of an airworthiness certificate, the owner, operator, or bailee of an aircraft must, upon request, surrender the certificate to the FAA.

[Amdt. 21–21, 33 FR 6858, May 7, 1968, as amended by Amdt. 21–49, 44 FR 46781, Aug. 9, 1979; Amdt. 21–70, 57 FR 41368, Sept. 9, 1992; Amdt. 21–85, 69 FR 44861, July 27, 2004]

§21.182 Aircraft identification.

(a) Except as provided in paragraph (b) of this section, each applicant for an airworthiness certificate under this subpart must show that his aircraft is identified as prescribed in §45.11.

(b) Paragraph (a) of this section does not apply to applicants for the following:

(1) A special flight permit.

(2) An experimental certificate for an aircraft not issued for the purpose of operating amateur-built aircraft, operating primary kit-built aircraft, or operating light-sport aircraft.

(3) A change from one airworthiness classification to another, for an aircraft already identified as prescribed in §45.11.

[Amdt. 21–13, 32 FR 188, Jan. 10, 1967, as amended by Amdt. 21–51, 45 FR 60170, Sept. 11, 1980; Amdt. 21–70, 57 FR 41368, Sept. 9, 1992; Amdt. 21–85, 69 FR 44862, July 27, 2004]

§21.183 Issue of standard airworthiness certificates for normal, utility, acrobatic, commuter, and transport category aircraft; manned free balloons; and special classes of aircraft.

(a) *New aircraft manufactured under a production certificate.* An applicant for a standard airworthiness certificate for a new aircraft manufactured under a production certificate is entitled to a standard airworthiness certificate without further showing, except that the FAA may inspect the aircraft to determine conformity to the type design and condition for safe operation.

(b) *New aircraft manufactured under type certificate.* An applicant for a standard airworthiness certificate for a new aircraft manufactured under a type certificate is entitled to a standard airworthiness certificate upon presentation, by the holder or licensee of the type certificate, of the statement of conformity prescribed in §21.130 if the FAA finds after inspection that the aircraft conforms to the type design and is in condition for safe operation.

(c) *Import aircraft.* An applicant for a standard airworthiness certificate for an import aircraft is entitled to that certificate if—

(1) The aircraft is type certificated in accordance with §21.21 or §21.29 and produced under the authority of another State of Manufacture;

(2) The State of Manufacture certifies, in accordance with the export provisions of an agreement with the United States for import of that aircraft, that the aircraft conforms to the type design and is in condition for safe operation; and

(3) The FAA finds that the aircraft conforms to the type design and is in condition for safe operation.

(d) *Used aircraft and surplus aircraft of the U.S. Armed Forces.* An applicant for a standard airworthiness certificate for a used aircraft or surplus aircraft of the U.S. Armed Forces is entitled to a standard airworthiness certificate if—

(1) The applicant presents evidence to the FAA that the aircraft conforms to a type design approved under a type certificate or a supplemental type certificate and to applicable Airworthiness Directives;

(2) The aircraft (except an experimentally certificated aircraft that previously had been issued a different airworthiness certificate under this section) has been inspected in accordance with the performance rules for 100-hour inspections set forth in § 43.15 of this chapter, or an equivalent performance standard acceptable to the FAA, and found airworthy by—

(i) The manufacturer;

(ii) The holder of a repair station certificate as provided in Part 145 of this chapter;

(iii) The holder of a mechanic certificate as authorized in Part 65 of this chapter; or

(iv) The holder of a certificate issued under part 121 of this chapter, and having a maintenance and inspection organization appropriate to the aircraft type; and

(3) The FAA finds after inspection, that the aircraft conforms to the type design, and is in condition for safe operation.

(e) *Noise requirements.* Notwithstanding all other provisions of this section, the following must be complied with for the original issuance of a standard airworthiness certificate:

(1) For transport category large airplanes and jet (turbojet powered) airplanes that have not had any flight time before the dates specified in § 36.1(d), no standard airworthiness certificate is originally issued under this section unless the FAA finds that the type design complies with the noise requirements in § 36.1(d) in addition to the applicable airworthiness requirements in this section. For import airplanes, compliance with this paragraph is shown if the country in which the airplane was manufactured certifies, and the FAA finds, that § 36.1(d) (or the applicable airplane noise requirements of the country in which the airplane was manufactured and any other requirements the FAA may prescribe to provide noise levels no greater than those provided by compliance with

§ 36.1(d)) and paragraph (c) of this section are complied with.

(2) For normal, utility, acrobatic, commuter, or transport category propeller driven small airplanes (except for those airplanes that are designed for "agricultural aircraft operations" (as defined in § 137.3 of this chapter, as effective on January 1, 1966) or for dispensing fire fighting materials to which § 36.1583 of this chapter does not apply) that have not had any flight time before the applicable date specified in part 36 of this chapter, no standard airworthiness certificate is originally issued under this section unless the applicant shows that the type design complies with the applicable noise requirements of part 36 of this chapter in addition to the applicable airworthiness requirements in this section. For import airplanes, compliance with this paragraph is shown if the country in which the airplane was manufactured certifies, and the FAA finds, that the applicable requirements of part of this chapter (or the applicable airplane noise requirements of the country in which the airplane was manufactured and any other requirements the FAA may prescribe to provide noise levels no greater than those provided by compliance with the applicable requirements of part 36 of this chapter) and paragraph (c) of this section are complied with.

(f) *Passenger emergency exit requirements.* Notwithstanding all other provisions of this section, each applicant for issuance of a standard airworthiness certificate for a transport category airplane manufactured after October 16, 1987, must show that the airplane meets the requirements of § 25.807(c)(7) in effect on July 24, 1989. For the purposes of this paragraph, the date of manufacture of an airplane is the date the inspection acceptance records reflect that the airplane is complete and meets the FAA-approved type design data.

(g) *Fuel venting and exhaust emission requirements.* Notwithstanding all other provisions of this section, and irrespective of the date of application, no airworthiness certificate is issued, on and after the dates specified in part 34 for the airplanes specified therein, unless the airplane complies with the applicable requirements of that part.

(h) *New aircraft manufactured under the provisions of § 21.6(b)*. An applicant for a standard airworthiness certificate for a new aircraft manufactured under the provisions of § 21.6(b) is entitled to a standard airworthiness certificate if—

(1) The applicant presents evidence to the FAA that the aircraft conforms to a type design approved under a type certificate or supplemental type certificate and to applicable Airworthiness Directives;

(2) The aircraft has been inspected in accordance with the performance rules for a 100-hour inspections set forth in § 43.15 of this chapter and found airworthy by a person specified in paragraph (d)(2) of this section; and

(3) The FAA finds after inspection, that the aircraft conforms to the type design, and is in condition for safe operation.

[Amdt. 21–17, 32 FR 14927, Oct. 28, 1967]

EDITORIAL NOTE: For FEDERAL REGISTER citations affecting § 21.183, see the List of CFR Sections Affected, which appears in the Finding Aids section of the printed volume and at *www.govinfo.gov.*

§ 21.184 Issue of special airworthiness certificates for primary category aircraft.

(a) *New primary category aircraft manufactured under a production certificate.* An applicant for an original, special airworthiness certificate-primary category for a new aircraft that meets the criteria of § 21.24(a)(1), manufactured under a production certificate, including aircraft assembled by another person from a kit provided by the holder of the production certificate and under the supervision and quality control of that holder, is entitled to a special airworthiness certificate without further showing, except that the FAA may inspect the aircraft to determine conformity to the type design and condition for safe operation.

(b) *Imported aircraft.* An applicant for a special airworthiness certificate-primary category for an imported aircraft type certificated under § 21.29 is entitled to a special airworthiness certificate if the civil airworthiness authority of the country in which the aircraft was manufactured certifies, and the FAA finds after inspection, that the aircraft conforms to an approved type design that meets the criteria of § 21.24(a)(1) and is in a condition for safe operation.

(c) *Aircraft having a current standard airworthiness certificate.* An applicant for a special airworthiness certificate-primary category, for an aircraft having a current standard airworthiness certificate that meets the criteria of § 21.24(a)(1), may obtain the primary category certificate in exchange for its standard airworthiness certificate through the supplemental type certification process. For the purposes of this paragraph, a current standard airworthiness certificate means that the aircraft conforms to its approved normal, utility, or acrobatic type design, complies with all applicable airworthiness directives, has been inspected and found airworthy within the last 12 calendar months in accordance with § 91.409(a)(1) of this chapter, and is found to be in a condition for safe operation by the FAA.

(d) *Other aircraft.* An applicant for a special airworthiness certificate-primary category for an aircraft that meets the criteria of § 21.24(a)(1), and is not covered by paragraph (a), (b), or (c) of this section, is entitled to a special airworthiness certificate if—

(1) The applicant presents evidence to the FAA that the aircraft conforms to an approved primary, normal, utility, or acrobatic type design, including compliance with all applicable airworthiness directives;

(2) The aircraft has been inspected and found airworthy within the past 12 calendar months in accordance with § 91.409(a)(1) of this chapter and;

(3) The aircraft is found by the FAA to conform to an approved type design and to be in a condition for safe operation.

(e) *Multiple-category airworthiness certificates* in the primary category and any other category will not be issued; a primary category aircraft may hold only one airworthiness certificate.

[Doc. No. 23345, 57 FR 41368, Sept. 9, 1992, as amended by Amdt. 21–70, 57 FR 43776, Sept. 22, 1992]

§ 21.185　Issue of airworthiness certificates for restricted category aircraft.

(a) *Aircraft manufactured under a production certificate or type certificate.* An applicant for the original issue of a restricted category airworthiness certificate for an aircraft type certificated in the restricted category, that was not previously type certificated in any other category, must comply with the appropriate provisions of § 21.183.

(b) *Other aircraft.* An applicant for a restricted category airworthiness certificate for an aircraft type certificated in the restricted category, that was either a surplus aircraft of the Armed Forces or previously type certificated in another category, is entitled to an airworthiness certificate if the aircraft has been inspected by the FAA and found by him to be in a good state of preservation and repair and in a condition for safe operation.

(c) *Import aircraft.* An applicant for the original issue of a special airworthiness certificate for a restricted category import aircraft is entitled to that certificate if—

(1) The aircraft is type-certificated in accordance with § 21.25 or § 21.29 and produced under the authority of another State of Manufacture;

(2) The State of Manufacture certifies, in accordance with the export provisions of an agreement with the United States for import of that aircraft that the aircraft conforms to the type design and is in condition for safe operation; and

(3) The FAA finds that the aircraft conforms to the type design and is in condition for safe operation.

(d) *Noise requirements.* For propeller-driven small airplanes (except airplanes designed for "agricultural aircraft operations," as defined in § 137.3 of this chapter, as effective on January 1, 1966, or for dispensing fire fighting materials) that have not had any flight time before the applicable date specified in Part 36 of this chapter, and notwithstanding the other provisions of this section, no original restricted category airworthiness certificate is issued under this section unless the FAA finds that the type design complies with the applicable noise requirements of Part 36 of this chapter in addition to the applicable airworthiness requirements of this section. For import airplanes, compliance with this paragraph is shown if the country in which the airplane was manufactured certifies, and the FAA finds, that the applicable requirements of Part 36 of this chapter (or the applicable airplane noise requirements of the country in which the airplane was manufactured and any other requirements the FAA may prescribe to provide noise levels no greater than those provided by compliance with the applicable requirements of Part 36 of this chapter) and paragraph (c) of this section are complied with.

[Amdt. 21–10, 31 FR 9211, July 6, 1966, as amended by Amdt. 21–32, 35 FR 10202, June 23, 1970; Amdt. 21–42, 40 FR 1034, Jan. 6, 1975; Amdt. 21–92, 74 FR 53389, Oct. 16, 2009; Amdt. 21–92, 74 FR 53389, Oct. 16, 2009; Amdt. 21–92A, 75 FR 9095, Mar. 1, 2010]

§ 21.187　Issue of multiple airworthiness certification.

(a) An applicant for an airworthiness certificate in the restricted category, and in one or more other categories except primary category, is entitled to the certificate, if—

(1) He shows compliance with the requirements for each category, when the aircraft is in the configuration for that category; and

(2) He shows that the aircraft can be converted from one category to another by removing or adding equipment by simple mechanical means.

(b) The operator of an aircraft certificated under this section must have the aircraft inspected by the FAA, or by a certificated mechanic with an appropriate airframe rating, to determine airworthiness each time the aircraft is converted from the restricted category to another category for the carriage of passengers for compensation or hire, unless the FAA finds this unnecessary for safety in a particular case.

(c) The aircraft complies with the applicable requirements of part 34.

[Doc. No. 5085, 29 FR 14569, Oct. 24, 1964, as amended by Amdt. 21–68, 55 FR 32860, Aug. 10, 1990; Amdt. 21–70, 57 FR 41369, Sept. 9, 1992]

§ 21.189 Issue of airworthiness certificate for limited category aircraft.

(a) An applicant for an airworthiness certificate for an aircraft in the limited category is entitled to the certificate when—

(1) He shows that the aircraft has been previously issued a limited category type certificate and that the aircraft conforms to that type certificate; and

(2) The FAA finds, after inspection (including a flight check by the applicant), that the aircraft is in a good state of preservation and repair and is in a condition for safe operation.

(b) The FAA prescribes limitations and conditions necessary for safe operation.

[Doc. No. 5085, 29 FR 14570, Oct. 24, 1964, as amended by Amdt. 21–4, 30 FR 9437, July 29, 1965]

§ 21.190 Issue of a special airworthiness certificate for a light-sport category aircraft.

(a) *Purpose.* The FAA issues a special airworthiness certificate in the light-sport category to operate a light-sport aircraft, other than a gyroplane.

(b) *Eligibility.* To be eligible for a special airworthiness certificate in the light-sport category:

(1) An applicant must provide the FAA with—

(i) The aircraft's operating instructions;

(ii) The aircraft's maintenance and inspection procedures;

(iii) The manufacturer's statement of compliance as described in paragraph (c) of this section; and

(iv) The aircraft's flight training supplement.

(2) The aircraft must not have been previously issued a standard, primary, restricted, limited, or provisional airworthiness certificate, or an equivalent airworthiness certificate issued by a foreign civil aviation authority.

(3) The aircraft must be inspected by the FAA and found to be in a condition for safe operation.

(c) *Manufacturer's statement of compliance for light-sport category aircraft.* The manufacturer's statement of compliance required in paragraph (b)(1)(iii) of this section must—

(1) Identify the aircraft by make and model, serial number, class, date of manufacture, and consensus standard used;

(2) State that the aircraft meets the provisions of the identified consensus standard;

(3) State that the aircraft conforms to the manufacturer's design data, using the manufacturer's quality assurance system that meets the identified consensus standard;

(4) State that the manufacturer will make available to any interested person the following documents that meet the identified consensus standard:

(i) The aircraft's operating instructions.

(ii) The aircraft's maintenance and inspection procedures.

(iii) The aircraft's flight training supplement.

(5) State that the manufacturer will monitor and correct safety-of-flight issues through the issuance of safety directives and a continued airworthiness system that meets the identified consensus standard;

(6) State that at the request of the FAA, the manufacturer will provide unrestricted access to its facilities; and

(7) State that the manufacturer, in accordance with a production acceptance test procedure that meets an applicable consensus standard has—

(i) Ground and flight tested the aircraft;

(ii) Found the aircraft performance acceptable; and

(iii) Determined that the aircraft is in a condition for safe operation.

(d) *Light-sport aircraft manufactured outside the United States.* For aircraft manufactured outside of the United States to be eligible for a special airworthiness certificate in the light-sport category, an applicant must meet the requirements of paragraph (b) of this section and provide to the FAA evidence that—

(1) The aircraft was manufactured in a country with which the United States has a Bilateral Airworthiness Agreement concerning airplanes or Bilateral Aviation Safety Agreement with associated Implementation Procedures for Airworthiness concerning airplanes, or an equivalent airworthiness agreement; and

(2) The aircraft is eligible for an airworthiness certificate, flight authorization, or other similar certification in its country of manufacture.

[Amdt. 21–85, 69 FR 44862, July 27, 2004]

§ 21.191   Experimental certificates.

Experimental certificates are issued for the following purposes:

(a) *Research and development.* Testing new aircraft design concepts, new aircraft equipment, new aircraft installations, new aircraft operating techniques, or new uses for aircraft.

(b) *Showing compliance with regulations.* Conducting flight tests and other operations to show compliance with the airworthiness regulations including flights to show compliance for issuance of type and supplemental type certificates, flights to substantiate major design changes, and flights to show compliance with the function and reliability requirements of the regulations.

(c) *Crew training.* Training of the applicant's flight crews.

(d) *Exhibition.* Exhibiting the aircraft's flight capabilities, performance, or unusual characteristics at air shows, motion picture, television, and similar productions, and the maintenance of exhibition flight proficiency, including (for persons exhibiting aircraft) flying to and from such air shows and productions.

(e) *Air racing.* Participating in air races, including (for such participants) practicing for such air races and flying to and from racing events.

(f) *Market surveys.* Use of aircraft for purposes of conducting market surveys, sales demonstrations, and customer crew training only as provided in § 21.195.

(g) *Operating amateur-built aircraft.* Operating an aircraft the major portion of which has been fabricated and assembled by persons who undertook the construction project solely for their own education or recreation.

(h) *Operating primary kit-built aircraft.* Operating a primary category aircraft that meets the criteria of § 21.24(a)(1) that was assembled by a person from a kit manufactured by the holder of a production certificate for that kit, without the supervision and quality control of the production certificate holder under § 21.184(a).

(i) *Operating light-sport aircraft.* Operating a light-sport aircraft that—

(1) Has not been issued a U.S. or foreign airworthiness certificate and does not meet the provisions of § 103.1 of this chapter. An experimental certificate will not be issued under this paragraph for these aircraft after January 31, 2008;

(2) Has been assembled—

(i) From an aircraft kit for which the applicant can provide the information required by § 21.193(e); and

(ii) In accordance with manufacturer's assembly instructions that meet an applicable consensus standard; or

(3) Has been previously issued a special airworthiness certificate in the light-sport category under § 21.190.

[Amdt. 21–21, 38 FR 6858, May 7, 1968, as amended by Amdt. 21–57, 49 FR 39651, Oct. 9, 1984; Amdt. 21–70, 57 FR 41369, Sept. 9, 1992; Amdt. 21–85, 69 FR 44862, July 27, 2004; Amdt. 21–85, 69 FR 53336, Sept. 1, 2004]

§ 21.193   Experimental certificates: general.

An applicant for an experimental certificate must submit the following information:

(a) A statement, in a form and manner prescribed by the FAA setting forth the purpose for which the aircraft is to be used.

(b) Enough data (such as photographs) to identify the aircraft.

(c) Upon inspection of the aircraft, any pertinent information found necessary by the FAA to safeguard the general public.

(d) In the case of an aircraft to be used for experimental purposes—

(1) The purpose of the experiment;

(2) The estimated time or number of flights required for the experiment;

(3) The areas over which the experiment will be conducted; and

(4) Except for aircraft converted from a previously certificated type without appreciable change in the external configuration, three-view drawings or three-view dimensioned photographs of the aircraft.

(e) In the case of a light-sport aircraft assembled from a kit to be certificated in accordance with

§21.191(i)(2), an applicant must provide the following:

(1) Evidence that an aircraft of the same make and model was manufactured and assembled by the aircraft kit manufacturer and issued a special airworthiness certificate in the light-sport category.

(2) The aircraft's operating instructions.

(3) The aircraft's maintenance and inspection procedures.

(4) The manufacturer's statement of compliance for the aircraft kit used in the aircraft assembly that meets §21.190(c), except that instead of meeting §21.190(c)(7), the statement must identify assembly instructions for the aircraft that meet an applicable consensus standard.

(5) The aircraft's flight training supplement.

(6) In addition to paragraphs (e)(1) through (e)(5) of this section, for an aircraft kit manufactured outside of the United States, evidence that the aircraft kit was manufactured in a country with which the United States has a Bilateral Airworthiness Agreement concerning airplanes or a Bilateral Aviation Safety Agreement with associated Implementation Procedures for Airworthiness concerning airplanes, or an equivalent airworthiness agreement.

[Doc. No. 5085, 29 FR 14569, Oct. 24, 1964, as amended by Amdt. 21–85, 69 FR 44862, July 27, 2004]

§21.195 Experimental certificates: Aircraft to be used for market surveys, sales demonstrations, and customer crew training.

(a) A manufacturer of aircraft manufactured within the United States may apply for an experimental certificate for an aircraft that is to be used for market surveys, sales demonstrations, or customer crew training.

(b) A manufacturer of aircraft engines who has altered a type certificated aircraft by installing different engines, manufactured by him within the United States, may apply for an experimental certificate for that aircraft to be used for market surveys, sales demonstrations, or customer crew training, if the basic aircraft, before alteration, was type certificated in the normal, acrobatic, commuter, or transport category.

(c) A person who has altered the design of a type certificated aircraft may apply for an experimental certificate for the altered aircraft to be used for market surveys, sales demonstrations, or customer crew training if the basic aircraft, before alteration, was type certificated in the normal, utility, acrobatic, or transport category.

(d) An applicant for an experimental certificate under this section is entitled to that certificate if, in addition to meeting the requirements of §21.193—

(1) He has established an inspection and maintenance program for the continued airworthiness of the aircraft; and

(2) The applicant shows that the aircraft has been flown for at least 50 hours, or for at least 5 hours if it is a type certificated aircraft which has been modified. The FAA may reduce these operational requirements if the applicant provides adequate justification.

[Amdt. 21–21, 33 FR 6858, May 7, 1968, as amended by Amdt. 21–28, 35 FR 2818, Feb. 11, 1970; Amdt. 21–57, 49 FR 39651, Oct. 9, 1984; Amdt. 21–59, 52 FR 1836, Jan. 15, 1987; Amdt. 21–92, 74 FR 53389, Oct. 16, 2009]

§21.197 Special flight permits.

(a) A special flight permit may be issued for an aircraft that may not currently meet applicable airworthiness requirements but is capable of safe flight, for the following purposes:

(1) Flying the aircraft to a base where repairs, alterations, or maintenance are to be performed, or to a point of storage.

(2) Delivering or exporting the aircraft.

(3) Production flight testing new production aircraft.

(4) Evacuating aircraft from areas of impending danger.

(5) Conducting customer demonstration flights in new production aircraft that have satisfactorily completed production flight tests.

(b) A special flight permit may also be issued to authorize the operation of an aircraft at a weight in excess of its maximum certificated takeoff weight for flight beyond the normal range over water, or over land areas where adequate landing facilities or appropriate fuel is not available. The excess weight that may be authorized under this paragraph is limited to the additional fuel, fuel-carrying facilities, and navigation equipment necessary for the flight.

(c) Upon application, as prescribed in §§ 91.1017 or 119.51 of this chapter, a special flight permit with a continuing authorization may be issued for aircraft that may not meet applicable airworthiness requirements, but are capable of safe flight for the purpose of flying aircraft to a base where maintenance or alterations are to be performed. The permit issued under this paragraph is an authorization, including conditions and limitations for flight, which is set forth in the certificate holder's operations specifications. The permit issued under this paragraph may be issued to—

(1) Certificate holders authorized to conduct operations under part 119 of this chapter, that have an approved program for continuing flight authorization; or

(2) Management specification holders authorized to conduct operations under part 91, subpart K of this chapter for those aircraft they operate and maintain under a continuous airworthiness maintenance program prescribed by § 91.1411 of this chapter.

[Doc. No. 5085, 29 FR 14570, Oct. 24, 1964, as amended by Amdt. 21–21, 33 FR 6859, May 7, 1968; Amdt. 21–51, 45 FR 60170, Sept. 11, 1980; Amdt. 21–54, 46 FR 37878, July 23, 1981; Amdt. 21–79, 66 FR 21066, Apr. 27, 2001; Amdt. 21–84, 68 FR 54559, Sept. 17, 2003; Amdt. 21–87, 71 FR 536, Jan. 4, 2006; Amdt. 21–92, 74 FR 53389, Oct. 16, 2009]

§ 21.199 Issue of special flight permits.

(a) Except as provided in § 21.197(c), an applicant for a special flight permit must submit a statement in a form and manner prescribed by the FAA, indicating—

(1) The purpose of the flight.

(2) The proposed itinerary.

(3) The crew required to operate the aircraft and its equipment, e.g., pilot, co-pilot, navigator, etc.

(4) The ways, if any, in which the aircraft does not comply with the applicable airworthiness requirements.

(5) Any restriction the applicant considers necessary for safe operation of the aircraft.

(6) Any other information considered necessary by the FAA for the purpose of prescribing operating limitations.

(b) The FAA may make, or require the applicant to make appropriate inspections or tests necessary for safety.

[Doc. No. 5085, 29 FR 14570, Oct. 24, 1964, as amended by Amdt. 21–21, 33 FR 6859, May 7, 1968; Amdt. 21–22, 33 FR 11901, Aug. 22, 1968]

Subpart I—Provisional Airworthiness Certificates

SOURCE: Docket No. 5085, 29 FR 14571, Oct. 24, 1964, unless otherwise noted.

§ 21.211 Applicability.

This subpart prescribes procedural requirements for the issue of provisional airworthiness certificates.

§ 21.213 Eligibility.

(a) A manufacturer who is a United States citizen may apply for a Class I or Class II provisional airworthiness certificate for aircraft manufactured by him within the U.S.

(b) Any holder of an air carrier operating certificate under Part 121 of this chapter who is a United States citizen may apply for a Class II provisional airworthiness certificate for transport category aircraft that meet either of the following:

(1) The aircraft has a current Class II provisional type certificate or an amendment thereto.

(2) The aircraft has a current provisional amendment to a type certificate that was preceded by a corresponding Class II provisional type certificate.

(c) An aircraft engine manufacturer who is a United States citizen and who has altered a type certificated aircraft by installing different type certificated engines, manufactured by him within the United States, may apply for a Class I provisional airworthiness certificate for that aircraft, if the basic aircraft, before alteration, was type certificated in the normal, utility, acrobatic, commuter, or transport category.

[Doc. No. 5085, 29 FR 14571, Oct. 24, 1964, as amended by Amdt. 21–59, 52 FR 1836, Jan. 15, 1987; Amdt. 21–79, 66 FR 21066, Apr. 27, 2001]

§21.215 Application.

Applications for provisional airworthiness certificates must be submitted to the FAA. The application must be accompanied by the pertinent information specified in this subpart.

[Amdt. 21–67, 54 FR 39291, Sept. 25, 1989; 54 FR 52872, Dec. 22, 1989; Doc. No. FAA–2018–0119, Amdt. 21–101, 83 FR 9169, Mar. 5, 2018]

§21.217 Duration.

Unless sooner surrendered, superseded, revoked, or otherwise terminated, provisional airworthiness certificates are effective for the duration of the corresponding provisional type certificate, amendment to a provisional type certificate, or provisional amendment to the type certificate.

§21.219 Transferability.

Class I provisional airworthiness certificates are not transferable. Class II provisional airworthiness certificates may be transferred to an air carrier eligible to apply for a certificate under §21.213(b).

§21.221 Class I provisional airworthiness certificates.

(a) Except as provided in §21.225, an applicant is entitled to a Class I provisional airworthiness certificate for an aircraft for which a Class I provisional type certificate has been issued if—

(1) He meets the eligibility requirements of §21.213 and he complies with this section; and

(2) The FAA finds that there is no feature, characteristic or condition of the aircraft that would make the aircraft unsafe when operated in accordance with the limitations established in §§21.81(e) and 91.317 of this subchapter.

(b) The manufacturer must hold a provisional type certificate for the aircraft.

(c) The manufacturer must submit a statement that the aircraft conforms to the type design corresponding to the provisional type certificate and has been found by him to be in safe operating condition under all applicable limitations.

(d) The aircraft must be flown at least five hours by the manufacturer.

(e) The aircraft must be supplied with a provisional aircraft flight manual or other document and appropriate placards containing the limitations established by §§21.81(e) and 91.317.

[Doc. No. 5085, 29 FR 14571, Oct. 24, 1964, as amended by Amdt. 21–66, 54 FR 34329, Aug. 18, 1989]

§21.223 Class II provisional airworthiness certificates.

(a) Except as provided in §21.225, an applicant is entitled to a Class II provisional airworthiness certificate for an aircraft for which a Class II provisional type certificate has been issued if—

(1) He meets the eligibility requirements of §21.213 and he complies with this section; and

(2) The FAA finds that there is no feature, characteristic, or condition of the aircraft that would make the aircraft unsafe when operated in accordance with the limitations established in §§21.83(h), 91.317, and 121.207 of this chapter.

(b) The applicant must show that a Class II provisional type certificate for the aircraft has been issued to the manufacturer.

(c) The applicant must submit a statement by the manufacturer that the aircraft has been manufactured under a quality system adequate to ensure that the aircraft conforms to the type design corresponding with the provisional type certificate.

(d) The applicant must submit a statement that the aircraft has been found by him to be in a safe operating condition under the applicable limitations.

(e) The aircraft must be flown at least five hours by the manufacturer.

(f) The aircraft must be supplied with a provisional aircraft flight manual containing the limitations established by §§21.83(h), 91.317, and 121.207 of this chapter.

[Doc. No. 5085, 29 FR 14571, Oct. 24, 1964, as amended by Amdt. 21–12, 31 FR 13389, Oct. 15, 1966; Amdt. 21–66, 54 FR 34329, Aug. 18, 1989; Amdt. 21–92, 74 FR 53390, Oct. 16, 2009]

§ 21.225 Provisional airworthiness certificates corresponding with provisional amendments to type certificates.

(a) An applicant is entitled to a Class I or a Class II provisional airworthiness certificate, for an aircraft, for which a provisional amendment to the type certificate has been issued, if—

(1) He meets the eligibility requirements of § 21.213 and he complies with this section; and

(2) The FAA finds that there is no feature, characteristic, or condition of the aircraft, as modified in accordance with the provisionally amended type certificate, that would make the aircraft unsafe when operated in accordance with the applicable limitations established in §§ 21.85(g), 91.317, and 121.207 of this chapter.

(b) The applicant must show that the modification was made under a quality system adequate to ensure that the modification conforms to the provisionally amended type certificate.

(c) The applicant must submit a statement that the aircraft has been found by him to be in a safe operating condition under the applicable limitations.

(d) The aircraft must be flown at least five hours by the manufacturer.

(e) The aircraft must be supplied with a provisional aircraft flight manual or other document and appropriate placards containing the limitations required by §§ 21.85(g), 91.317, and 121.207 of this chapter.

[Doc. No. 5085, 29 FR 14571, Oct. 24, 1964, as amended by Amdt. 21-12, 31 FR 13389, Oct. 15, 1966; Amdt. 21-66, 54 FR 34329, Aug. 18, 1989; Amdt. 21-92, 74 FR 53390, Oct. 16, 2009]

Subpart J [Reserved]

Subpart K—Parts Manufacturer Approvals

SOURCE: Docket No. FAA-2006-25877, Amdt. 21-92, 74 FR 53390, Oct. 16, 2009, unless otherwise noted.

§ 21.301 Applicability.

This subpart prescribes—

(a) Procedural requirements for issuing PMAs; and

(b) Rules governing holders of PMAs.

§ 21.303 Application.

(a) The applicant for a PMA must apply in a form and manner prescribed by the FAA, and include the following:

(1) The identity of the product on which the article is to be installed.

(2) The name and address of the manufacturing facilities at which these articles are to be manufactured.

(3) The design of the article, which consists of—

(i) Drawings and specifications necessary to show the configuration of the article; and

(ii) Information on dimensions, materials, and processes necessary to define the structural strength of the article.

(4) Test reports and computations necessary to show that the design of the article meets the airworthiness requirements of this subchapter. The test reports and computations must be applicable to the product on which the article is to be installed, unless the applicant shows that the design of the article is identical to the design of a article that is covered under a type certificate. If the design of the article was obtained by a licensing agreement, the applicant must provide evidence of that agreement.

(5) An applicant for a PMA based on test reports and computations must provide a statement certifying that the applicant has complied with the airworthiness requirements of this subchapter.

(b) Each applicant for a PMA must make all inspections and tests necessary to determine—

(1) Compliance with the applicable airworthiness requirements;

(2) That materials conform to the specifications in the design;

(3) That the article conforms to its approved design; and

(4) That the manufacturing processes, construction, and assembly conform to those specified in the design.

§ 21.305 Organization.

(a) Each applicant for or holder of a PMA must provide the FAA with a document—

(1) Describing how its organization will ensure compliance with the provisions of this subpart;

(2) Describing assigned responsibilities, delegated authorities, and the

functional relationship of those responsible for quality to management and other organizational components; and

(3) Identifying an accountable manager.

(b) The accountable manager specified in paragraph (a) of this section must be responsible within the applicant's or production approval holder's organization for, and have authority over, all production operations conducted under this part. The accountable manager must confirm that the procedures described in the quality manual required by §21.308 are in place and that the production approval holder satisfies the requirements of the applicable regulations of subchapter C, Aircraft. The accountable manager must serve as the primary contact with the FAA.

[Doc. No. FAA–2013–0933, Amdt. 21–98, 80 FR 59031, Oct. 1, 2015]

§21.307   Quality system.

Each applicant for or holder of a PMA must establish a quality system that meets the requirements of §21.137.

§21.308   Quality manual.

Each applicant for or holder of a PMA must provide a manual describing its quality system to the FAA for approval. The manual must be in the English language and retrievable in a form acceptable to the FAA.

§21.309   Location of or change to manufacturing facilities.

(a) An applicant may obtain a PMA for manufacturing facilities located outside of the United States if the FAA finds no undue burden in administering the applicable requirements of Title 49 U.S.C. and this subchapter.

(b) The PMA holder must obtain FAA approval before making any changes to the location of any of its manufacturing facilities.

(c) The PMA holder must immediately notify the FAA, in writing, of any change to the manufacturing facilities that may affect the inspection, conformity, or airworthiness of its PMA article.

§21.310   Inspections and tests.

(a) Each applicant for or holder of a PMA must allow the FAA to inspect its quality system, facilities, technical data, and any manufactured articles and witness any tests, including any inspections or tests at a supplier facility, necessary to determine compliance with this subchapter.

(b) Unless otherwise authorized by the FAA, the applicant or holder—

(1) May not present any article to the FAA for an inspection or test unless compliance with §21.303(b)(2) through (4) has been shown for that article; and

(2) May not make any change to an article between the time that compliance with §21.303(b)(2) through (4) is shown for that article and the time that the article is presented to the FAA for the inspection or test.

§21.311   Issuance.

The FAA issues a PMA after finding that the applicant complies with the requirements of this subpart and the design complies with the requirements of this chapter applicable to the product on which the article is to be installed.

§21.313   Duration.

A PMA is effective until surrendered, withdrawn, or the FAA otherwise terminates it.

§21.314   Transferability.

The holder of a PMA may not transfer the PMA.

§21.316   Responsibility of holder.

Each holder of a PMA must—

(a) Amend the document required by §21.305 as necessary to reflect changes in the organization and provide these amendments to the FAA;

(b) Maintain the quality system in compliance with the data and procedures approved for the PMA;

(c) Ensure that each PMA article conforms to its approved design and is in a condition for safe operation;

(d) Mark the PMA article for which an approval has been issued. Marking must be in accordance with part 45 of this chapter, including any critical parts;

(e) Identify any portion of the PMA article (*e.g.*, sub-assemblies, component parts, or replacement articles) that leave the manufacturer's facility as FAA approved with the manufacturer's

169

part number and name, trademark, symbol, or other FAA approved manufacturer's identification;

(f) Have access to design data necessary to determine conformity and airworthiness for each article produced under the PMA;

(g) Retain each document granting PMA and make it available to the FAA upon request; and

(h) Make available to the FAA information regarding all delegation of authority to suppliers.

§ 21.319 Design changes.

(a) *Classification of design changes.* (1) A "minor change" to the design of an article produced under a PMA is one that has no appreciable effect on the approval basis.

(2) A "major change" to the design of an article produced under a PMA is any change that is not minor.

(b) *Approval of design changes.* (1) Minor changes to the basic design of a PMA may be approved using a method acceptable to the FAA.

(2) The PMA holder must obtain FAA approval of any major change before including it in the design of an article produced under a PMA.

§ 21.320 Changes in quality system.

After the issuance of a PMA—

(a) Each change to the quality system is subject to review by the FAA; and

(b) The holder of the PMA must immediately notify the FAA, in writing, of any change that may affect the inspection, conformity, or airworthiness of its article.

Subpart L—Export Airworthiness Approvals

SOURCE: Docket No. FAA–2006–25877, Amdt. 21–92, 74 FR 53391, Oct. 16, 2009, unless otherwise noted.

§ 21.321 Applicability.

This subpart prescribes—

(a) Procedural requirements for issuing export airworthiness approvals; and

(b) Rules governing the holders of those approvals.

§ 21.325 Export airworthiness approvals.

(a) An export airworthiness approval for an aircraft is issued in the form of an export certificate of airworthiness. This certificate does not authorize operation of that aircraft.

(b) The FAA prescribes the form and manner in which an export airworthiness approval for an aircraft engine, propeller, or article is issued.

(c) If the FAA finds no undue burden in administering the applicable requirements of Title 49 U.S.C. and this subchapter, an export airworthiness approval may be issued for a product or article located outside of the United States.

§ 21.327 Application.

Any person may apply for an export airworthiness approval. Each applicant must apply in a form and manner prescribed by the FAA.

§ 21.329 Issuance of export certificates of airworthiness.

(a) A person may obtain from the FAA an export certificate of airworthiness for an aircraft if—

(1) A new or used aircraft manufactured under subpart F or G of this part meets the airworthiness requirements under subpart H of this part for a—

(i) Standard airworthiness certificate; or

(ii) Special airworthiness certificate in either the "primary" or the "restricted" category; or

(2) A new or used aircraft not manufactured under subpart F or G of this part has a valid—

(i) Standard airworthiness certificate; or

(ii) Special airworthiness certificate in either the "primary" or the "restricted" category.

(b) An aircraft need not meet a requirement specified in paragraph (a) of this section, as applicable, if—

(1) The importing country or jurisdiction accepts, in a form and manner acceptable to the FAA, a deviation from that requirement; and

(2) The export certificate of airworthiness lists as an exception any difference between the aircraft to be exported and its type design.

§21.331 Issuance of export airworthiness approvals for aircraft engines, propellers, and articles.

(a) A person may obtain from the FAA an export airworthiness approval to export a new aircraft engine, propeller, or article that is manufactured under this part if it conforms to its approved design and is in a condition for safe operation.

(b) A new aircraft engine, propeller, or article need not meet a requirement of paragraph (a) of this section if—

(1) The importing country or jurisdiction accepts, in a form and manner acceptable to the FAA, a deviation from that requirement; and

(2) The export airworthiness approval lists as an exception any difference between the aircraft engine, propeller, or article to be exported and its approved design.

(c) A person may obtain from the FAA an export airworthiness approval to export a used aircraft engine, propeller, or article if it conforms to its approved design and is in a condition for safe operation.

(d) A used aircraft engine or propeller need not meet a requirement of paragraph (c) of this section if—

(1) The importing country or jurisdiction accepts, in a form and manner acceptable to the FAA, a deviation from that requirement; and

(2) The export airworthiness approval lists as an exception any difference between the used aircraft engine or propeller to be exported and its approved design.

§21.335 Responsibilities of exporters.

Unless otherwise agreed to by the importing country or jurisdiction, each exporter must—

(a) Forward to the importing country or jurisdiction all documents specified by that country or jurisdiction;

(b) Preserve and package products and articles as necessary to protect them against corrosion and damage during transit or storage and state the duration of effectiveness of such preservation and packaging;

(c) Remove or cause to be removed any temporary installation incorporated on an aircraft for the purpose of export delivery and restore the aircraft to the approved configuration upon completion of the delivery flight;

(d) Secure all proper foreign entry clearances from all the countries or jurisdictions involved when conducting sales demonstrations or delivery flights; and

(e) When title to an aircraft passes or has passed to a foreign purchaser—

(1) Request cancellation of the U.S. registration and airworthiness certificates from the FAA, giving the date of transfer of title, and the name and address of the foreign owner;

(2) Return the Registration and Airworthiness Certificates to the FAA; and

(3) Provide a statement to the FAA certifying that the U.S. identification and registration numbers have been removed from the aircraft in compliance with §45.33.

Subpart M [Reserved]

Subpart N—Acceptance of Aircraft Engines, Propellers, and Articles for Import

SOURCE: Docket No. FAA-2006-25877, 74 FR 53392, Amdt. 21-92, Oct. 16, 2009, unless otherwise noted.

§21.500 Acceptance of aircraft engines and propellers.

An aircraft engine or propeller manufactured in a foreign country or jurisdiction meets the requirements for acceptance under this subchapter if—

(a) That country or jurisdiction is subject to the provisions of an agreement with the United States for the acceptance of that product;

(b) That product is marked in accordance with part 45 of this chapter; and

(c) The holder or licensee of a U.S. type certificate for that product furnishes with each such aircraft engine or propeller imported into the United States, an export airworthiness approval issued in accordance with the provisions of that agreement certifying that the individual aircraft engine or propeller—

(1) Conforms to its U.S. type certificate and is in condition for safe operation; and

(2) Has been subjected by the manufacturer to a final operational check.

§ 21.502   Acceptance of articles.

An article (including an article produced under a letter of TSO design approval) manufactured in a foreign country or jurisdiction meets the requirements for acceptance under this subchapter if—

(a) That country or jurisdiction is subject to the provisions of an agreement with the United States for the acceptance of that article;

(b) That article is marked in accordance with part 45 of this chapter; and

(c) An export airworthiness approval has been issued in accordance with the provisions of that agreement for that article for import into the United States.

Subpart O—Technical Standard Order Approvals

Source: Docket No. FAA-2006-25877, Amdt. 21-92, 74 FR 53392, Oct. 16, 2009, unless otherwise noted.

§ 21.601   Applicability and definitions.

(a) This subpart prescribes—

(1) Procedural requirements for issuing TSO authorizations;

(2) Rules governing the holders of TSO authorizations; and

(3) Procedural requirements for issuing letters of TSO design approval.

(b) For the purposes of this subpart—

(1) A TSO issued by the FAA is a minimum performance standard for specified articles used on civil aircraft;

(2) A TSO authorization is an FAA design and production approval issued to the manufacturer of an article that has been found to meet a specific TSO;

(3) A letter of TSO design approval is an FAA design approval for an article that has been found to meet a specific TSO in accordance with the procedures of § 21.621;

(4) An article manufactured under a TSO authorization, an FAA letter of acceptance as described in § 21.613(b), or an article manufactured under a letter of TSO design approval described in § 21.621 is an approved article for the purpose of meeting the regulations of this chapter that require the article to be approved; and

(5) An article manufacturer is the person who controls the design and quality of the article produced (or to be produced, in the case of an application), including any related parts, processes, or services procured from an outside source.

§ 21.603   Application.

(a) An applicant for a TSO authorization must apply in the form and manner prescribed by the FAA. The applicant must include the following documents in the application:

(1) A statement of conformance certifying that the applicant has met the requirements of this subpart and that the article concerned meets the applicable TSO that is effective on the date of application for that article.

(2) One copy of the technical data required in the applicable TSO.

(b) If the applicant anticipates a series of minor changes in accordance with § 21.619, the applicant may set forth in its application the basic model number of the article and the part number of the components with open brackets after it to denote that suffix change letters or numbers (or combinations of them) will be added from time to time.

(c) If the application is deficient, the applicant must, when requested by the FAA, provide any additional information necessary to show compliance with this part. If the applicant fails to provide the additional information within 30 days after the FAA's request, the FAA denies the application and notifies the applicant.

[Docket No. FAA-2006-25877, Amdt. 21-92, 74 FR 53392, Oct. 16, 2009, as amended by Doc. No. FAA-2018-0119, Amdt. 21-101, 83 FR 9169, Mar. 5, 2018]

§ 21.605   Organization.

(a) Each applicant for or holder of a TSO authorization must provide the FAA with a document—

(1) Describing how its organization will ensure compliance with the provisions of this subpart;

(2) Describing assigned responsibilities, delegated authorities, and the functional relationship of those responsible for quality to management and other organizational components; and

(3) Identifying an accountable manager.

(b) The accountable manager specified in paragraph (a) of this section must be responsible within the applicant's or production approval holder's organization for, and have authority over, all production operations conducted under this part. The accountable manager must confirm that the procedures described in the quality manual required by §21.608 are in place and that the production approval holder satisfies the requirements of the applicable regulations of subchapter C, Aircraft. The accountable manager must serve as the primary contact with the FAA.

[Doc. No. FAA–2013–0933, Amdt. 21–98, 80 FR 59032, Oct. 1, 2015]

§21.607 Quality system.

Each applicant for or holder of a TSO authorization must establish a quality system that meets the requirements of §21.137.

§21.608 Quality manual.

Each applicant for or holder of a TSO authorization must provide a manual describing its quality system to the FAA for approval. The manual must be in the English language and retrievable in a form acceptable to the FAA.

§21.609 Location of or change to manufacturing facilities.

(a) An applicant may obtain a TSO authorization for manufacturing facilities located outside of the United States if the FAA finds no undue burden in administering the applicable requirements of Title 49 U.S.C. and this subchapter.

(b) The TSO authorization holder must obtain FAA approval before making any changes to the location of any of its manufacturing facilities.

(c) The TSO authorization holder must immediately notify the FAA, in writing, of any change to the manufacturing facilities that may affect the inspection, conformity, or airworthiness of its product or article.

§21.610 Inspections and tests.

Each applicant for or holder of a TSO authorization must allow the FAA to inspect its quality system, facilities, technical data, and any manufactured articles and witness any tests, including any inspections or tests at a supplier facility, necessary to determine compliance with this subchapter.

§21.611 Issuance.

If the FAA finds that the applicant complies with the requirements of this subchapter, the FAA issues a TSO authorization to the applicant (including all TSO deviations granted to the applicant).

§21.613 Duration.

(a) A TSO authorization or letter of TSO design approval is effective until surrendered, withdrawn, or otherwise terminated by the FAA.

(b) If a TSO is revised or canceled, the holder of an affected FAA letter of acceptance of a statement of conformance, TSO authorization, or letter of TSO design approval may continue to manufacture articles that meet the original TSO without obtaining a new acceptance, authorization, or approval but must comply with the requirements of this chapter.

§21.614 Transferability.

The holder of a TSO authorization or letter of TSO design approval may not transfer the TSO authorization or letter of TSO design approval.

§21.616 Responsibility of holder.

Each holder of a TSO authorization must—

(a) Amend the document required by §21.605 as necessary to reflect changes in the organization and provide these amendments to the FAA.

(b) Maintain a quality system in compliance with the data and procedures approved for the TSO authorization;

(c) Ensure that each manufactured article conforms to its approved design, is in a condition for safe operation, and meets the applicable TSO;

(d) Mark the TSO article for which an approval has been issued. Marking must be in accordance with part 45 of this chapter, including any critical parts;

(e) Identify any portion of the TSO article (e.g., sub-assemblies, component parts, or replacement articles) that leave the manufacturer's facility as FAA approved with the manufacturer's part number and name, trademark, symbol, or other FAA approved manufacturer's identification;

(f) Have access to design data necessary to determine conformity and airworthiness for each article produced under the TSO authorization. The manufacturer must retain this data until it no longer manufactures the article. At that time, copies of the data must be sent to the FAA;

(g) Retain its TSO authorization and make it available to the FAA upon request; and

(h) Make available to the FAA information regarding all delegation of authority to suppliers.

§ 21.618   Approval for deviation.

(a) Each manufacturer who requests approval to deviate from any performance standard of a TSO must show that factors or design features providing an equivalent level of safety compensate for the standards from which a deviation is requested.

(b) The manufacturer must send requests for approval to deviate, together with all pertinent data, to the FAA. If the article is manufactured under the authority of a foreign country or jurisdiction, the manufacturer must send requests for approval to deviate, together with all pertinent data, through the civil aviation authority of that country or jurisdiction to the FAA.

[Docket No. FAA-2006-25877, Amdt. 21-92, 74 FR 53392, Oct. 16, 2009, as amended by Doc. No. FAA-2018-0119, Amdt. 21-101, 83 FR 9169, Mar. 5, 2018]

§ 21.619   Design changes.

(a) *Minor changes by the manufacturer holding a TSO authorization.* The manufacturer of an article under an authorization issued under this part may make minor design changes (any change other than a major change) without further approval by the FAA. In this case, the changed article keeps the original model number (part numbers may be used to identify minor changes) and the manufacturer must forward to the FAA, any revised data that are necessary for compliance with § 21.603(a).

(b) *Major changes by the manufacturer holding a TSO authorization.* Any design change by the manufacturer extensive enough to require a substantially complete investigation to determine compliance with a TSO is a major change. Before making a major change, the manufacturer must assign a new type or model designation to the article and apply for an authorization under § 21.603.

(c) *Changes by persons other than the manufacturer.* No design change by any person (other than the manufacturer who provided the statement of conformance for the article) is eligible for approval under this part unless the person seeking the approval is a manufacturer and applies under § 21.603(a) for a separate TSO authorization. Persons other than a manufacturer may obtain approval for design changes under part 43 or under the applicable airworthiness regulations of this chapter.

[Docket No. FAA-2006-25877, Amdt. 21-92, 74 FR 53392, Oct. 16, 2009, as amended by Doc. No. FAA-2018-0119, Amdt. 21-101, 83 FR 9169, Mar. 5, 2018; Doc. No. FAA-2022-1355, Amdt. 21-106, 87 FR 75710, Dec. 9, 2022]

§ 21.620   Changes in quality system.

After the issuance of a TSO authorization—

(a) Each change to the quality system is subject to review by the FAA; and

(b) The holder of the TSO authorization must immediately notify the FAA, in writing, of any change that may affect the inspection, conformity, or airworthiness of its article.

§ 21.621   Issue of letters of TSO design approval: Import articles.

(a) The FAA may issue a letter of TSO design approval for an article—

(1) Designed and manufactured in a foreign country or jurisdiction subject to the export provisions of an agreement with the United States for the acceptance of these articles for import; and

(2) For import into the United States if—

(i) The State of Design certifies that the article has been examined, tested, and found to meet the applicable TSO or the applicable performance standards of the State of Design and any other performance standards the FAA may prescribe to provide a level of safety equivalent to that provided by the TSO; and

(ii) The manufacturer has provided to the FAA one copy of the technical data required in the applicable performance standard through its State of Design.

(b) The FAA issues the letter of TSO design approval that lists any deviation granted under § 21.618.

[Doc. No. FAA–2006–25877, Amdt. 21–92, 74 FR 53392, Oct. 16, 2009, as amended by Amdt. 21–92A, 75 FR 9095, Mar. 1, 2010]

Subpart P—Special Federal Aviation Regulations

SOURCE: Docket No. FAA–2011–0186, Amdt. 21–92, 76 FR 12555, Mar. 8, 2011, unless otherwise noted.

§ 21.700 SFAR No. 111—Lavatory Oxygen Systems.

The requirements of § 121.1500 of this chapter also apply to this part.

PART 23—AIRWORTHINESS STANDARDS: NORMAL CATEGORY AIRPLANES

AUTHORITY: 49 U.S.C. 106(f), 106(g), 40113, 44701–44702, 44704, Pub. L. 113–53, 127 Stat. 584 (49 U.S.C. 44704) note.

SOURCE: Doc. No. FAA–2015–1621, Amdt. 23–64, 81 FR 96689, Dec. 30, 2016, unless otherwise noted.

§ 23.1457 Cockpit voice recorders.

(a) Each cockpit voice recorder required by the operating rules of this chapter must be approved and must be installed so that it will record the following:

(1) Voice communications transmitted from or received in the airplane by radio.

(2) Voice communications of flightcrew members on the flight deck.

(3) Voice communications of flightcrew members on the flight deck, using the airplane's interphone system.

(4) Voice or audio signals identifying navigation or approach aids introduced into a headset or speaker.

(5) Voice communications of flightcrew members using the passenger loudspeaker system, if there is such a system and if the fourth channel is available in accordance with the requirements of paragraph (c)(4)(ii) of this section.

(6) If datalink communication equipment is installed, all datalink communications, using an approved data message set. Datalink messages must be recorded as the output signal from the communications unit that translates the signal into usable data.

(b) The recording requirements of paragraph (a)(2) of this section must be met by installing a cockpit-mounted area microphone, located in the best position for recording voice communications originating at the first and second pilot stations and voice communications of other crewmembers on the flight deck when directed to those stations. The microphone must be so located and, if necessary, the pre-amplifiers and filters of the recorder must be so adjusted or supplemented, so that the intelligibility of the recorded communications is as high as practicable when recorded under flight cockpit noise conditions and played back. Repeated aural or visual playback of the record may be used in evaluating intelligibility.

(c) Each cockpit voice recorder must be installed so that the part of the communication or audio signals specified in paragraph (a) of this section obtained from each of the following sources is recorded on a separate channel:

(1) For the first channel, from each boom, mask, or handheld microphone, headset, or speaker used at the first pilot station.

(2) For the second channel from each boom, mask, or handheld microphone, headset, or speaker used at the second pilot station.

(3) For the third channel—from the cockpit-mounted area microphone.

(4) For the fourth channel from:

(i) Each boom, mask, or handheld microphone, headset, or speaker used at the station for the third and fourth crewmembers.

(ii) If the stations specified in paragraph (c)(4)(i) of this section are not required or if the signal at such a station is picked up by another channel, each microphone on the flight deck that is used with the passenger loudspeaker system, if its signals are not picked up by another channel.

(5) And that as far as is practicable all sounds received by the microphone

176

listed in paragraphs (c)(1), (2), and (4) of this section must be recorded without interruption irrespective of the position of the interphone-transmitter key switch. The design shall ensure that sidetone for the flightcrew is produced only when the interphone, public address system, or radio transmitters are in use.

(d) Each cockpit voice recorder must be installed so that:

(1)(i) It receives its electrical power from the bus that provides the maximum reliability for operation of the cockpit voice recorder without jeopardizing service to essential or emergency loads.

(ii) It remains powered for as long as possible without jeopardizing emergency operation of the airplane.

(2) There is an automatic means to simultaneously stop the recorder and prevent each erasure feature from functioning, within 10 minutes after crash impact.

(3) There is an aural or visual means for preflight checking of the recorder for proper operation.

(4) Any single electrical failure external to the recorder does not disable both the cockpit voice recorder and the flight data recorder.

(5) It has an independent power source—

(i) That provides 10 ±1 minutes of electrical power to operate both the cockpit voice recorder and cockpit-mounted area microphone;

(ii) That is located as close as practicable to the cockpit voice recorder; and

(iii) To which the cockpit voice recorder and cockpit-mounted area microphone are switched automatically in the event that all other power to the cockpit voice recorder is interrupted either by normal shutdown or by any other loss of power to the electrical power bus.

(6) It is in a separate container from the flight data recorder when both are required. If used to comply with only the cockpit voice recorder requirements, a combination unit may be installed.

(e) The recorder container must be located and mounted to minimize the probability of rupture of the container as a result of crash impact and con-

sequent heat damage to the recorder from fire.

(1) Except as provided in paragraph (e)(2) of this section, the recorder container must be located as far aft as practicable, but need not be outside of the pressurized compartment, and may not be located where aft-mounted engines may crush the container during impact.

(2) If two separate combination digital flight data recorder and cockpit voice recorder units are installed instead of one cockpit voice recorder and one digital flight data recorder, the combination unit that is installed to comply with the cockpit voice recorder requirements may be located near the cockpit.

(f) If the cockpit voice recorder has a bulk erasure device, the installation must be designed to minimize the probability of inadvertent operation and actuation of the device during crash impact.

(g) Each recorder container must—

(1) Be either bright orange or bright yellow;

(2) Have reflective tape affixed to its external surface to facilitate its location under water; and

(3) Have an underwater locating device, when required by the operating rules of this chapter, on or adjacent to the container, which is secured in such manner that they are not likely to be separated during crash impact.

§ 23.1459 **Flight data recorders.**

(a) Each flight recorder required by the operating rules of this chapter must be installed so that—

(1) It is supplied with airspeed, altitude, and directional data obtained from sources that meet the aircraft level system requirements and the functionality specified in § 23.2500;

(2) The vertical acceleration sensor is rigidly attached, and located longitudinally either within the approved center of gravity limits of the airplane, or at a distance forward or aft of these limits that does not exceed 25 percent of the airplane's mean aerodynamic chord;

(3)(i) It receives its electrical power from the bus that provides the maximum reliability for operation of the flight data recorder without jeopardizing service to essential or emergency loads;

(ii) It remains powered for as long as possible without jeopardizing emergency operation of the airplane;

(4) There is an aural or visual means for preflight checking of the recorder for proper recording of data in the storage medium;

(5) Except for recorders powered solely by the engine-driven electrical generator system, there is an automatic means to simultaneously stop a recorder that has a data erasure feature and prevent each erasure feature from functioning, within 10 minutes after crash impact;

(6) Any single electrical failure external to the recorder does not disable both the cockpit voice recorder and the flight data recorder; and

(7) It is in a separate container from the cockpit voice recorder when both are required. If used to comply with only the flight data recorder requirements, a combination unit may be installed. If a combination unit is installed as a cockpit voice recorder to comply with § 23.1457(e)(2), a combination unit must be used to comply with this flight data recorder requirement.

(b) Each non-ejectable record container must be located and mounted so as to minimize the probability of container rupture resulting from crash impact and subsequent damage to the record from fire. In meeting this requirement, the record container must be located as far aft as practicable, but need not be aft of the pressurized compartment, and may not be where aft-mounted engines may crush the container upon impact.

(c) A correlation must be established between the flight recorder readings of airspeed, altitude, and heading and the corresponding readings (taking into account correction factors) of the first pilot's instruments. The correlation must cover the airspeed range over which the airplane is to be operated, the range of altitude to which the airplane is limited, and 360 degrees of heading. Correlation may be established on the ground as appropriate.

(d) Each recorder container must—

(1) Be either bright orange or bright yellow;

(2) Have reflective tape affixed to its external surface to facilitate its location under water; and

(3) Have an underwater locating device, when required by the operating rules of this chapter, on or adjacent to the container, which is secured in such a manner that they are not likely to be separated during crash impact.

(e) Any novel or unique design or operational characteristics of the aircraft shall be evaluated to determine if any dedicated parameters must be recorded on flight recorders in addition to or in place of existing requirements.

§ 23.1529 Instructions for continued airworthiness.

The applicant must prepare Instructions for Continued Airworthiness, in accordance with appendix A of this part, that are acceptable to the Administrator. The instructions may be incomplete at type certification if a program exists to ensure their completion prior to delivery of the first airplane or issuance of a standard certificate of airworthiness, whichever occurs later.

Subpart A—General

§ 23.2000 Applicability and definitions.

(a) This part prescribes airworthiness standards for the issuance of type certificates, and changes to those certificates, for airplanes in the normal category.

(b) For the purposes of this part, the following definition applies:

Continued safe flight and landing means an airplane is capable of continued controlled flight and landing, possibly using emergency procedures, without requiring exceptional pilot skill or strength. Upon landing, some airplane damage may occur as a result of a failure condition.

§ 23.2005 Certification of normal category airplanes.

(a) Certification in the normal category applies to airplanes with a passenger-seating configuration of 19 or less and a maximum certificated takeoff weight of 19,000 pounds or less.

(b) Airplane certification levels are:

(1) Level 1—for airplanes with a maximum seating configuration of 0 to 1 passengers.

(2) Level 2—for airplanes with a maximum seating configuration of 2 to 6 passengers.

(3) Level 3—for airplanes with a maximum seating configuration of 7 to 9 passengers.

(4) Level 4—for airplanes with a maximum seating configuration of 10 to 19 passengers.

(c) Airplane performance levels are:

(1) Low speed—for airplanes with a V_{NO} and $V_{MO} \leq 250$ Knots Calibrated Airspeed (KCAS) and a $M_{MO} \leq 0.6$.

(2) High speed—for airplanes with a V_{NO} or $V_{MO} > 250$ KCAS or a $M_{MO} > 0.6$.

(d) Airplanes not certified for aerobatics may be used to perform any maneuver incident to normal flying, including—

(1) Stalls (except whip stalls); and

(2) Lazy eights, chandelles, and steep turns, in which the angle of bank is not more than 60 degrees.

(e) Airplanes certified for aerobatics may be used to perform maneuvers without limitations, other than those limitations established under subpart G of this part.

§23.2010 Accepted means of compliance.

(a) An applicant must comply with this part using a means of compliance, which may include consensus standards, accepted by the Administrator.

(b) An applicant requesting acceptance of a means of compliance must provide the means of compliance to the FAA in a form and manner acceptable to the Administrator.

Subpart B—Flight

PERFORMANCE

§23.2100 Weight and center of gravity.

(a) The applicant must determine limits for weights and centers of gravity that provide for the safe operation of the airplane.

(b) The applicant must comply with each requirement of this subpart at critical combinations of weight and center of gravity within the airplane's range of loading conditions using tolerances acceptable to the Administrator.

(c) The condition of the airplane at the time of determining its empty weight and center of gravity must be well defined and easily repeatable.

§23.2105 Performance data.

(a) Unless otherwise prescribed, an airplane must meet the performance requirements of this subpart in—

(1) Still air and standard atmospheric conditions at sea level for all airplanes; and

(2) Ambient atmospheric conditions within the operating envelope for levels 1 and 2 high-speed and levels 3 and 4 airplanes.

(b) Unless otherwise prescribed, the applicant must develop the performance data required by this subpart for the following conditions:

(1) Airport altitudes from sea level to 10,000 feet (3,048 meters); and

(2) Temperatures above and below standard day temperature that are within the range of operating limitations, if those temperatures could have a negative effect on performance.

(c) The procedures used for determining takeoff and landing distances must be executable consistently by pilots of average skill in atmospheric conditions expected to be encountered in service.

(d) Performance data determined in accordance with paragraph (b) of this section must account for losses due to atmospheric conditions, cooling needs, and other demands on power sources.

§23.2110 Stall speed.

The applicant must determine the airplane stall speed or the minimum steady flight speed for each flight configuration used in normal operations, including takeoff, climb, cruise, descent, approach, and landing. The stall speed or minimum steady flight speed determination must account for the most adverse conditions for each flight configuration with power set at—

(a) Idle or zero thrust for propulsion systems that are used primarily for thrust; and

(b) A nominal thrust for propulsion systems that are used for thrust, flight control, and/or high-lift systems.

§23.2115 Takeoff performance.

(a) The applicant must determine airplane takeoff performance accounting for—

(1) Stall speed safety margins;

(2) Minimum control speeds; and

(3) Climb gradients.

(b) For single engine airplanes and levels 1, 2, and 3 low-speed multiengine airplanes, takeoff performance includes the determination of ground roll and initial climb distance to 50 feet (15 meters) above the takeoff surface.

(c) For levels 1, 2, and 3 high-speed multiengine airplanes, and level 4 multiengine airplanes, takeoff performance includes a determination of the following distances after a sudden critical loss of thrust—

(1) An aborted takeoff at critical speed;

(2) Ground roll and initial climb to 35 feet (11 meters) above the takeoff surface; and

(3) Net takeoff flight path.

[Doc. No. FAA–2015–1621, Amdt. 23–64, 81 FR 96689, Dec. 30, 2016, as amended by Doc. No. FAA–2022–1355, Amdt. 23–65, 87 FR 75710, Dec. 9, 2022]

§ 23.2120 Climb requirements.

The design must comply with the following minimum climb performance out of ground effect:

(a) With all engines operating and in the initial climb configuration(s)—

(1) For levels 1 and 2 low-speed airplanes, a climb gradient of 8.3 percent for landplanes and 6.7 percent for seaplanes and amphibians; and

(2) For levels 1 and 2 high-speed airplanes, all level 3 airplanes, and level 4 single-engines a climb gradient after takeoff of 4 percent.

(b) After a critical loss of thrust on multiengine airplanes—

(1) For levels 1 and 2 low-speed airplanes that do not meet single-engine crashworthiness requirements, a climb gradient of 1.5 percent at a pressure altitude of 5,000 feet (1,524 meters) in the cruise configuration(s);

(2) For levels 1 and 2 high-speed airplanes, and level 3 low-speed airplanes, a 1 percent climb gradient at 400 feet (122 meters) above the takeoff surface with the landing gear retracted and flaps in the takeoff configuration(s); and

(3) For level 3 high-speed airplanes and all level 4 airplanes, a 2 percent climb gradient at 400 feet (122 meters) above the takeoff surface with the

landing gear retracted and flaps in the approach configuration(s).

(c) For a balked landing, a climb gradient of 3 percent without creating undue pilot workload with the landing gear extended and flaps in the landing configuration(s).

[Doc. No. FAA–2015–1621, Amdt. 23–64, 81 FR 96689, Dec. 30, 2016, as amended by Doc. No. FAA–2022–1355, Amdt. 23–65, 87 FR 75710, Dec. 9, 2022]

§ 23.2125 Climb information.

(a) The applicant must determine climb performance at each weight, altitude, and ambient temperature within the operating limitations—

(1) For all single-engine airplanes;

(2) For levels 1 and 2 high-speed multiengine airplanes and level 3 multiengine airplanes, following a critical loss of thrust on takeoff in the initial climb configuration; and

(3) For all multiengine airplanes, during the enroute phase of flight with all engines operating and after a critical loss of thrust in the cruise configuration.

(b) The applicant must determine the glide performance for single-engine airplanes after a complete loss of thrust.

§ 23.2130 Landing.

The applicant must determine the following, for standard temperatures at critical combinations of weight and altitude within the operational limits:

(a) The distance, starting from a height of 50 feet (15 meters) above the landing surface, required to land and come to a stop.

(b) The approach and landing speeds, configurations, and procedures, which allow a pilot of average skill to land within the published landing distance consistently and without causing damage or injury, and which allow for a safe transition to the balked landing conditions of this part accounting for:

(1) Stall speed safety margin; and

(2) Minimum control speeds.

FLIGHT CHARACTERISTICS

§ 23.2135 Controllability.

(a) The airplane must be controllable and maneuverable, without requiring exceptional piloting skill, alertness, or strength, within the operating envelope—

(1) At all loading conditions for which certification is requested;

(2) During all phases of flight;

(3) With likely reversible flight control or propulsion system failure; and

(4) During configuration changes.

(b) The airplane must be able to complete a landing without causing substantial damage or serious injury using the steepest approved approach gradient procedures and providing a reasonable margin below V_{ref} or above approach angle of attack.

(c) V_{MC} is the calibrated airspeed at which, following the sudden critical loss of thrust, it is possible to maintain control of the airplane. For multiengine airplanes, the applicant must determine V_{MC}, if applicable, for the most critical configurations used in takeoff and landing operations.

(d) If the applicant requests certification of an airplane for aerobatics, the applicant must demonstrate those aerobatic maneuvers for which certification is requested and determine entry speeds.

§23.2140 Trim.

(a) The airplane must maintain lateral and directional trim without further force upon, or movement of, the primary flight controls or corresponding trim controls by the pilot, or the flight control system, under the following conditions:

(1) For levels 1, 2, and 3 airplanes in cruise.

(2) For level 4 airplanes in normal operations.

(b) The airplane must maintain longitudinal trim without further force upon, or movement of, the primary flight controls or corresponding trim controls by the pilot, or the flight control system, under the following conditions:

(1) Climb.

(2) Level flight.

(3) Descent.

(4) Approach.

(c) Residual control forces must not fatigue or distract the pilot during normal operations of the airplane and likely abnormal or emergency operations, including a critical loss of thrust on multiengine airplanes.

§23.2145 Stability.

(a) Airplanes not certified for aerobatics must—

(1) Have static longitudinal, lateral, and directional stability in normal operations;

(2) Have dynamic short period and Dutch roll stability in normal operations; and

(3) Provide stable control force feedback throughout the operating envelope.

(b) No airplane may exhibit any divergent longitudinal stability characteristic so unstable as to increase the pilot's workload or otherwise endanger the airplane and its occupants.

§23.2150 Stall characteristics, stall warning, and spins.

(a) The airplane must have controllable stall characteristics in straight flight, turning flight, and accelerated turning flight with a clear and distinctive stall warning that provides sufficient margin to prevent inadvertent stalling.

(b) Single-engine airplanes, not certified for aerobatics, must not have a tendency to inadvertently depart controlled flight.

(c) Levels 1 and 2 multiengine airplanes, not certified for aerobatics, must not have a tendency to inadvertently depart controlled flight from thrust asymmetry after a critical loss of thrust.

(d) Airplanes certified for aerobatics that include spins must have controllable stall characteristics and the ability to recover within one and one-half additional turns after initiation of the first control action from any point in a spin, not exceeding six turns or any greater number of turns for which certification is requested, while remaining within the operating limitations of the airplane.

(e) Spin characteristics in airplanes certified for aerobatics that includes spins must recover without exceeding limitations and may not result in unrecoverable spins—

(1) With any typical use of the flight or engine power controls; or

(2) Due to pilot disorientation or incapacitation.

§ 23.2155 Ground and water handling characteristics.

For airplanes intended for operation on land or water, the airplane must have controllable longitudinal and directional handling characteristics during taxi, takeoff, and landing operations.

§ 23.2160 Vibration, buffeting, and high-speed characteristics.

(a) Vibration and buffeting, for operations up to V_D/M_D, must not interfere with the control of the airplane or cause excessive fatigue to the flightcrew. Stall warning buffet within these limits is allowable.

(b) For high-speed airplanes and all airplanes with a maximum operating altitude greater than 25,000 feet (7,620 meters) pressure altitude, there must be no perceptible buffeting in cruise configuration at 1g and at any speed up to V_{MO}/M_{MO}, except stall buffeting.

(c) For high-speed airplanes, the applicant must determine the positive maneuvering load factors at which the onset of perceptible buffet occurs in the cruise configuration within the operational envelope. Likely inadvertent excursions beyond this boundary must not result in structural damage.

(d) High-speed airplanes must have recovery characteristics that do not result in structural damage or loss of control, beginning at any likely speed up to V_{MO}/M_{MO}, following—

(1) An inadvertent speed increase; and

(2) A high-speed trim upset for airplanes where dynamic pressure can impair the longitudinal trim system operation.

§ 23.2165 Performance and flight characteristics requirements for flight in icing conditions.

(a) An applicant who requests certification for flight in icing conditions defined in part 1 of appendix C to part 25 of this chapter, or an applicant who requests certification for flight in these icing conditions and any additional atmospheric icing conditions, must show the following in the icing conditions for which certification is requested under normal operation of the ice protection system(s):

(1) Compliance with each requirement of this subpart, except those applicable to spins and any that must be demonstrated at speeds in excess of—

(i) 250 KCAS;

(ii) V_{MO}/M_{MO} or V_{NE}; or

(iii) A speed at which the applicant demonstrates the airframe will be free of ice accretion.

(2) The means by which stall warning is provided to the pilot for flight in icing conditions and non-icing conditions is the same.

(b) If an applicant requests certification for flight in icing conditions, the applicant must provide a means to detect any icing conditions for which certification is not requested and show the airplane's ability to avoid or exit those conditions.

(c) The applicant must develop an operating limitation to prohibit intentional flight, including takeoff and landing, into icing conditions for which the airplane is not certified to operate.

[Doc. No. FAA–2015–1621, Amdt. 23–64, 81 FR 96689, Dec. 30, 2016, as amended by Doc. No. FAA–2022–1355, Amdt. 23–65, 87 FR 75710, Dec. 9, 2022]

Subpart C—Structures

§ 23.2200 Structural design envelope.

The applicant must determine the structural design envelope, which describes the range and limits of airplane design and operational parameters for which the applicant will show compliance with the requirements of this subpart. The applicant must account for all airplane design and operational parameters that affect structural loads, strength, durability, and aeroelasticity, including:

(a) Structural design airspeeds, landing descent speeds, and any other airspeed limitation at which the applicant must show compliance to the requirements of this subpart. The structural design airspeeds must—

(1) Be sufficiently greater than the stalling speed of the airplane to safeguard against loss of control in turbulent air; and

(2) Provide sufficient margin for the establishment of practical operational limiting airspeeds.

(b) Design maneuvering load factors not less than those, which service history shows, may occur within the structural design envelope.

(c) Inertial properties including weight, center of gravity, and mass moments of inertia, accounting for—

(1) Each critical weight from the airplane empty weight to the maximum weight; and

(2) The weight and distribution of occupants, payload, and fuel.

(d) Characteristics of airplane control systems, including range of motion and tolerances for control surfaces, high-lift devices, or other moveable surfaces.

(e) Each critical altitude up to the maximum altitude.

[Doc. No. FAA–2015–1621, Amdt. 23–64, 81 FR 96689, Dec. 30, 2016, as amended by Doc. No. FAA–2022–1355, Amdt. 23–65, 87 FR 75710, Dec. 9, 2022]

§ 23.2205 Interaction of systems and structures.

For airplanes equipped with systems that modify structural performance, alleviate the impact of this subpart's requirements, or provide a means of compliance with this subpart, the applicant must account for the influence and failure of these systems when showing compliance with the requirements of this subpart.

STRUCTURAL LOADS

§ 23.2210 Structural design loads.

(a) The applicant must:

(1) Determine the applicable structural design loads resulting from likely externally or internally applied pressures, forces, or moments that may occur in flight, ground and water operations, ground and water handling, and while the airplane is parked or moored.

(2) Determine the loads required by paragraph (a)(1) of this section at all critical combinations of parameters, on and within the boundaries of the structural design envelope.

(b) The magnitude and distribution of the applicable structural design loads required by this section must be based on physical principles.

§ 23.2215 Flight load conditions.

The applicant must determine the structural design loads resulting from the following flight conditions:

(a) Atmospheric gusts where the magnitude and gradient of these gusts are based on measured gust statistics.

(b) Symmetric and asymmetric maneuvers.

(c) Asymmetric thrust resulting from the failure of a powerplant unit.

§ 23.2220 Ground and water load conditions.

The applicant must determine the structural design loads resulting from taxi, takeoff, landing, and handling conditions on the applicable surface in normal and adverse attitudes and configurations.

§ 23.2225 Component loading conditions.

The applicant must determine the structural design loads acting on:

(a) Each engine mount and its supporting structure such that both are designed to withstand loads resulting from—

(1) Powerplant operation combined with flight gust and maneuver loads; and

(2) For non-reciprocating powerplants, sudden powerplant stoppage.

(b) Each flight control and high-lift surface, their associated system and supporting structure resulting from—

(1) The inertia of each surface and mass balance attachment;

(2) Flight gusts and maneuvers;

(3) Pilot or automated system inputs;

(4) System induced conditions, including jamming and friction; and

(5) Taxi, takeoff, and landing operations on the applicable surface, including downwind taxi and gusts occurring on the applicable surface.

(c) A pressurized cabin resulting from the pressurization differential—

(1) From zero up to the maximum relief pressure combined with gust and maneuver loads;

(2) From zero up to the maximum relief pressure combined with ground and water loads if the airplane may land with the cabin pressurized; and

(3) At the maximum relief pressure multiplied by 1.33, omitting all other loads.

§ 23.2230 Limit and ultimate loads.

The applicant must determine—

(a) The limit loads, which are equal to the structural design loads unless otherwise specified elsewhere in this part; and

(b) The ultimate loads, which are equal to the limit loads multiplied by a 1.5 factor of safety unless otherwise specified elsewhere in this part.

STRUCTURAL PERFORMANCE

§ 23.2235 Structural strength.

The structure must support:

(a) Limit loads without—

(1) Interference with the safe operation of the airplane; and

(2) Detrimental permanent deformation.

(b) Ultimate loads.

§ 23.2240 Structural durability.

(a) The applicant must develop and implement inspections or other procedures to prevent structural failures due to foreseeable causes of strength degradation, which could result in serious or fatal injuries, or extended periods of operation with reduced safety margins. Each of the inspections or other procedures developed under this section must be included in the Airworthiness Limitations Section of the Instructions for Continued Airworthiness required by § 23.1529.

(b) For Level 4 airplanes, the procedures developed for compliance with paragraph (a) of this section must be capable of detecting structural damage before the damage could result in structural failure.

(c) For pressurized airplanes:

(1) The airplane must be capable of continued safe flight and landing following a sudden release of cabin pressure, including sudden releases caused by door and window failures.

(2) For airplanes with maximum operating altitude greater than 41,000 feet, the procedures developed for compliance with paragraph (a) of this section must be capable of detecting damage to the pressurized cabin structure before the damage could result in rapid decompression that would result in serious or fatal injuries.

(d) The airplane must be designed to minimize hazards to the airplane due to structural damage caused by high-energy fragments from an uncontained engine or rotating machinery failure.

§ 23.2245 Aeroelasticity.

(a) The airplane must be free from flutter, control reversal, and divergence—

(1) At all speeds within and sufficiently beyond the structural design envelope;

(2) For any configuration and condition of operation;

(3) Accounting for critical degrees of freedom; and

(4) Accounting for any critical failures or malfunctions.

(b) The applicant must establish tolerances for all quantities that affect flutter.

DESIGN

§ 23.2250 Design and construction principles.

(a) The applicant must design each part, article, and assembly for the expected operating conditions of the airplane.

(b) Design data must adequately define the part, article, or assembly configuration, its design features, and any materials and processes used.

(c) The applicant must determine the suitability of each design detail and part having an important bearing on safety in operations.

(d) The control system must be free from jamming, excessive friction, and excessive deflection when the airplane is subjected to expected limit airloads.

(e) Doors, canopies, and exits must be protected against inadvertent opening in flight, unless shown to create no hazard when opened in flight.

§ 23.2255 Protection of structure.

(a) The applicant must protect each part of the airplane, including small parts such as fasteners, against deterioration or loss of strength due to any cause likely to occur in the expected operational environment.

(b) Each part of the airplane must have adequate provisions for ventilation and drainage.

(c) For each part that requires maintenance, preventive maintenance, or servicing, the applicant must incorporate a means into the airplane design to allow such actions to be accomplished.

[Doc. No. FAA–2015–1621, Amdt. 23–64, 81 FR 96689, Dec. 30, 2016, as amended by Doc. No. FAA–2022–1355, Amdt. 23–65, 87 FR 75710, Dec. 9, 2022]

§23.2260 Materials and processes.

(a) The applicant must determine the suitability and durability of materials used for parts, articles, and assemblies, accounting for the effects of likely environmental conditions expected in service, the failure of which could prevent continued safe flight and landing.

(b) The methods and processes of fabrication and assembly used must produce consistently sound structures. If a fabrication process requires close control to reach this objective, the applicant must perform the process under an approved process specification.

(c) Except as provided in paragraphs (f) and (g) of this section, the applicant must select design values that ensure material strength with probabilities that account for the criticality of the structural element. Design values must account for the probability of structural failure due to material variability.

(d) If material strength properties are required, a determination of those properties must be based on sufficient tests of material meeting specifications to establish design values on a statistical basis.

(e) If thermal effects are significant on a critical component or structure under normal operating conditions, the applicant must determine those effects on allowable stresses used for design.

(f) Design values, greater than the minimums specified by this section, may be used, where only guaranteed minimum values are normally allowed, if a specimen of each individual item is tested before use to determine that the actual strength properties of that particular item will equal or exceed those used in the design.

(g) An applicant may use other material design values if approved by the Administrator.

§23.2265 Special factors of safety.

(a) The applicant must determine a special factor of safety for each critical design value for each part, article, or assembly for which that critical design value is uncertain, and for each part, article, or assembly that is—

(1) Likely to deteriorate in service before normal replacement; or

(2) Subject to appreciable variability because of uncertainties in manufacturing processes or inspection methods.

(b) The applicant must determine a special factor of safety using quality controls and specifications that account for each—

(1) Type of application;

(2) Inspection method;

(3) Structural test requirement;

(4) Sampling percentage; and

(5) Process and material control.

(c) The applicant must multiply the highest pertinent special factor of safety in the design for each part of the structure by each limit and ultimate load, or ultimate load only, if there is no corresponding limit load, such as occurs with emergency condition loading.

STRUCTURAL OCCUPANT PROTECTION

§23.2270 Emergency conditions.

(a) The airplane, even when damaged in an emergency landing, must protect each occupant against injury that would preclude egress when—

(1) Properly using safety equipment and features provided for in the design;

(2) The occupant experiences ultimate static inertia loads likely to occur in an emergency landing; and

(3) Items of mass, including engines or auxiliary power units (APUs), within or aft of the cabin, that could injure an occupant, experience ultimate static inertia loads likely to occur in an emergency landing.

(b) The emergency landing conditions specified in paragraph (a)(1) and (a)(2) of this section, must—

(1) Include dynamic conditions that are likely to occur in an emergency landing; and

(2) Not generate loads experienced by the occupants, which exceed established human injury criteria for human tolerance due to restraint or contact with objects in the airplane.

(c) The airplane must provide protection for all occupants, accounting for likely flight, ground, and emergency landing conditions.

(d) Each occupant protection system must perform its intended function and not create a hazard that could cause a secondary injury to an occupant. The occupant protection system must not prevent occupant egress or interfere with the operation of the airplane when not in use.

(e) Each baggage and cargo compartment must—

(1) Be designed for its maximum weight of contents and for the critical load distributions at the maximum load factors corresponding to the flight and ground load conditions determined under this part;

(2) Have a means to prevent the contents of the compartment from becoming a hazard by impacting occupants or shifting; and

(3) Protect any controls, wiring, lines, equipment, or accessories whose damage or failure would affect safe operations.

Subpart D—Design and Construction

§ 23.2300 Flight control systems.

(a) The applicant must design airplane flight control systems to:

(1) Operate easily, smoothly, and positively enough to allow proper performance of their functions.

(2) Protect against likely hazards.

(b) The applicant must design trim systems, if installed, to:

(1) Protect against inadvertent, incorrect, or abrupt trim operation.

(2) Provide a means to indicate—

(i) The direction of trim control movement relative to airplane motion;

(ii) The trim position with respect to the trim range;

(iii) The neutral position for lateral and directional trim; and

(iv) The range for takeoff for all applicant requested center of gravity ranges and configurations.

§ 23.2305 Landing gear systems.

(a) The landing gear must be designed to—

(1) Provide stable support and control to the airplane during surface operation; and

(2) Account for likely system failures and likely operation environments (including anticipated limitation exceedances and emergency procedures).

(b) All airplanes must have a reliable means of stopping the airplane with sufficient kinetic energy absorption to account for landing. Airplanes that are required to demonstrate aborted takeoff capability must account for this additional kinetic energy.

(c) For airplanes that have a system that actuates the landing gear, there is—

(1) A positive means to keep the landing gear in the landing position; and

(2) An alternative means available to bring the landing gear in the landing position when a non-deployed system position would be a hazard.

§ 23.2310 Buoyancy for seaplanes and amphibians.

Airplanes intended for operations on water, must—

(a) Provide buoyancy of 80 percent in excess of the buoyancy required to support the maximum weight of the airplane in fresh water; and

(b) Have sufficient margin so the airplane will stay afloat at rest in calm water without capsizing in case of a likely float or hull flooding.

OCCUPANT SYSTEM DESIGN PROTECTION

§ 23.2315 Means of egress and emergency exits.

(a) With the cabin configured for takeoff or landing, the airplane is designed to:

(1) Facilitate rapid and safe evacuation of the airplane in conditions likely to occur following an emergency landing, excluding ditching for level 1, level 2, and single-engine level 3 airplanes.

(2) Have means of egress (openings, exits, or emergency exits), that can be readily located and opened from the inside and outside. The means of opening must be simple and obvious and marked inside and outside the airplane.

(3) Have easy access to emergency exits when present.

(b) Airplanes approved for aerobatics must have a means to egress the airplane in flight.

[Doc. No. FAA–2015–1621, Amdt. 23–64, 81 FR 96689, Dec. 30, 2016, as amended by Doc. No. FAA–2022–1355, Amdt. 23–65, 87 FR 75710, Dec. 9, 2022]

§ 23.2320 Occupant physical environment.

(a) The applicant must design the airplane to—

(1) Allow clear communication between the flightcrew and passengers;

(2) Protect the pilot and flight controls from propellers; and

(3) Protect the occupants from serious injury due to damage to windshields, windows, and canopies.

(b) For level 4 airplanes, each windshield and its supporting structure directly in front of the pilot must withstand, without penetration, the impact equivalent to a two-pound bird when the velocity of the airplane is equal to the airplane's maximum approach flap speed.

(c) The airplane must provide each occupant with air at a breathable pressure, free of hazardous concentrations of gases, vapors, and smoke during normal operations and likely failures.

(d) If a pressurization system is installed in the airplane, it must be designed to protect against—

(1) Decompression to an unsafe level; and

(2) Excessive differential pressure.

(e) If an oxygen system is installed in the airplane, it must—

(1) Effectively provide oxygen to each user to prevent the effects of hypoxia; and

(2) Be free from hazards in itself, in its method of operation, and its effect upon other components.

Fire and High Energy Protection

§ 23.2325 Fire protection.

(a) The following materials must be self-extinguishing—

(1) Insulation on electrical wire and electrical cable;

(2) For levels 1, 2, and 3 airplanes, materials in the baggage and cargo compartments inaccessible in flight; and

(3) For level 4 airplanes, materials in the cockpit, cabin, baggage, and cargo compartments.

(b) The following materials must be flame resistant—

(1) For levels 1, 2 and 3 airplanes, materials in each compartment accessible in flight; and

(2) Any equipment associated with any electrical cable installation and that would overheat in the event of circuit overload or fault.

(c) Thermal/acoustic materials in the fuselage, if installed, must not be a flame propagation hazard.

(d) Sources of heat within each baggage and cargo compartment that are capable of igniting adjacent objects must be shielded and insulated to prevent such ignition.

(e) For level 4 airplanes, each baggage and cargo compartment must—

(1) Be located where a fire would be visible to the pilots, or equipped with a fire detection system and warning system; and

(2) Be accessible for the manual extinguishing of a fire, have a built-in fire extinguishing system, or be constructed and sealed to contain any fire within the compartment.

(f) There must be a means to extinguish any fire in the cabin such that—

(1) The pilot, while seated, can easily access the fire extinguishing means; and

(2) For levels 3 and 4 airplanes, passengers have a fire extinguishing means available within the passenger compartment.

(g) Each area where flammable fluids or vapors might escape by leakage of a fluid system must—

(1) Be defined; and

(2) Have a means to minimize the probability of fluid and vapor ignition, and the resultant hazard, if ignition occurs.

(h) Combustion heater installations must be protected from uncontained fire.

§ 23.2330 Fire protection in designated fire zones and adjacent areas.

(a) Flight controls, engine mounts, and other flight structures within or adjacent to designated fire zones must be capable of withstanding the effects of a fire.

(b) Engines in a designated fire zone must remain attached to the airplane in the event of a fire.

(c) In designated fire zones, terminals, equipment, and electrical cables used during emergency procedures must be fire-resistant.

§ 23.2335　Lightning protection.

The airplane must be protected against catastrophic effects from lightning.

Subpart E—Powerplant

§ 23.2400　Powerplant installation.

(a) For the purpose of this subpart, the airplane powerplant installation must include each component necessary for propulsion, which affects propulsion safety, or provides auxiliary power to the airplane.

(b) Each airplane engine and propeller must be type certificated, except for engines and propellers installed on level 1 low-speed airplanes, which may be approved under the airplane type certificate in accordance with a standard accepted by the Administrator that contains airworthiness criteria the Administrator has found appropriate and applicable to the specific design and intended use of the engine or propeller and provides a level of safety acceptable to the Administrator.

(c) The applicant must construct and arrange each powerplant installation to account for—

(1) Likely operating conditions, including foreign object threats;

(2) Sufficient clearance of moving parts to other airplane parts and their surroundings;

(3) Likely hazards in operation including hazards to ground personnel; and

(4) Vibration and fatigue.

(d) Hazardous accumulations of fluids, vapors, or gases must be isolated from the airplane and personnel compartments, and be safely contained or discharged.

(e) Powerplant components must comply with their component limitations and installation instructions or be shown not to create a hazard.

[Doc. No. FAA–2015–1621, Amdt. 23–64, 81 FR 96689, Dec. 30, 2016, as amended by Doc. No. FAA–2022–1355, Amdt. 23–65, 87 FR 75710, Dec. 9, 2022]

§ 23.2405　Automatic power or thrust control systems.

(a) An automatic power or thrust control system intended for in-flight use must be designed so no unsafe condition will result during normal operation of the system.

(b) Any single failure or likely combination of failures of an automatic power or thrust control system must not prevent continued safe flight and landing of the airplane.

(c) Inadvertent operation of an automatic power or thrust control system by the flightcrew must be prevented, or if not prevented, must not result in an unsafe condition.

(d) Unless the failure of an automatic power or thrust control system is extremely remote, the system must—

(1) Provide a means for the flightcrew to verify the system is in an operating condition;

(2) Provide a means for the flightcrew to override the automatic function; and

(3) Prevent inadvertent deactivation of the system.

§ 23.2410　Powerplant installation hazard assessment.

The applicant must assess each powerplant separately and in relation to other airplane systems and installations to show that any hazard resulting from the likely failure of any powerplant system, component, or accessory will not—

(a) Prevent continued safe flight and landing or, if continued safe flight and landing cannot be ensured, the hazard has been minimized;

(b) Cause serious injury that may be avoided; and

(c) Require immediate action by any crewmember for continued operation of any remaining powerplant system.

§ 23.2415　Powerplant ice protection.

(a) The airplane design, including the induction and inlet system, must prevent foreseeable accumulation of ice or snow that adversely affects powerplant operation.

(b) The powerplant installation design must prevent any accumulation of ice or snow that adversely affects powerplant operation, in those icing conditions for which certification is requested.

§ 23.2420 Reversing systems.

Each reversing system must be designed so that—

(a) No unsafe condition will result during normal operation of the system; and

(b) The airplane is capable of continued safe flight and landing after any single failure, likely combination of failures, or malfunction of the reversing system.

§ 23.2425 Powerplant operational characteristics.

(a) The installed powerplant must operate without any hazardous characteristics during normal and emergency operation within the range of operating limitations for the airplane and the engine.

(b) The pilot must have the capability to stop the powerplant in flight and restart the powerplant within an established operational envelope.

§ 23.2430 Fuel systems.

(a) Each fuel system must—

(1) Be designed and arranged to provide independence between multiple fuel storage and supply systems so that failure of any one component in one system will not result in loss of fuel storage or supply of another system;

(2) Be designed and arranged to prevent ignition of the fuel within the system by direct lightning strikes or swept lightning strokes to areas where such occurrences are highly probable, or by corona or streamering at fuel vent outlets;

(3) Provide the fuel necessary to ensure each powerplant and auxiliary power unit functions properly in all likely operating conditions;

(4) Provide the flightcrew with a means to determine the total useable fuel available and provide uninterrupted supply of that fuel when the system is correctly operated, accounting for likely fuel fluctuations;

(5) Provide a means to safely remove or isolate the fuel stored in the system from the airplane;

(6) Be designed to retain fuel under all likely operating conditions and minimize hazards to the occupants during any survivable emergency landing. For level 4 airplanes, failure due to overload of the landing system must be taken into account; and

(7) Prevent hazardous contamination of the fuel supplied to each powerplant and auxiliary power unit.

(b) Each fuel storage system must—

(1) Withstand the loads under likely operating conditions without failure;

(2) Be isolated from personnel compartments and protected from hazards due to unintended temperature influences;

(3) Be designed to prevent significant loss of stored fuel from any vent system due to fuel transfer between fuel storage or supply systems, or under likely operating conditions;

(4) Provide fuel for at least one-half hour of operation at maximum continuous power or thrust; and

(5) Be capable of jettisoning fuel safely if required for landing.

(c) Each fuel storage refilling or recharging system must be designed to—

(1) Prevent improper refilling or recharging;

(2) Prevent contamination of the fuel stored during likely operating conditions; and

(3) Prevent the occurrence of any hazard to the airplane or to persons during refilling or recharging.

§ 23.2435 Powerplant induction and exhaust systems.

(a) The air induction system for each powerplant or auxiliary power unit and their accessories must—

(1) Supply the air required by that powerplant or auxiliary power unit and its accessories under likely operating conditions;

(2) Be designed to prevent likely hazards in the event of fire or backfire;

(3) Minimize the ingestion of foreign matter; and

(4) Provide an alternate intake if blockage of the primary intake is likely.

(b) The exhaust system, including exhaust heat exchangers for each powerplant or auxiliary power unit, must—

(1) Provide a means to safely discharge potential harmful material; and

(2) Be designed to prevent likely hazards from heat, corrosion, or blockage.

§ 23.2440 Powerplant fire protection.

(a) A powerplant, auxiliary power unit, or combustion heater that includes a flammable fluid and an ignition source for that fluid must be installed in a designated fire zone.

(b) Each designated fire zone must provide a means to isolate and mitigate hazards to the airplane in the event of fire or overheat within the zone.

(c) Each component, line, fitting, and control subject to fire conditions must—

(1) Be designed and located to prevent hazards resulting from a fire, including any located adjacent to a designated fire zone that may be affected by fire within that zone;

(2) Be fire-resistant if carrying flammable fluid, gas or air, or is required to operate in the event of a fire; and

(3) Be fireproof or enclosed by a fire proof shield if storing concentrated flammable fluids.

(d) The applicant must provide a means to prevent hazardous quantities of flammable fluids from flowing into, within or through each designated fire zone. This means must—

(1) Not restrict flow or limit operation of any remaining powerplant or auxiliary power unit, or equipment necessary for safety;

(2) Prevent inadvertent operation; and

(3) Be located outside the fire zone unless an equal degree of safety is provided with a means inside the fire zone.

(e) A means to ensure the prompt detection of fire must be provided for each designated fire zone—

(1) On a multiengine airplane where detection will mitigate likely hazards to the airplane; or

(2) That contains a fire extinguisher.

(f) A means to extinguish fire within a fire zone, except a combustion heater fire zone, must be provided for—

(1) Any fire zone located outside the pilot's view;

(2) Any fire zone embedded within the fuselage, which must also include a redundant means to extinguish fire; and

(3) Any fire zone on a level 4 airplane.

[Doc. No. FAA–2015–1621, Amdt. 23–64, 81 FR 96689, Dec. 30, 2016, as amended by Doc. No. FAA–2022–1355, Amdt. 23–65, 87 FR 75710, Dec. 9, 2022]

Subpart F—Equipment

§ 23.2500 Airplane level systems requirements.

This section applies generally to installed equipment and systems unless a section of this part imposes requirements for a specific piece of equipment, system, or systems.

(a) The equipment and systems required for an airplane to operate safely in the kinds of operations for which certification is requested (Day VFR, Night VFR, IFR) must be designed and installed to—

(1) Meet the level of safety applicable to the certification and performance level of the airplane; and

(2) Perform their intended function throughout the operating and environmental limits for which the airplane is certificated.

(b) The systems and equipment not covered by paragraph (a) of this section—considered separately and in relation to other systems—must be designed and installed so their operation does not have an adverse effect on the airplane or its occupants.

[Doc. No. FAA–2015–1621, Amdt. 23–64, 81 FR 96689, Dec. 30, 2016, as amended by Doc. No. FAA–2022–1355, Amdt. 23–65, 87 FR 75710, Dec. 9, 2022]

§ 23.2505 Function and installation.

When installed, each item of equipment must function as intended.

§ 23.2510 Equipment, systems, and installations.

For any airplane system or equipment whose failure or abnormal operation has not been specifically addressed by another requirement in this part, the applicant must design and install each system and equipment, such that there is a logical and acceptable inverse relationship between the average probability and the severity of failure conditions to the extent that:

(a) Each catastrophic failure condition is extremely improbable;

(b) Each hazardous failure condition is extremely remote; and

(c) Each major failure condition is remote.

§ 23.2515 Electrical and electronic system lightning protection.

An airplane approved for IFR operations must meet the following requirements, unless an applicant shows that exposure to lightning is unlikely:

(a) Each electrical or electronic system that performs a function, the failure of which would prevent the continued safe flight and landing of the airplane, must be designed and installed such that—

(1) The function at the airplane level is not adversely affected during and after the time the airplane is exposed to lightning; and

(2) The system recovers normal operation of that function in a timely manner after the airplane is exposed to lightning unless the system's recovery conflicts with other operational or functional requirements of the system.

(b) Each electrical and electronic system that performs a function, the failure of which would significantly reduce the capability of the airplane or the ability of the flightcrew to respond to an adverse operating condition, must be designed and installed such that the system recovers normal operation of that function in a timely manner after the airplane is exposed to lightning.

§ 23.2520 High-intensity Radiated Fields (HIRF) protection.

(a) Each electrical and electronic system that performs a function, the failure of which would prevent the continued safe flight and landing of the airplane, must be designed and installed such that—

(1) The function at the airplane level is not adversely affected during and after the time the airplane is exposed to the HIRF environment; and

(2) The system recovers normal operation of that function in a timely manner after the airplane is exposed to the HIRF environment, unless the system's recovery conflicts with other operational or functional requirements of the system.

(b) For airplanes approved for IFR operations, each electrical and electronic system that performs a function, the failure of which would significantly reduce the capability of the airplane or the ability of the flightcrew to respond to an adverse operating condition, must be designed and installed such that the system recovers normal operation of that function in a timely manner after the airplane is exposed to the HIRF environment.

[Doc. No. FAA–2015–1621, Amdt. 23–64, 81 FR 96689, Dec. 30, 2016, as amended by Doc. No. FAA–2022–1355, Amdt. 23–65, 87 FR 75710, Dec. 9, 2022]

§ 23.2525 System power generation, storage, and distribution.

The power generation, storage, and distribution for any system must be designed and installed to—

(a) Supply the power required for operation of connected loads during all intended operating conditions;

(b) Ensure no single failure or malfunction of any one power supply, distribution system, or other utilization system will prevent the system from supplying the essential loads required for continued safe flight and landing; and

(c) Have enough capacity, if the primary source fails, to supply essential loads, including non-continuous essential loads for the time needed to complete the function required for continued safe flight and landing.

§ 23.2530 External and cockpit lighting.

(a) The applicant must design and install all lights to minimize any adverse effects on the performance of flightcrew duties.

(b) Any position and anti-collision lights, if required by part 91 of this chapter, must have the intensities, flash rate, colors, fields of coverage, and other characteristics to provide sufficient time for another aircraft to avoid a collision.

(c) Any position lights, if required by part 91 of this chapter, must include a red light on the left side of the airplane, a green light on the right side of the airplane, spaced laterally as far

191

apart as practicable, and a white light facing aft, located on an aft portion of the airplane or on the wing tips.

(d) Any taxi and landing lights must be designed and installed so they provide sufficient light for night operations.

(e) For seaplanes or amphibian airplanes, riding lights must provide a white light visible in clear atmospheric conditions.

§ 23.2535   Safety equipment.

Safety and survival equipment, required by the operating rules of this chapter, must be reliable, readily accessible, easily identifiable, and clearly marked to identify its method of operation.

§ 23.2540   Flight in icing conditions.

An applicant who requests certification for flight in icing conditions defined in part 1 of appendix C to part 25 of this chapter, or an applicant who requests certification for flight in these icing conditions and any additional atmospheric icing conditions, must show the following in the icing conditions for which certification is requested:

(a) The ice protection system provides for safe operation.

(b) The airplane design must provide protection from stalling when the autopilot is operating.

§ 23.2545   Pressurized   systems   elements.

Pressurized systems must withstand appropriate proof and burst pressures.

§ 23.2550   Equipment containing high-energy rotors.

Equipment containing high-energy rotors must be designed or installed to protect the occupants and airplane from uncontained fragments.

Subpart G—Flightcrew Interface and Other Information

§ 23.2600   Flightcrew interface.

(a) The pilot compartment, its equipment, and its arrangement to include pilot view, must allow each pilot to perform his or her duties, including taxi, takeoff, climb, cruise, descent, approach, landing, and perform any maneuvers within the operating envelope of the airplane, without excessive concentration, skill, alertness, or fatigue.

(b) The applicant must install flight, navigation, surveillance, and powerplant controls and displays so flightcrew members can monitor and perform defined tasks associated with the intended functions of systems and equipment. The system and equipment design must minimize flightcrew errors, which could result in additional hazards.

(c) For level 4 airplanes, the flightcrew interface design must allow for continued safe flight and landing after the loss of vision through any one of the windshield panels.

[Doc. No. FAA–2015–1621, Amdt. 23–64, 81 FR 96689, Dec. 30, 2016, as amended by Doc. No. FAA–2022–1355, Amdt. 23–65, 87 FR 75710, Dec. 9, 2022]

§ 23.2605   Installation and operation.

(a) Each item of installed equipment related to the flightcrew interface must be labelled, if applicable, as to it identification, function, or operating limitations, or any combination of these factors.

(b) There must be a discernible means of providing system operating parameters required to operate the airplane, including warnings, cautions, and normal indications to the responsible crewmember.

(c) Information concerning an unsafe system operating condition must be provided in a timely manner to the crewmember responsible for taking corrective action. The information must be clear enough to avoid likely crewmember errors.

§ 23.2610   Instrument markings, control markings, and placards.

(a) Each airplane must display in a conspicuous manner any placard and instrument marking necessary for operation.

(b) The design must clearly indicate the function of each cockpit control, other than primary flight controls.

(c) The applicant must include instrument marking and placard information in the Airplane Flight Manual.

§ 23.2615 Flight, navigation, and powerplant instruments.

(a) Installed systems must provide the flightcrew member who sets or monitors parameters for the flight, navigation, and powerplant, the information necessary to do so during each phase of flight. This information must—

(1) Be presented in a manner that the crewmember can monitor the parameter and determine trends, as needed, to operate the airplane; and

(2) Include limitations, unless the limitation cannot be exceeded in all intended operations.

(b) Indication systems that integrate the display of flight or powerplant parameters to operate the airplane or are required by the operating rules of this chapter must—

(1) Not inhibit the primary display of flight or powerplant parameters needed by any flightcrew member in any normal mode of operation; and

(2) In combination with other systems, be designed and installed so information essential for continued safe flight and landing will be available to the flightcrew in a timely manner after any single failure or probable combination of failures.

§ 23.2620 Airplane flight manual.

The applicant must provide an Airplane Flight Manual that must be delivered with each airplane.

(a) The Airplane Flight Manual must contain the following information—

(1) Airplane operating limitations;

(2) Airplane operating procedures;

(3) Performance information;

(4) Loading information; and

(5) Other information that is necessary for safe operation because of design, operating, or handling characteristics.

(b) The following sections of the Airplane Flight Manual must be approved by the FAA in a manner specified by the Administrator—

(1) For low-speed, level 1 and 2 airplanes, those portions of the Airplane Flight Manual containing the information specified in paragraph (a)(1) of this section; and

(2) For high-speed level 1 and 2 airplanes and all level 3 and 4 airplanes, those portions of the Airplane Flight Manual containing the information specified in paragraphs (a)(1) thru (a)(4) of this section.

[Doc. No. FAA–2015–1621, Amdt. 23–64, 81 FR 96689, Dec. 30, 2016, as amended by Doc. No. FAA–2022–1355, Amdt. 23–65, 87 FR 75710, Dec. 9, 2022]

APPENDIX A TO PART 23—INSTRUCTIONS FOR CONTINUED AIRWORTHINESS

A23.1 General

(a) This appendix specifies requirements for the preparation of Instructions for Continued Airworthiness as required by this part.

(b) The Instructions for Continued Airworthiness for each airplane must include the Instructions for Continued Airworthiness for each engine and propeller (hereinafter designated "products"), for each appliance required by this chapter, and any required information relating to the interface of those appliances and products with the airplane. If Instructions for Continued Airworthiness are not supplied by the manufacturer of an appliance or product installed in the airplane, the Instructions for Continued Airworthiness for the airplane must include the information essential to the continued airworthiness of the airplane.

(c) The applicant must submit to the FAA a program to show how changes to the Instructions for Continued Airworthiness made by the applicant or by the manufacturers of products and appliances installed in the airplane will be distributed.

A23.2 Format

(a) The Instructions for Continued Airworthiness must be in the form of a manual or manuals as appropriate for the quantity of data to be provided.

(b) The format of the manual or manuals must provide for a practical arrangement.

A23.3 Content

The contents of the manual or manuals must be prepared in the English language. The Instructions for Continued Airworthiness must contain the following manuals or sections and information:

(a) Airplane maintenance manual or section.

(1) Introduction information that includes an explanation of the airplane's features and data to the extent necessary for maintenance or preventive maintenance.

(2) A description of the airplane and its systems and installations including its engines, propellers, and appliances.

(3) Basic control and operation information describing how the airplane components and systems are controlled and how they operate, including any special procedures and limitations that apply.

(4) Servicing information that covers details regarding servicing points, capacities of tanks, reservoirs, types of fluids to be used, pressures applicable to the various systems, location of access panels for inspection and servicing, locations of lubrication points, lubricants to be used, equipment required for servicing, tow instructions and limitations, mooring, jacking, and leveling information.

(b) Maintenance Instructions.

(1) Scheduling information for each part of the airplane and its engines, auxiliary power units, propellers, accessories, instruments, and equipment that provides the recommended periods at which they should be cleaned, inspected, adjusted, tested, and lubricated, and the degree of inspection, the applicable wear tolerances, and work recommended at these periods. However, the applicant may refer to an accessory, instrument, or equipment manufacturer as the source of this information if the applicant shows that the item has an exceptionally high degree of complexity requiring specialized maintenance techniques, test equipment, or expertise. The recommended overhaul periods and necessary cross reference to the Airworthiness Limitations section of the manual must also be included. In addition, the applicant must include an inspection program that includes the frequency and extent of the inspections necessary to provide for the continued airworthiness of the airplane.

(2) Troubleshooting information describing probable malfunctions, how to recognize those malfunctions, and the remedial action for those malfunctions.

(3) Information describing the order and method of removing and replacing products and parts with any necessary precautions to be taken.

(4) Other general procedural instructions including procedures for system testing during ground running, symmetry checks, weighing and determining the center of gravity, lifting and shoring, and storage limitations.

(c) Diagrams of structural access plates and information needed to gain access for inspections when access plates are not provided.

(d) Details for the application of special inspection techniques including radiographic and ultrasonic testing where such processes are specified by the applicant.

(e) Information needed to apply protective treatments to the structure after inspection.

(f) All data relative to structural fasteners such as identification, discard recommendations, and torque values.

(g) A list of special tools needed.

(h) In addition, for level 4 airplanes, the following information must be furnished—

(1) Electrical loads applicable to the various systems;

(2) Methods of balancing control surfaces;

(3) Identification of primary and secondary structures; and

(4) Special repair methods applicable to the airplane.

A23.4 Airworthiness limitations section.

The Instructions for Continued Airworthiness must contain a section titled Airworthiness Limitations that is segregated and clearly distinguishable from the rest of the document. This section must set forth each mandatory replacement time, structural inspection interval, and related structural inspection procedure required for type certification. If the Instructions for Continued Airworthiness consist of multiple documents, the section required by this paragraph must be included in the principal manual. This section must contain a legible statement in a prominent location that reads ''The Airworthiness Limitations section is FAA approved and specifies maintenance required under §§ 43.16 and 91.403 of Title 14 of the Code of Federal Regulations unless an alternative program has been FAA approved.''

PART 25—AIRWORTHINESS STANDARDS: TRANSPORT CATEGORY AIRPLANES

SPECIAL FEDERAL AVIATION REGULATION NO. 13

SPECIAL FEDERAL AVIATION REGULATION NO. 109

Subpart A—General

Subpart B—Flight

GENERAL

PERFORMANCE

AUTHORITY: 49 U.S.C. 106(f), 106(g), 40113, 44701, 44702 and 44704; Pub. L. 115–254, 132 Stat 3281 (49 U.S.C. 44903 note).

SOURCE: Docket No. 5066, 29 FR 18291, Dec. 24, 1964, unless otherwise noted.

SPECIAL FEDERAL AVIATION REGULATION NO. 13

1. *Applicability.* Contrary provisions of the Civil Air Regulations regarding certification notwithstanding,[1] this regulation shall provide the basis for approval by the Administrator of modifications of individual Douglas DC–3 and Lockheed L–18 airplanes subsequent to the effective date of this regulation.

2. *General modifications.* Except as modified in sections 3 and 4 of this regulation, an applicant for approval of modifications to a DC–3 or L–18 airplane which result in changes in design or in changes to approved limitations shall show that the modifications were accomplished in accordance with the rules of either Part 4a or Part 4b in effect on September 1, 1953, which are applicable to the modification being made: *Provided,* That an applicant may elect to accomplish a modification in accordance with the rules of Part 4b in effect on the date of application for the modification in lieu of Part 4a or Part 4b as in effect on September 1, 1953: *And provided further,* That each specific modification must be accomplished in accordance with all of the provisions contained in the elected rules relating to the particular modification.

3. *Specific conditions for approval.* An applicant for any approval of the following specific changes shall comply with section 2 of this regulation as modified by the applicable provisions of this section.

(a) *Increase in take-off power limitation—1,200 to 1,350 horsepower.* The engine take-off power limitation for the airplane may be increased to more than 1,200 horsepower but not to more than 1,350 horsepower per engine if the increase in power does not adversely affect the flight characteristics of the airplane.

---

[1] It is not intended to waive compliance with such airworthiness requirements as are included in the operating parts of the Civil Air Regulations for specific types of operation.

(b) *Increase in take-off power limitation to more than 1,350 horsepower.* The engine take-off power limitation for the airplane may be increased to more than 1,350 horsepower per engine if compliance is shown with the flight characteristics and ground handling requirements of Part 4b.

(c) *Installation of engines of not more than 1,830 cubic inches displacement and not having a certificated take-off rating of more than 1,350 horsepower.* Engines of not more than 1,830 cubic inches displacement and not having a certificated take-off rating of more than 1,350 horsepower which necessitate a major modification of redesign of the engine installation may be installed, if the engine fire prevention and fire protection are equivalent to that on the prior engine installation.

(d) *Installation of engines of more than 1,830 cubic inches displacement or having certificated take-off rating of more than 1,350 horsepower.* Engines of more than 1,830 cubic inches displacement or having certificated take-off rating of more than 1,350 horsepower may be installed if compliance is shown with the engine installation requirements of Part 4b: *Provided,* That where literal compliance with the engine installation requirements of Part 4b is extremely difficult to accomplish and would not contribute materially to the objective sought, and the Administrator finds that the experience with the DC-3 or L-18 airplanes justifies it, he is authorized to accept such measures of compliance as he finds will effectively accomplish the basic objective.

4. *Establishment of new maximum certificated weights.* An applicant for approval of new maximum certificated weights shall apply for an amendment of the airworthiness certificate of the airplane and shall show that the weights sought have been established, and the appropriate manual material obtained, as provided in this section.

NOTE: Transport category performance requirements result in the establishment of maximum certificated weights for various altitudes.

(a) *Weights–25,200 to 26,900 for the DC-3 and 18,500 to 19,500 for the L-18.* New maximum certificated weights of more than 25,200 but not more than 26,900 pounds for DC-3 and more than 18,500 but not more than 19,500 pounds for L-18 airplanes may be established in accordance with the transport category performance requirements of either Part 4a or Part 4b, if the airplane at the new maximum weights can meet the structural requirements of the elected part.

(b) *Weights of more than 26,900 for the DC-3 and 19,500 for the L-18.* New maximum certificated weights of more than 26,900 pounds for DC-3 and 19,500 pounds for L-18 airplanes shall be established in accordance with the structural performance, flight characteristics, and ground handling requirements of Part 4b: *Provided,* That where literal compliance with the structural requirements of Part 4b is extremely difficult to accomplish and would not contribute materially to the objective sought, and the Administrator finds that the experience with the DC-3 or L-18 airplanes justifies it, he is authorized to accept such measures of compliance as he finds will effectively accomplish the basic objective.

(c) *Airplane flight manual-performance operating information.* An approved airplane flight manual shall be provided for each DC-3 and L-18 airplane which has had new maximum certificated weights established under this section. The airplane flight manual shall contain the applicable performance information prescribed in that part of the regulations under which the new certificated weights were established and such additional information as may be necessary to enable the application of the take-off, en route, and landing limitations prescribed for transport category airplanes in the operating parts of the Civil Air Regulations.

(d) *Performance operating limitations.* Each airplane for which new maximum certificated weights are established in accordance with paragraphs (a) or (b) of this section shall be considered a transport category airplane for the purpose of complying with the performance operating limitations applicable to the operations in which it is utilized.

5. *Reference.* Unless otherwise provided, all references in this regulation to Part 4a and Part 4b are those parts of the Civil Air Regulations in effect on September 1, 1953.

This regulation supersedes Special Civil Air Regulation SR-398 and shall remain effective until superseded or rescinded by the Board.

[19 FR 5039, Aug. 11, 1954. Redesignated at 29 FR 19099, Dec. 30, 1964]

SPECIAL FEDERAL AVIATION REGULATION NO. 109

1. *Applicability.* Contrary provisions of 14 CFR parts 21, 25, and 119 of this chapter notwithstanding, an applicant is entitled to an amended type certificate or supplemental type certificate in the transport category, if the applicant complies with all applicable provisions of this SFAR.

Operations

2. *General.*

(a) The passenger capacity may not exceed 60. If more than 60 passenger seats are installed, then:

(1) If the extra seats are not suitable for occupancy during taxi, takeoff and landing, each extra seat must be clearly marked (e.g., a placard on the top of an armrest, or a

placard sewn into the top of the back cushion) that the seat is not to be occupied during taxi, takeoff and landing.

(2) If the extra seats are suitable for occupancy during taxi, takeoff and landing (*i.e.*, meet all the strength and passenger injury criteria in part 25), then a note must be included in the Limitations Section of the Airplane Flight Manual that there are extra seats installed but that the number of passengers on the airplane must not exceed 60. Additionally, there must be a placard installed adjacent to each door that can be used as a passenger boarding door that states that the maximum passenger capacity is 60. The placard must be clearly legible to passengers entering the airplane.

(b) For airplanes outfitted with interior doors under paragraph 10 of this SFAR, the airplane flight manual (AFM) must include an appropriate limitation that the airplane must be staffed with at least the following number of flight attendants who meet the requirements of 14 CFR 91.533(b):

(1) The number of flight attendants required by § 91.533(a)(1) and (2) of this chapter, and

(2) At least one flight attendant if the airplane model was originally certified for 75 passengers or more.

(c) The AFM must include appropriate limitation(s) to require a preflight passenger briefing describing the appropriate functions to be performed by the passengers and the relevant features of the airplane to ensure the safety of the passengers and crew.

(d) The airplane may not be offered for common carriage or operated for hire. The operating limitations section of the AFM must be revised to prohibit any operations involving the carriage of persons or property for compensation or hire. The operators may receive remuneration to the extent consistent with parts 125 and 91, subpart F, of this chapter.

(e) A placard stating that "Operations involving the carriage of persons or property for compensation or hire are prohibited," must be located in the area of the Airworthiness Certificate holder at the entrance to the flightdeck.

(f) For passenger capacities of 45 to 60 passengers, analysis must be submitted that demonstrates that the airplane can be evacuated in less than 90 seconds under the conditions specified in § 25.803 and appendix J to part 25.

(g) In order for any airplane certified under this SFAR to be placed in part 135 or part 121 operations, the airplane must be brought back into full compliance with the applicable operational part.

Equipment and Design

3. *General.* Unless otherwise noted, compliance is required with the applicable certification basis for the airplane. Some provisions of this SFAR impose alternative requirements to certain airworthiness standards that do not apply to airplanes certificated to earlier standards. Those airplanes with an earlier certification basis are not required to comply with those alternative requirements.

4. *Occupant Protection.*

(a) Firm Handhold. In lieu of the requirements of § 25.785(j), there must be means provided to enable persons to steady themselves in moderately rough air while occupying aisles that are along the cabin sidewall, or where practicable, bordered by seats (seat backs providing a 25-pound minimum breakaway force are an acceptable means of compliance).

(b) Injury criteria for multiple occupancy side-facing seats. The following requirements are only applicable to airplanes that are subject to § 25.562.

(1) *Existing Criteria.* All injury protection criteria of § 25.562(c)(1) through (c)(6) apply to the occupants of side-facing seating. The Head Injury Criterion (HIC) assessments are only required for head contact with the seat and/or adjacent structures.

(2) *Body-to-Body Contact.* Contact between the head, pelvis, torso or shoulder area of one Anthropomorphic Test Dummy (ATD) with the head, pelvis, torso or shoulder area of the ATD in the adjacent seat is not allowed during the tests conducted in accordance with § 25.562(b)(1) and (b)(2). Contact during rebound is allowed.

(3) *Thoracic Trauma.* If the torso of an ATD at the forward-most seat place impacts the seat and/or adjacent structure during testing, compliance with the Thoracic Trauma Index (TTI) injury criterion must be substantiated by dynamic test or by rational analysis based on previous test(s) of a similar seat installation. TTI data must be acquired with a Side Impact Dummy (SID), as defined by 49 CFR part 572, subpart F, or an equivalent ATD or a more appropriate ATD and must be processed as defined in Federal Motor Vehicle Safety Standards (FMVSS) part 571.214, section S6.13.5 (49 CFR 571.214). The TTI must be less than 85, as defined in 49 CFR part 572, subpart F. Torso contact during rebound is acceptable and need not be measured.

(4) *Pelvis.* If the pelvis of an ATD at any seat place impacts seat and/or adjacent structure during testing, pelvic lateral acceleration injury criteria must be substantiated by dynamic test or by rational analysis based on previous test(s) of a similar seat installation. Pelvic lateral acceleration may not exceed 130g. Pelvic acceleration data must be processed as defined in FMVSS part 571.214, section S6.13.5 (49 CFR 571.214).

(5) *Body-to-Wall/Furnishing Contact.* If the seat is installed aft of a structure—such as an interior wall or furnishing that may contact the pelvis, upper arm, chest, or head of

an occupant seated next to the structure—the structure or a conservative representation of the structure and its stiffness must be included in the tests. It is recommended, but not required, that the contact surface of the actual structure be covered with at least two inches of energy absorbing protective padding (foam or equivalent) such as Ensolite.

(6) *Shoulder Strap Loads.* Where upper torso straps (shoulder straps) are used for sofa occupants, the tension loads in individual straps may not exceed 1,750 pounds. If dual straps are used for restraining the upper torso, the total strap tension loads may not exceed 2,000 pounds.

(7) *Occupant Retention.* All side-facing seats require end closures or other means to prevent the ATD's pelvis from translating beyond the end of the seat at any time during testing.

(8) *Test Parameters.*

(i) All seat positions need to be occupied by ATDs for the longitudinal tests.

(ii) A minimum of one longitudinal test, conducted in accordance with the conditions specified in § 25.562(b)(2), is required to assess the injury criteria as follows. Note that if a seat is installed aft of structure (such as an interior wall or furnishing) that does not have a homogeneous surface, an additional test or tests may be required to demonstrate that the injury criteria are met for the area which an occupant could contact. For example, different yaw angles could result in different injury considerations and may require separate tests to evaluate:

(A) For configurations without structure (such as a wall or bulkhead) installed directly forward of the forward seat place, Hybrid II ATDs or equivalent must be in all seat places.

(B) For configurations with structure (such as a wall or bulkhead) installed directly forward of the forward seat place, a side impact dummy or equivalent ATD or more appropriate ATD must be in the forward seat place and a Hybrid II ATD or equivalent must be in all other seat places.

(C) The test may be conducted with or without deformed floor.

(D) The test must be conducted with either no yaw or 10 degrees yaw for evaluating occupant injury. Deviating from the no yaw condition may not result in the critical area of contact not being evaluated. The upper torso restraint straps, where installed, must remain on the occupant's shoulder during the impact condition of § 25.562(b)(2).

(c) For the vertical test, conducted in accordance with the conditions specified in § 25.562(b)(1), Hybrid II ATDs or equivalent must be used in all seat positions.

5. *Direct View.* In lieu of the requirements of § 25.785(h)(2), to the extent practical without compromising proximity to a required floor level emergency exit, the majority of installed flight attendant seats must be located to face the cabin area for which the flight attendant is responsible.

6. *Passenger Information Signs.* Compliance with § 25.791 is required except that for § 25.791(a), when smoking is to be prohibited, notification to the passengers may be provided by a single placard so stating, to be conspicuously located inside the passenger compartment, easily visible to all persons entering the cabin in the immediate vicinity of each passenger entry door.

7. *Distance Between Exits.* For an airplane that is required to comply with § 25.807(f)(4), in effect as of July 24, 1989, which has more than one passenger emergency exit on each side of the fuselage, no passenger emergency exit may be more than 60 feet from any adjacent passenger emergency exit on the same side of the same deck of the fuselage, as measured parallel to the airplane's longitudinal axis between the nearest exit edges, unless the following conditions are met:

(a) Each passenger seat must be located within 30 feet from the nearest exit on each side of the fuselage, as measured parallel to the airplane's longitudinal axis, between the nearest exit edge and the front of the seat bottom cushion.

(b) The number of passenger seats located between two adjacent pairs of emergency exits (commonly referred to as a passenger zone) or between a pair of exits and a bulkhead or a compartment door (commonly referred to as a "dead-end zone"), may not exceed the following:

(1) For zones between two pairs of exits, 50 percent of the combined rated capacity of the two pairs of emergency exits.

(2) For zones between one pair of exits and a bulkhead, 40 percent of the rated capacity of the pair of emergency exits.

(c) The total number of passenger seats in the airplane may not exceed 33 percent of the maximum seating capacity for the airplane model using the exit ratings listed in § 25.807(g) for the original certified exits or the maximum allowable after modification when exits are deactivated, whichever is less.

(d) A distance of more than 60 feet between adjacent passenger emergency exits on the same side of the same deck of the fuselage, as measured parallel to the airplane's longitudinal axis between the nearest exit edges, is allowed only once on each side of the fuselage.

8. *Emergency Exit Signs.* In lieu of the requirements of § 25.811(d)(1) and (2) a single sign at each exit may be installed provided:

(a) The sign can be read from the aisle while directly facing the exit, and

(b) The sign can be read from the aisle adjacent to the passenger seat that is farthest from the exit and that does not have an intervening bulkhead/divider or exit.

9. *Emergency Lighting.*

(a) *Exit Signs.* In lieu of the requirements of §25.812(b)(1), for airplanes that have a passenger seating configuration, excluding pilot seats, of 19 seats or less, the emergency exit signs required by §25.811(d)(1), (2), and (3) must have red letters at least 1-inch high on a white background at least 2 inches high. These signs may be internally electrically illuminated, or self illuminated by other than electrical means, with an initial brightness of at least 160 microlamberts. The color may be reversed in the case of a sign that is self-illuminated by other than electrical means.

(b) *Floor Proximity Escape Path Marking.* In lieu of the requirements of §25.812(e)(1), for cabin seating compartments that do not have the main cabin aisle entering and exiting the compartment, the following are applicable:

(1) After a passenger leaves any passenger seat in the compartment, he/she must be able to exit the compartment to the main cabin aisle using only markings and visual features not more that 4 feet above the cabin floor, and

(2) Proceed to the exits using the marking system necessary to accomplish the actions in §25.812(e)(1) and (e)(2).

(c) *Transverse Separation of the Fuselage.* In the event of a transverse separation of the fuselage, compliance must be shown with §25.812(l) except as follows:

(1) For each airplane type originally type certificated with a maximum passenger seating capacity of 9 or less, not more than 50 percent of all electrically illuminated emergency lights required by §25.812 may be rendered inoperative in addition to the lights that are directly damaged by the separation.

(2) For each airplane type originally type certificated with a maximum passenger seating capacity of 10 to 19, not more than 33 percent of all electrically illuminated emergency lights required by §25.812 may be rendered inoperative in addition to the lights that are directly damaged by the separation.

10. *Interior doors.* In lieu of the requirements of §25.813(e), interior doors may be installed between passenger seats and exits, provided the following requirements are met.

(a) Each door between any passenger seat, occupiable for taxi, takeoff, and landing, and any emergency exit must have a means to signal to the flightcrew, at the flightdeck, that the door is in the open position for taxi, takeoff and landing.

(b) Appropriate procedures/limitations must be established to ensure that any such door is in the open configuration for takeoff and landing.

(c) Each door between any passenger seat and any exit must have dual means to retain it in the open position, each of which is capable of reacting the inertia loads specified in §25.561.

(d) Doors installed across a longitudinal aisle must translate laterally to open and close, e.g., pocket doors.

(e) Each door between any passenger seat and any exit must be frangible in either direction.

(f) Each door between any passenger seat and any exit must be operable from either side, and if a locking mechanism is installed, it must be capable of being unlocked from either side without the use of special tools.

11. *Width of Aisle.* Compliance is required with §25.815, except that aisle width may be reduced to 0 inches between passenger seats during in-flight operations only, provided that the applicant demonstrates that all areas of the cabin are easily accessible by a crew member in the event of an emergency (e.g., in-flight fire, decompression). Additionally, instructions must be provided at each passenger seat for restoring the aisle width required by §25.815. Procedures must be established and documented in the AFM to ensure that the required aisle widths are provided during taxi, takeoff, and landing.

12. *Materials for Compartment Interiors.* Compliance is required with the applicable provisions of §25.853, except that compliance with appendix F, parts IV and V, to part 25, need not be demonstrated if it can be shown by test or a combination of test and analysis that the maximum time for evacuation of all occupants does not exceed 45 seconds under the conditions specified in appendix J to part 25.

13. *Fire Detection.* For airplanes with a type certificated passenger capacity of 20 or more, there must be means that meet the requirements of §25.858(a) through (d) to signal the flightcrew in the event of a fire in any isolated room not occupiable for taxi, takeoff and landing, which can be closed off from the rest of the cabin by a door. The indication must identify the compartment where the fire is located. This does not apply to lavatories, which continue to be governed by §25.854.

14. *Cooktops.* Each cooktop must be designed and installed to minimize any potential threat to the airplane, passengers, and crew. Compliance with this requirement must be found in accordance with the following criteria:

(a) Means, such as conspicuous burner-on indicators, physical barriers, or handholds, must be installed to minimize the potential for inadvertent personnel contact with hot surfaces of both the cooktop and cookware. Conditions of turbulence must be considered.

(b) Sufficient design means must be included to restrain cookware while in place on the cooktop, as well as representative contents, e.g., soup, sauces, etc., from the effects of flight loads and turbulence. Restraints must be provided to preclude hazardous movement of cookware and contents. These restraints must accommodate any cookware that is identified for use with the cooktop. Restraints must be designed to be easily utilized and effective in service. The cookware restraint system should also be designed so that it will not be easily disabled, thus rendering it unusable. Placarding must be installed which prohibits the use of cookware that cannot be accommodated by the restraint system.

(c) Placarding must be installed which prohibits the use of cooktops (i.e., power on any burner) during taxi, takeoff, and landing.

(d) Means must be provided to address the possibility of a fire occurring on or in the immediate vicinity of the cooktop. Two acceptable means of complying with this requirement are as follows:

(1) Placarding must be installed that prohibits any burner from being powered when the cooktop is unattended. (NOTE: This would prohibit a single person from cooking on the cooktop and intermittently serving food to passengers while any burner is powered.) A fire detector must be installed in the vicinity of the cooktop which provides an audible warning in the passenger cabin, and a fire extinguisher of appropriate size and extinguishing agent must be installed in the immediate vicinity of the cooktop. Access to the extinguisher may not be blocked by a fire on or around the cooktop.

(2) An automatic, thermally activated fire suppression system must be installed to extinguish a fire at the cooktop and immediately adjacent surfaces. The agent used in the system must be an approved total flooding agent suitable for use in an occupied area. The fire suppression system must have a manual override. The automatic activation of the fire suppression system must also automatically shut off power to the cooktop.

(e) The surfaces of the galley surrounding the cooktop which would be exposed to a fire on the cooktop surface or in cookware on the cooktop must be constructed of materials that comply with the flammability requirements of part III of appendix F to part 25. This requirement is in addition to the flammability requirements typically required of the materials in these galley surfaces. During the selection of these materials, consideration must also be given to ensure that the flammability characteristics of the materials will not be adversely affected by the use of cleaning agents and utensils used to remove cooking stains.

(f) The cooktop must be ventilated with a system independent of the airplane cabin and cargo ventilation system. Procedures and time intervals must be established to inspect and clean or replace the ventilation system to prevent a fire hazard from the accumulation of flammable oils and be included in the instructions for continued airworthiness. The ventilation system ducting must be protected by a flame arrestor. [NOTE: The applicant may find additional useful information in Society of Automotive Engineers, Aerospace Recommended Practice 85, Rev. E, entitled "Air Conditioning Systems for Subsonic Airplanes," dated August 1, 1991.]

(g) Means must be provided to contain spilled foods or fluids in a manner that will prevent the creation of a slipping hazard to occupants and will not lead to the loss of structural strength due to airplane corrosion.

(h) Cooktop installations must provide adequate space for the user to immediately escape a hazardous cooktop condition.

(i) A means to shut off power to the cooktop must be provided at the galley containing the cooktop and in the cockpit. If additional switches are introduced in the cockpit, revisions to smoke or fire emergency procedures of the AFM will be required.

(j) If the cooktop is required to have a lid to enclose the cooktop there must be a means to automatically shut off power to the cooktop when the lid is closed.

15. *Hand-Held Fire Extinguishers.*

(a) For airplanes that were originally type certificated with more than 60 passengers, the number of hand-held fire extinguishers must be the greater of—

(1) That provided in accordance with the requirements of § 25.851, or

(2) A number equal to the number of originally type certificated exit pairs, regardless of whether the exits are deactivated for the proposed configuration.

(b) Extinguishers must be evenly distributed throughout the cabin. These extinguishers are in addition to those required by paragraph 14 of this SFAR, unless it can be shown that the cooktop was installed in the immediate vicinity of the original exits.

16. *Security.* The requirements of § 25.795 are not applicable to airplanes approved in accordance with this SFAR.

[Doc. No. FAA–2007–28250, 74 FR 21541, May 8, 2009]

Subpart A—General

§ 25.1 Applicability.

(a) This part prescribes airworthiness standards for the issue of type certificates, and changes to those certificates, for transport category airplanes.

(b) Each person who applies under Part 21 for such a certificate or change must show compliance with the applicable requirements in this part.

§25.2 Special retroactive requirements.

The following special retroactive requirements are applicable to an airplane for which the regulations referenced in the type certificate predate the sections specified below—

(a) Irrespective of the date of application, each applicant for a supplemental type certificate (or an amendment to a type certificate) involving an increase in passenger seating capacity to a total greater than that for which the airplane has been type certificated must show that the airplane concerned meets the requirements of:

(1) Sections 25.721(d), 25.783(g), 25.785(c), 25.803(c)(2) through (9), 25.803 (d) and (e), 25.807 (a), (c), and (d), 25.809 (f) and (h), 25.811, 25.812, 25.813 (a), (b), and (c), 25.815, 25.817, 25.853 (a) and (b), 25.855(a), 25.993(f), and 25.1359(c) in effect on October 24, 1967, and

(2) Sections 25.803(b) and 25.803(c)(1) in effect on April 23, 1969.

(b) Irrespective of the date of application, each applicant for a supplemental type certificate (or an amendment to a type certificate) for an airplane manufactured after October 16, 1987, must show that the airplane meets the requirements of §25.807(c)(7) in effect on July 24, 1989.

(c) Compliance with subsequent revisions to the sections specified in paragraph (a) or (b) of this section may be elected or may be required in accordance with §21.101(a) of this chapter.

[Amdt. 25–72, 55 FR 29773, July 20, 1990, as amended by Amdt. 25–99, 65 FR 36266, June 7, 2000]

§25.3 Special provisions for ETOPS type design approvals.

(a) *Applicability.* This section applies to an applicant for ETOPS type design approval of an airplane:

(1) That has an existing type certificate on February 15, 2007; or

(2) For which an application for an original type certificate was submitted before February 15, 2007.

(b) *Airplanes with two engines.* (1) For ETOPS type design approval of an airplane up to and including 180 minutes, an applicant must comply with §25.1535, except that it need not comply with the following provisions of Appendix K, K25.1.4, of this part:

(i) K25.1.4(a), fuel system pressure and flow requirements;

(ii) K25.1.4(a)(3), low fuel alerting; and

(iii) K25.1.4(c), engine oil tank design.

(2) For ETOPS type design approval of an airplane beyond 180 minutes an applicant must comply with §25.1535.

(c) *Airplanes with more than two engines.* An applicant for ETOPS type design approval must comply with §25.1535 for an airplane manufactured on or after February 17, 2015, except that, for an airplane configured for a three person flight crew, the applicant need not comply with Appendix K, K25.1.4(a)(3), of this part, low fuel alerting.

[Doc. No. FAA–2002–6717, 72 FR 1873, Jan. 16, 2007]

§25.5 Incorporations by reference.

(a) The materials listed in this section are incorporated by reference in the corresponding sections noted. These incorporations by reference were approved by the Director of the Federal Register in accordance with 5 U.S.C. 552(a) and 1 CFR part 51. These materials are incorporated as they exist on the date of the approval, and notice of any change in these materials will be published in the FEDERAL REGISTER. The materials are available for purchase at the corresponding addresses noted below, and all are available for inspection at the National Archives and Records Administration (NARA). For information on the availability of this material at NARA, call 202–741–6030, or go to: *http://www.archives.gov/ federal-register/cfr/ibr-locations.html.*

(b) The following materials are available for purchase from the following address: The National Technical Information Services (NTIS), Springfield, Virginia 22166.

(1) Fuel Tank Flammability Assessment Method User's Manual, dated May 2008, document number DOT/FAA/ AR–05/8, IBR approved for §25.981 and Appendix N. It can also be obtained at the following Web site: *http:// www.fire.tc.faa.gov/systems/fueltank/ FTFAM.stm.*

(2) [Reserved]

[73 FR 42494, July 21, 2008, as amended by Doc. No. FAA–2018–0119, Amdt. 21–101, 83 FR 9169, Mar. 5, 2018]

Subpart B—Flight

GENERAL

§ 25.21   Proof of compliance.

(a) Each requirement of this subpart must be met at each appropriate combination of weight and center of gravity within the range of loading conditions for which certification is requested. This must be shown—

(1) By tests upon an airplane of the type for which certification is requested, or by calculations based on, and equal in accuracy to, the results of testing; and

(2) By systematic investigation of each probable combination of weight and center of gravity, if compliance cannot be reasonably inferred from combinations investigated.

(b) [Reserved]

(c) The controllability, stability, trim, and stalling characteristics of the airplane must be shown for each altitude up to the maximum expected in operation.

(d) Parameters critical for the test being conducted, such as weight, loading (center of gravity and inertia), airspeed, power, and wind, must be maintained within acceptable tolerances of the critical values during flight testing.

(e) If compliance with the flight characteristics requirements is dependent upon a stability augmentation system or upon any other automatic or power-operated system, compliance must be shown with §§ 25.671 and 25.672.

(f) In meeting the requirements of §§ 25.105(d), 25.125, 25.233, and 25.237, the wind velocity must be measured at a height of 10 meters above the surface, or corrected for the difference between the height at which the wind velocity is measured and the 10-meter height.

(g) The requirements of this subpart associated with icing conditions apply only if the applicant is seeking certification for flight in icing conditions.

(1) Paragraphs (g)(3) and (4) of this section apply only to airplanes with one or both of the following attributes:

(i) Maximum takeoff gross weight is less than 60,000 lbs; or

(ii) The airplane is equipped with reversible flight controls.

(2) Each requirement of this subpart, except §§ 25.121(a), 25.123(c), 25.143(b)(1) and (2), 25.149, 25.201(c)(2), 25.239, and 25.251(b) through (e), must be met in the icing conditions specified in Appendix C of this part. Section 25.207(c) and (d) must be met in the landing configuration in the icing conditions specified in Appendix C, but need not be met for other configurations. Compliance must be shown using the ice accretions defined in part II of Appendix C of this part, assuming normal operation of the airplane and its ice protection system in accordance with the operating limitations and operating procedures established by the applicant and provided in the airplane flight manual.

(3) If the applicant does not seek certification for flight in all icing conditions defined in Appendix O of this part, each requirement of this subpart, except §§ 25.105, 25.107, 25.109, 25.111, 25.113, 25.115, 25.121, 25.123, 25.143(b)(1), (b)(2), and (c)(1), 25.149, 25.201(c)(2), 25.207(c), (d), and (e)(1), 25.239, and 25.251(b) through (e), must be met in the Appendix O icing conditions for which certification is not sought in order to allow a safe exit from those conditions. Compliance must be shown using the ice accretions defined in part II, paragraphs (b) and (d) of Appendix O, assuming normal operation of the airplane and its ice protection system in accordance with the operating limitations and operating procedures established by the applicant and provided in the airplane flight manual.

(4) If the applicant seeks certification for flight in any portion of the icing conditions of Appendix O of this part, each requirement of this subpart, except §§ 25.121(a), 25.123(c), 25.143(b)(1) and (2), 25.149, 25.201(c)(2), 25.239, and 25.251(b) through (e), must be met in the Appendix O icing conditions for which certification is sought. Section 25.207(c) and (d) must be met in the landing configuration in the Appendix O icing conditions for which certification is sought, but need not be met for other configurations. Compliance must be shown using the ice accretions defined in part II, paragraphs (c) and (d) of Appendix O, assuming normal operation of the airplane and its ice protection system in accordance with the operating limitations and operating

procedures established by the applicant and provided in the airplane flight manual.

[Doc. No. 5066, 29 FR 18291, Dec. 24, 1964, as amended by Amdt. 25–23, 35 FR 5671, Apr. 8, 1970; Amdt. 25–42, 43 FR 2320, Jan. 16, 1978; Amdt. 25–72, 55 FR 29774, July 20, 1990; Amdt. 25–121, 72 FR 44665, Aug. 8, 2007 Amdt. 25–135, 76 FR 74654, Dec. 1, 2011; Amdt. 25–140, 79 FR 65524, Nov. 4, 2014]

§ 25.23 Load distribution limits.

(a) Ranges of weights and centers of gravity within which the airplane may be safely operated must be established. If a weight and center of gravity combination is allowable only within certain load distribution limits (such as spanwise) that could be inadvertently exceeded, these limits and the corresponding weight and center of gravity combinations must be established.

(b) The load distribution limits may not exceed—

(1) The selected limits;

(2) The limits at which the structure is proven; or

(3) The limits at which compliance with each applicable flight requirement of this subpart is shown.

§ 25.25 Weight limits.

(a) *Maximum weights.* Maximum weights corresponding to the airplane operating conditions (such as ramp, ground or water taxi, takeoff, en route, and landing), environmental conditions (such as altitude and temperature), and loading conditions (such as zero fuel weight, center of gravity position and weight distribution) must be established so that they are not more than—

(1) The highest weight selected by the applicant for the particular conditions; or

(2) The highest weight at which compliance with each applicable structural loading and flight requirement is shown, except that for airplanes equipped with standby power rocket engines the maximum weight must not be more than the highest weight established in accordance with appendix E of this part; or

(3) The highest weight at which compliance is shown with the certification requirements of Part 36 of this chapter.

(b) *Minimum weight.* The minimum weight (the lowest weight at which compliance with each applicable requirement of this part is shown) must be established so that it is not less than—

(1) The lowest weight selected by the applicant;

(2) The design minimum weight (the lowest weight at which compliance with each structural loading condition of this part is shown); or

(3) The lowest weight at which compliance with each applicable flight requirement is shown.

[Doc. No. 5066, 29 FR 18291, Dec. 24, 1964, as amended by Amdt. 25–23, 35 FR 5671, Apr. 8, 1970; Amdt. 25–63, 53 FR 16365, May 6, 1988]

§ 25.27 Center of gravity limits.

The extreme forward and the extreme aft center of gravity limitations must be established for each practicably separable operating condition. No such limit may lie beyond—

(a) The extremes selected by the applicant;

(b) The extremes within which the structure is proven; or

(c) The extremes within which compliance with each applicable flight requirement is shown.

§ 25.29 Empty weight and corresponding center of gravity.

(a) The empty weight and corresponding center of gravity must be determined by weighing the airplane with—

(1) Fixed ballast;

(2) Unusable fuel determined under § 25.959; and

(3) Full operating fluids, including—

(i) Oil;

(ii) Hydraulic fluid; and

(iii) Other fluids required for normal operation of airplane systems, except potable water, lavatory precharge water, and fluids intended for injection in the engine.

(b) The condition of the airplane at the time of determining empty weight must be one that is well defined and can be easily repeated.

[Doc. No. 5066, 29 FR 18291, Dec. 24, 1964, as amended by Amdt. 25–42, 43 FR 2320, Jan. 16, 1978; Amdt. 25–72, 55 FR 29774, July 20, 1990]

§ 25.31 Removable ballast.

Removable ballast may be used on showing compliance with the flight requirements of this subpart.

§ 25.33 Propeller speed and pitch limits.

(a) The propeller speed and pitch must be limited to values that will ensure—

(1) Safe operation under normal operating conditions; and

(2) Compliance with the performance requirements of §§ 25.101 through 25.125.

(b) There must be a propeller speed limiting means at the governor. It must limit the maximum possible governed engine speed to a value not exceeding the maximum allowable r.p.m.

(c) The means used to limit the low pitch position of the propeller blades must be set so that the engine does not exceed 103 percent of the maximum allowable engine rpm or 99 percent of an approved maximum overspeed, whichever is greater, with—

(1) The propeller blades at the low pitch limit and governor inoperative;

(2) The airplane stationary under standard atmospheric conditions with no wind; and

(3) The engines operating at the takeoff manifold pressure limit for reciprocating engine powered airplanes or the maximum takeoff torque limit for turbopropeller engine-powered airplanes.

[Doc. No. 5066, 29 FR 18291, Dec. 24, 1964, as amended by Amdt. 25–57, 49 FR 6848, Feb. 23, 1984; Amdt. 25–72, 55 FR 29774, July 20, 1990]

PERFORMANCE

§ 25.101 General.

(a) Unless otherwise prescribed, airplanes must meet the applicable performance requirements of this subpart for ambient atmospheric conditions and still air.

(b) The performance, as affected by engine power or thrust, must be based on the following relative humidities;

(1) For turbine engine powered airplanes, a relative humidity of—

(i) 80 percent, at and below standard temperatures; and

(ii) 34 percent, at and above standard temperatures plus 50 °F.

Between these two temperatures, the relative humidity must vary linearly.

(2) For reciprocating engine powered airplanes, a relative humidity of 80 percent in a standard atmosphere. Engine power corrections for vapor pressure must be made in accordance with the following table:

Altitude H (ft.)	Vapor pressure e (In. Hg.)	Specific humidity w (Lb. moisture per lb. dry air)	Density ratio $\rho / \sigma =$ 0.0023769
0	0.403	0.00849	0.99508
1,000	.354	.00773	.96672
2,000	.311	.00703	.93895
3,000	.272	.00638	.91178
4,000	.238	.00578	.88514
5,000	.207	.00523	.85910
6,000	.1805	.00472	.83361
7,000	.1566	.00425	.80870
8,000	.1356	.00382	.78434
9,000	.1172	.00343	.76053
10,000	.1010	.00307	.73722
15,000	.0463	.001710	.62868
20,000	.01978	.000896	.53263
25,000	.00778	.000436	.44806

(c) The performance must correspond to the propulsive thrust available under the particular ambient atmospheric conditions, the particular flight condition, and the relative humidity specified in paragraph (b) of this section. The available propulsive thrust must correspond to engine power or thrust, not exceeding the approved power or thrust less—

(1) Installation losses; and

(2) The power or equivalent thrust absorbed by the accessories and services appropriate to the particular ambient atmospheric conditions and the particular flight condition.

(d) Unless otherwise prescribed, the applicant must select the takeoff, en route, approach, and landing configurations for the airplane.

(e) The airplane configurations may vary with weight, altitude, and temperature, to the extent they are compatible with the operating procedures required by paragraph (f) of this section.

(f) Unless otherwise prescribed, in determining the accelerate-stop distances, takeoff flight paths, takeoff distances, and landing distances, changes in the airplane's configuration, speed, power, and thrust, must be made in accordance with procedures established by the applicant for operation in service.

(g) Procedures for the execution of balked landings and missed approaches associated with the conditions prescribed in §§ 25.119 and 25.121(d) must be established.

(h) The procedures established under paragraphs (f) and (g) of this section must—

(1) Be able to be consistently executed in service by crews of average skill;

(2) Use methods or devices that are safe and reliable; and

(3) Include allowance for any time delays, in the execution of the procedures, that may reasonably be expected in service.

(i) The accelerate-stop and landing distances prescribed in §§ 25.109 and 25.125, respectively, must be determined with all the airplane wheel brake assemblies at the fully worn limit of their allowable wear range.

[Doc. No. 5066, 29 FR 18291, Dec. 24, 1964, as amended by Amdt. 25–38, 41 FR 55466, Dec. 20, 1976; Amdt. 25–92, 63 FR 8318, Feb. 18, 1998]

§ 25.103 Stall speed.

(a) The reference stall speed, V_{SR}, is a calibrated airspeed defined by the applicant. V_{SR} may not be less than a 1-g stall speed. V_{SR} is expressed as:

$$V_{SR} \geq \frac{V_{CL_{MAX}}}{\sqrt{n_{ZW}}}$$

where:

$V_{CL_{MAX}}$ = Calibrated airspeed obtained when the load factor-corrected lift coefficient

$$\left(\frac{n_{ZW}W}{qS}\right)$$

is first a maximum during the maneuver prescribed in paragraph (c) of this section. In addition, when the maneuver is limited by a device that abruptly pushes the nose down at a selected angle of attack (e.g., a stick pusher), $V_{CL_{MAX}}$ may not be less than the speed existing at the instant the device operates;

n_{ZW} = Load factor normal to the flight path at $V_{CL_{MAX}}$

W = Airplane gross weight;

S = Aerodynamic reference wing area; and

q = Dynamic pressure.

(b) $V_{CL_{MAX}}$ is determined with:

(1) Engines idling, or, if that resultant thrust causes an appreciable decrease in stall speed, not more than zero thrust at the stall speed;

(2) Propeller pitch controls (if applicable) in the takeoff position;

(3) The airplane in other respects (such as flaps, landing gear, and ice accretions) in the condition existing in the test or performance standard in which V_{SR} is being used;

(4) The weight used when V_{SR} is being used as a factor to determine compliance with a required performance standard;

(5) The center of gravity position that results in the highest value of reference stall speed; and

(6) The airplane trimmed for straight flight at a speed selected by the applicant, but not less than $1.13V_{SR}$ and not greater than $1.3V_{SR}$.

(c) Starting from the stabilized trim condition, apply the longitudinal control to decelerate the airplane so that the speed reduction does not exceed one knot per second.

(d) In addition to the requirements of paragraph (a) of this section, when a device that abruptly pushes the nose down at a selected angle of attack (e.g., a stick pusher) is installed, the reference stall speed, V_{SR}, may not be less than 2 knots or 2 percent, whichever is greater, above the speed at which the device operates.

[Doc. No. 28404, 67 FR 70825, Nov. 26, 2002, as amended by Amdt. 25–121, 72 FR 44665, Aug. 8, 2007]

§ 25.105 Takeoff.

(a) The takeoff speeds prescribed by § 25.107, the accelerate-stop distance prescribed by § 25.109, the takeoff path prescribed by § 25.111, the takeoff distance and takeoff run prescribed by § 25.113, and the net takeoff flight path prescribed by § 25.115, must be determined in the selected configuration for takeoff at each weight, altitude, and ambient temperature within the operational limits selected by the applicant—

(1) In non-icing conditions; and

(2) In icing conditions, if in the configuration used to show compliance with § 25.121(b), and with the most critical of the takeoff ice accretion(s) defined in appendices C and O of this part, as applicable, in accordance with § 25.21(g):

(i) The stall speed at maximum takeoff weight exceeds that in non-icing conditions by more than the greater of 3 knots CAS or 3 percent of V_{SR}; or

(ii) The degradation of the gradient of climb determined in accordance with § 25.121(b) is greater than one-half of the applicable actual-to-net takeoff flight path gradient reduction defined in § 25.115(b).

(b) No takeoff made to determine the data required by this section may require exceptional piloting skill or alertness.

(c) The takeoff data must be based on—

(1) In the case of land planes and amphibians:

(i) Smooth, dry and wet, hard-surfaced runways; and

(ii) At the option of the applicant, grooved or porous friction course wet, hard-surfaced runways.

(2) Smooth water, in the case of seaplanes and amphibians; and

(3) Smooth, dry snow, in the case of skiplanes.

(d) The takeoff data must include, within the established operational limits of the airplane, the following operational correction factors:

(1) Not more than 50 percent of nominal wind components along the takeoff path opposite to the direction of takeoff, and not less than 150 percent of nominal wind components along the takeoff path in the direction of takeoff.

(2) Effective runway gradients.

[Doc. No. 5066, 29 FR 18291, Dec. 24, 1964, as amended by Amdt. 25–92, 63 FR 8318, Feb. 18, 1998; Amdt. 25–121, 72 FR 44665, Aug. 8, 2007; Amdt. 25–140, 79 FR 65525, Nov. 4, 2014]

§ 25.107   Takeoff speeds.

(a) V_1 must be established in relation to V_{EF} as follows:

(1) V_{EF} is the calibrated airspeed at which the critical engine is assumed to fail. V_{EF} must be selected by the applicant, but may not be less than V_{MCG} determined under § 25.149(e).

(2) V_1, in terms of calibrated airspeed, is selected by the applicant; however, V_1 may not be less than V_{EF} plus the speed gained with critical engine inoperative during the time interval between the instant at which the critical engine is failed, and the instant at which the pilot recognizes and reacts to the engine failure, as indicated by the pilot's initiation of the first action (e.g., applying brakes, reducing thrust, deploying speed brakes) to stop the airplane during accelerate-stop tests.

(b) V_{2MIN}, in terms of calibrated airspeed, may not be less than—

(1) 1.13 V_{SR} for—

(i) Two-engine and three-engine turbopropeller and reciprocating engine powered airplanes; and

(ii) Turbojet powered airplanes without provisions for obtaining a significant reduction in the one-engine-inoperative power-on stall speed;

(2) 1.08 V_{SR} for—

(i) Turbopropeller and reciprocating engine powered airplanes with more than three engines; and

(ii) Turbojet powered airplanes with provisions for obtaining a significant reduction in the one-engine-inoperative power-on stall speed; and

(3) 1.10 times V_{MC} established under § 25.149.

(c) V_2, in terms of calibrated airspeed, must be selected by the applicant to provide at least the gradient of climb required by § 25.121(b) but may not be less than—

(1) V_{2MIN};

(2) V_R plus the speed increment attained (in accordance with § 25.111(c)(2)) before reaching a height of 35 feet above the takeoff surface; and

(3) A speed that provides the maneuvering capability specified in § 25.143(h).

(d) V_{MU} is the calibrated airspeed at and above which the airplane can safely lift off the ground, and con- tinue the takeoff. V_{MU} speeds must be selected by the applicant throughout the range of thrust-to-weight ratios to be certificated. These speeds may be established from free air data if these data are verified by ground takeoff tests.

(e) V_R, in terms of calibrated airspeed, must be selected in accordance with the conditions of paragraphs (e)(1) through (4) of this section:

(1) V_R may not be less than—

(i) V_1;

(ii) 105 percent of V_{MC};

(iii) The speed (determined in accordance with § 25.111(c)(2)) that allows reaching V_2 before reaching a height of 35 feet above the takeoff surface; or

(iv) A speed that, if the airplane is rotated at its maximum practicable rate, will result in a V_{LOF} of not less than —

(A) 110 percent of V_{MU} in the all-engines-operating condition, and 105 percent of V_{MU} determined at the thrust-to-weight ratio corresponding to the one-engine-inoperative condition; or

(B) If the V_{MU} attitude is limited by the geometry of the airplane (*i.e.*, tail contact with the runway), 108 percent of V_{MU} in the all-engines-operating condition, and 104 percent of V_{MU} determined at the thrust-to-weight ratio corresponding to the one-engine-inoperative condition.

(2) For any given set of conditions (such as weight, configuration, and temperature), a single value of V_R obtained in accordance with this paragraph, must be used to show compliance with both the one-engine-inoperative and the all-engines-operating takeoff provisions.

(3) It must be shown that the one-engine-inoperative takeoff distance, using a rotation speed of 5 knots less than V_R established in accordance with paragraphs (e)(1) and (2) of this section, does not exceed the corresponding one-engine-inoperative takeoff distance using the established V_R. The takeoff distances must be determined in accordance with §25.113(a)(1).

(4) Reasonably expected variations in service from the established takeoff procedures for the operation of the airplane (such as over-rotation of the airplane and out-of-trim conditions) may not result in unsafe flight characteristics or in marked increases in the scheduled takeoff distances established in accordance with §25.113(a).

(f) V_{LOF} is the calibrated airspeed at which the airplane first becomes airborne.

(g) V_{FTO}, in terms of calibrated airspeed, must be selected by the applicant to provide at least the gradient of climb required by §25.121(c), but may not be less than—

(1) 1.18 V_{SR}; and

(2) A speed that provides the maneuvering capability specified in §25.143(h).

(h) In determining the takeoff speeds V_1, V_R, and V_2 for flight in icing conditions, the values of V_{MCG}, V_{MC}, and V_{MU} determined for non-icing conditions may be used.

[Doc. No. 5066, 29 FR 18291, Dec. 24, 1964, as amended by Amdt. 25–38, 41 FR 55466, Dec. 20, 1976; Amdt. 25–42, 43 FR 2320, Jan. 16, 1978; Amdt. 25–92, 63 FR 8318, Feb. 18, 1998; Amdt. 25–94, 63 FR 8848, Feb. 23, 1998; Amdt. 25–108, 67 FR 70826, Nov. 26, 2002; Amdt. 25–121, 72 FR 44665, Aug. 8, 2007; Amdt. 25–135, 76 FR 74654, Dec. 1, 2011]

§25.109 Accelerate-stop distance.

(a) The accelerate-stop distance on a dry runway is the greater of the following distances:

(1) The sum of the distances necessary to—

(i) Accelerate the airplane from a standing start with all engines operating to V_{EF} for takeoff from a dry runway;

(ii) Allow the airplane to accelerate from V_{EF} to the highest speed reached during the rejected takeoff, assuming the critical engine fails at V_{EF} and the pilot takes the first action to reject the takeoff at the V_1 for takeoff from a dry runway; and

(iii) Come to a full stop on a dry runway from the speed reached as prescribed in paragraph (a)(1)(ii) of this section; plus

(iv) A distance equivalent to 2 seconds at the V_1 for takeoff from a dry runway.

(2) The sum of the distances necessary to—

(i) Accelerate the airplane from a standing start with all engines operating to the highest speed reached during the rejected takeoff, assuming the pilot takes the first action to reject the takeoff at the V_1 for takeoff from a dry runway; and

(ii) With all engines still operating, come to a full stop on dry runway from the speed reached as prescribed in paragraph (a)(2)(i) of this section; plus

(iii) A distance equivalent to 2 seconds at the V_1 for takeoff from a dry runway.

(b) The accelerate-stop distance on a wet runway is the greater of the following distances:

(1) The accelerate-stop distance on a dry runway determined in accordance with paragraph (a) of this section; or

(2) The accelerate-stop distance determined in accordance with paragraph

(a) of this section, except that the runway is wet and the corresponding wet runway values of V_{EF} and V_1 are used. In determining the wet runway accelerate-stop distance, the stopping force from the wheel brakes may never exceed:

(i) The wheel brakes stopping force determined in meeting the requirements of § 25.101(i) and paragraph (a) of this section; and

(ii) The force resulting from the wet runway braking coefficient of friction determined in accordance with paragraphs (c) or (d) of this section, as applicable, taking into account the distribution of the normal load between braked and unbraked wheels at the most adverse center-of-gravity position approved for takeoff.

(c) The wet runway braking coefficient of friction for a smooth wet runway is defined as a curve of friction coefficient versus ground speed and must be computed as follows:

(1) The maximum tire-to-ground wet runway braking coefficient of friction is defined as:

Tire Pressure (psi)	Maximum Braking Coefficient (tire-to-ground)
50	$\mu_{t/g_{MAX}} = -0.0350\left(\dfrac{V}{100}\right)^3 + 0.306\left(\dfrac{V}{100}\right)^2 - 0.851\left(\dfrac{V}{100}\right) + 0.883$
100	$\mu_{t/g_{MAX}} = -0.0437\left(\dfrac{V}{100}\right)^3 + 0.320\left(\dfrac{V}{100}\right)^2 - 0.805\left(\dfrac{V}{100}\right) + 0.804$
200	$\mu_{t/g_{MAX}} = -0.0331\left(\dfrac{V}{100}\right)^3 + 0.252\left(\dfrac{V}{100}\right)^2 - 0.658\left(\dfrac{V}{100}\right) + 0.692$
300	$\mu_{t/g_{MAX}} = -0.0401\left(\dfrac{V}{100}\right)^3 + 0.263\left(\dfrac{V}{100}\right)^2 - 0.611\left(\dfrac{V}{100}\right) + 0.614$

Where—

Tire Pressure = maximum airplane operating tire pressure (psi);

$\mu_{t/g_{MAX}}$ = maximum tire-to-ground braking coefficient;

V = airplane true ground speed (knots); and

Linear interpolation may be used for tire pressures other than those listed.

(2) The maximum tire-to-ground wet runway braking coefficient of friction must be adjusted to take into account the efficiency of the anti-skid system on a wet runway. Anti-skid system operation must be demonstrated by flight testing on a smooth wet runway, and its efficiency must be determined. Unless a specific anti-skid system efficiency is determined from a quantitative analysis of the flight testing on a smooth wet runway, the maximum tire-to-ground wet runway braking coefficient of friction determined in paragraph (c)(1) of this section must be multiplied by the efficiency value associated with the type of anti-skid system installed on the airplane:

Type of anti-skid system	Efficiency value
On-Off	0.30
Quasi-Modulating	0.50
Fully Modulating	0.80

(d) At the option of the applicant, a higher wet runway braking coefficient of friction may be used for runway surfaces that have been grooved or treated with a porous friction course material. For grooved and porous friction course runways, the wet runway braking coefficient of friction is defined as either:

(1) 70 percent of the dry runway braking coefficient of friction used to determine the dry runway accelerate-stop distance; or

(2) The wet runway braking coefficient defined in paragraph (c) of this section, except that a specific anti-skid

system efficiency, if determined, is appropriate for a grooved or porous friction course wet runway, and the max-

imum tire-to-ground wet runway braking coefficient of friction is defined as:

Tire Pressure (psi)   Maximum Braking Coefficient (tire-to-ground)

50 $\quad \mu_{t/gMAX} = 0.1470\left(\dfrac{V}{100}\right)^5 - 1.050\left(\dfrac{V}{100}\right)^4 + 2.673\left(\dfrac{V}{100}\right)^3 - 2.683\left(\dfrac{V}{100}\right)^2 + 0.403\left(\dfrac{V}{100}\right) + 0.859$

100 $\quad \mu_{t/gMAX} = 0.1106\left(\dfrac{V}{100}\right)^5 - 0.813\left(\dfrac{V}{100}\right)^4 + 2.130\left(\dfrac{V}{100}\right)^3 - 2.200\left(\dfrac{V}{100}\right)^2 + 0.317\left(\dfrac{V}{100}\right) + 0.807$

200 $\quad \mu_{t/gMAX} = 0.0498\left(\dfrac{V}{100}\right)^5 - 0.398\left(\dfrac{V}{100}\right)^4 + 1.140\left(\dfrac{V}{100}\right)^3 - 1.285\left(\dfrac{V}{100}\right)^2 + 0.140\left(\dfrac{V}{100}\right) + 0.701$

300 $\quad \mu_{t/gMAX} = 0.0314\left(\dfrac{V}{100}\right)^5 - 0.247\left(\dfrac{V}{100}\right)^4 + 0.703\left(\dfrac{V}{100}\right)^3 - 0.779\left(\dfrac{V}{100}\right)^2 - 0.00954\left(\dfrac{V}{100}\right) + 0.614$

Where—

Tire Pressure = maximum airplane operating tire pressure (psi);

$\mu_{t/gMAX}$ = maximum tire-to-ground braking coefficient;

V = airplane true ground speed (knots); and

Linear interpolation may be used for tire pressures other than those listed.

(e) Except as provided in paragraph (f)(1) of this section, means other than wheel brakes may be used to determine the accelerate-stop distance if that means—

(1) Is safe and reliable;

(2) Is used so that consistent results can be expected under normal operating conditions; and

(3) Is such that exceptional skill is not required to control the airplane.

(f) The effects of available reverse thrust—

(1) Shall not be included as an additional means of deceleration when determining the accelerate-stop distance on a dry runway; and

(2) May be included as an additional means of deceleration using recommended reverse thrust procedures when determining the accelerate-stop distance on a wet runway, provided the requirements of paragraph (e) of this section are met.

(g) The landing gear must remain extended throughout the accelerate-stop distance.

(h) If the accelerate-stop distance includes a stopway with surface characteristics substantially different from

those of the runway, the takeoff data must include operational correction factors for the accelerate-stop distance. The correction factors must account for the particular surface characteristics of the stopway and the variations in these characteristics with seasonal weather conditions (such as temperature, rain, snow, and ice) within the established operational limits.

(i) A flight test demonstration of the maximum brake kinetic energy accelerate-stop distance must be conducted with not more than 10 percent of the allowable brake wear range remaining on each of the airplane wheel brakes.

[Doc. No. 5066, 29 FR 18291, Dec. 24, 1964, as amended by Amdt. 25–42, 43 FR 2321, Jan. 16, 1978; Amdt. 25–92, 63 FR 8318, Feb. 18, 1998]

§25.111 Takeoff path.

(a) The takeoff path extends from a standing start to a point in the takeoff at which the airplane is 1,500 feet above the takeoff surface, or at which the transition from the takeoff to the en route configuration is completed and V_{FTO} is reached, whichever point is higher. In addition—

(1) The takeoff path must be based on the procedures prescribed in §25.101(f);

(2) The airplane must be accelerated on the ground to V_{EF} at which point the critical engine must be made inoperative and remain inoperative for the rest of the takeoff; and

(3) After reaching V_{EF} the airplane must be accelerated to V_2.

(b) During the acceleration to speed V_2, the nose gear may be raised off the ground at a speed not less than V_R. However, landing gear retraction may not be begun until the airplane is airborne.

(c) During the takeoff path determination in accordance with paragraphs (a) and (b) of this section—

(1) The slope of the airborne part of the takeoff path must be positive at each point;

(2) The airplane must reach V_2 before it is 35 feet above the takeoff surface and must continue at a speed as close as practical to, but not less than V_2, until it is 400 feet above the takeoff surface;

(3) At each point along the takeoff path, starting at the point at which the airplane reaches 400 feet above the takeoff surface, the available gradient of climb may not be less than—

(i) 1.2 percent for two-engine airplanes;

(ii) 1.5 percent for three-engine airplanes; and

(iii) 1.7 percent for four-engine airplanes.

(4) The airplane configuration may not be changed, except for gear retraction and automatic propeller feathering, and no change in power or thrust that requires action by the pilot may be made until the airplane is 400 feet above the takeoff surface; and

(5) If § 25.105(a)(2) requires the takeoff path to be determined for flight in icing conditions, the airborne part of the takeoff must be based on the airplane drag:

(i) With the most critical of the takeoff ice accretion(s) defined in Appendices C and O of this part, as applicable, in accordance with § 25.21(g), from a height of 35 feet above the takeoff surface up to the point where the airplane is 400 feet above the takeoff surface; and

(ii) With the most critical of the final takeoff ice accretion(s) defined in Appendices C and O of this part, as applicable, in accordance with § 25.21(g), from the point where the airplane is 400 feet above the takeoff surface to the end of the takeoff path.

(d) The takeoff path must be determined by a continuous demonstrated takeoff or by synthesis from segments. If the takeoff path is determined by the segmental method—

(1) The segments must be clearly defined and must be related to the distinct changes in the configuration, power or thrust, and speed;

(2) The weight of the airplane, the configuration, and the power or thrust must be constant throughout each segment and must correspond to the most critical condition prevailing in the segment;

(3) The flight path must be based on the airplane's performance without ground effect; and

(4) The takeoff path data must be checked by continuous demonstrated takeoffs up to the point at which the airplane is out of ground effect and its speed is stabilized, to ensure that the path is conservative relative to the continous path.

The airplane is considered to be out of the ground effect when it reaches a height equal to its wing span.

(e) For airplanes equipped with standby power rocket engines, the takeoff path may be determined in accordance with section II of appendix E.

[Doc. No. 5066, 29 FR 18291, Dec. 24, 1964, as amended by Amdt. 25–6, 30 FR 8468, July 2, 1965; Amdt. 25–42, 43 FR 2321, Jan. 16, 1978; Amdt. 25–54, 45 FR 60172, Sept. 11, 1980; Amdt. 25–72, 55 FR 29774, July 20, 1990; Amdt. 25–94, 63 FR 8848, Feb. 23, 1998; Amdt. 25–108, 67 FR 70826, Nov. 26, 2002; Amdt. 25–115, 69 FR 40527, July 2, 2004; Amdt. 25–121, 72 FR 44666; Aug. 8, 2007; Amdt. 25–140, 79 FR 65525, Nov. 4, 2014]

§ 25.113   Takeoff distance and takeoff run.

(a) Takeoff distance on a dry runway is the greater of—

(1) The horizontal distance along the takeoff path from the start of the takeoff to the point at which the airplane is 35 feet above the takeoff surface, determined under § 25.111 for a dry runway; or

(2) 115 percent of the horizontal distance along the takeoff path, with all engines operating, from the start of the takeoff to the point at which the airplane is 35 feet above the takeoff surface, as determined by a procedure consistent with § 25.111.

(b) Takeoff distance on a wet runway is the greater of—

(1) The takeoff distance on a dry runway determined in accordance with paragraph (a) of this section; or

(2) The horizontal distance along the takeoff path from the start of the takeoff to the point at which the airplane is 15 feet above the takeoff surface, achieved in a manner consistent with the achievement of V_2 before reaching 35 feet above the takeoff surface, determined under § 25.111 for a wet runway.

(c) If the takeoff distance does not include a clearway, the takeoff run is equal to the takeoff distance. If the takeoff distance includes a clearway—

(1) The takeoff run on a dry runway is the greater of—

(i) The horizontal distance along the takeoff path from the start of the takeoff to a point equidistant between the point at which V_{LOF} is reached and the point at which the airplane is 35 feet above the takeoff surface, as determined under § 25.111 for a dry runway; or

(ii) 115 percent of the horizontal distance along the takeoff path, with all engines operating, from the start of the takeoff to a point equidistant between the point at which V_{LOF} is reached and the point at which the airplane is 35 feet above the takeoff surface, determined by a procedure consistent with § 25.111.

(2) The takeoff run on a wet runway is the greater of—

(i) The horizontal distance along the takeoff path from the start of the takeoff to the point at which the airplane is 15 feet above the takeoff surface, achieved in a manner consistent with the achievement of V_2 before reaching 35 feet above the takeoff surface, as determined under § 25.111 for a wet runway; or

(ii) 115 percent of the horizontal distance along the takeoff path, with all engines operating, from the start of the takeoff to a point equidistant between the point at which V_{LOF} is reached and the point at which the airplane is 35 feet above the takeoff surface, determined by a procedure consistent with § 25.111.

[Doc. No. 5066, 29 FR 18291, Dec. 24, 1964, as amended by Amdt. 25–23, 35 FR 5671, Apr. 8, 1970; Amdt. 25–92, 63 FR 8320, Feb. 18, 1998]

§ 25.115 Takeoff flight path.

(a) The takeoff flight path shall be considered to begin 35 feet above the takeoff surface at the end of the takeoff distance determined in accordance with § 25.113(a) or (b), as appropriate for the runway surface condition.

(b) The net takeoff flight path data must be determined so that they represent the actual takeoff flight paths (determined in accordance with § 25.111 and with paragraph (a) of this section) reduced at each point by a gradient of climb equal to—

(1) 0.8 percent for two-engine airplanes;

(2) 0.9 percent for three-engine airplanes; and

(3) 1.0 percent for four-engine airplanes.

(c) The prescribed reduction in climb gradient may be applied as an equivalent reduction in acceleration along that part of the takeoff flight path at which the airplane is accelerated in level flight.

[Doc. No. 5066, 29 FR 18291, Dec. 24, 1964, as amended by Amdt. 25–92, 63 FR 8320, Feb. 18, 1998]

§ 25.117 Climb: general.

Compliance with the requirements of §§ 25.119 and 25.121 must be shown at each weight, altitude, and ambient temperature within the operational limits established for the airplane and with the most unfavorable center of gravity for each configuration.

§ 25.119 Landing climb: All-engines-operating.

In the landing configuration, the steady gradient of climb may not be less than 3.2 percent, with the engines at the power or thrust that is available 8 seconds after initiation of movement of the power or thrust controls from the minimum flight idle to the go-around power or thrust setting—

(a) In non-icing conditions, with a climb speed of V_{REF} determined in accordance with § 25.125(b)(2)(i); and

(b) In icing conditions with the most critical of the landing ice accretion(s) defined in Appendices C and O of this part, as applicable, in accordance with § 25.21(g), and with a climb speed of

V_{REF} determined in accordance with § 25.125(b)(2)(ii).

[Amdt. 25–121, 72 FR 44666; Aug. 8, 2007, as amended by Amdt. 25–,140, 79 FR 65525, Nov. 4, 2014]

§ 25.121 Climb: One-engine-inoperative.

(a) *Takeoff; landing gear extended.* In the critical takeoff configuration existing along the flight path (between the points at which the airplane reaches V_{LOF} and at which the landing gear is fully retracted) and in the configuration used in § 25.111 but without ground effect, the steady gradient of climb must be positive for two-engine airplanes, and not less than 0.3 percent for three-engine airplanes or 0.5 percent for four-engine airplanes, at V_{LOF} and with—

(1) The critical engine inoperative and the remaining engines at the power or thrust available when retraction of the landing gear is begun in accordance with § 25.111 unless there is a more critical power operating condition existing later along the flight path but before the point at which the landing gear is fully retracted; and

(2) The weight equal to the weight existing when retraction of the landing gear is begun, determined under § 25.111.

(b) *Takeoff; landing gear retracted.* In the takeoff configuration existing at the point of the flight path at which the landing gear is fully retracted, and in the configuration used in § 25.111 but without ground effect:

(1) The steady gradient of climb may not be less than 2.4 percent for two-engine airplanes, 2.7 percent for three-engine airplanes, and 3.0 percent for four-engine airplanes, at V_2 with:

(i) The critical engine inoperative, the remaining engines at the takeoff power or thrust available at the time the landing gear is fully retracted, determined under § 25.111, unless there is a more critical power operating condition existing later along the flight path but before the point where the airplane reaches a height of 400 feet above the takeoff surface; and

(ii) The weight equal to the weight existing when the airplane's landing gear is fully retracted, determined under § 25.111.

(2) The requirements of paragraph (b)(1) of this section must be met:

(i) In non-icing conditions; and

(ii) In icing conditions with the most critical of the takeoff ice accretion(s) defined in Appendices C and O of this part, as applicable, in accordance with § 25.21(g), if in the configuration used to show compliance with § 25.121(b) with this takeoff ice accretion:

(A) The stall speed at maximum takeoff weight exceeds that in non-icing conditions by more than the greater of 3 knots CAS or 3 percent of V_{SR}; or

(B) The degradation of the gradient of climb determined in accordance with § 25.121(b) is greater than one-half of the applicable actual-to-net takeoff flight path gradient reduction defined in § 25.115(b).

(c) *Final takeoff.* In the en route configuration at the end of the takeoff path determined in accordance with § 25.111:

(1) The steady gradient of climb may not be less than 1.2 percent for two-engine airplanes, 1.5 percent for three-engine airplanes, and 1.7 percent for four-engine airplanes, at V_{FTO} with—

(i) The critical engine inoperative and the remaining engines at the available maximum continuous power or thrust; and

(ii) The weight equal to the weight existing at the end of the takeoff path, determined under § 25.111.

(2) The requirements of paragraph (c)(1) of this section must be met:

(i) In non-icing conditions; and

(ii) In icing conditions with the most critical of the final takeoff ice accretion(s) defined in Appendices C and O of this part, as applicable, in accordance with § 25.21(g), if in the configuration used to show compliance with § 25.121(b) with the takeoff ice accretion used to show compliance with § 25.111(c)(5)(i):

(A) The stall speed at maximum takeoff weight exceeds that in non-icing conditions by more than the greater of 3 knots CAS or 3 percent of V_{SR}; or

(B) The degradation of the gradient of climb determined in accordance with § 25.121(b) is greater than one-half of the applicable actual-to-net takeoff flight path gradient reduction defined in § 25.115(b).

(d) *Approach.* In a configuration corresponding to the normal all-engines-operating procedure in which V_{SR} for this configuration does not exceed 110 percent of the V_{SR} for the related all-engines-operating landing configuration:

(1) The steady gradient of climb may not be less than 2.1 percent for two-engine airplanes, 2.4 percent for three-engine airplanes, and 2.7 percent for four-engine airplanes, with—

(i) The critical engine inoperative, the remaining engines at the go-around power or thrust setting;

(ii) The maximum landing weight;

(iii) A climb speed established in connection with normal landing procedures, but not exceeding 1.4 V_{SR}; and

(iv) Landing gear retracted.

(2) The requirements of paragraph (d)(1) of this section must be met:

(i) In non-icing conditions; and

(ii) In icing conditions with the most critical of the approach ice accretion(s) defined in Appendices C and O of this part, as applicable, in accordance with § 25.21(g). The climb speed selected for non-icing conditions may be used if the climb speed for icing conditions, computed in accordance with paragraph (d)(1)(iii) of this section, does not exceed that for non-icing conditions by more than the greater of 3 knots CAS or 3 percent.

[Doc. No. 5066, 29 FR 18291, Dec. 24, 1964, as amended by Amdt. 25–84, 60 FR 30749, June 9, 1995; Amdt. 25–108, 67 FR 70826, Nov. 26, 2002; Amdt. 25–121, 72 FR 44666; Aug. 8, 2007; Amdt. 25–140, 79 FR 65525, Nov. 4, 2014]

§ 25.123 En route flight paths.

(a) For the en route configuration, the flight paths prescribed in paragraph (b) and (c) of this section must be determined at each weight, altitude, and ambient temperature, within the operating limits established for the airplane. The variation of weight along the flight path, accounting for the progressive consumption of fuel and oil by the operating engines, may be included in the computation. The flight paths must be determined at a speed not less than V_{FTO}, with—

(1) The most unfavorable center of gravity;

(2) The critical engines inoperative;

(3) The remaining engines at the available maximum continuous power or thrust; and

(4) The means for controlling the engine-cooling air supply in the position that provides adequate cooling in the hot-day condition.

(b) The one-engine-inoperative net flight path data must represent the actual climb performance diminished by a gradient of climb of 1.1 percent for two-engine airplanes, 1.4 percent for three-engine airplanes, and 1.6 percent for four-engine airplanes—

(1) In non-icing conditions; and

(2) In icing conditions with the most critical of the en route ice accretion(s) defined in Appendices C and O of this part, as applicable, in accordance with § 25.21(g), if:

(i) A speed of 1.18 "V_{SR0} with the en route ice accretion exceeds the en route speed selected for non-icing conditions by more than the greater of 3 knots CAS or 3 percent of V_{SR}; or

(ii) The degradation of the gradient of climb is greater than one-half of the applicable actual-to-net flight path reduction defined in paragraph (b) of this section.

(c) For three- or four-engine airplanes, the two-engine-inoperative net flight path data must represent the actual climb performance diminished by a gradient of climb of 0.3 percent for three-engine airplanes and 0.5 percent for four-engine airplanes.

[Doc. No. 5066, 29 FR 18291, Dec. 24, 1964, as amended by Amdt. 25–121, 72 FR 44666; Aug. 8, 2007; Amdt. 25–140, 79 FR 65525, Nov. 4, 2014]

§ 25.125 Landing.

(a) The horizontal distance necessary to land and to come to a complete stop (or to a speed of approximately 3 knots for water landings) from a point 50 feet above the landing surface must be determined (for standard temperatures, at each weight, altitude, and wind within the operational limits established by the applicant for the airplane):

(1) In non-icing conditions; and

(2) In icing conditions with the most critical of the landing ice accretion(s) defined in Appendices C and O of this part, as applicable, in accordance with § 25.21(g), if V_{REF} for icing conditions exceeds V_{REF} for non-icing conditions by

more than 5 knots CAS at the maximum landing weight.

(d) In determining the distance in paragraph (a) of this section:

(1) The airplane must be in the landing configuration.

(2) A stabilized approach, with a calibrated airspeed of not less than V_{REF}, must be maintained down to the 50-foot height.

(i) In non-icing conditions, V_{REF} may not be less than:

(A) 1.23 $V_{SR}0$;

(B) V_{MCL} established under § 25.149(f); and

(C) A speed that provides the maneuvering capability specified in § 25.143(h).

(ii) In icing conditions, V_{REF} may not be less than:

(A) The speed determined in paragraph (b)(2)(i) of this section;

(B) 1.23 V_{SR0} with the most critical of the landing ice accretion(s) defined in Appendices C and O of this part, as applicable, in accordance with § 25.21(g), if that speed exceeds V_{REF} selected for non-icing conditions by more than 5 knots CAS; and

(C) A speed that provides the maneuvering capability specified in § 25.143(h) with the most critical of the landing ice accretion(s) defined in Appendices C and O of this part, as applicable, in accordance with § 25.21(g).

(3) Changes in configuration, power or thrust, and speed, must be made in accordance with the established procedures for service operation.

(4) The landing must be made without excessive vertical acceleration, tendency to bounce, nose over, ground loop, porpoise, or water loop.

(5) The landings may not require exceptional piloting skill or alertness.

(c) For landplanes and amphibians, the landing distance on land must be determined on a level, smooth, dry, hard-surfaced runway. In addition—

(1) The pressures on the wheel braking systems may not exceed those specified by the brake manufacturer;

(2) The brakes may not be used so as to cause excessive wear of brakes or tires; and

(3) Means other than wheel brakes may be used if that means—

(i) Is safe and reliable;

(ii) Is used so that consistent results can be expected in service; and

(iii) Is such that exceptional skill is not required to control the airplane.

(d) For seaplanes and amphibians, the landing distance on water must be determined on smooth water.

(e) For skiplanes, the landing distance on snow must be determined on smooth, dry, snow.

(f) The landing distance data must include correction factors for not more than 50 percent of the nominal wind components along the landing path opposite to the direction of landing, and not less than 150 percent of the nominal wind components along the landing path in the direction of landing.

(g) If any device is used that depends on the operation of any engine, and if the landing distance would be noticeably increased when a landing is made with that engine inoperative, the landing distance must be determined with that engine inoperative unless the use of compensating means will result in a landing distance not more than that with each engine operating.

[Amdt. 25–121, 72 FR 44666; Aug. 8, 2007; 72 FR 50467, Aug. 31, 2007; Amdt. 25–140, 79 FR 65525, Nov. 4, 2014]

CONTROLLABILITY AND
MANEUVERABILITY

§ 25.143 General.

(a) The airplane must be safely controllable and maneuverable during—

(1) Takeoff;

(2) Climb;

(3) Level flight;

(4) Descent; and

(5) Landing.

(b) It must be possible to make a smooth transition from one flight condition to any other flight condition without exceptional piloting skill, alertness, or strength, and without danger of exceeding the airplane limit-load factor under any probable operating conditions, including—

(1) The sudden failure of the critical engine;

(2) For airplanes with three or more engines, the sudden failure of the second critical engine when the airplane is in the en route, approach, or landing configuration and is trimmed with the critical engine inoperative; and

(3) Configuration changes, including deployment or retraction of deceleration devices.

(c) The airplane must be shown to be safely controllable and maneuverable with the most critical of the ice accretion(s) appropriate to the phase of flight as defined in Appendices C and O of this part, as applicable, in accordance with §25.21(g), and with the critical engine inoperative and its propeller (if applicable) in the minimum drag position:

(1) At the minimum V_2 for takeoff;

(2) During an approach and go-around; and

(3) During an approach and landing.

(d) The following table prescribes, for conventional wheel type controls, the maximum control forces permitted during the testing required by paragraph (a) through (c) of this section:

Force, in pounds, applied to the control wheel or rudder pedals	Pitch	Roll	Yaw
For short term application for pitch and roll control—two hands available for control	75	50	
For short term application for pitch and roll control—one hand available for control	50	25	
For short term application for yaw control			150
For long term application	10	5	20

(e) Approved operating procedures or conventional operating practices must be followed when demonstrating compliance with the control force limitations for short term application that are prescribed in paragraph (d) of this section. The airplane must be in trim, or as near to being in trim as practical, in the preceding steady flight condition. For the takeoff condition, the airplane must be trimmed according to the approved operating procedures.

(f) When demonstrating compliance with the control force limitations for long term application that are prescribed in paragraph (d) of this section, the airplane must be in trim, or as near to being in trim as practical.

(g) When maneuvering at a constant airspeed or Mach number (up to V_{FC}/M_{FC}), the stick forces and the gradient of the stick force versus maneuvering load factor must lie within satisfactory limits. The stick forces must not be so great as to make excessive demands on the pilot's strength when maneuvering the airplane, and must not be so low that the airplane can easily be overstressed inadvertently. Changes of gradient that occur with changes of load factor must not cause undue difficulty in maintaining control of the airplane, and local gradients must not be so low as to result in a danger of overcontrolling.

(h) The maneuvering capabilities in a constant speed coordinated turn at forward center of gravity, as specified in the following table, must be free of stall warning or other characteristics that might interfere with normal maneuvering:

Configuration	Speed	Maneuvering bank angle in a coordinated turn	Thrust/power setting
Takeoff	V_2	30°	Asymmetric WAT-Limited. [1]
Takeoff	[2]V_2 + XX	40°	All-engines-operating climb. [3]
En route	V_{FTO}	40°	Asymmetric WAT-Limited. [1]
Landing	V_{REF}	40°	Symmetric for −3° flight path angle.

[1] A combination of weight, altitude, and temperature (WAT) such that the thrust or power setting produces the minimum climb gradient specified in §25.121 for the flight condition.

[2] Airspeed approved for all-engines-operating initial climb.

[3] That thrust or power setting which, in the event of failure of the critical engine and without any crew action to adjust the thrust or power of the remaining engines, would result in the thrust or power specified for the takeoff condition at V_2, or any lesser thrust or power setting that is used for all-engines-operating initial climb procedures.

(i) When demonstrating compliance with §25.143 in icing conditions—

(1) Controllability must be demonstrated with the most critical of the ice accretion(s) for the particular flight phase as defined in Appendices C and O of this part, as applicable, in accordance with §25.21(g);

(2) It must be shown that a push force is required throughout a pushover maneuver down to a zero g load factor, or the lowest load factor obtainable if limited by elevator power or other design characteristic of the flight control system. It must be possible to promptly recover from the maneuver without

exceeding a pull control force of 50 pounds; and

(3) Any changes in force that the pilot must apply to the pitch control to maintain speed with increasing sideslip angle must be steadily increasing with no force reversals, unless the change in control force is gradual and easily controllable by the pilot without using exceptional piloting skill, alertness, or strength.

(j) For flight in icing conditions before the ice protection system has been activated and is performing its intended function, it must be demonstrated in flight with the most critical of the ice accretion(s) defined in Appendix C, part II, paragraph (e) of this part and Appendix O, part II, paragraph (d) of this part, as applicable, in accordance with § 25.21(g), that:

(1) The airplane is controllable in a pull-up maneuver up to 1.5 g load factor; and

(2) There is no pitch control force reversal during a pushover maneuver down to 0.5 g load factor.

[Doc. No. 5066, 29 FR 18291, Dec. 24, 1964, as amended by Amdt. 25–42, 43 FR 2321, Jan. 16, 1978; Amdt. 25–84, 60 FR 30749, June 9, 1995; Amdt. 25–108, 67 FR 70826, Nov. 26, 2002; Amdt. 25–121, 72 FR 44667, Aug. 8, 2007; Amdt. 25–129, 74 FR 38339, Aug. 3, 2009; Amdt. 25–140, 79 FR 65525, Nov. 4, 2014]

§ 25.145 Longitudinal control.

(a) It must be possible, at any point between the trim speed prescribed in § 25.103(b)(6) and stall identification (as defined in § 25.201(d)), to pitch the nose downward so that the acceleration to this selected trim speed is prompt with

(1) The airplane trimmed at the trim speed prescribed in § 25.103(b)(6);

(2) The landing gear extended;

(3) The wing flaps (i) retracted and (ii) extended; and

(4) Power (i) off and (ii) at maximum continuous power on the engines.

(b) With the landing gear extended, no change in trim control, or exertion of more than 50 pounds control force (representative of the maximum short term force that can be applied readily by one hand) may be required for the following maneuvers:

(1) With power off, flaps retracted, and the airplane trimmed at 1.3 V_{SR1}, extend the flaps as rapidly as possible while maintaining the airspeed at approximately 30 percent above the reference stall speed existing at each instant throughout the maneuver.

(2) Repeat paragraph (b)(1) except initially extend the flaps and then retract them as rapidly as possible.

(3) Repeat paragraph (b)(2), except at the go-around power or thrust setting.

(4) With power off, flaps retracted, and the airplane trimmed at 1.3 V_{SR1}, rapidly set go-around power or thrust while maintaining the same airspeed.

(5) Repeat paragraph (b)(4) except with flaps extended.

(6) With power off, flaps extended, and the airplane trimmed at 1.3 V_{SR1}, obtain and maintain airspeeds between V_{SW} and either 1.6 V_{SR1} or V_{FE}, whichever is lower.

(c) It must be possible, without exceptional piloting skill, to prevent loss of altitude when complete retraction of the high lift devices from any position is begun during steady, straight, level flight at 1.08 V_{SR1} for propeller powered airplanes, or 1.13 V_{SR1} for turbojet powered airplanes, with—

(1) Simultaneous movement of the power or thrust controls to the go-around power or thrust setting;

(2) The landing gear extended; and

(3) The critical combinations of landing weights and altitudes.

(d) If gated high-lift device control positions are provided, paragraph (c) of this section applies to retractions of the high-lift devices from any position from the maximum landing position to the first gated position, between gated positions, and from the last gated position to the fully retracted position. The requirements of paragraph (c) of this section also apply to retractions from each approved landing position to the control position(s) associated with the high-lift device configuration(s) used to establish the go-around procedure(s) from that landing position. In addition, the first gated control position from the maximum landing position must correspond with a configuration of the high-lift devices used to establish a go-around procedure from a landing configuration. Each gated control position must require a separate and distinct motion of the control to pass through the gated position and must have features to prevent inadvertent movement of the control through the gated position. It must only be possible to make this separate and distinct motion once the control has reached the gated position.

[Doc. No. 5066, 29 FR 18291, Dec. 24, 1964, as amended by Amdt. 25–23, 35 FR 5671, Apr. 8, 1970; Amdt. 25–72, 55 FR 29774, July 20, 1990; Amdt. 25–84, 60 FR 30749, June 9, 1995; Amdt. 25–98, 64 FR 6164, Feb. 8, 1999; 64 FR 10740, Mar. 5, 1999; Amdt. 25–108, 67 FR 70827, Nov. 26, 2002]

§ 25.147 Directional and lateral control.

(a) *Directional control; general.* It must be possible, with the wings level, to yaw into the operative engine and to safely make a reasonably sudden change in heading of up to 15 degrees in the direction of the critical inoperative engine. This must be shown at 1.3 V_SR1 for heading changes up to 15 degrees (except that the heading change at which the rudder pedal force is 150 pounds need not be exceeded), and with—

(1) The critical engine inoperative and its propeller in the minimum drag position;

(2) The power required for level flight at 1.3 V_SR1, but not more than maximum continuous power;

(3) The most unfavorable center of gravity;

(4) Landing gear retracted;

(5) Flaps in the approach position; and

(6) Maximum landing weight.

(b) *Directional control; airplanes with four or more engines.* Airplanes with four or more engines must meet the requirements of paragraph (a) of this section except that—

(1) The two critical engines must be inoperative with their propellers (if applicable) in the minimum drag position;

(2) [Reserved]

(3) The flaps must be in the most favorable climb position.

(c) *Lateral control; general.* It must be possible to make 20° banked turns, with and against the inoperative engine, from steady flight at a speed equal to 1.3 V_SR1, with—

(1) The critical engine inoperative and its propeller (if applicable) in the minimum drag position;

(2) The remaining engines at maximum continuous power;

(3) The most unfavorable center of gravity;

(4) Landing gear (i) retracted and (ii) extended;

(5) Flaps in the most favorable climb position; and

(6) Maximum takeoff weight.

(d) *Lateral control; roll capability.* With the critical engine inoperative, roll response must allow normal maneuvers. Lateral control must be sufficient, at the speeds likely to be used with one engine inoperative, to provide a roll rate necessary for safety without excessive control forces or travel.

(e) *Lateral control; airplanes with four or more engines.* Airplanes with four or more engines must be able to make 20° banked turns, with and against the inoperative engines, from steady flight at a speed equal to 1.3 V_SR1, with maximum continuous power, and with the airplane in the configuration prescribed by paragraph (b) of this section.

(f) *Lateral control; all engines operating.* With the engines operating, roll response must allow normal maneuvers (such as recovery from upsets produced by gusts and the initiation of evasive maneuvers). There must be enough excess lateral control in sideslips (up to sideslip angles that might be required in normal operation), to allow a limited amount of maneuvering and to correct for gusts. Lateral control must be enough at any speed up to V_{FC}/M_{FC} to provide a peak roll rate necessary for safety, without excessive control forces or travel.

[Doc. No. 5066, 29 FR 18291, Dec. 24, 1964, as amended by Amdt. 25–42, 43 FR 2321, Jan. 16, 1978; Amdt. 25–72, 55 FR 29774, July 20, 1990; Amdt. 25–108, 67 FR 70827, Nov. 26, 2002; Amdt. 25–115, 69 FR 40527, July 2, 2004]

§ 25.149 Minimum control speed.

(a) In establishing the minimum control speeds required by this section, the method used to simulate critical engine failure must represent the most critical mode of powerplant failure with respect to controllability expected in service.

(b) V_{MC} is the calibrated airspeed at which, when the critical engine is suddenly made inoperative, it is possible to maintain control of the airplane with that engine still inoperative and maintain straight flight with an angle of bank of not more than 5 degrees.

(c) V_{MC} may not exceed 1.13 V_{SR} with—

(1) Maximum available takeoff power or thrust on the engines;

(2) The most unfavorable center of gravity;

(3) The airplane trimmed for takeoff;

(4) The maximum sea level takeoff weight (or any lesser weight necessary to show V_{MC});

(5) The airplane in the most critical takeoff configuration existing along the flight path after the airplane becomes airborne, except with the landing gear retracted;

(6) The airplane airborne and the ground effect negligible; and

(7) If applicable, the propeller of the inoperative engine—

(i) Windmilling;

(ii) In the most probable position for the specific design of the propeller control; or

(iii) Feathered, if the airplane has an automatic feathering device acceptable for showing compliance with the climb requirements of § 25.121.

(d) The rudder forces required to maintain control at V_{MC} may not exceed 150 pounds nor may it be necessary to reduce power or thrust of the operative engines. During recovery, the airplane may not assume any dangerous attitude or require exceptional piloting skill, alertness, or strength to prevent a heading change of more than 20 degrees.

(e) V_{MCG}, the minimum control speed on the ground, is the calibrated airspeed during the takeoff run at which, when the critical engine is suddenly made inoperative, it is possible to maintain control of the airplane using the rudder control alone (without the use of nosewheel steering), as limited by 150 pounds of force, and the lateral control to the extent of keeping the wings level to enable the takeoff to be safely continued using normal piloting skill. In the determination of V_{MCG}, assuming that the path of the airplane accelerating with all engines operating is along the centerline of the runway, its path from the point at which the critical engine is made inoperative to the point at which recovery to a direc-

tion parallel to the centerline is completed may not deviate more than 30 feet laterally from the centerline at any point. V_{MCG} must be established with—

(1) The airplane in each takeoff configuration or, at the option of the applicant, in the most critical takeoff configuration;

(2) Maximum available takeoff power or thrust on the operating engines;

(3) The most unfavorable center of gravity;

(4) The airplane trimmed for takeoff; and

(5) The most unfavorable weight in the range of takeoff weights.

(f) V_{MCL}, the minimum control speed during approach and landing with all engines operating, is the calibrated airspeed at which, when the critical engine is suddenly made inoperative, it is possible to maintain control of the airplane with that engine still inoperative, and maintain straight flight with an angle of bank of not more than 5 degrees. V_{MCL} must be established with—

(1) The airplane in the most critical configuration (or, at the option of the applicant, each configuration) for approach and landing with all engines operating;

(2) The most unfavorable center of gravity;

(3) The airplane trimmed for approach with all engines operating;

(4) The most favorable weight, or, at the option of the applicant, as a function of weight;

(5) For propeller airplanes, the propeller of the inoperative engine in the position it achieves without pilot action, assuming the engine fails while at the power or thrust necessary to maintain a three degree approach path angle; and

(6) Go-around power or thrust setting on the operating engine(s).

(g) For airplanes with three or more engines, V_{MCL-2}, the minimum control speed during approach and landing with one critical engine inoperative, is the calibrated airspeed at which, when a second critical engine is suddenly made inoperative, it is possible to maintain control of the airplane with both engines still inoperative, and maintain straight flight with an angle of bank of not more than 5 degrees. V_{MCL-2} must be established with—

(1) The airplane in the most critical configuration (or, at the option of the applicant, each configuration) for approach and landing with one critical engine inoperative;

(2) The most unfavorable center of gravity;

(3) The airplane trimmed for approach with one critical engine inoperative;

(4) The most unfavorable weight, or, at the option of the applicant, as a function of weight;

(5) For propeller airplanes, the propeller of the more critical inoperative engine in the position it achieves without pilot action, assuming the engine fails while at the power or thrust necessary to maintain a three degree approach path angle, and the propeller of the other inoperative engine feathered;

(6) The power or thrust on the operating engine(s) necessary to maintain an approach path angle of three degrees when one critical engine is inoperative; and

(7) The power or thrust on the operating engine(s) rapidly changed, immediately after the second critical engine is made inoperative, from the power or thrust prescribed in paragraph (g)(6) of this section to—

(i) Minimum power or thrust; and

(ii) Go-around power or thrust setting.

(h) In demonstrations of V_{MCL} and V_{MCL-2}—

(1) The rudder force may not exceed 150 pounds;

(2) The airplane may not exhibit hazardous flight characteristics or require exceptional piloting skill, alertness, or strength;

(3) Lateral control must be sufficient to roll the airplane, from an initial condition of steady flight, through an angle of 20 degrees in the direction necessary to initiate a turn away from the inoperative engine(s), in not more than 5 seconds; and

(4) For propeller airplanes, hazardous flight characteristics must not be exhibited due to any propeller position achieved when the engine fails or during any likely subsequent movements of the engine or propeller controls.

[Doc. No. 5066, 29 FR 18291, Dec. 24, 1964, as amended by Amdt. 25–42, 43 FR 2321, Jan. 16, 1978; Amdt. 25–72, 55 FR 29774, July 20, 1990; 55 FR 37607, Sept. 12, 1990; Amdt. 25–84, 60 FR 30749, June 9, 1995; Amdt. 25–108, 67 FR 70827, Nov. 26, 2002]

TRIM

§25.161   Trim.

(a) *General.* Each airplane must meet the trim requirements of this section after being trimmed, and without further pressure upon, or movement of, either the primary controls or their corresponding trim controls by the pilot or the automatic pilot.

(b) *Lateral and directional trim.* The airplane must maintain lateral and directional trim with the most adverse lateral displacement of the center of gravity within the relevant operating limitations, during normally expected conditions of operation (including operation at any speed from 1.3 V_{SR1} to V_{MO}/M_{MO}).

(c) *Longitudinal trim.* The airplane must maintain longitudinal trim during—

(1) A climb with maximum continuous power at a speed not more than 1.3 V_{SR1}, with the landing gear retracted, and the flaps (i) retracted and (ii) in the takeoff position;

(2) Either a glide with power off at a speed not more than 1.3 V_{SR1}, or an approach within the normal range of approach speeds appropriate to the weight and configuration with power settings corresponding to a 3 degree glidepath, whichever is the most severe, with the landing gear extended, the wing flaps (i) retracted and (ii) extended, and with the most unfavorable combination of center of gravity position and weight approved for landing; and

(3) Level flight at any speed from 1.3 V_{SR1}, to V_{MO}/M_{MO}, with the landing gear and flaps retracted, and from 1.3 V_{SR1} to V_{LE} with the landing gear extended.

(d) *Longitudinal, directional, and lateral trim.* The airplane must maintain longitudinal, directional, and lateral trim (and for the lateral trim, the

223

angle of bank may not exceed five degrees) at 1.3 V_{SR1} during climbing flight with—

(1) The critical engine inoperative;

(2) The remaining engines at maximum continuous power; and

(3) The landing gear and flaps retracted.

(e) *Airplanes with four or more engines.* Each airplane with four or more engines must also maintain trim in rectilinear flight with the most unfavorable center of gravity and at the climb speed, configuration, and power required by § 25.123(a) for the purpose of establishing the en route flight paths with two engines inoperative.

[Doc. No. 5066, 29 FR 18291, Dec. 24, 1964, as amended by Amdt. 25–23, 35 FR 5671, Apr. 8, 1970; Amdt. 25–38, 41 FR 55466, Dec. 20, 1976; Amdt. 25–108, 67 FR 70827, Nov. 26, 2002; Amdt. 25–115, 69 FR 40527, July 2, 2004]

STABILITY

§ 25.171   General.

The airplane must be longitudinally, directionally, and laterally stable in accordance with the provisions of §§ 25.173 through 25.177. In addition, suitable stability and control feel (static stability) is required in any condition normally encountered in service, if flight tests show it is necessary for safe operation.

[Doc. No. 5066, 29 FR 18291, Dec. 24, 1964, as amended by Amdt. 25–7, 30 FR 13117, Oct. 15, 1965]

§ 25.173   Static longitudinal stability.

Under the conditions specified in § 25.175, the characteristics of the elevator control forces (including friction) must be as follows:

(a) A pull must be required to obtain and maintain speeds below the specified trim speed, and a push must be required to obtain and maintain speeds above the specified trim speed. This must be shown at any speed that can be obtained except speeds higher than the landing gear or wing flap operating limit speeds or V_{FC}/M_{FC}, whichever is appropriate, or lower than the minimum speed for steady unstalled flight.

(b) The airspeed must return to within 10 percent of the original trim speed for the climb, approach, and landing conditions specified in § 25.175 (a), (c),

and (d), and must return to within 7.5 percent of the original trim speed for the cruising condition specified in § 25.175(b), when the control force is slowly released from any speed within the range specified in paragraph (a) of this section.

(c) The average gradient of the stable slope of the stick force versus speed curve may not be less than 1 pound for each 6 knots.

(d) Within the free return speed range specified in paragraph (b) of this section, it is permissible for the airplane, without control forces, to stabilize on speeds above or below the desired trim speeds if exceptional attention on the part of the pilot is not required to return to and maintain the desired trim speed and altitude.

[Amdt. 25–7, 30 FR 13117, Oct. 15, 1965]

§ 25.175   Demonstration of static longitudinal stability.

Static longitudinal stability must be shown as follows:

(a) *Climb.* The stick force curve must have a stable slope at speeds between 85 and 115 percent of the speed at which the airplane—

(1) Is trimmed, with—

(i) Wing flaps retracted;

(ii) Landing gear retracted;

(iii) Maximum takeoff weight; and

(iv) 75 percent of maximum continuous power for reciprocating engines or the maximum power or thrust selected by the applicant as an operating limitation for use during climb for turbine engines; and

(2) Is trimmed at the speed for best rate-of-climb except that the speed need not be less than 1.3 V_{SR1}.

(b) *Cruise.* Static longitudinal stability must be shown in the cruise condition as follows:

(1) With the landing gear retracted at high speed, the stick force curve must have a stable slope at all speeds within a range which is the greater of 15 percent of the trim speed plus the resulting free return speed range, or 50 knots plus the resulting free return speed range, above and below the trim speed (except that the speed range need not include speeds less than 1.3 V_{SR1}, nor speeds greater than V_{FC}/M_{FC} nor speeds that require a stick force of more than 50 pounds), with—

224

(i) The wing flaps retracted;

(ii) The center of gravity in the most adverse position (see § 25.27);

(iii) The most critical weight between the maximum takeoff and maximum landing weights;

(iv) 75 percent of maximum continuous power for reciprocating engines or for turbine engines, the maximum cruising power selected by the applicant as an operating limitation (see § 25.1521), except that the power need not exceed that required at V_{MO}/M_{MO}; and

(v) The airplane trimmed for level flight with the power required in paragraph (b)(1)(iv) of this section.

(2) With the landing gear retracted at low speed, the stick force curve must have a stable slope at all speeds within a range which is the greater of 15 percent of the trim speed plus the resulting free return speed range, or 50 knots plus the resulting free return speed range, above and below the trim speed (except that the speed range need not include speeds less than 1.3 V_{SR1}, nor speeds greater than the minimum speed of the applicable speed range prescribed in paragraph (b)(1), nor speeds that require a stick force of more than 50 pounds), with—

(i) Wing flaps, center of gravity position, and weight as specified in paragraph (b)(1) of this section;

(ii) Power required for level flight at a speed equal to $(V_{MO} + 1.3\ V_{SR1})/2$; and

(iii) The airplane trimmed for level flight with the power required in paragraph (b)(2)(ii) of this section.

(3) With the landing gear extended, the stick force curve must have a stable slope at all speeds within a range which is the greater of 15 percent of the trim speed plus the resulting free return speed range, or 50 knots plus the resulting free return speed range, above and below the trim speed (except that the speed range need not include speeds less than 1.3 V_{SR1}, nor speeds greater than V_{LE} nor speeds that require a stick force of more than 50 pounds), with—

(i) Wing flap, center of gravity position, and weight as specified in paragraph (b)(1) of this section;

(ii) 75 percent of maximum continuous power for reciprocating engines or, for turbine engines, the maximum cruising power selected by the applicant as an operating limitation, except that the power need not exceed that required for level flight at V_{LE}; and

(iii) The aircraft trimmed for level flight with the power required in paragraph (b)(3)(ii) of this section.

(c) *Approach.* The stick force curve must have a stable slope at speeds between V_{SW} and 1.7 V_{SR1}, with—

(1) Wing flaps in the approach position;

(2) Landing gear retracted;

(3) Maximum landing weight; and

(4) The airplane trimmed at 1.3 V_{SR1} with enough power to maintain level flight at this speed.

(d) *Landing.* The stick force curve must have a stable slope, and the stick force may not exceed 80 pounds, at speeds between V_{SW} and 1.7 V_{SR0} with—

(1) Wing flaps in the landing position;

(2) Landing gear extended;

(3) Maximum landing weight;

(4) The airplane trimmed at 1.3 V_{SR0} with—

(i) Power or thrust off, and

(ii) Power or thrust for level flight.

(5) The airplane trimmed at 1.3 V_{SR0} with power or thrust off.

[Doc. No. 5066, 29 FR 18291, Dec. 24, 1964, as amended by Amdt. 25–7, 30 FR 13117, Oct. 15, 1965; Amdt. 25–108, 67 FR 70827, Nov. 26, 2002; Amdt. 25–115, 69 FR 40527, July 2, 2004]

§ 25.177 Static lateral-directional stability.

(a) The static directional stability (as shown by the tendency to recover from a skid with the rudder free) must be positive for any landing gear and flap position and symmetric power condition, at speeds from 1.13 V_{SR1}, up to V_{FE}, V_{LE}, or V_{FC}/M_{FC} (as appropriate for the airplane configuration).

(b) The static lateral stability (as shown by the tendency to raise the low wing in a sideslip with the aileron controls free) for any landing gear and flap position and symmetric power condition, may not be negative at any airspeed (except that speeds higher than V_{FE} need not be considered for flaps extended configurations nor speeds higher than V_{LE} for landing gear extended configurations) in the following airspeed ranges:

(1) From 1.13 V_{SR1} to V_{MO}/M_{MO}.

(2) From V_{MO}/M_{MO} to V_{FC}/M_{FC}, unless the divergence is—

(i) Gradual;

(ii) Easily recognizable by the pilot; and

(iii) Easily controllable by the pilot.

(c) The following requirement must be met for the configurations and speed specified in paragraph (a) of this section. In straight, steady sideslips over the range of sideslip angles appropriate to the operation of the airplane, the aileron and rudder control movements and forces must be substantially proportional to the angle of sideslip in a stable sense. This factor of proportionality must lie between limits found necessary for safe operation. The range of sideslip angles evaluated must include those sideslip angles resulting from the lesser of:

(1) One-half of the available rudder control input; and

(2) A rudder control force of 180 pounds.

(d) For sideslip angles greater than those prescribed by paragraph (c) of this section, up to the angle at which full rudder control is used or a rudder control force of 180 pounds is obtained, the rudder control forces may not reverse, and increased rudder deflection must be needed for increased angles of sideslip. Compliance with this requirement must be shown using straight, steady sideslips, unless full lateral control input is achieved before reaching either full rudder control input or a rudder control force of 180 pounds; a straight, steady sideslip need not be maintained after achieving full lateral control input. This requirement must be met at all approved landing gear and flap positions for the range of operating speeds and power conditions appropriate to each landing gear and flap position with all engines operating.

[Amdt. 25–135, 76 FR 74654, Dec. 1, 2011]

§ 25.181 Dynamic stability.

(a) Any short period oscillation, not including combined lateral-directional oscillations, occurring between 1.13 V_{SR} and maximum allowable speed appropriate to the configuration of the airplane must be heavily damped with the primary controls—

(1) Free; and

(2) In a fixed position.

(b) Any combined lateral-directional oscillations ("Dutch roll") occurring between 1.13 V_{SR} and maximum allowable speed appropriate to the configuration of the airplane must be positively damped with controls free, and must be controllable with normal use of the primary controls without requiring exceptional pilot skill.

[Amdt. 25–42, 43 FR 2322, Jan. 16, 1978, as amended by Amdt. 25–72, 55 FR 29775, July 20, 1990; 55 FR 37607, Sept. 12, 1990; Amdt. 25–108, 67 FR 70827, Nov. 26, 2002]

STALLS

§ 25.201 Stall demonstration.

(a) Stalls must be shown in straight flight and in 30 degree banked turns with—

(1) Power off; and

(2) The power necessary to maintain level flight at 1.5 V_{SR1} (where V_{SR1} corresponds to the reference stall speed at maximum landing weight with flaps in the approach position and the landing gear retracted).

(b) In each condition required by paragraph (a) of this section, it must be possible to meet the applicable requirements of § 25.203 with—

(1) Flaps, landing gear, and deceleration devices in any likely combination of positions approved for operation;

(2) Representative weights within the range for which certification is requested;

(3) The most adverse center of gravity for recovery; and

(4) The airplane trimmed for straight flight at the speed prescribed in § 25.103(b)(6).

(c) The following procedures must be used to show compliance with § 25.203;

(1) Starting at a speed sufficiently above the stalling speed to ensure that a steady rate of speed reduction can be established, apply the longitudinal control so that the speed reduction does not exceed one knot per second until the airplane is stalled.

(2) In addition, for turning flight stalls, apply the longitudinal control to achieve airspeed deceleration rates up to 3 knots per second.

(3) As soon as the airplane is stalled, recover by normal recovery techniques.

(d) The airplane is considered stalled when the behavior of the airplane gives the pilot a clear and distinctive indication of an acceptable nature that the airplane is stalled. Acceptable indications of a stall, occurring either individually or in combination, are—

(1) A nose-down pitch that cannot be readily arrested;

(2) Buffeting, of a magnitude and severity that is a strong and effective deterrent to further speed reduction; or

(3) The pitch control reaches the aft stop and no further increase in pitch attitude occurs when the control is held full aft for a short time before recovery is initiated.

[Doc. No. 5066, 29 FR 18291, Dec. 24, 1964, as amended by Amdt. 25–84, 60 FR 30750, June 9, 1995; Amdt. 25–108, 67 FR 70827, Nov. 26, 2002]

§ 25.203 Stall characteristics.

(a) It must be possible to produce and to correct roll and yaw by unreversed use of the aileron and rudder controls, up to the time the airplane is stalled. No abnormal nose-up pitching may occur. The longitudinal control force must be positive up to and throughout the stall. In addition, it must be possible to promptly prevent stalling and to recover from a stall by normal use of the controls.

(b) For level wing stalls, the roll occurring between the stall and the completion of the recovery may not exceed approximately 20 degrees.

(c) For turning flight stalls, the action of the airplane after the stall may not be so violent or extreme as to make it difficult, with normal piloting skill, to effect a prompt recovery and to regain control of the airplane. The maximum bank angle that occurs during the recovery may not exceed—

(1) Approximately 60 degrees in the original direction of the turn, or 30 degrees in the opposite direction, for deceleration rates up to 1 knot per second; and

(2) Approximately 90 degrees in the original direction of the turn, or 60 degrees in the opposite direction, for deceleration rates in excess of 1 knot per second.

[Doc. No. 5066, 29 FR 18291, Dec. 24, 1964, as amended by Amdt. 25–84, 60 FR 30750, June 9, 1995]

§ 25.207 Stall warning.

(a) Stall warning with sufficient margin to prevent inadvertent stalling with the flaps and landing gear in any normal position must be clear and distinctive to the pilot in straight and turning flight.

(b) The warning must be furnished either through the inherent aerodynamic qualities of the airplane or by a device that will give clearly distinguishable indications under expected conditions of flight. However, a visual stall warning device that requires the attention of the crew within the cockpit is not acceptable by itself. If a warning device is used, it must provide a warning in each of the airplane configurations prescribed in paragraph (a) of this section at the speed prescribed in paragraphs (c) and (d) of this section. Except for the stall warning prescribed in paragraph (h)(3)(ii) of this section, the stall warning for flight in icing conditions must be provided by the same means as the stall warning for flight in non-icing conditions.

(c) When the speed is reduced at rates not exceeding one knot per second, stall warning must begin, in each normal configuration, at a speed, V_{SW}, exceeding the speed at which the stall is identified in accordance with § 25.201(d) by not less than five knots or five percent CAS, whichever is greater. Once initiated, stall warning must continue until the angle of attack is reduced to approximately that at which stall warning began.

(d) In addition to the requirement of paragraph (c) of this section, when the speed is reduced at rates not exceeding one knot per second, in straight flight with engines idling and at the center-of-gravity position specified in § 25.103(b)(5), V_{SW}, in each normal configuration, must exceed V_{SR} by not less than three knots or three percent CAS, whichever is greater.

(e) In icing conditions, the stall warning margin in straight and turning flight must be sufficient to allow the pilot to prevent stalling (as defined in § 25.201(d)) when the pilot starts a recovery maneuver not less than three seconds after the onset of stall warning. When demonstrating compliance with this paragraph, the pilot must perform the recovery maneuver in the same way as for the airplane in non-icing conditions. Compliance with this requirement must be demonstrated in flight with the speed reduced at rates not exceeding one knot per second, with—

(1) The most critical of the takeoff ice and final takeoff ice accretions defined in Appendices C and O of this part, as applicable, in accordance with § 25.21(g), for each configuration used in the takeoff phase of flight;

(2) The most critical of the en route ice accretion(s) defined in Appendices C and O of this part, as applicable, in accordance with § 25.21(g), for the en route configuration;

(3) The most critical of the holding ice accretion(s) defined in Appendices C and O of this part, as applicable, in accordance with § 25.21(g), for the holding configuration(s);

(4) The most critical of the approach ice accretion(s) defined in Appendices C and O of this part, as applicable, in accordance with § 25.21(g), for the approach configuration(s); and

(5) The most critical of the landing ice accretion(s) defined in Appendices C and O of this part, as applicable, in accordance with § 25.21(g), for the landing and go-around configuration(s).

(f) The stall warning margin must be sufficient in both non-icing and icing conditions to allow the pilot to prevent stalling when the pilot starts a recovery maneuver not less than one second after the onset of stall warning in slowdown turns with at least 1.5 g load factor normal to the flight path and airspeed deceleration rates of at least 2 knots per second. When demonstrating compliance with this paragraph for icing conditions, the pilot must perform the recovery maneuver in the same way as for the airplane in non-icing conditions. Compliance with this requirement must be demonstrated in flight with—

(1) The flaps and landing gear in any normal position;

(2) The airplane trimmed for straight flight at a speed of 1.3 V_{SR}; and

(3) The power or thrust necessary to maintain level flight at 1.3 V_{SR}.

(g) Stall warning must also be provided in each abnormal configuration of the high lift devices that is likely to be used in flight following system fail-

ures (including all configurations covered by Airplane Flight Manual procedures).

(h) The following stall warning margin is required for flight in icing conditions before the ice protection system has been activated and is performing its intended function. Compliance must be shown using the most critical of the ice accretion(s) defined in Appendix C, part II, paragraph (e) of this part and Appendix O, part II, paragraph (d) of this part, as applicable, in accordance with § 25.21(g). The stall warning margin in straight and turning flight must be sufficient to allow the pilot to prevent stalling without encountering any adverse flight characteristics when:

(1) The speed is reduced at rates not exceeding one knot per second;

(2) The pilot performs the recovery maneuver in the same way as for flight in non-icing conditions; and

(3) The recovery maneuver is started no earlier than:

(i) One second after the onset of stall warning if stall warning is provided by the same means as for flight in non-icing conditions; or

(ii) Three seconds after the onset of stall warning if stall warning is provided by a different means than for flight in non-icing conditions.

(i) In showing compliance with paragraph (h) of this section, if stall warning is provided by a different means in icing conditions than for non-icing conditions, compliance with § 25.203 must be shown using the accretion defined in appendix C, part II(e) of this part. Compliance with this requirement must be shown using the demonstration prescribed by § 25.201, except that the deceleration rates of § 25.201(c)(2) need not be demonstrated.

[Doc. No. 5066, 29 FR 18291, Dec. 24, 1964, as amended by Amdt. 25-7, 30 FR 13118, Oct. 15, 1965; Amdt. 25-42, 43 FR 2322, Jan. 16, 1978; Amdt. 25-108, 67 FR 70827, Nov. 26, 2002; Amdt. 25-121, 72 FR 44668, Aug. 8, 2007; Amdt. 25-129, 74 FR 38339, Aug. 3, 2009; Amdt. 25-140, 79 FR 65526, Nov. 4, 2014]

GROUND AND WATER HANDLING CHARACTERISTICS

§ 25.231 Longitudinal stability and control.

(a) Landplanes may have no uncontrollable tendency to nose over in any reasonably expected operating condition or when rebound occurs during landing or takeoff. In addition—

(1) Wheel brakes must operate smoothly and may not cause any undue tendency to nose over; and

(2) If a tail-wheel landing gear is used, it must be possible, during the takeoff ground run on concrete, to maintain any attitude up to thrust line level, at 75 percent of V_{SR1}.

(b) For seaplanes and amphibians, the most adverse water conditions safe for takeoff, taxiing, and landing, must be established.

[Doc. No. 5066, 29 FR 18291, Dec. 24, 1964, as amended by Amdt. 25-108, 67 FR 70828, Nov. 26, 2002]

§25.233 Directional stability and control.

(a) There may be no uncontrollable ground-looping tendency in 90° cross winds, up to a wind velocity of 20 knots or 0.2 V_{SR0}, whichever is greater, except that the wind velocity need not exceed 25 knots at any speed at which the airplane may be expected to be operated on the ground. This may be shown while establishing the 90° cross component of wind velocity required by §25.237.

(b) Landplanes must be satisfactorily controllable, without exceptional piloting skill or alertness, in power-off landings at normal landing speed, without using brakes or engine power to maintain a straight path. This may be shown during power-off landings made in conjunction with other tests.

(c) The airplane must have adequate directional control during taxiing. This may be shown during taxiing prior to takeoffs made in conjunction with other tests.

[Doc. No. 5066, 29 FR 18291, Dec. 24, 1964, as amended by Amdt. 25-23, 35 FR 5671, Apr. 8, 1970; Amdt. 25-42, 43 FR 2322, Jan. 16, 1978; Amdt. 25-94, 63 FR 8848, Feb. 23, 1998; Amdt. 25-108, 67 FR 70828, Nov. 26, 2002]

§25.235 Taxiing condition.

The shock absorbing mechanism may not damage the structure of the airplane when the airplane is taxied on the roughest ground that may reasonably be expected in normal operation.

§25.237 Wind velocities.

(a) For land planes and amphibians, the following applies:

(1) A 90-degree cross component of wind velocity, demonstrated to be safe for takeoff and landing, must be established for dry runways and must be at least 20 knots or 0.2 V_{SR0}, whichever is greater, except that it need not exceed 25 knots.

(2) The crosswind component for takeoff established without ice accretions is valid in icing conditions.

(3) The landing crosswind component must be established for:

(i) Non-icing conditions, and

(ii) Icing conditions with the most critical of the landing ice accretion(s) defined in Appendices C and O of this part, as applicable, in accordance with §25.21(g).

(b) For seaplanes and amphibians, the following applies:

(1) A 90-degree cross component of wind velocity, up to which takeoff and landing is safe under all water conditions that may reasonably be expected in normal operation, must be established and must be at least 20 knots or 0.2 V_{SR0}, whichever is greater, except that it need not exceed 25 knots.

(2) A wind velocity, for which taxiing is safe in any direction under all water conditions that may reasonably be expected in normal operation, must be established and must be at least 20 knots or 0.2 V_{SR0}, whichever is greater, except that it need not exceed 25 knots.

[Amdt. 25-42, 43 FR 2322, Jan. 16, 1978, as amended by Amdt. 25-108, 67 FR 70827, Nov. 26, 2002; Amdt. 25-121, 72 FR 44668, Aug. 8, 2007; Amdt. 25-140, 79 FR 65525, Nov. 4, 2014]

§25.239 Spray characteristics, control, and stability on water.

(a) For seaplanes and amphibians, during takeoff, taxiing, and landing, and in the conditions set forth in paragraph (b) of this section, there may be no—

(1) Spray characteristics that would impair the pilot's view, cause damage, or result in the taking in of an undue quantity of water;

(2) Dangerously uncontrollable porpoising, bounding, or swinging tendency; or

(3) Immersion of auxiliary floats or sponsons, wing tips, propeller blades,

or other parts not designed to withstand the resulting water loads.

(b) Compliance with the requirements of paragraph (a) of this section must be shown—

(1) In water conditions, from smooth to the most adverse condition established in accordance with § 25.231;

(2) In wind and cross-wind velocities, water currents, and associated waves and swells that may reasonably be expected in operation on water;

(3) At speeds that may reasonably be expected in operation on water;

(4) With sudden failure of the critical engine at any time while on water; and

(5) At each weight and center of gravity position, relevant to each operating condition, within the range of loading conditions for which certification is requested.

(c) In the water conditions of paragraph (b) of this section, and in the corresponding wind conditions, the seaplane or amphibian must be able to drift for five minutes with engines inoperative, aided, if necessary, by a sea anchor.

MISCELLANEOUS FLIGHT REQUIREMENTS

§ 25.251 Vibration and buffeting.

(a) The airplane must be demonstrated in flight to be free from any vibration and buffeting that would prevent continued safe flight in any likely operating condition.

(b) Each part of the airplane must be demonstrated in flight to be free from excessive vibration under any appropriate speed and power conditions up to V_{DF}/M_{DF}. The maximum speeds shown must be used in establishing the operating limitations of the airplane in accordance with § 25.1505.

(c) Except as provided in paragraph (d) of this section, there may be no buffeting condition, in normal flight, including configuration changes during cruise, severe enough to interfere with the control of the airplane, to cause excessive fatigue to the crew, or to cause structural damage. Stall warning buffeting within these limits is allowable.

(d) There may be no perceptible buffeting condition in the cruise configuration in straight flight at any speed up to V_{MO}/M_{MO}, except that stall warning buffeting is allowable.

(e) For an airplane with M_D greater than .6 or with a maximum operating altitude greater than 25,000 feet, the positive maneuvering load factors at which the onset of perceptible buffeting occurs must be determined with the airplane in the cruise configuration for the ranges of airspeed or Mach number, weight, and altitude for which the airplane is to be certificated. The envelopes of load factor, speed, altitude, and weight must provide a sufficient range of speeds and load factors for normal operations. Probable inadvertent excursions beyond the boundaries of the buffet onset envelopes may not result in unsafe conditions.

[Doc. No. 5066, 29 FR 18291, Dec. 24, 1964, as amended by Amdt. 25–23, 35 FR 5671, Apr. 8, 1970; Amdt. 25–72, 55 FR 29775, July 20, 1990; Amdt. 25–77, 57 FR 28949, June 29, 1992]

§ 25.253 High-speed characteristics.

(a) *Speed increase and recovery characteristics.* The following speed increase and recovery characteristics must be met:

(1) Operating conditions and characteristics likely to cause inadvertent speed increases (including upsets in pitch and roll) must be simulated with the airplane trimmed at any likely cruise speed up to V_{MO}/M_{MO}. These conditions and characteristics include gust upsets, inadvertent control movements, low stick force gradient in relation to control friction, passenger movement, leveling off from climb, and descent from Mach to airspeed limit altitudes.

(2) Allowing for pilot reaction time after effective inherent or artificial speed warning occurs, it must be shown that the airplane can be recovered to a normal attitude and its speed reduced to V_{MO}/M_{MO}, without—

(i) Exceptional piloting strength or skill;

(ii) Exceeding V_D/M_D, V_{DF}/M_{DF}, or the structural limitations; and

(iii) Buffeting that would impair the pilot's ability to read the instruments or control the airplane for recovery.

(3) With the airplane trimmed at any speed up to V_{MO}/M_{MO}, there must be no reversal of the response to control input about any axis at any speed up to V_{DF}/M_{DF}. Any tendency to pitch, roll, or

yaw must be mild and readily controllable, using normal piloting techniques. When the airplane is trimmed at V_{MO}/M_{MO}, the slope of the elevator control force versus speed curve need not be stable at speeds greater than V_{FC}/M_{FC}, but there must be a push force at all speeds up to V_{DF}/M_{DF} and there must be no sudden or excessive reduction of elevator control force as V_{DF}/M_{DF} is reached.

(4) Adequate roll capability to assure a prompt recovery from a lateral upset condition must be available at any speed up to V_{DF}/M_{DF}.

(5) With the airplane trimmed at V_{MO}/M_{MO}, extension of the speedbrakes over the available range of movements of the pilot's control, at all speeds above V_{MO}/M_{MO}, but not so high that V_{DF}/M_{DF} would be exceeded during the maneuver, must not result in:

(i) An excessive positive load factor when the pilot does not take action to counteract the effects of extension;

(ii) Buffeting that would impair the pilot's ability to read the instruments or control the airplane for recovery; or

(iii) A nose down pitching moment, unless it is small.

(b) *Maximum speed for stability characteristics, V_{FC}/M_{FC}.* V_{FC}/M_{FC} is the maximum speed at which the requirements of §§ 25.143(g), 25.147(f), 25.175(b)(1), 25.177(a) through (c), and 25.181 must be met with flaps and landing gear retracted. Except as noted in § 25.253(c), V_{FC}/M_{FC} may not be less than a speed midway between V_{MO}/M_{MO} and V_{DF}/M_{DF}, except that, for altitudes where Mach number is the limiting factor, M_{FC} need not exceed the Mach number at which effective speed warning occurs.

(c) *Maximum speed for stability characteristics in icing conditions.* The maximum speed for stability characteristics with the most critical of the ice accretions defined in Appendices C and O of this part, as applicable, in accordance with § 25.21(g), at which the requirements of §§ 25.143(g), 25.147(f), 25.175(b)(1), 25.177(a) through (c), and 25.181 must be met, is the lower of:

(1) 300 knots CAS;

(2) V_{FC}; or

(3) A speed at which it is demonstrated that the airframe will be free of ice accretion due to the effects of increased dynamic pressure.

[Doc. No. 5066, 29 FR 18291, Dec. 24, 1964, as amended by Amdt. 25–23, 35 FR 5671, Apr. 8, 1970; Amdt. 25–54, 45 FR 60172, Sept. 11, 1980; Amdt. 25–72, 55 FR 29775, July 20, 1990; Amdt. 25–84, 60 FR 30750, June 9, 1995; Amdt. 25–121, 72 FR 44668, Aug. 8, 2007; Amdt. 25–135, 76 FR 74654, Dec. 1, 2011; Amdt. 25–140, 79 FR 65525, Nov. 4, 2014]

§ 25.255 Out-of-trim characteristics.

(a) From an initial condition with the airplane trimmed at cruise speeds up to V_{MO}/M_{MO}, the airplane must have satisfactory maneuvering stability and controllability with the degree of out-of-trim in both the airplane nose-up and nose-down directions, which results from the greater of—

(1) A three-second movement of the longitudinal trim system at its normal rate for the particular flight condition with no aerodynamic load (or an equivalent degree of trim for airplanes that do not have a power-operated trim system), except as limited by stops in the trim system, including those required by § 25.655(b) for adjustable stabilizers; or

(2) The maximum mistrim that can be sustained by the autopilot while maintaining level flight in the high speed cruising condition.

(b) In the out-of-trim condition specified in paragraph (a) of this section, when the normal acceleration is varied from + 1 g to the positive and negative values specified in paragraph (c) of this section—

(1) The stick force vs. g curve must have a positive slope at any speed up to and including V_{FC}/M_{FC}; and

(2) At speeds between V_{FC}/M_{FC} and V_{DF}/M_{DF} the direction of the primary longitudinal control force may not reverse.

(c) Except as provided in paragraphs (d) and (e) of this section, compliance with the provisions of paragraph (a) of this section must be demonstrated in flight over the acceleration range—

(1) − 1 g to + 2.5 g; or

(2) 0 g to 2.0 g, and extrapolating by an acceptable method to − 1 g and + 2.5 g.

(d) If the procedure set forth in paragraph (c)(2) of this section is used to demonstrate compliance and marginal conditions exist during flight test with regard to reversal of primary longitudinal control force, flight tests must be accomplished from the normal acceleration at which a marginal condition is found to exist to the applicable limit specified in paragraph (b)(1) of this section.

(e) During flight tests required by paragraph (a) of this section, the limit maneuvering load factors prescribed in §§ 25.333(b) and 25.337, and the maneuvering load factors associated with probable inadvertent excursions beyond the boundaries of the buffet onset envelopes determined under § 25.251(e), need not be exceeded. In addition, the entry speeds for flight test demonstrations at normal acceleration values less than 1 g must be limited to the extent necessary to accomplish a recovery without exceeding V_{DF}/M_{DF}.

(f) In the out-of-trim condition specified in paragraph (a) of this section, it must be possible from an overspeed condition at V_{DF}/M_{DF} to produce at least 1.5 g for recovery by applying not more than 125 pounds of longitudinal control force using either the primary longitudinal control alone or the primary longitudinal control and the longitudinal trim system. If the longitudinal trim is used to assist in producing the required load factor, it must be shown at V_{DF}/M_{DF} that the longitudinal trim can be actuated in the airplane nose-up direction with the primary surface loaded to correspond to the least of the following airplane nose-up control forces:

(1) The maximum control forces expected in service as specified in §§ 25.301 and 25.397.

(2) The control force required to produce 1.5 g.

(3) The control force corresponding to buffeting or other phenomena of such intensity that it is a strong deterrent to further application of primary longitudinal control force.

[Amdt. 25–42, 43 FR 2322, Jan. 16, 1978]

Subpart C—Structure

GENERAL

§ 25.301 Loads.

(a) Strength requirements are specified in terms of limit loads (the maximum loads to be expected in service) and ultimate loads (limit loads multiplied by prescribed factors of safety). Unless otherwise provided, prescribed loads are limit loads.

(b) Unless otherwise provided, the specified air, ground, and water loads must be placed in equilibrium with inertia forces, considering each item of mass in the airplane. These loads must be distributed to conservatively approximate or closely represent actual conditions. Methods used to determine load intensities and distribution must be validated by flight load measurement unless the methods used for determining those loading conditions are shown to be reliable.

(c) If deflections under load would significantly change the distribution of external or internal loads, this redistribution must be taken into account.

[Doc. No. 5066, 29 FR 18291, Dec. 24, 1964, as amended by Amdt. 25–23, 35 FR 5672, Apr. 8, 1970]

§ 25.303 Factor of safety.

Unless otherwise specified, a factor of safety of 1.5 must be applied to the prescribed limit load which are considered external loads on the structure. When a loading condition is prescribed in terms of ultimate loads, a factor of safety need not be applied unless otherwise specified.

[Amdt. 25–23, 35 FR 5672, Apr. 8, 1970]

§ 25.305 Strength and deformation.

(a) The structure must be able to support limit loads without detrimental permanent deformation. At any load up to limit loads, the deformation may not interfere with safe operation.

(b) The structure must be able to support ultimate loads without failure for at least 3 seconds. However, when proof of strength is shown by dynamic tests simulating actual load conditions, the 3-second limit does not apply. Static tests conducted to ultimate load must include the ultimate deflections and ultimate deformation induced by the loading. When analytical methods are used to show compliance with the ultimate load strength requirements, it must be shown that—

(1) The effects of deformation are not significant;

(2) The deformations involved are fully accounted for in the analysis; or

(3) The methods and assumptions used are sufficient to cover the effects of these deformations.

(c) Where structural flexibility is such that any rate of load application likely to occur in the operating conditions might produce transient stresses appreciably higher than those corresponding to static loads, the effects of this rate of application must be considered.

(d) [Reserved]

(e) The airplane must be designed to withstand any vibration and buffeting that might occur in any likely operating condition up to V_D/M_D, including stall and probable inadvertent excursions beyond the boundaries of the buffet onset envelope. This must be shown by analysis, flight tests, or other tests found necessary by the Administrator.

(f) Unless shown to be extremely improbable, the airplane must be designed to withstand any forced structural vibration resulting from any failure, malfunction or adverse condition in the flight control system. These must be considered limit loads and must be investigated at airspeeds up to V_C/M_C.

[Doc. No. 5066, 29 FR 18291, Dec. 24, 1964, as amended by Amdt. 25–23, 35 FR 5672, Apr. 8, 1970; Amdt. 25–54, 45 FR 60172, Sept. 11, 1980; Amdt. 25–77, 57 FR 28949, June 29, 1992; Amdt. 25–86, 61 FR 5220, Feb. 9, 1996]

§25.307 Proof of structure.

(a) Compliance with the strength and deformation requirements of this subpart must be shown for each critical loading condition. Structural analysis may be used only if the structure conforms to that for which experience has shown this method to be reliable. In other cases, substantiating tests must be made to load levels that are sufficient to verify structural behavior up to loads specified in §25.305.

(b)–(c) [Reserved]

(d) When static or dynamic tests are used to show compliance with the requirements of §25.305(b) for flight structures, appropriate material correction factors must be applied to the test results, unless the structure, or part thereof, being tested has features such that a number of elements contribute to the total strength of the structure and the failure of one element results in the redistribution of the load through alternate load paths.

[Doc. No. 5066, 29 FR 18291, Dec. 24, 1964, as amended by Amdt. 25–23, 35 FR 5672, Apr. 8, 1970; Amdt. 25–54, 45 FR 60172, Sept. 11, 1980; Amdt. 25–72, 55 FR 29775, July 20, 1990; Amdt. 25–139, 79 FR 59429, Oct. 2, 2014]

FLIGHT LOADS

§25.321 General.

(a) Flight load factors represent the ratio of the aerodynamic force component (acting normal to the assumed longitudinal axis of the airplane) to the weight of the airplane. A positive load factor is one in which the aerodynamic force acts upward with respect to the airplane.

(b) Considering compressibility effects at each speed, compliance with the flight load requirements of this subpart must be shown—

(1) At each critical altitude within the range of altitudes selected by the applicant;

(2) At each weight from the design minimum weight to the design maximum weight appropriate to each particular flight load condition; and

(3) For each required altitude and weight, for any practicable distribution of disposable load within the operating limitations recorded in the Airplane Flight Manual.

(c) Enough points on and within the boundaries of the design envelope must be investigated to ensure that the maximum load for each part of the airplane structure is obtained.

(d) The significant forces acting on the airplane must be placed in equilibrium in a rational or conservative manner. The linear inertia forces must be considered in equilibrium with the thrust and all aerodynamic loads, while the angular (pitching) inertia forces must be considered in equilibrium with thrust and all aerodynamic moments, including moments due to loads on components such as tail surfaces and nacelles. Critical thrust values in the range from zero to maximum continuous thrust must be considered.

[Doc. No. 5066, 29 FR 18291, Dec. 24, 1964, as amended by Amdt. 25–23, 35 FR 5672, Apr. 8, 1970; Amdt. 25–86, 61 FR 5220, Feb. 9, 1996]

FLIGHT MANEUVER AND GUST
CONDITIONS

§ 25.331 Symmetric maneuvering conditions.

(a) *Procedure.* For the analysis of the maneuvering flight conditions specified in paragraphs (b) and (c) of this section, the following provisions apply:

(1) Where sudden displacement of a control is specified, the assumed rate of control surface displacement may not be less than the rate that could be applied by the pilot through the control system.

(2) In determining elevator angles and chordwise load distribution in the maneuvering conditions of paragraphs (b) and (c) of this section, the effect of corresponding pitching velocities must be taken into account. The in-trim and out-of-trim flight conditions specified in § 25.255 must be considered.

(b) *Maneuvering balanced conditions.* Assuming the airplane to be in equilibrium with zero pitching acceleration, the maneuvering conditions A through I on the maneuvering envelope in § 25.333(b) must be investigated.

(c) *Maneuvering pitching conditions.* The following conditions must be investigated:

(1) *Maximum pitch control displacement at* V_A. The airplane is assumed to be flying in steady level flight (point A_1, § 25.333(b)) and the cockpit pitch control is suddenly moved to obtain extreme nose up pitching acceleration. In defining the tail load, the response of the airplane must be taken into account. Airplane loads that occur subsequent to the time when normal acceleration at the c.g. exceeds the positive limit maneuvering load factor (at point A_2 in § 25.333(b)), or the resulting tailplane normal load reaches its maximum, whichever occurs first, need not be considered.

(2) *Checked maneuver between* V_A *and* V_D. Nose-up checked pitching maneuvers must be analyzed in which the positive limit load factor prescribed in § 25.337 is achieved. As a separate condition, nose-down checked pitching maneuvers must be analyzed in which a limit load factor of 0g is achieved. In defining the airplane loads, the flight deck pitch control motions described in paragraphs (c)(2)(i) through (iv) of this section must be used:

(i) The airplane is assumed to be flying in steady level flight at any speed between V_A and V_D and the flight deck pitch control is moved in accordance with the following formula:

$$\delta(t) = \delta_1 \sin(\omega t) \text{ for } 0 \leq t \leq t_{max}$$

Where—

δ_1 = the maximum available displacement of the flight deck pitch control in the initial direction, as limited by the control system stops, control surface stops, or by pilot effort in accordance with § 25.397(b);

$\delta(t)$ = the displacement of the flight deck pitch control as a function of time. In the initial direction, $\delta(t)$ is limited to δ_1. In the reverse direction, $\delta(t)$ may be truncated at the maximum available displacement of the flight deck pitch control as limited by the control system stops, control surface stops, or by pilot effort in accordance with 25.397(b);

$t_{max} = 3\pi/2\omega$;

ω = the circular frequency (radians/second) of the control deflection taken equal to the undamped natural frequency of the short period rigid mode of the airplane, with active control system effects included where appropriate; but not less than:

$$\omega = \frac{\pi V}{2 V_A} \text{ radians per second;}$$

Where

V = the speed of the airplane at entry to the maneuver.

V_A = the design maneuvering speed prescribed in § 25.335(c).

(ii) For nose-up pitching maneuvers, the complete flight deck pitch control displacement history may be scaled down in amplitude to the extent necessary to ensure that the positive limit load factor prescribed in § 25.337 is not exceeded. For nose-down pitching maneuvers, the complete flight deck control displacement history may be scaled down in amplitude to the extent necessary to ensure that the normal acceleration at the center of gravity does not go below 0g.

(iii) In addition, for cases where the airplane response to the specified flight deck pitch control motion does not achieve the prescribed limit load factors, then the following flight deck pitch control motion must be used:

$\delta(t) = \delta_1 \sin(\omega t)$ for $0 \leq t \leq t_1$

$\delta(t) = \delta_1$ for $t_1 \leq t \leq t_2$

$\delta(t) = \delta_1 \sin(\omega[t + t_1 - t_2])$ for $t_2 \leq t \leq t_{max}$

Where—

$t_1 = \pi/2\omega$

$t_2 = t_1 + \Delta t$

$t_{max} = t_2 + \pi/\omega$;

Δt = the minimum period of time necessary to allow the prescribed limit load factor to be achieved in the initial direction, but it need not exceed five seconds (see figure below).

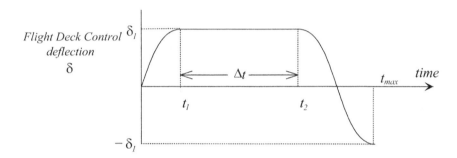

(iv) In cases where the flight deck pitch control motion may be affected by inputs from systems (for example, by a stick pusher that can operate at high load factor as well as at 1g), then the effects of those systems shall be taken into account.

(v) Airplane loads that occur beyond the following times need not be considered:

(A) For the nose-up pitching maneuver, the time at which the normal acceleration at the center of gravity goes below 0g;

(B) For the nose-down pitching maneuver, the time at which the normal acceleration at the center of gravity goes above the positive limit load factor prescribed in §25.337;

(C) t_{max}.

[Doc. No. 5066, 29 FR 18291, Dec. 24, 1964, as amended by Amdt. 25–23, 35 FR 5672, Apr. 8, 1970; Amdt. 25–46, 43 FR 50594, Oct. 30, 1978; 43 FR 52495, Nov. 13, 1978; 43 FR 54082, Nov. 20, 1978; Amdt. 25–72, 55 FR 29775, July 20, 1990; 55 FR 37607, Sept. 12, 1990; Amdt. 25–86, 61 FR 5220, Feb. 9, 1996; Amdt. 25–91, 62 FR 40704, July 29, 1997; Amdt. 25–141, 79 FR 73466, Dec. 11, 2014]

§25.333 Flight maneuvering envelope.

(a) *General.* The strength requirements must be met at each combination of airspeed and load factor on and within the boundaries of the representative maneuvering envelope (*V-n* diagram) of paragraph (b) of this section. This envelope must also be used in determining the airplane structural operating limitations as specified in §25.1501.

(b) *Maneuvering envelope.*

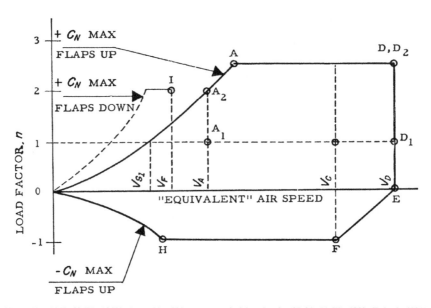

[Doc. No. 5066, 29 FR 18291, Dec. 24, 1964, as amended by Amdt. 25–86, 61 FR 5220, Feb. 9, 1996]

§ 25.335 Design airspeeds.

The selected design airspeeds are equivalent airspeeds (EAS). Estimated values of V_{S0} and V_{S1} must be conservative.

(a) *Design cruising speed, V_C.* For V_C, the following apply:

(1) The minimum value of V_C must be sufficiently greater than V_B to provide for inadvertent speed increases likely to occur as a result of severe atmospheric turbulence.

(2) Except as provided in § 25.335(d)(2), V_C may not be less than V_B + 1.32 U_{REF} (with U_{REF} as specified in § 25.341(a)(5)(i)). However V_C need not exceed the maximum speed in level flight at maximum continuous power for the corresponding altitude.

(3) At altitudes where V_D is limited by Mach number, V_C may be limited to a selected Mach number.

(b) *Design dive speed, V_D.* V_D must be selected so that V_C/M_C is not greater than 0.8 V_D/M_D, or so that the minimum speed margin between V_C/M_C and V_D/M_D is the greater of the following values:

(1) From an initial condition of stabilized flight at V_C/M_C, the airplane is upset, flown for 20 seconds along a flight path 7.5° below the initial path, and then pulled up at a load factor of 1.5g (0.5g acceleration increment). The speed increase occurring in this maneuver may be calculated if reliable or conservative aerodynamic data is used. Power as specified in § 25.175(b)(1)(iv) is assumed until the pullup is initiated, at which time power reduction and the use of pilot controlled drag devices may be assumed;

(2) The minimum speed margin must be enough to provide for atmospheric variations (such as horizontal gusts, and penetration of jet streams and cold fronts) and for instrument errors and airframe production variations. These factors may be considered on a probability basis. The margin at altitude where M_C is limited by compressibility effects must not be less than 0.07M unless a lower margin is determined using a rational analysis that includes the effects of any automatic systems. In any

case, the margin may not be reduced to less than 0.05M.

(c) *Design maneuvering speed V_A.* For V_A, the following apply:

(1) V_A may not be less than $V_{S1} \sqrt{n}$ where—

(i) n is the limit positive maneuvering load factor at V_C; and

(ii) V_{S1} is the stalling speed with flaps retracted.

(2) V_A and V_S must be evaluated at the design weight and altitude under consideration.

(3) V_A need not be more than V_C or the speed at which the positive $C_{N \, max}$ curve intersects the positive maneuver load factor line, whichever is less.

(d) *Design speed for maximum gust intensity,* V_B.

(1) V_B may not be less than

$$ V_{S1} \left[1 + \frac{K_g U_{ref} V_c a}{498w} \right]^{1/2} $$

where—

V_{S1} = the 1-g stalling speed based on C_{NAmax} with the flaps retracted at the particular weight under consideration;

V_c = design cruise speed (knots equivalent airspeed);

U_{ref} = the reference gust velocity (feet per second equivalent airspeed) from §25.341(a)(5)(i);

w = average wing loading (pounds per square foot) at the particular weight under consideration.

$$ K_g = \frac{.88\mu}{5.3 + \mu} $$

$$ \mu = \frac{2w}{\rho cag} $$

ρ = density of air (slugs/ft³);
c = mean geometric chord of the wing (feet);
g = acceleration due to gravity (ft/sec²);
a = slope of the airplane normal force coefficient curve, C_{NA} per radian;

(2) At altitudes where V_C is limited by Mach number—

(i) V_B may be chosen to provide an optimum margin between low and high speed buffet boundaries; and,

(ii) V_B need not be greater than V_C.

(e) *Design flap speeds,* V_F. For V_F, the following apply:

(1) The design flap speed for each flap position (established in accordance with §25.697(a)) must be sufficiently

greater than the operating speed recommended for the corresponding stage of flight (including balked landings) to allow for probable variations in control of airspeed and for transition from one flap position to another.

(2) If an automatic flap positioning or load limiting device is used, the speeds and corresponding flap positions programmed or allowed by the device may be used.

(3) V_F may not be less than—

(i) 1.6 V_{S1} with the flaps in takeoff position at maximum takeoff weight;

(ii) 1.8 V_{S1} with the flaps in approach position at maximum landing weight, and

(iii) 1.8 V_{S0} with the flaps in landing position at maximum landing weight.

(f) *Design drag device speeds,* V_{DD}. The selected design speed for each drag device must be sufficiently greater than the speed recommended for the operation of the device to allow for probable variations in speed control. For drag devices intended for use in high speed descents, V_{DD} may not be less than V_D. When an automatic drag device positioning or load limiting means is used, the speeds and corresponding drag device positions programmed or allowed by the automatic means must be used for design.

[Doc. No. 5066, 29 FR 18291, Dec. 24, 1964, as amended by Amdt. 25–23, 35 FR 5672, Apr. 8, 1970; Amdt. 25–86, 61 FR 5220, Feb. 9, 1996; Amdt. 25–91, 62 FR 40704, July 29, 1997]

§25.337 Limit maneuvering load factors.

(a) Except where limited by maximum (static) lift coefficients, the airplane is assumed to be subjected to symmetrical maneuvers resulting in the limit maneuvering load factors prescribed in this section. Pitching velocities appropriate to the corresponding pull-up and steady turn maneuvers must be taken into account.

(b) The positive limit maneuvering load factor n for any speed up to Vn may not be less than 2.1 + 24,000/ (W + 10,000) except that n may not be less than 2.5 and need not be greater than 3.8—where W is the design maximum takeoff weight.

(c) The negative limit maneuvering load factor—

(1) May not be less than −1.0 at speeds up to V_C; and

(2) Must vary linearly with speed from the value at V_C to zero at V_D.

(d) Maneuvering load factors lower than those specified in this section may be used if the airplane has design features that make it impossible to exceed these values in flight.

[Doc. No. 5066, 29 FR 18291, Dec. 24, 1964, as amended by Amdt. 25–23, 35 FR 5672, Apr. 8, 1970]

§ 25.341 Gust and turbulence loads.

(a) *Discrete Gust Design Criteria.* The airplane is assumed to be subjected to symmetrical vertical and lateral gusts in level flight. Limit gust loads must be determined in accordance with the provisions:

(1) Loads on each part of the structure must be determined by dynamic analysis. The analysis must take into account unsteady aerodynamic characteristics and all significant structural degrees of freedom including rigid body motions.

(2) The shape of the gust must be:

$$U = \frac{U_{ds}}{2}\left[1 - \cos\left(\frac{\pi s}{H}\right)\right]$$

for $0 \leq s \leq 2H$

where—

s = distance penetrated into the gust (feet);

U_{ds} = the design gust velocity in equivalent airspeed specified in paragraph (a)(4) of this section; and

H = the gust gradient which is the distance (feet) parallel to the airplane's flight path for the gust to reach its peak velocity.

(3) A sufficient number of gust gradient distances in the range 30 feet to 350 feet must be investigated to find the critical response for each load quantity.

(4) The design gust velocity must be:

$$U_{ds} = U_{ref} F_g \left(\frac{H}{350}\right)^{1/6}$$

where—

U_{ref} = the reference gust velocity in equivalent airspeed defined in paragraph (a)(5) of this section.

F_g = the flight profile alleviation factor defined in paragraph (a)(6) of this section.

(5) The following reference gust velocities apply:

(i) At airplane speeds between V_B and V_C: Positive and negative gusts with reference gust velocities of 56.0 ft/sec EAS must be considered at sea level. The reference gust velocity may be reduced linearly from 56.0 ft/sec EAS at sea level to 44.0 ft/sec EAS at 15,000 feet. The reference gust velocity may be further reduced linearly from 44.0 ft/sec EAS at 15,000 feet to 20.86 ft/sec EAS at 60,000 feet.

(ii) At the airplane design speed V_D: The reference gust velocity must be 0.5 times the value obtained under § 25.341(a)(5)(i).

(6) The flight profile alleviation factor, F_g, must be increased linearly from the sea level value to a value of 1.0 at the maximum operating altitude defined in § 25.1527. At sea level, the flight profile alleviation factor is determined by the following equation:

$$F_g = 0.5\left(F_{gz} + F_{gm}\right)$$

Where:

$$F_{gz} = 1 - \frac{Z_{mo}}{250000};$$

$$F_{gm} = \sqrt{R_2 \tan\left(\frac{\pi R_1}{4}\right)};$$

$$R_1 = \frac{\text{Maximum Landing Weight}}{\text{Maximum Take-off Weight}};$$

$$R_2 = \frac{\text{Maximum Zero Fuel Weight}}{\text{Maximum Take-off Weight}};$$

Z_{mo} = Maximum operating altitude defined in § 25.1527 (feet).

(7) When a stability augmentation system is included in the analysis, the effect of any significant system nonlinearities should be accounted for when deriving limit loads from limit gust conditions.

(b) *Continuous turbulence design criteria.* The dynamic response of the airplane to vertical and lateral continuous turbulence must be taken into account. The dynamic analysis must take into account unsteady aerodynamic characteristics and all significant structural degrees of freedom including rigid body motions. The limit loads must be determined for all critical altitudes, weights, and weight distributions as specified in §25.321(b), and all critical speeds within the ranges indicated in §25.341(b)(3).

(1) Except as provided in paragraphs (b)(4) and (5) of this section, the following equation must be used:

$$P_L = P_{L-1g} \pm U_\sigma \bar{A}$$

Where—

P_L = limit load;
P_{L-1g} = steady 1g load for the condition;
$\bar{A}$ = ratio of root-mean-square incremental load for the condition to root-mean-square turbulence velocity; and
U_σ = limit turbulence intensity in true airspeed, specified in paragraph (b)(3) of this section.

(2) Values of $\bar{A}$ must be determined according to the following formula:

$$\bar{A} = \sqrt{\int_0^\infty |H(\Omega)|^2 \Phi(\Omega) d\Omega}$$

Where—

$H(\Omega)$ = the frequency response function, determined by dynamic analysis, that relates the loads in the aircraft structure to the atmospheric turbulence; and
$\Phi(\Omega)$ = normalized power spectral density of atmospheric turbulence given by—

$$\Phi(\Omega) = \frac{L}{\pi} \frac{1 + \frac{8}{3}(1.339\Omega L)^2}{\left[1 + (1.339\Omega L)^2\right]^{11/6}}$$

Where—

Ω = reduced frequency, radians per foot; and
L = scale of turbulence = 2,500 ft.

(3) The limit turbulence intensities, U_σ, in feet per second true airspeed required for compliance with this paragraph are—

(i) At airplane speeds between V_B and V_C:

$$U_\sigma = U_{\sigma ref} F_g$$

Where—

$U_{\sigma ref}$ is the reference turbulence intensity that varies linearly with altitude from 90 fps (TAS) at sea level to 79 fps (TAS) at 24,000 feet and is then constant at 79 fps (TAS) up to the altitude of 60,000 feet.
F_g is the flight profile alleviation factor defined in paragraph (a)(6) of this section;

(ii) At speed V_D: U_σ is equal to ½ the values obtained under paragraph (b)(3)(i) of this section.

(iii) At speeds between V_C and V_D: U_σ is equal to a value obtained by linear interpolation.

(iv) At all speeds, both positive and negative incremental loads due to continuous turbulence must be considered.

(4) When an automatic system affecting the dynamic response of the airplane is included in the analysis, the effects of system non-linearities on loads at the limit load level must be taken into account in a realistic or conservative manner.

(5) If necessary for the assessment of loads on airplanes with significant non-linearities, it must be assumed that the turbulence field has a root-mean-square velocity equal to 40 percent of the U_σ values specified in paragraph (b)(3) of this section. The value of limit load is that load with the same probability of exceedance in the turbulence field as $\bar{A}U_\sigma$ of the same load quantity in a linear approximated model.

(c) *Supplementary gust conditions for wing-mounted engines.* For airplanes equipped with wing-mounted engines, the engine mounts, pylons, and wing supporting structure must be designed

239

for the maximum response at the nacelle center of gravity derived from the following dynamic gust conditions applied to the airplane:

(1) A discrete gust determined in accordance with § 25.341(a) at each angle normal to the flight path, and separately,

(2) A pair of discrete gusts, one vertical and one lateral. The length of each of these gusts must be independently tuned to the maximum response in accordance with § 25.341(a). The penetration of the airplane in the combined gust field and the phasing of the vertical and lateral component gusts must be established to develop the maximum response to the gust pair. In the absence of a more rational analysis, the following formula must be used for each of the maximum engine loads in all six degrees of freedom:

$$P_L = P_{L-1g} \pm 0.85\sqrt{L_V^2 + L_L^2}$$

Where—

P_L = limit load;
P_{L-1g} = steady 1g load for the condition;
L_V = peak incremental response load due to a vertical gust according to § 25.341(a); and
L_L = peak incremental response load due to a lateral gust according to § 25.341(a).

[Doc. No. 27902, 61 FR 5221, Feb. 9, 1996; 61 FR 9533, Mar. 8, 1996; Doc. No. FAA–2013–0142; 79 FR 73467, Dec. 11, 2014; Amdt. 25–141, 80 FR 4762, Jan. 29, 2015; 80 FR 6435, Feb. 5, 2015]

§ 25.343 Design fuel and oil loads.

(a) The disposable load combinations must include each fuel and oil load in the range from zero fuel and oil to the selected maximum fuel and oil load. A structural reserve fuel condition, not exceeding 45 minutes of fuel under the operating conditions in § 25.1001(e) and (f), as applicable, may be selected.

(b) If a structural reserve fuel condition is selected, it must be used as the minimum fuel weight condition for showing compliance with the flight load requirements as prescribed in this subpart. In addition—

(1) The structure must be designed for a condition of zero fuel and oil in the wing at limit loads corresponding to—

(i) A maneuvering load factor of + 2.25; and

(ii) The gust and turbulence conditions of § 25.341(a) and (b), but assuming 85% of the gust velocities prescribed in § 25.341(a)(4) and 85% of the turbulence intensities prescribed in § 25.341(b)(3).

(2) Fatigue evaluation of the structure must account for any increase in operating stresses resulting from the design condition of paragraph (b)(1) of this section; and

(3) The flutter, deformation, and vibration requirements must also be met with zero fuel.

[Doc. No. 5066, 29 FR 18291, Dec. 24, 1964, as amended by Amdt. 25–18, 33 FR 12226, Aug. 30, 1968; Amdt. 25–72, 55 FR 37607, Sept. 12, 1990; Amdt. 25–86, 61 FR 5221, Feb. 9, 1996; Amdt. 25–141, 79 FR 73468, Dec. 11, 2014]

§ 25.345 High lift devices.

(a) If wing flaps are to be used during takeoff, approach, or landing, at the design flap speeds established for these stages of flight under § 25.335(e) and with the wing flaps in the corresponding positions, the airplane is assumed to be subjected to symmetrical maneuvers and gusts. The resulting limit loads must correspond to the conditions determined as follows:

(1) Maneuvering to a positive limit load factor of 2.0; and

(2) Positive and negative gusts of 25 ft/sec EAS acting normal to the flight path in level flight. Gust loads resulting on each part of the structure must be determined by rational analysis. The analysis must take into account the unsteady aerodynamic characteristics and rigid body motions of the aircraft. The shape of the gust must be as described in § 25.341(a)(2) except that—

U_{ds} = 25 ft/sec EAS;
H = 12.5 c; and
c = mean geometric chord of the wing (feet).

(b) The airplane must be designed for the conditions prescribed in paragraph (a) of this section, except that the airplane load factor need not exceed 1.0, taking into account, as separate conditions, the effects of—

(1) Propeller slipstream corresponding to maximum continuous power at the design flap speeds V_F, and with takeoff power at not less than 1.4 times the stalling speed for the particular flap position and associated maximum weight; and

(2) A head-on gust of 25 feet per second velocity (EAS).

(c) If flaps or other high lift devices are to be used in en route conditions, and with flaps in the appropriate position at speeds up to the flap design speed chosen for these conditions, the airplane is assumed to be subjected to symmetrical maneuvers and gusts within the range determined by—

(1) Maneuvering to a positive limit load factor as prescribed in §25.337(b); and

(2) The vertical gust and turbulence conditions prescribed in §25.341(a) and (b).

(d) The airplane must be designed for a maneuvering load factor of 1.5 g at the maximum take-off weight with the wing-flaps and similar high lift devices in the landing configurations.

[Doc. No. 5066, 29 FR 18291, Dec. 24, 1964, as amended by Amdt. 25–46, 43 FR 50595, Oct. 30, 1978; Amdt. 25–72, 55 FR 37607, Sept. 17, 1990; Amdt. 25–86, 61 FR 5221, Feb. 9, 1996; Amdt. 25–91, 62 FR 40704, July 29, 1997; Amdt. 25–141, 79 FR 73468, Dec. 11, 2014]

§25.349 Rolling conditions.

The airplane must be designed for loads resulting from the rolling conditions specified in paragraphs (a) and (b) of this section. Unbalanced aerodynamic moments about the center of gravity must be reacted in a rational or conservative manner, considering the principal masses furnishing the reacting inertia forces.

(a) *Maneuvering.* The following conditions, speeds, and aileron deflections (except as the deflections may be limited by pilot effort) must be considered in combination with an airplane load factor of zero and of two-thirds of the positive maneuvering factor used in design. In determining the required aileron deflections, the torsional flexibility of the wing must be considered in accordance with §25.301(b):

(1) Conditions corresponding to steady rolling velocities must be investigated. In addition, conditions corresponding to maximum angular acceleration must be investigated for airplanes with engines or other weight concentrations outboard of the fuselage. For the angular acceleration conditions, zero rolling velocity may be assumed in the absence of a rational time history investigation of the maneuver.

(2) At V_A, a sudden deflection of the aileron to the stop is assumed.

(3) At V_C, the aileron deflection must be that required to produce a rate of roll not less than that obtained in paragraph (a)(2) of this section.

(4) At V_D, the aileron deflection must be that required to produce a rate of roll not less than one-third of that in paragraph (a)(2) of this section.

(b) *Unsymmetrical gusts.* The airplane is assumed to be subjected to unsymmetrical vertical gusts in level flight. The resulting limit loads must be determined from either the wing maximum airload derived directly from §25.341(a), or the wing maximum airload derived indirectly from the vertical load factor calculated from §25.341(a). It must be assumed that 100 percent of the wing air load acts on one side of the airplane and 80 percent of the wing air load acts on the other side.

[Doc. No. 5066, 29 FR 18291, Dec. 24, 1964, as amended by Amdt. 25–23, 35 FR 5672, Apr. 8, 1970; Amdt. 25–86, 61 FR 5222, Feb. 9, 1996; Amdt. 25–94, 63 FR 8848, Feb. 23, 1998]

§25.351 Yaw maneuver conditions.

The airplane must be designed for loads resulting from the yaw maneuver conditions specified in paragraphs (a) through (d) of this section at speeds from V_{MC} to V_D. Unbalanced aerodynamic moments about the center of gravity must be reacted in a rational or conservative manner considering the airplane inertia forces. In computing the tail loads the yawing velocity may be assumed to be zero.

(a) With the airplane in unaccelerated flight at zero yaw, it is assumed that the cockpit rudder control is suddenly displaced to achieve the resulting rudder deflection, as limited by:

(1) The control system on control surface stops; or

(2) A limit pilot force of 300 pounds from V_{MC} to V_A and 200 pounds from V_C/M_C to V_D/M_D, with a linear variation between V_A and V_C/M_C.

(b) With the cockpit rudder control deflected so as always to maintain the maximum rudder deflection available within the limitations specified in paragraph (a) of this section, it is assumed that the airplane yaws to the overswing sideslip angle.

(c) With the airplane yawed to the static equilibrium sideslip angle, it is assumed that the cockpit rudder control is held so as to achieve the maximum rudder deflection available within the limitations specified in paragraph (a) of this section.

(d) With the airplane yawed to the static equilibrium sideslip angle of paragraph (c) of this section, it is assumed that the cockpit rudder control is suddenly returned to neutral.

[Amdt. 25-91, 62 FR 40704, July 29, 1997]

§ 25.353 Rudder control reversal conditions.

Airplanes with a powered rudder control surface or surfaces must be designed for loads, considered to be ultimate, resulting from the yaw maneuver conditions specified in paragraphs (a) through (e) of this section at speeds from V_{MC} to V_C/M_C. Any permanent deformation resulting from these ultimate load conditions must not prevent continued safe flight and landing. The applicant must evaluate these conditions with the landing gear retracted and speed brakes (and spoilers when used as speed brakes) retracted. The applicant must evaluate the effects of flaps, flaperons, or any other aerodynamic devices when used as flaps, and slats-extended configurations, if they are used in en route conditions. Unbalanced aerodynamic moments about the center of gravity must be reacted in a rational or conservative manner considering the airplane inertia forces. In computing the loads on the airplane, the yawing velocity may be assumed to be zero. The applicant must assume a pilot force of 200 pounds when evaluating each of the following conditions:

(a) With the airplane in unaccelerated flight at zero yaw, the flightdeck rudder control is suddenly and fully displaced to achieve the resulting rudder deflection, as limited by the control system or the control surface stops.

(b) With the airplane yawed to the overswing sideslip angle, the flightdeck rudder control is suddenly and fully displaced in the opposite direction, as limited by the control system or control surface stops.

(c) With the airplane yawed to the opposite overswing sideslip angle, the flightdeck rudder control is suddenly and fully displaced in the opposite direction, as limited by the control system or control surface stops.

(d) With the airplane yawed to the subsequent overswing sideslip angle, the flightdeck rudder control is suddenly and fully displaced in the opposite direction, as limited by the control system or control surface stops.

(e) With the airplane yawed to the opposite overswing sideslip angle, the flightdeck rudder control is suddenly returned to neutral.

[Amdt. No. 25-147, 87 FR 71210, Nov. 22, 2022]

SUPPLEMENTARY CONDITIONS

§ 25.361 Engine and auxiliary power unit torque.

(a) For engine installations—

(1) Each engine mount, pylon, and adjacent supporting airframe structures must be designed for the effects of—

(i) A limit engine torque corresponding to takeoff power/thrust and, if applicable, corresponding propeller speed, acting simultaneously with 75% of the limit loads from flight condition A of § 25.333(b);

(ii) A limit engine torque corresponding to the maximum continuous power/thrust and, if applicable, corresponding propeller speed, acting simultaneously with the limit loads from flight condition A of § 25.333(b); and

(iii) For turbopropeller installations only, in addition to the conditions specified in paragraphs (a)(1)(i) and (ii) of this section, a limit engine torque corresponding to takeoff power and propeller speed, multiplied by a factor accounting for propeller control system malfunction, including quick feathering, acting simultaneously with 1g level flight loads. In the absence of a rational analysis, a factor of 1.6 must be used.

(2) The limit engine torque to be considered under paragraph (a)(1) of this section must be obtained by—

(i) For turbopropeller installations, multiplying mean engine torque for the specified power/thrust and speed by a factor of 1.25;

(ii) For other turbine engines, the limit engine torque must be equal to the maximum accelerating torque for the case considered.

(3) The engine mounts, pylons, and adjacent supporting airframe structure must be designed to withstand 1g level flight loads acting simultaneously with the limit engine torque loads imposed by each of the following conditions to be considered separately:

(i) Sudden maximum engine deceleration due to malfunction or abnormal condition; and

(ii) The maximum acceleration of engine.

(b) For auxiliary power unit installations, the power unit mounts and adjacent supporting airframe structure must be designed to withstand 1g level flight loads acting simultaneously with the limit torque loads imposed by each of the following conditions to be considered separately:

(1) Sudden maximum auxiliary power unit deceleration due to malfunction, abnormal condition, or structural failure; and

(2) The maximum acceleration of the auxiliary power unit.

[Amdt. 25–141, 79 FR 73468, Dec. 11, 2014]

§ 25.362 Engine failure loads.

(a) For engine mounts, pylons, and adjacent supporting airframe structure, an ultimate loading condition must be considered that combines 1g flight loads with the most critical transient dynamic loads and vibrations, as determined by dynamic analysis, resulting from failure of a blade, shaft, bearing or bearing support, or bird strike event. Any permanent deformation from these ultimate load conditions must not prevent continued safe flight and landing.

(b) The ultimate loads developed from the conditions specified in paragraph (a) of this section are to be—

(1) Multiplied by a factor of 1.0 when applied to engine mounts and pylons; and

(2) Multiplied by a factor of 1.25 when applied to adjacent supporting airframe structure.

[Amdt. 25–141, 79 FR 73468, Dec. 11, 2014]

§ 25.363 Side load on engine and auxiliary power unit mounts.

(a) Each engine and auxiliary power unit mount and its supporting structure must be designed for a limit load factor in lateral direction, for the side load on the engine and auxiliary power unit mount, at least equal to the maximum load factor obtained in the yawing conditions but not less than—

(1) 1.33; or

(2) One-third of the limit load factor for flight condition A as prescribed in § 25.333(b).

(b) The side load prescribed in paragraph (a) of this section may be assumed to be independent of other flight conditions.

[Doc. No. 5066, 29 FR 18291, Dec. 24, 1964, as amended by Amdt. 25–23, 35 FR 5672, Apr. 8, 1970; Amdt. 25–91, 62 FR 40704, July 29, 1997]

§ 25.365 Pressurized compartment loads.

For airplanes with one or more pressurized compartments the following apply:

(a) The airplane structure must be strong enough to withstand the flight loads combined with pressure differential loads from zero up to the maximum relief valve setting.

(b) The external pressure distribution in flight, and stress concentrations and fatigue effects must be accounted for.

(c) If landings may be made with the compartment pressurized, landing loads must be combined with pressure differential loads from zero up to the maximum allowed during landing.

(d) The airplane structure must be designed to be able to withstand the pressure differential loads corresponding to the maximum relief valve setting multiplied by a factor of 1.33 for airplanes to be approved for operation to 45,000 feet or by a factor of

1.67 for airplanes to be approved for operation above 45,000 feet, omitting other loads.

(e) Any structure, component or part, inside or outside a pressurized compartment, the failure of which could interfere with continued safe flight and landing, must be designed to withstand the effects of a sudden release of pressure through an opening in any compartment at any operating altitude resulting from each of the following conditions:

(1) The penetration of the compartment by a portion of an engine following an engine disintegration;

(2) Any opening in any pressurized compartment up to the size H_o in square feet; however, small compartments may be combined with an adjacent pressurized compartment and both considered as a single compartment for openings that cannot reasonably be expected to be confined to the small compartment. The size H_o must be computed by the following formula:

$$H_o = PA_s$$

where,

H_o = Maximum opening in square feet, need not exceed 20 square feet.
$P = (A_s/6240) + .024$
A_s = Maximum cross-sectional area of the pressurized shell normal to the longitudinal axis, in square feet; and

(3) The maximum opening caused by airplane or equipment failures not shown to be extremely improbable.

(f) In complying with paragraph (e) of this section, the fail-safe features of the design may be considered in determining the probability of failure or penetration and probable size of openings, provided that possible improper operation of closure devices and inadvertent door openings are also considered. Furthermore, the resulting differential pressure loads must be combined in a rational and conservative manner with 1–g level flight loads and any loads arising from emergency depressurization conditions. These loads may be considered as ultimate conditions; however, any deformations associated with these conditions must not interfere with continued safe flight and landing. The pressure relief provided by intercompartment venting may also be considered.

(g)(1) Except as provided in paragraph (g)(2) of this section, bulkheads, floors, and partitions in pressurized compartments for occupants must be designed to withstand the conditions specified in paragraph (e) of this section. In addition, reasonable design precautions must be taken to minimize the probability of parts becoming detached and injuring occupants while in their seats.

(2) Partitions adjacent to the opening specified in paragraph (e)(2) of this section need not be designed to withstand that condition provided—

(i) Failure of the partition would not interfere with continued safe flight and landing; and

(ii) Designing the partition to withstand the condition specified in paragraph (e)(2) of this section would be impractical.

[Doc. No. 5066, 29 FR 18291, Dec. 24, 1964, as amended by Amdt. 25–54, 45 FR 60172, Sept. 11, 1980; Amdt. 25–71, 55 FR 13477, Apr. 10, 1990; Amdt. 25–72, 55 FR 29776, July 20, 1990; Amdt. 25–87, 61 FR 28695, June 5, 1996; Amdt. No. 25–149, 88 FR 38382, June 13, 2023]

§ 25.367 Unsymmetrical loads due to engine failure.

(a) The airplane must be designed for the unsymmetrical loads resulting from the failure of the critical engine. Turbopropeller airplanes must be designed for the following conditions in combination with a single malfunction of the propeller drag limiting system, considering the probable pilot corrective action on the flight controls:

(1) At speeds between V_{MC} and V_D, the loads resulting from power failure because of fuel flow interruption are considered to be limit loads.

(2) At speeds between V_{MC} and V_C, the loads resulting from the disconnection of the engine compressor from the turbine or from loss of the turbine blades are considered to be ultimate loads.

(3) The time history of the thrust decay and drag build-up occurring as a result of the prescribed engine failures must be substantiated by test or other data applicable to the particular engine-propeller combination.

(4) The timing and magnitude of the probable pilot corrective action must

be conservatively estimated, considering the characteristics of the particular engine-propeller-airplane combination.

(b) Pilot corrective action may be assumed to be initiated at the time maximum yawing velocity is reached, but not earlier than two seconds after the engine failure. The magnitude of the corrective action may be based on the control forces specified in § 25.397(b) except that lower forces may be assumed where it is shown by anaylsis or test that these forces can control the yaw and roll resulting from the prescribed engine failure conditions.

§ 25.371 Gyroscopic loads.

The structure supporting any engine or auxiliary power unit must be designed for the loads, including gyroscopic loads, arising from the conditions specified in §§ 25.331, 25.341, 25.349, 25.351, 25.473, 25.479, and 25.481, with the engine or auxiliary power unit at the maximum rotating speed appropriate to the condition. For the purposes of compliance with this paragraph, the pitch maneuver in § 25.331(c)(1) must be carried out until the positive limit maneuvering load factor (point A_2 in § 25.333(b)) is reached.

[Amdt. 25–141, 79 FR 73468, Dec. 11, 2014]

§ 25.373 Speed control devices.

If speed control devices (such as spoilers and drag flaps) are installed for use in en route conditions—

(a) The airplane must be designed for the symmetrical maneuvers prescribed in §§ 25.333 and 25.337, the yawing maneuvers in § 25.351, and the vertical and lateral gust and turbulence conditions prescribed in § 25.341(a) and (b) at each setting and the maximum speed associated with that setting; and

(b) If the device has automatic operating or load limiting features, the airplane must be designed for the maneuver and gust conditions prescribed in paragraph (a) of this section, at the speeds and corresponding device positions that the mechanism allows.

[Doc. No. 5066, 29 FR 18291, Dec. 24, 1964, as amended by Amdt. 25–72, 55 FR 29776, July 20, 1990; Amdt. 25–86, 61 FR 5222, Feb. 9, 1996; Amdt. 25–141, 79 FR 73468, Dec. 11, 2014]

CONTROL SURFACE AND SYSTEM LOADS

§ 25.391 Control surface loads: General.

The control surfaces must be designed for the limit loads resulting from the flight conditions in §§ 25.331, 25.341(a) and (b), 25.349, and 25.351, considering the requirements for—

(a) Loads parallel to hinge line, in § 25.393;

(b) Pilot effort effects, in § 25.397;

(c) Trim tab effects, in § 25.407;

(d) Unsymmetrical loads, in § 25.427; and

(e) Auxiliary aerodynamic surfaces, in § 25.445.

[Doc. No. 5066, 29 FR 18291, Dec. 24, 1964, as amended by Amdt. 25–86, 61 FR 5222, Feb. 9, 1996; Amdt. 25–141, 79 FR 73468, Dec. 11, 2014]

§ 25.393 Loads parallel to hinge line.

(a) Control surfaces and supporting hinge brackets must be designed for inertia loads acting parallel to the hinge line.

(b) In the absence of more rational data, the inertia loads may be assumed to be equal to KW, where—

(1) K = 24 for vertical surfaces;

(2) K = 12 for horizontal surfaces; and

(3) W = weight of the movable surfaces.

§ 25.395 Control system.

(a) Longitudinal, lateral, directional, and drag control system and their supporting structures must be designed for loads corresponding to 125 percent of the computed hinge moments of the movable control surface in the conditions prescribed in § 25.391.

(b) The system limit loads of paragraph (a) of this section need not exceed the loads that can be produced by the pilot (or pilots) and by automatic or power devices operating the controls.

(c) The loads must not be less than those resulting from application of the minimum forces prescribed in § 25.397(c).

[Doc. No. 5066, 29 FR 18291, Dec. 24, 1964, as amended by Amdt. 25–23, 35 FR 5672, Apr. 8, 1970; Amdt. 25–72, 55 FR 29776, July 20, 1990; Amdt. 25–141, 79 FR 73468, Dec. 11, 2014]

§ 25.397 Control system loads.

(a) *General.* The maximum and minimum pilot forces, specified in paragraph (c) of this section, are assumed to act at the appropriate control grips or pads (in a manner simulating flight conditions) and to be reacted at the attachment of the control system to the control surface horn.

(b) *Pilot effort effects.* In the control surface flight loading condition, the air loads on movable surfaces and the corresponding deflections need not exceed those that would result in flight from the application of any pilot force within the ranges specified in paragraph (c) of this section. Two-thirds of the maximum values specified for the aileron and elevator may be used if control surface hinge moments are based on reliable data. In applying this criterion, the effects of servo mechanisms, tabs, and automatic pilot systems, must be considered.

(c) *Limit pilot forces and torques.* The limit pilot forces and torques are as follows:

Control	Maximum forces or torques	Minimum forces or torques
Aileron:		
Stick	100 lbs	40 lbs.
Wheel [1]	80 D in.-lbs [2] ...	40 D in.-lbs.
Elevator:		
Stick	250 lbs	100 lbs.
Wheel (symmetrical)	300 lbs	100 lbs.
Wheel (unsymmetrical) [3]	....	100 lbs.
Rudder	300 lbs	130 lbs.

[1] The critical parts of the aileron control system must be designed for a single tangential force with a limit value equal to 1.25 times the couple force determined from these criteria.

[2] D = wheel diameter (inches).

[3] The unsymmetrical forces must be applied at one of the normal handgrip points on the periphery of the control wheel.

[Doc. No. 5066, 29 FR 18291, Dec. 24, 1964, as amended by Amdt. 25–38, 41 FR 55466, Dec. 20, 1976; Amdt. 25–72, 55 FR 29776, July 20, 1990]

§ 25.399 Dual control system.

(a) Each dual control system must be designed for the pilots operating in opposition, using individual pilot forces not less than—

(1) 0.75 times those obtained under § 25.395; or

(2) The minimum forces specified in § 25.397(c).

(b) The control system must be designed for pilot forces applied in the same direction, using individual pilot forces not less than 0.75 times those obtained under § 25.395.

§ 25.405 Secondary control system.

Secondary controls, such as wheel brake, spoiler, and tab controls, must be designed for the maximum forces that a pilot is likely to apply to those controls. The following values may be used:

PILOT CONTROL FORCE LIMITS (SECONDARY CONTROLS)

Control	Limit pilot forces
Miscellaneous:	
*Crank, wheel, or lever ..	((1 + R) / 3) × 50 lbs., but not less than 50 lbs. nor more than 150 lbs. (R = radius). (Applicable to any angle within 20° of plane of control).
Twist	133 in.-lbs.
Push-pull	To be chosen by applicant.

*Limited to flap, tab, stabilizer, spoiler, and landing gear operation controls.

§ 25.407 Trim tab effects.

The effects of trim tabs on the control surface design conditions must be accounted for only where the surface loads are limited by maximum pilot effort. In these cases, the tabs are considered to be deflected in the direction that would assist the pilot, and the deflections are—

(a) For elevator trim tabs, those required to trim the airplane at any point within the positive portion of the pertinent flight envelope in § 25.333(b), except as limited by the stops; and

(b) For aileron and rudder trim tabs, those required to trim the airplane in the critical unsymmetrical power and loading conditions, with appropriate allowance for rigging tolerances.

§ 25.409 Tabs.

(a) *Trim tabs.* Trim tabs must be designed to withstand loads arising from all likely combinations of tab setting, primary control position, and airplane speed (obtainable without exceeding the flight load conditions prescribed for the airplane as a whole), when the effect of the tab is opposed by pilot effort forces up to those specified in § 25.397(b).

(b) *Balancing tabs.* Balancing tabs must be designed for deflections consistent with the primary control surface loading conditions.

(c) *Servo tabs.* Servo tabs must be designed for deflections consistent with the primary control surface loading conditions obtainable within the pilot maneuvering effort, considering possible opposition from the trim tabs.

§ 25.415 Ground gust conditions.

(a) The flight control systems and surfaces must be designed for the limit loads generated when the airplane is subjected to a horizontal 65-knot ground gust from any direction while taxiing and while parked. For airplanes equipped with control system gust locks, the taxiing condition must be evaluated with the controls locked and unlocked, and the parked condition must be evaluated with the controls locked.

(b) The control system and surface loads due to ground gust may be assumed to be static loads, and the hinge moments H must be computed from the formula:

$$H = K \, (1/2) \, \rho_o \, V^2 \, c \, S$$

Where—

K = hinge moment factor for ground gusts derived in paragraph (c) of this section;
ρ_o = density of air at sea level;
V = 65 knots relative to the aircraft;
S = area of the control surface aft of the hinge line;
c = mean aerodynamic chord of the control surface aft of the hinge line.

(c) The hinge moment factor K for ground gusts must be taken from the following table:

Surface	K	Position of controls
(1) Aileron	0.75	Control column locked or lashed in mid-position.
(2) Aileron	*±0.50	Ailerons at full throw.
(3) Elevator	*±0.75	Elevator full down.
(4) Elevator	*±0.75	Elevator full up.
(5) Rudder	0.75	Rudder in neutral.
(6) Rudder	0.75	Rudder at full throw.

*A positive value of K indicates a moment tending to depress the surface, while a negative value of K indicates a moment tending to raise the surface.

(d) The computed hinge moment of paragraph (b) of this section must be used to determine the limit loads due to ground gust conditions for the control surface. A 1.25 factor on the computed hinge moments must be used in calculating limit control system loads.

(e) Where control system flexibility is such that the rate of load application in the ground gust conditions might produce transient stresses appreciably higher than those corresponding to static loads, in the absence of a rational analysis substantiating a different dynamic factor, an additional factor of 1.6 must be applied to the control system loads of paragraph (d) of this section to obtain limit loads. If a rational analysis is used, the additional factor must not be less than 1.2.

(f) For the condition of the control locks engaged, the control surfaces, the control system locks, and the parts of any control systems between the surfaces and the locks must be designed to the resultant limit loads. Where control locks are not provided, then the control surfaces, the control system stops nearest the surfaces, and the parts of any control systems between the surfaces and the stops must be designed to the resultant limit loads. If the control system design is such as to allow any part of the control system to impact with the stops due to flexibility, then the resultant impact loads must be taken into account in deriving the limit loads due to ground gust.

(g) For the condition of taxiing with the control locks disengaged, or where control locks are not provided, the following apply:

(1) The control surfaces, the control system stops nearest the surfaces, and the parts of any control systems between the surfaces and the stops must be designed to the resultant limit loads.

(2) The parts of the control systems between the stops nearest the surfaces and the flight deck controls must be designed to the resultant limit loads, except that the parts of the control system where loads are eventually reacted by the pilot need not exceed:

(i) The loads corresponding to the maximum pilot loads in § 25.397(c) for each pilot alone; or

(ii) 0.75 times these maximum loads for each pilot when the pilot forces are applied in the same direction.

[Amdt. 25–141, 79 FR 73468, Dec. 11, 2014]

§ 25.427 Unsymmetrical loads.

(a) In designing the airplane for lateral gust, yaw maneuver and roll maneuver conditions, account must be taken of unsymmetrical loads on the empennage arising from effects such as slipstream and aerodynamic interference with the wing, vertical fin and other aerodynamic surfaces.

(b) The horizontal tail must be assumed to be subjected to unsymmetrical loading conditions determined as follows:

(1) 100 percent of the maximum loading from the symmetrical maneuver conditions of § 25.331 and the vertical gust conditions of § 25.341(a) acting separately on the surface on one side of the plane of symmetry; and

(2) 80 percent of these loadings acting on the other side.

(c) For empennage arrangements where the horizontal tail surfaces have dihedral angles greater than plus or minus 10 degrees, or are supported by the vertical tail surfaces, the surfaces and the supporting structure must be designed for gust velocities specified in § 25.341(a) acting in any orientation at right angles to the flight path.

(d) Unsymmetrical loading on the empennage arising from buffet conditions of § 25.305(e) must be taken into account.

[Doc. No. 27902, 61 FR 5222, Feb. 9, 1996]

§ 25.445 Auxiliary aerodynamic surfaces.

(a) When significant, the aerodynamic influence between auxiliary aerodynamic surfaces, such as outboard fins and winglets, and their supporting aerodynamic surfaces, must be taken into account for all loading conditions including pitch, roll, and yaw maneuvers, and gusts as specified in § 25.341(a) acting at any orientation at right angles to the flight path.

(b) To provide for unsymmetrical loading when outboard fins extend above and below the horizontal surface, the critical vertical surface loading (load per unit area) determined under § 25.391 must also be applied as follows:

(1) 100 percent to the area of the vertical surfaces above (or below) the horizontal surface.

(2) 80 percent to the area below (or above) the horizontal surface.

[Doc. No. 5066, 29 FR 18291, Dec. 24, 1964, as amended by Amdt. 25–86, 61 FR 5222, Feb. 9, 1996]

§ 25.457 Wing flaps.

Wing flaps, their operating mechanisms, and their supporting structures must be designed for critical loads occurring in the conditions prescribed in § 25.345, accounting for the loads occurring during transition from one flap position and airspeed to another.

§ 25.459 Special devices.

The loading for special devices using aerodynamic surfaces (such as slots, slats and spoilers) must be determined from test data.

[Doc. No. 5066, 29 FR 18291, Dec. 24, 1964, as amended by Amdt. 25–72, 55 FR 29776, July 20, 1990]

GROUND LOADS

§ 25.471 General.

(a) *Loads and equilibrium.* For limit ground loads—

(1) Limit ground loads obtained under this subpart are considered to be external forces applied to the airplane structure; and

(2) In each specified ground load condition, the external loads must be placed in equilibrium with the linear and angular inertia loads in a rational or conservative manner.

(b) *Critical centers of gravity.* The critical centers of gravity within the range for which certification is requested must be selected so that the maximum design loads are obtained in each landing gear element. Fore and aft, vertical, and lateral airplane centers of gravity must be considered. Lateral displacements of the c.g. from the airplane centerline which would result in main gear loads not greater than 103 percent of the critical design load for symmetrical loading conditions may be selected without considering the effects of these lateral c.g. displacements on the loading of the main gear elements, or on the airplane structure provided—

(1) The lateral displacement of the c.g. results from random passenger or cargo disposition within the fuselage or from random unsymmetrical fuel loading or fuel usage; and

(2) Appropriate loading instructions for random disposable loads are included under the provisions of § 25.1583(c)(2) to ensure that the lateral

displacement of the center of gravity is maintained within these limits.

(c) *Landing gear dimension data.* Figure 1 of appendix A contains the basic landing gear dimension data.

[Amdt. 25–23, 35 FR 5673, Apr. 8, 1970, as amended by Doc. No. FAA–2022–1355, Amdt. 25–148, 87 FR 75710, Dec. 9, 2022; 88 FR 2813, Jan. 18, 2023]

§ 25.473 Landing load conditions and assumptions.

(a) For the landing conditions specified in § 25.479 to § 25.485 the airplane is assumed to contact the ground—

(1) In the attitudes defined in § 25.479 and § 25.481;

(2) With a limit descent velocity of 10 fps at the design landing weight (the maximum weight for landing conditions at maximum descent velocity); and

(3) With a limit descent velocity of 6 fps at the design take-off weight (the maximum weight for landing conditions at a reduced descent velocity).

(4) The prescribed descent velocities may be modified if it is shown that the airplane has design features that make it impossible to develop these velocities.

(b) Airplane lift, not exceeding airplane weight, may be assumed unless the presence of systems or procedures significantly affects the lift.

(c) The method of analysis of airplane and landing gear loads must take into account at least the following elements:

(1) Landing gear dynamic characteristics.

(2) Spin-up and springback.

(3) Rigid body response.

(4) Structural dynamic response of the airframe, if significant.

(d) The landing gear dynamic characteristics must be validated by tests as defined in § 25.723(a).

(e) The coefficient of friction between the tires and the ground may be established by considering the effects of skidding velocity and tire pressure. However, this coefficient of friction need not be more than 0.8.

[Amdt. 25–91, 62 FR 40705, July 29, 1997; Amdt. 25–91, 62 FR 45481, Aug. 27, 1997; Amdt. 25–103, 66 FR 27394, May 16, 2001]

§ 25.477 Landing gear arrangement.

Sections 25.479 through 25.485 apply to airplanes with conventional arrangements of main and nose gears, or main and tail gears, when normal operating techniques are used.

§ 25.479 Level landing conditions.

(a) In the level attitude, the airplane is assumed to contact the ground at forward velocity components, ranging from V_{L1} to 1.25 V_{L2} parallel to the ground under the conditions prescribed in § 25.473 with—

(1) V_{L1} equal to V_{S0} (TAS) at the appropriate landing weight and in standard sea level conditions; and

(2) V_{L2} equal to V_{S0} (TAS) at the appropriate landing weight and altitudes in a hot day temperature of 41 degrees F. above standard.

(3) The effects of increased contact speed must be investigated if approval of downwind landings exceeding 10 knots is requested.

(b) For the level landing attitude for airplanes with tail wheels, the conditions specified in this section must be investigated with the airplane horizontal reference line horizontal in accordance with Figure 2 of Appendix A of this part.

(c) For the level landing attitude for airplanes with nose wheels, shown in Figure 2 of Appendix A of this part, the conditions specified in this section must be investigated assuming the following attitudes:

(1) An attitude in which the main wheels are assumed to contact the ground with the nose wheel just clear of the ground; and

(2) If reasonably attainable at the specified descent and forward velocities, an attitude in which the nose and main wheels are assumed to contact the ground simultaneously.

(d) In addition to the loading conditions prescribed in paragraph (a) of this section, but with maximum vertical ground reactions calculated from paragraph (a), the following apply:

(1) The landing gear and directly affected attaching structure must be designed for the maximum vertical ground reaction combined with an aft acting drag component of not less than 25% of this maximum vertical ground reaction.

(2) The most severe combination of loads that are likely to arise during a lateral drift landing must be taken into account. In absence of a more rational analysis of this condition, the following must be investigated:

(i) A vertical load equal to 75% of the maximum ground reaction of § 25.473 must be considered in combination with a drag and side load of 40% and 25% respectively of that vertical load.

(ii) The shock absorber and tire deflections must be assumed to be 75% of the deflection corresponding to the maximum ground reaction of § 25.473(a)(2). This load case need not be considered in combination with flat tires.

(3) The combination of vertical and drag components is considered to be acting at the wheel axle centerline.

[Amdt. 25-91, 62 FR 40705, July 29, 1997; Amdt. 25-91, 62 FR 45481, Aug. 27, 1997]

§ 25.481 Tail-down landing conditions.

(a) In the tail-down attitude, the airplane is assumed to contact the ground at forward velocity components, ranging from V_{L1} to V_{L2} parallel to the ground under the conditions prescribed in § 25.473 with—

(1) V_{L1} equal to V_{S0} (TAS) at the appropriate landing weight and in standard sea level conditions; and

(2) V_{L2} equal to V_{S0} (TAS) at the appropriate landing weight and altitudes in a hot day temperature of 41 degrees F. above standard.

(3) The combination of vertical and drag components considered to be acting at the main wheel axle centerline.

(b) For the tail-down landing condition for airplanes with tail wheels, the main and tail wheels are assumed to contact the ground simultaneously, in accordance with figure 3 of appendix A. Ground reaction conditions on the tail wheel are assumed to act—

(1) Vertically; and

(2) Up and aft through the axle at 45 degrees to the ground line.

(c) For the tail-down landing condition for airplanes with nose wheels, the airplane is assumed to be at an attitude corresponding to either the stalling angle or the maximum angle allowing clearance with the ground by each part of the airplane other than the main wheels, in accordance with figure 3 of appendix A, whichever is less.

[Doc. No. 5066, 29 FR 18291, Dec. 24, 1964, as amended by Amdt. 25-91, 62 FR 40705, July 29, 1997; Amdt. 25-94, 63 FR 8848, Feb. 23, 1998]

§ 25.483 One-gear landing conditions.

For the one-gear landing conditions, the airplane is assumed to be in the level attitude and to contact the ground on one main landing gear, in accordance with Figure 4 of Appendix A of this part. In this attitude—

(a) The ground reactions must be the same as those obtained on that side under § 25.479(d)(1), and

(b) Each unbalanced external load must be reacted by airplane inertia in a rational or conservative manner.

[Doc. No. 5066, 29 FR 18291, Dec. 24, 1964, as amended by Amdt. 25-91, 62 FR 40705, July 29, 1997]

§ 25.485 Side load conditions.

In addition to § 25.479(d)(2) the following conditions must be considered:

(a) For the side load condition, the airplane is assumed to be in the level attitude with only the main wheels contacting the ground, in accordance with figure 5 of appendix A.

(b) Side loads of 0.8 of the vertical reaction (on one side) acting inward and 0.6 of the vertical reaction (on the other side) acting outward must be combined with one-half of the maximum vertical ground reactions obtained in the level landing conditions. These loads are assumed to be applied at the ground contact point and to be resisted by the inertia of the airplane. The drag loads may be assumed to be zero.

[Doc. No. 5066, 29 FR 18291, Dec. 24, 1964, as amended by Amdt. 25-91, 62 FR 40705, July 29, 1997]

§ 25.487 Rebound landing condition.

(a) The landing gear and its supporting structure must be investigated for the loads occurring during rebound of the airplane from the landing surface.

(b) With the landing gear fully extended and not in contact with the ground, a load factor of 20.0 must act on the unsprung weights of the landing gear. This load factor must act in the direction of motion of the unsprung weights as they reach their limiting positions in extending with relation to the sprung parts of the landing gear.

§25.489 Ground handling conditions.

Unless otherwise prescribed, the landing gear and airplane structure must be investigated for the conditions in §§25.491 through 25.509 with the airplane at the design ramp weight (the maximum weight for ground handling conditions). No wing lift may be considered. The shock absorbers and tires may be assumed to be in their static position.

[Doc. No. 5066, 29 FR 18291, Dec. 24, 1964, as amended by Amdt. 25–23, 35 FR 5673, Apr. 8, 1970]

§25.491 Taxi, takeoff and landing roll.

Within the range of appropriate ground speeds and approved weights, the airplane structure and landing gear are assumed to be subjected to loads not less than those obtained when the aircraft is operating over the roughest ground that may reasonably be expected in normal operation.

[Amdt. 25–91, 62 FR 40705, July 29, 1997]

§25.493 Braked roll conditions.

(a) An airplane with a tail wheel is assumed to be in the level attitude with the load on the main wheels, in accordance with figure 6 of appendix A. The limit vertical load factor is 1.2 at the design landing weight and 1.0 at the design ramp weight. A drag reaction equal to the vertical reaction multiplied by a coefficient of friction of 0.8, must be combined with the vertical ground reaction and applied at the ground contact point.

(b) For an airplane with a nose wheel the limit vertical load factor is 1.2 at the design landing weight, and 1.0 at the design ramp weight. A drag reaction equal to the vertical reaction, multiplied by a coefficient of friction of 0.8, must be combined with the vertical reaction and applied at the ground contact point of each wheel with brakes. The following two attitudes, in accordance with figure 6 of appendix A, must be considered:

(1) The level attitude with the wheels contacting the ground and the loads distributed between the main and nose gear. Zero pitching acceleration is assumed.

(2) The level attitude with only the main gear contacting the ground and with the pitching moment resisted by angular acceleration.

(c) A drag reaction lower than that prescribed in this section may be used if it is substantiated that an effective drag force of 0.8 times the vertical reaction cannot be attained under any likely loading condition.

(d) An airplane equipped with a nose gear must be designed to withstand the loads arising from the dynamic pitching motion of the airplane due to sudden application of maximum braking force. The airplane is considered to be at design takeoff weight with the nose and main gears in contact with the ground, and with a steady-state vertical load factor of 1.0. The steady-state nose gear reaction must be combined with the maximum incremental nose gear vertical reaction caused by the sudden application of maximum braking force as described in paragraphs (b) and (c) of this section.

(e) In the absence of a more rational analysis, the nose gear vertical reaction prescribed in paragraph (d) of this section must be calculated according to the following formula:

$$V_N = \frac{W_T}{A+B}\left[B + \frac{f\mu AE}{A+B+\mu E}\right]$$

Where:

V_N = Nose gear vertical reaction.
W_T = Design takeoff weight.
A = Horizontal distance between the c.g. of the airplane and the nose wheel.
B = Horizontal distance between the c.g. of the airplane and the line joining the centers of the main wheels.
E = Vertical height of the c.g. of the airplane above the ground in the 1.0 g static condition.
μ = Coefficient of friction of 0.80.
f = Dynamic response factor; 2.0 is to be used unless a lower factor is substantiated. In the absence of other information, the dynamic response factor f may be defined by the equation:

$$f = 1 + \exp\left(\frac{-\pi\xi}{\sqrt{1-\xi^2}}\right)$$

Where:

251

ξ is the effective critical damping ratio of the rigid body pitching mode about the main landing gear effective ground contact point.

[Doc. No. 5066, 29 FR 18291, Dec. 24, 1964, as amended by Amdt. 25–23, 35 FR 5673, Apr. 8, 1970; Amdt. 25–97, 63 FR 29072, May 27, 1998]

§ 25.495   Turning.

In the static position, in accordance with figure 7 of appendix A, the airplane is assumed to execute a steady turn by nose gear steering, or by application of sufficient differential power, so that the limit load factors applied at the center of gravity are 1.0 vertically and 0.5 laterally. The side ground reaction of each wheel must be 0.5 of the vertical reaction.

§ 25.497   Tail-wheel yawing.

(a) A vertical ground reaction equal to the static load on the tail wheel, in combination with a side component of equal magnitude, is assumed.

(b) If there is a swivel, the tail wheel is assumed to be swiveled 90° to the airplane longitudinal axis with the resultant load passing through the axle.

(c) If there is a lock, steering device, or shimmy damper the tail wheel is also assumed to be in the trailing position with the side load acting at the ground contact point.

§ 25.499   Nose-wheel yaw and steering.

(a) A vertical load factor of 1.0 at the airplane center of gravity, and a side component at the nose wheel ground contact equal to 0.8 of the vertical ground reaction at that point are assumed.

(b) With the airplane assumed to be in static equilibrium with the loads resulting from the use of brakes on one side of the main landing gear, the nose gear, its attaching structure, and the fuselage structure forward of the center of gravity must be designed for the following loads:

(1) A vertical load factor at the center of gravity of 1.0.

(2) A forward acting load at the airplane center of gravity of 0.8 times the vertical load on one main gear.

(3) Side and vertical loads at the ground contact point on the nose gear that are required for static equilibrium.

(4) A side load factor at the airplane center of gravity of zero.

(c) If the loads prescribed in paragraph (b) of this section result in a nose gear side load higher than 0.8 times the vertical nose gear load, the design nose gear side load may be limited to 0.8 times the vertical load, with unbalanced yawing moments assumed to be resisted by airplane inertia forces.

(d) For other than the nose gear, its attaching structure, and the forward fuselage structure, the loading conditions are those prescribed in paragraph (b) of this section, except that—

(1) A lower drag reaction may be used if an effective drag force of 0.8 times the vertical reaction cannot be reached under any likely loading condition; and

(2) The forward acting load at the center of gravity need not exceed the maximum drag reaction on one main gear, determined in accordance with § 25.493(b).

(e) With the airplane at design ramp weight, and the nose gear in any steerable position, the combined application of full normal steering torque and vertical force equal to 1.33 times the maximum static reaction on the nose gear must be considered in designing the nose gear, its attaching structure, and the forward fuselage structure.

[Doc. No. 5066, 29 FR 18291, Dec. 24, 1964, as amended by Amdt. 25–23, 35 FR 5673, Apr. 8, 1970; Amdt. 25–46, 43 FR 50595, Oct. 30, 1978; Amdt. 25–91, 62 FR 40705, July 29, 1997]

§ 25.503   Pivoting.

(a) The airplane is assumed to pivot about one side of the main gear with the brakes on that side locked. The limit vertical load factor must be 1.0 and the coefficient of friction 0.8.

(b) The airplane is assumed to be in static equilibrium, with the loads being applied at the ground contact points, in accordance with figure 8 of appendix A.

§ 25.507   Reversed braking.

(a) The airplane must be in a three point static ground attitude. Horizontal reactions parallel to the ground and directed forward must be applied at the ground contact point of each wheel with brakes. The limit loads must be equal to 0.55 times the vertical load at each wheel or to the load developed by 1.2 times the nominal maximum static brake torque, whichever is less.

(b) For airplanes with nose wheels, the pitching moment must be balanced by rotational inertia.

(c) For airplanes with tail wheels, the resultant of the ground reactions must pass through the center of gravity of the airplane.

§ 25.509 Towing loads.

(a) The towing loads specified in paragraph (d) of this section must be considered separately. These loads must be applied at the towing fittings and must act parallel to the ground. In addition—

(1) A vertical load factor equal to 1.0 must be considered acting at the center of gravity;

(2) The shock struts and tires must be in their static positions; and

(3) With W_T as the design ramp weight, the towing load, F_{TOW} is—

(i) $0.3\ W_T$ for W_T less than 30,000 pounds;

(ii) $(6W_T + 450,000)/70$ for W_T between 30,000 and 100,000 pounds; and

(iii) $0.15\ W_T$ for W_T over 100,000 pounds.

(b) For towing points not on the landing gear but near the plane of symmetry of the airplane, the drag and side tow load components specified for the auxiliary gear apply. For towing points located outboard of the main gear, the drag and side tow load components specified for the main gear apply. Where the specified angle of swivel cannot be reached, the maximum obtainable angle must be used.

(c) The towing loads specified in paragraph (d) of this section must be reacted as follows:

(1) The side component of the towing load at the main gear must be reacted by a side force at the static ground line of the wheel to which the load is applied.

(2) The towing loads at the auxiliary gear and the drag components of the towing loads at the main gear must be reacted as follows:

(i) A reaction with a maximum value equal to the vertical reaction must be applied at the axle of the wheel to which the load is applied. Enough airplane inertia to achieve equilibrium must be applied.

(ii) The loads must be reacted by airplane inertia.

(d) The prescribed towing loads are as follows:

Tow point	Position	Load		
		Magnitude	No.	Direction
Main gear	....	0.75 F_{TOW} per main gear unit.	1	Forward, parallel to drag axis.
			2	Forward, at 30° to drag axis.
			3	Aft, parallel to drag axis.
			4	Aft, at 30° to drag axis.
Auxiliary gear	Swiveled forward	1.0 F_{TOW}	5	Forward.
			6	Aft.
	Swiveled aft	......do	7	Forward.
			8	Aft.
	Swiveled 45° from forward	0.5 F_{TOW}	9	Forward, in plane of wheel.
			10	Aft, in plane of wheel.
	Swiveled 45° from aft	......do	11	Forward, in plane of wheel.
			12	Aft, in plane of wheel.

[Doc. No. 5066, 29 FR 18291, Dec. 24, 1964, as amended by Amdt. 25–23, 35 FR 5673, Apr. 8, 1970]

§ 25.511 Ground load: unsymmetrical loads on multiple-wheel units.

(a) *General.* Multiple-wheel landing gear units are assumed to be subjected to the limit ground loads prescribed in this subpart under paragraphs (b) through (f) of this section. In addition—

(1) A tandem strut gear arrangement is a multiple-wheel unit; and

(2) In determining the total load on a gear unit with respect to the provisions of paragraphs (b) through (f) of this section, the transverse shift in the load centroid, due to unsymmetrical load

distribution on the wheels, may be neglected.

(b) *Distribution of limit loads to wheels; tires inflated.* The distribution of the limit loads among the wheels of the landing gear must be established for each landing, taxiing, and ground handling condition, taking into account the effects of the following factors:

(1) The number of wheels and their physical arrangements. For truck type landing gear units, the effects of any seesaw motion of the truck during the landing impact must be considered in determining the maximum design loads for the fore and aft wheel pairs.

(2) Any differentials in tire diameters resulting from a combination of manufacturing tolerances, tire growth, and tire wear. A maximum tire-diameter differential equal to ⅔ of the most unfavorable combination of diameter variations that is obtained when taking into account manufacturing tolerances, tire growth, and tire wear, may be assumed.

(3) Any unequal tire inflation pressure, assuming the maximum variation to be ±5 percent of the nominal tire inflation pressure.

(4) A runway crown of zero and a runway crown having a convex upward shape that may be approximated by a slope of 1½ percent with the horizontal. Runway crown effects must be considered with the nose gear unit on either slope of the crown.

(5) The airplane attitude.

(6) Any structural deflections.

(c) *Deflated tires.* The effect of deflated tires on the structure must be considered with respect to the loading conditions specified in paragraphs (d) through (f) of this section, taking into account the physical arrangement of the gear components. In addition—

(1) The deflation of any one tire for each multiple wheel landing gear unit, and the deflation of any two critical tires for each landing gear unit using four or more wheels per unit, must be considered; and

(2) The ground reactions must be applied to the wheels with inflated tires except that, for multiple-wheel gear units with more than one shock strut, a rational distribution of the ground reactions between the deflated and inflated tires, accounting for the differences in shock strut extensions resulting from a deflated tire, may be used.

(d) *Landing conditions.* For one and for two deflated tires, the applied load to each gear unit is assumed to be 60 percent and 50 percent, respectively, of the limit load applied to each gear for each of the prescribed landing conditions. However, for the drift landing condition of § 25.485, 100 percent of the vertical load must be applied.

(e) *Taxiing and ground handling conditions.* For one and for two deflated tires—

(1) The applied side or drag load factor, or both factors, at the center of gravity must be the most critical value up to 50 percent and 40 percent, respectively, of the limit side or drag load factors, or both factors, corresponding to the most severe condition resulting from consideration of the prescribed taxiing and ground handling conditions;

(2) For the braked roll conditions of § 25.493 (a) and (b)(2), the drag loads on each inflated tire may not be less than those at each tire for the symmetrical load distribution with no deflated tires;

(3) The vertical load factor at the center of gravity must be 60 percent and 50 percent, respectively, of the factor with no deflated tires, except that it may not be less than 1g; and

(4) Pivoting need not be considered.

(f) *Towing conditions.* For one and for two deflated tires, the towing load, F_{Tow} must be 60 percent and 50 percent, respectively, of the load prescribed.

§ 25.519 Jacking and tie-down provisions.

(a) General. The airplane must be designed to withstand the limit load conditions resulting from the static ground load conditions of paragraph (b) of this section and, if applicable, paragraph (c) of this section at the most critical combinations of airplane weight and center of gravity. The maximum allowable load at each jack pad must be specified.

(b) Jacking. The airplane must have provisions for jacking and must withstand the following limit loads when the airplane is supported on jacks—

(1) For jacking by the landing gear at the maximum ramp weight of the airplane, the airplane structure must be designed for a vertical load of 1.33 times the vertical static reaction at each jacking point acting singly and in combination with a horizontal load of 0.33 times the vertical static reaction applied in any direction.

(2) For jacking by other airplane structure at maximum approved jacking weight:

(i) The airplane structure must be designed for a vertical load of 1.33 times the vertical reaction at each jacking point acting singly and in combination with a horizontal load of 0.33 times the vertical static reaction applied in any direction.

(ii) The jacking pads and local structure must be designed for a vertical load of 2.0 times the vertical static reaction at each jacking point, acting singly and in combination with a horizontal load of 0.33 times the vertical static reaction applied in any direction.

(c) Tie-down. If tie-down points are provided, the main tie-down points and local structure must withstand the limit loads resulting from a 65-knot horizontal wind from any direction.

[Doc. No. 26129, 59 FR 22102, Apr. 28, 1994]

WATER LOADS

§25.521 General.

(a) Seaplanes must be designed for the water loads developed during takeoff and landing, with the seaplane in any attitude likely to occur in normal operation, and at the appropriate forward and sinking velocities under the most severe sea conditions likely to be encountered.

(b) Unless a more rational analysis of the water loads is made, or the standards in ANC–3 are used, §§25.523 through 25.537 apply.

(c) The requirements of this section and §§25.523 through 25.537 apply also to amphibians.

§25.523 Design weights and center of gravity positions.

(a) Design weights. The water load requirements must be met at each operating weight up to the design landing weight except that, for the takeoff condition prescribed in §25.531, the design water takeoff weight (the maximum weight for water taxi and takeoff run) must be used.

(b) Center of gravity positions. The critical centers of gravity within the limits for which certification is requested must be considered to reach maximum design loads for each part of the seaplane structure.

[Doc. No. 5066, 29 FR 18291, Dec. 24, 1964, as amended by Amdt. 25–23, 35 FR 5673, Apr. 8, 1970]

§25.525 Application of loads.

(a) Unless otherwise prescribed, the seaplane as a whole is assumed to be subjected to the loads corresponding to the load factors specified in §25.527.

(b) In applying the loads resulting from the load factors prescribed in §25.527, the loads may be distributed over the hull or main float bottom (in order to avoid excessive local shear loads and bending moments at the location of water load application) using pressures not less than those prescribed in §25.533(c).

(c) For twin float seaplanes, each float must be treated as an equivalent hull on a fictitious seaplane with a weight equal to one-half the weight of the twin float seaplane.

(d) Except in the takeoff condition of §25.531, the aerodynamic lift on the seaplane during the impact is assumed to be ⅔ of the weight of the seaplane.

[Doc. No. 5066, 29 FR 18291, Dec. 24, 1964, as amended by Doc. No. FAA–2022–1355, Amdt. 25–148, 87 FR 75710, Dec. 9, 2022; 88 FR 2813, Jan. 18, 2023]

§25.527 Hull and main float load factors.

(a) Water reaction load factors n_W must be computed in the following manner:

(1) For the step landing case

$$n_w = \frac{C_1 V_{S0}{}^2}{\left(\mathrm{Tan}^{\frac{2}{3}}\beta\right) W^{\frac{1}{3}}}$$

(2) For the bow and stern landing cases

255

$$n_w = \frac{C_1 V_{S0}{}^2}{\left(\operatorname{Tan}^{\frac{2}{3}}\beta\right) W^{\frac{1}{3}}} \times \frac{K_1}{\left(1 + r_x{}^2\right)^{\frac{2}{3}}}$$

(b) The following values are used:

(1) n_w = water reaction load factor (that is, the water reaction divided by seaplane weight).

(2) C_1 = empirical seaplane operations factor equal to 0.012 (except that this factor may not be less than that necessary to obtain the minimum value of step load factor of 2.33).

(3) V_{S0} = seaplane stalling speed in knots with flaps extended in the appropriate landing position and with no slipstream effect.

(4) β = angle of dead rise at the longitudinal station at which the load factor is being determined in accordance with figure 1 of appendix B.

(5) W = seaplane design landing weight in pounds.

(6) K_1 = empirical hull station weighing factor, in accordance with figure 2 of appendix B.

(7) r_x = ratio of distance, measured parallel to hull reference axis, from the center of gravity of the seaplane to the hull longitudinal station at which the load factor is being computed to the radius of gyration in pitch of the seaplane, the hull reference axis being a straight line, in the plane of symmetry, tangential to the keel at the main step.

(c) For a twin float seaplane, because of the effect of flexibility of the attachment of the floats to the seaplane, the factor K_1 may be reduced at the bow and stern to 0.8 of the value shown in figure 2 of appendix B. This reduction applies only to the design of the carry-through and seaplane structure.

[Doc. No. 5066, 29 FR 18291, Dec. 24, 1964, as amended by Amdt. 25–23, 35 FR 5673, Apr. 8, 1970]

§ 25.529 Hull and main float landing conditions.

(a) *Symmetrical step, bow, and stern landing.* For symmetrical step, bow, and stern landings, the limit water reaction load factors are those computed under § 25.527. In addition—

(1) For symmetrical step landings, the resultant water load must be applied at the keel, through the center of gravity, and must be directed perpendicularly to the keel line;

(2) For symmetrical bow landings, the resultant water load must be applied at the keel, one-fifth of the longitudinal distance from the bow to the step, and must be directed perpendicularly to the keel line; and

(3) For symmetrical stern landings, the resultant water load must be applied at the keel, at a point 85 percent of the longitudinal distance from the step to the stern post, and must be directed perpendicularly to the keel line.

(b) *Unsymmetrical landing for hull and single float seaplanes.* Unsymmetrical step, bow, and stern landing conditions must be investigated. In addition—

(1) The loading for each condition consists of an upward component and a side component equal, respectively, to 0.75 and 0.25 tan β times the resultant load in the corresponding symmetrical landing condition; and

(2) The point of application and direction of the upward component of the load is the same as that in the symmetrical condition, and the point of application of the side component is at the same longitudinal station as the upward component but is directed inward perpendicularly to the plane of symmetry at a point midway between the keel and chine lines.

(c) *Unsymmetrical landing; twin float seaplanes.* The unsymmetrical loading consists of an upward load at the step of each float of 0.75 and a side load of 0.25 tan β at one float times the step landing load reached under § 25.527. The side load is directed inboard, perpendicularly to the plane of symmetry midway between the keel and chine lines of the float, at the same longitudinal station as the upward load.

§ 25.531 Hull and main float takeoff condition.

For the wing and its attachment to the hull or main float—

(a) The aerodynamic wing lift is assumed to be zero; and

(b) A downward inertia load, corresponding to a load factor computed from the following formula, must be applied:

$$n = \frac{C_{TO} V_{S1}^2}{\left(\tan^{\frac{2}{3}} \beta\right) W^{\frac{1}{3}}}$$

where—

n = inertia load factor;
C_{TO} = empirical seaplane operations factor equal to 0.004;
V_{S1} = seaplane stalling speed (knots) at the design takeoff weight with the flaps extended in the appropriate takeoff position;
β = angle of dead rise at the main step (degrees); and
W = design water takeoff weight in pounds.

[Doc. No. 5066, 29 FR 18291, Dec. 24, 1964, as amended by Amdt. 25–23, 35 FR 5673, Apr. 8, 1970]

§25.533 Hull and main float bottom pressures.

(a) *General.* The hull and main float structure, including frames and bulkheads, stringers, and bottom plating, must be designed under this section.

(b) *Local pressures.* For the design of the bottom plating and stringers and their attachments to the supporting structure, the following pressure distributions must be applied:

(1) For an unflared bottom, the pressure at the chine is 0.75 times the pressure at the keel, and the pressures between the keel and chine vary linearly, in accordance with figure 3 of appendix B. The pressure at the keel (psi) is computed as follows:

$$P_k = C_2 \times \frac{K_2 V_{S1}^2}{\tan \beta_k}$$

where—

P_k = pressure (p.s.i.) at the keel;
C_2 = 0.00213;
K_2 = hull station weighing factor, in accordance with figure 2 of appendix B;
V_{S1} = seaplane stalling speed (Knots) at the design water takeoff weight with flaps extended in the appropriate takeoff position; and
β_K = angle of dead rise at keel, in accordance with figure 1 of appendix B.

(2) For a flared bottom, the pressure at the beginning of the flare is the same as that for an unflared bottom, and the pressure between the chine and the beginning of the flare varies linearly, in accordance with figure 3 of appendix B. The pressure distribution is the same as that prescribed in paragraph (b)(1) of this section for an unflared bottom except that the pressure at the chine is computed as follows:

$$P_{ch} = C_3 \times \frac{K_2 V_{S1}^2}{\tan \beta}$$

where—

P_{ch} = pressure (p.s.i.) at the chine;
C_3 = 0.0016;
K_2 = hull station weighing factor, in accordance with figure 2 of appendix B;
V_{S1} = seaplane stalling speed at the design water takeoff weight with flaps extended in the appropriate takeoff position; and
β = angle of dead rise at appropriate station.

The area over which these pressures are applied must simulate pressures occurring during high localized impacts on the hull or float, but need not extend over an area that would induce critical stresses in the frames or in the overall structure.

(c) *Distributed pressures.* For the design of the frames, keel, and chine structure, the following pressure distributions apply:

(1) Symmetrical pressures are computed as follows:

$$P = C_4 \times \frac{K_2 V_{S0}^2}{\tan \beta}$$

where—

P = pressure (p.s.i.);
C_4 = 0.078 C_1 (with C_1 computed under §25.527);
K_2 = hull station weighing factor, determined in accordance with figure 2 of appendix B;
V_{S0} = seaplane stalling speed (Knots) with landing flaps extended in the appropriate position and with no slipstream effect; and
V_{S0} = seaplane stalling speed with landing flaps extended in the appropriate position and with no slipstream effect; and β = angle of dead rise at appropriate station.

(2) The unsymmetrical pressure distribution consists of the pressures prescribed in paragraph (c)(1) of this section on one side of the hull or main float centerline and one-half of that pressure on the other side of the hull or main float centerline, in accordance with figure 3 of appendix B.

These pressures are uniform and must be applied simultaneously over the entire hull or main float bottom. The loads obtained must be carried into the sidewall structure of the hull proper, but need not be transmitted in a fore and aft direction as shear and bending loads.

[Doc. No. 5066, 29 FR 18291, Dec. 24, 1964, as amended by Amdt. 25–23, 35 FR 5673, Apr. 8, 1970]

§ 25.535 Auxiliary float loads.

(a) *General.* Auxiliary floats and their attachments and supporting structures must be designed for the conditions prescribed in this section. In the cases specified in paragraphs (b) through (e) of this section, the prescribed water loads may be distributed over the float bottom to avoid excessive local loads, using bottom pressures not less than those prescribed in paragraph (g) of this section.

(b) *Step loading.* The resultant water load must be applied in the plane of symmetry of the float at a point three-fourths of the distance from the bow to the step and must be perpendicular to the keel. The resultant limit load is computed as follows, except that the value of L need not exceed three times the weight of the displaced water when the float is completely submerged:

$$L = \frac{C_5\, V_{S_0}^2\, W^{\frac{2}{3}}}{\tan^{\frac{2}{3}} \beta_s \left(1 + r_y^2\right)^{\frac{2}{3}}}$$

where—

L = limit load (lbs.);
C_5 = 0.0053;
V_{S0} = seaplane stalling speed (knots) with landing flaps extended in the appropriate position and with no slipstream effect;
W = seaplane design landing weight in pounds;
β_s = angle of dead rise at a station ¾ of the distance from the bow to the step, but need not be less than 15 degrees; and
r_y = ratio of the lateral distance between the center of gravity and the plane of symmetry of the float to the radius of gyration in roll.

(c) *Bow loading.* The resultant limit load must be applied in the plane of symmetry of the float at a point one-fourth of the distance from the bow to the step and must be perpendicular to

the tangent to the keel line at that point. The magnitude of the resultant load is that specified in paragraph (b) of this section.

(d) *Unsymmetrical step loading.* The resultant water load consists of a component equal to 0.75 times the load specified in paragraph (a) of this section and a side component equal to 0.25 tan β times the load specified in paragraph (b) of this section. The side load must be applied perpendicularly to the plane of symmetry of the float at a point midway between the keel and the chine.

(e) *Unsymmetrical bow loading.* The resultant water load consists of a component equal to 0.75 times the load specified in paragraph (b) of this section and a side component equal to 0.25 tan β times the load specified in paragraph (c) of this section. The side load must be applied perpendicularly to the plane of symmetry at a point midway between the keel and the chine.

(f) *Immersed float condition.* The resultant load must be applied at the centroid of the cross section of the float at a point one-third of the distance from the bow to the step. The limit load components are as follows:

$$\text{vertical} = \rho g V$$

$$\text{aft} = C_{x_2}\, \rho\, V^{\frac{2}{3}} \left(K V_{S_0} \right)^2$$

$$\text{side} = C_{y_2}\, \rho\, V^{\frac{2}{3}} \left(K V_{S_0} \right)^2$$

where—

ρ = mass density of water (slugs/ft.²);
V = volume of float (ft.²);
C_x = coefficient of drag force, equal to 0.133;
C_y = coefficient of side force, equal to 0.106;
K = 0.8, except that lower values may be used if it is shown that the floats are incapable of submerging at a speed of 0.8 V_{S0} in normal operations;
V_{S0} = seaplane stalling speed (knots) with landing flaps extended in the appropriate position and with no slipstream effect; and
g = acceleration due to gravity (ft./sec.²).

(g) *Float bottom pressures.* The float bottom pressures must be established under § 25.533, except that the value of K_2 in the formulae may be taken as 1.0. The angle of dead rise to be used in determining the float bottom pressures is set forth in paragraph (b) of this section.

[Doc. No. 5066, 29 FR 18291, Dec. 24, 1964, as amended by Amdt. 25–23, 35 FR 5673, Apr. 8, 1970; Amdt. 25–148, 87 FR 75710, Dec. 9, 2022; 88 FR 2813, Jan. 18, 2023]

§25.537 Seawing loads.

Seawing design loads must be based on applicable test data.

EMERGENCY LANDING CONDITIONS

§25.561 General.

(a) The airplane, although it may be damaged in emergency landing conditions on land or water, must be designed as prescribed in this section to protect each occupant under those conditions.

(b) The structure must be designed to give each occupant every reasonable chance of escaping serious injury in a minor crash landing when—

(1) Proper use is made of seats, belts, and all other safety design provisions;

(2) The wheels are retracted (where applicable); and

(3) The occupant experiences the following ultimate inertia forces acting separately relative to the surrounding structure:

(i) Upward, 3.0g

(ii) Forward, 9.0g

(iii) Sideward, 3.0g on the airframe; and 4.0g on the seats and their attachments.

(iv) Downward, 6.0g

(v) Rearward, 1.5g

(c) For equipment, cargo in the passenger compartments and any other large masses, the following apply:

(1) Except as provided in paragraph (c)(2) of this section, these items must be positioned so that if they break loose they will be unlikely to:

(i) Cause direct injury to occupants;

(ii) Penetrate fuel tanks or lines or cause fire or explosion hazard by damage to adjacent systems; or

(iii) Nullify any of the escape facilities provided for use after an emergency landing.

(2) When such positioning is not practical (e.g. fuselage mounted engines or auxiliary power units) each such item of mass shall be restrained under all loads up to those specified in paragraph (b)(3) of this section. The local attach-

ments for these items should be designed to withstand 1.33 times the specified loads if these items are subject to severe wear and tear through frequent removal (e.g. quick change interior items).

(d) Seats and items of mass (and their supporting structure) must not deform under any loads up to those specified in paragraph (b)(3) of this section in any manner that would impede subsequent rapid evacuation of occupants.

[Doc. No. 5066, 29 FR 18291, Dec. 24, 1964, as amended by Amdt. 25–23, 35 FR 5673, Apr. 8, 1970; Amdt. 25–64, 53 FR 17646, May 17, 1988; Amdt. 25–91, 62 FR 40706, July 29, 1997]

§25.562 Emergency landing dynamic conditions.

(a) The seat and restraint system in the airplane must be designed as prescribed in this section to protect each occupant during an emergency landing condition when—

(1) Proper use is made of seats, safety belts, and shoulder harnesses provided for in the design; and

(2) The occupant is exposed to loads resulting from the conditions prescribed in this section.

(b) Each seat type design approved for crew or passenger occupancy during takeoff and landing must successfully complete dynamic tests or be demonstrated by rational analysis based on dynamic tests of a similar type seat, in accordance with each of the following emergency landing conditions. The tests must be conducted with an occupant simulated by a 170-pound anthropomorphic test dummy, as defined by 49 CFR Part 572, Subpart B, or its equivalent, sitting in the normal upright position.

(1) A change in downward vertical velocity (Δ v) of not less than 35 feet per second, with the airplane's longitudinal axis canted downward 30 degrees with respect to the horizontal plane and with the wings level. Peak floor deceleration must occur in not more than 0.08 seconds after impact and must reach a minimum of 14g.

(2) A change in forward longitudinal velocity (Δ v) of not less than 44 feet per second, with the airplane's longitudinal axis horizontal and yawed 10 degrees either right or left, whichever

would cause the greatest likelihood of the upper torso restraint system (where installed) moving off the occupant's shoulder, and with the wings level. Peak floor deceleration must occur in not more than 0.09 seconds after impact and must reach a minimum of 16g. Where floor rails or floor fittings are used to attach the seating devices to the test fixture, the rails or fittings must be misaligned with respect to the adjacent set of rails or fittings by at least 10 degrees vertically (*i.e.*, out of Parallel) with one rolled 10 degrees.

(c) The following performance measures must not be exceeded during the dynamic tests conducted in accordance with paragraph (b) of this section:

(1) Where upper torso straps are used for crewmembers, tension loads in individual straps must not exceed 1,750 pounds. If dual straps are used for re-straining the upper torso, the total strap tension loads must not exceed 2,000 pounds.

(2) The maximum compressive load measured between the pelvis and the lumbar column of the anthropomorphic dummy must not exceed 1,500 pounds.

(3) The upper torso restraint straps (where installed) must remain on the occupant's shoulder during the impact.

(4) The lap safety belt must remain on the occupant's pelvis during the impact.

(5) Each occupant must be protected from serious head injury under the conditions prescribed in paragraph (b) of this section. Where head contact with seats or other structure can occur, protection must be provided so that the head impact does not exceed a Head Injury Criterion (HIC) of 1,000 units. The level of HIC is defined by the equation:

$$HIC = \left\{ (t_2 - t_1) \left[\frac{1}{(t_2 - t_1)} \int_{t_1}^{t_2} a(t)dt \right]^{2.5} \right\}_{max}$$

Where:

t_1 is the initial integration time,
t_2 is the final integration time, and
$a(t)$ is the total acceleration vs. time curve for the head strike, and where
(t) is in seconds, and (a) is in units of gravity (g).

(6) Where leg injuries may result from contact with seats or other structure, protection must be provided to prevent axially compressive loads exceeding 2,250 pounds in each femur.

(7) The seat must remain attached at all points of attachment, although the structure may have yielded.

(8) Seats must not yield under the tests specified in paragraphs (b)(1) and (b)(2) of this section to the extent they would impede rapid evacuation of the airplane occupants.

[Amdt. 25–64, 53 FR 17646, May 17, 1988]

§ 25.563   Structural ditching provisions.

Structural strength considerations of ditching provisions must be in accordance with § 25.801(e).

FATIGUE EVALUATION

§ 25.571   Damage-tolerance and fatigue evaluation of structure.

(a) *General.* An evaluation of the strength, detail design, and fabrication must show that catastrophic failure due to fatigue, corrosion, manufacturing defects, or accidental damage, will be avoided throughout the operational life of the airplane. This evaluation must be conducted in accordance with the provisions of paragraphs (b) and (e) of this section, except as specified in paragraph (c) of this section, for each part of the structure that could contribute to a catastrophic failure (such as wing, empennage, control surfaces and their systems, the fuselage, engine mounting, landing gear, and their related primary attachments). For turbojet powered airplanes, those parts that could contribute to a catastrophic failure must also be evaluated under paragraph (d) of this section. In addition, the following apply:

(1) Each evaluation required by this section must include—

(i) The typical loading spectra, temperatures, and humidities expected in service;

(ii) The identification of principal structural elements and detail design points, the failure of which could cause catastrophic failure of the airplane; and

(iii) An analysis, supported by test evidence, of the principal structural elements and detail design points identified in paragraph (a)(1)(ii) of this section.

(2) The service history of airplanes of similar structural design, taking due account of differences in operating conditions and procedures, may be used in the evaluations required by this section.

(3) Based on the evaluations required by this section, inspections or other procedures must be established, as necessary, to prevent catastrophic failure, and must be included in the Airworthiness Limitations section of the Instructions for Continued Airworthiness required by § 25.1529. The limit of validity of the engineering data that supports the structural maintenance program (hereafter referred to as LOV), stated as a number of total accumulated flight cycles or flight hours or both, established by this section must also be included in the Airworthiness Limitations section of the Instructions for Continued Airworthiness required by § 25.1529. Inspection thresholds for the following types of structure must be established based on crack growth analyses and/or tests, assuming the structure contains an initial flaw of the maximum probable size that could exist as a result of manufacturing or service-induced damage:

(i) Single load path structure, and

(ii) Multiple load path "fail-safe" structure and crack arrest "fail-safe" structure, where it cannot be demonstrated that load path failure, partial failure, or crack arrest will be detected and repaired during normal maintenance, inspection, or operation of an airplane prior to failure of the remaining structure.

(b) *Damage-tolerance evaluation.* The evaluation must include a determination of the probable locations and modes of damage due to fatigue, corrosion, or accidental damage. Repeated load and static analyses supported by test evidence and (if available) service experience must also be incorporated in the evaluation. Special consideration for widespread fatigue damage must be included where the design is such that this type of damage could occur. An LOV must be established that corresponds to the period of time, stated as a number of total accumulated flight cycles or flight hours or both, during which it is demonstrated that widespread fatigue damage will not occur in the airplane structure. This demonstration must be by full-scale fatigue test evidence. The type certificate may be issued prior to completion of full-scale fatigue testing, provided the Administrator has approved a plan for completing the required tests. In that case, the Airworthiness Limitations section of the Instructions for Continued Airworthiness required by § 25.1529 must specify that no airplane may be operated beyond a number of cycles equal to ½ the number of cycles accumulated on the fatigue test article, until such testing is completed. The extent of damage for residual strength evaluation at any time within the operational life of the airplane must be consistent with the initial detectability and subsequent growth under repeated loads. The residual strength evaluation must show that the remaining structure is able to withstand loads (considered as static ultimate loads) corresponding to the following conditions:

(1) The limit symmetrical maneuvering conditions specified in § 25.337 at all speeds up to V_c and in § 25.345.

(2) The limit gust conditions specified in § 25.341 at the specified speeds up to V_C and in § 25.345.

(3) The limit rolling conditions specified in § 25.349 and the limit unsymmetrical conditions specified in §§ 25.367 and 25.427 (a) through (c), at speeds up to V_C.

(4) The limit yaw maneuvering conditions specified in § 25.351(a) at the specified speeds up to V_C.

(5) For pressurized cabins, the following conditions:

(i) The normal operating differential pressure combined with the expected

261

external aerodynamic pressures applied simultaneously with the flight loading conditions specified in paragraphs (b)(1) through (4) of this section, if they have a significant effect.

(ii) The maximum value of normal operating differential pressure (including the expected external aerodynamic pressures during 1 g level flight) multiplied by a factor of 1.15, omitting other loads.

(6) For landing gear and directly-affected airframe structure, the limit ground loading conditions specified in §§ 25.473, 25.491, and 25.493.

If significant changes in structural stiffness or geometry, or both, follow from a structural failure, or partial failure, the effect on damage tolerance must be further investigated.

(c) *Fatigue (safe-life) evaluation.* Compliance with the damage-tolerance requirements of paragraph (b) of this section is not required if the applicant establishes that their application for particular structure is impractical. This structure must be shown by analysis, supported by test evidence, to be able to withstand the repeated loads of variable magnitude expected during its service life without detectable cracks. Appropriate safe-life scatter factors must be applied.

(d) *Sonic fatigue strength.* It must be shown by analysis, supported by test evidence, or by the service history of airplanes of similar structural design and sonic excitation environment, that—

(1) Sonic fatigue cracks are not probable in any part of the flight structure subject to sonic excitation; or

(2) Catastrophic failure caused by sonic cracks is not probable assuming that the loads prescribed in paragraph (b) of this section are applied to all areas affected by those cracks.

(e) *Damage-tolerance (discrete source) evaluation.* The airplane must be capable of successfully completing a flight during which likely structural damage occurs as a result of—

(1) Impact with a 4-pound bird when the velocity of the airplane relative to the bird along the airplane's flight path is equal to V_c at sea level or $0.85V_c$ at 8,000 feet, whichever is more critical;

(2) Uncontained fan blade impact;

(3) Uncontained engine failure; or

(4) Uncontained high energy rotating machinery failure.

The damaged structure must be able to withstand the static loads (considered as ultimate loads) which are reasonably expected to occur on the flight. Dynamic effects on these static loads need not be considered. Corrective action to be taken by the pilot following the incident, such as limiting maneuvers, avoiding turbulence, and reducing speed, must be considered. If significant changes in structural stiffness or geometry, or both, follow from a structural failure or partial failure, the effect on damage tolerance must be further investigated.

[Amdt. 25–45, 43 FR 46242, Oct. 5, 1978, as amended by Amdt. 25–54, 45 FR 60173, Sept. 11, 1980; Amdt. 25–72, 55 FR 29776, July 20, 1990; Amdt. 25–86, 61 FR 5222, Feb. 9, 1996; Amdt. 25–96, 63 FR 15714, Mar. 31, 1998; 63 FR 23338, Apr. 28, 1998; Amdt. 25–132, 75 FR 69781, Nov. 15, 2010; Amdt. No. 25–148, 87 FR 75710, Dec. 9, 2022; 88 FR 2813, Jan. 18, 2023]

LIGHTNING PROTECTION

§ 25.581 Lightning protection.

(a) The airplane must be protected against catastrophic effects from lightning.

(b) For metallic components, compliance with paragraph (a) of this section may be shown by—

(1) Bonding the components properly to the airframe; or

(2) Designing the components so that a strike will not endanger the airplane.

(c) For nonmetallic components, compliance with paragraph (a) of this section may be shown by—

(1) Designing the components to minimize the effect of a strike; or

(2) Incorporating acceptable means of diverting the resulting electrical current so as not to endanger the airplane.

[Amdt. 25–23, 35 FR 5674, Apr. 8, 1970]

Subpart D—Design and Construction

GENERAL

§ 25.601 General.

The airplane may not have design features or details that experience has shown to be hazardous or unreliable. The suitability of each questionable design detail and part must be established by tests.

§ 25.603 Materials.

The suitability and durability of materials used for parts, the failure of which could adversely affect safety, must—

(a) Be established on the basis of experience or tests;

(b) Conform to approved specifications (such as industry or military specifications, or Technical Standard Orders) that ensure their having the strength and other properties assumed in the design data; and

(c) Take into account the effects of environmental conditions, such as temperature and humidity, expected in service.

[Doc. No. 5066, 29 FR 18291, Dec. 24, 1964, as amended by Amdt. 25–38, 41 FR 55466, Dec. 20, 1976; Amdt. 25–46, 43 FR 50595, Oct. 30, 1978]

§ 25.605 Fabrication methods.

(a) The methods of fabrication used must produce a consistently sound structure. If a fabrication process (such as gluing, spot welding, or heat treating) requires close control to reach this objective, the process must be performed under an approved process specification.

(b) Each new aircraft fabrication method must be substantiated by a test program.

[Doc. No. 5066, 29 FR 18291, Dec. 24, 1964, as amended by Amdt. 25–46, 43 FR 50595, Oct. 30, 1978]

§ 25.607 Fasteners.

(a) Each removable bolt, screw, nut, pin, or other removable fastener must incorporate two separate locking devices if—

(1) Its loss could preclude continued flight and landing within the design limitations of the airplane using normal pilot skill and strength; or

(2) Its loss could result in reduction in pitch, yaw, or roll control capability or response below that required by Subpart B of this chapter.

(b) The fasteners specified in paragraph (a) of this section and their locking devices may not be adversely affected by the environmental conditions associated with the particular installation.

(c) No self-locking nut may be used on any bolt subject to rotation in operation unless a nonfriction locking device is used in addition to the self-locking device.

[Amdt. 25–23, 35 FR 5674, Apr. 8, 1970]

§ 25.609 Protection of structure.

Each part of the structure must—

(a) Be suitably protected against deterioration or loss of strength in service due to any cause, including—

(1) Weathering;

(2) Corrosion; and

(3) Abrasion; and

(b) Have provisions for ventilation and drainage where necessary for protection.

§ 25.611 Accessibility provisions.

(a) Means must be provided to allow inspection (including inspection of principal structural elements and control systems), replacement of parts normally requiring replacement, adjustment, and lubrication as necessary for continued airworthiness. The inspection means for each item must be practicable for the inspection interval for the item. Nondestructive inspection aids may be used to inspect structural elements where it is impracticable to provide means for direct visual inspection if it is shown that the inspection is effective and the inspection procedures are specified in the maintenance manual required by § 25.1529.

(b) EWIS must meet the accessibility requirements of § 25.1719.

[Amdt. 25–23, 35 FR 5674, Apr. 8, 1970, as amended by Amdt. 25–123, 72 FR 63404, Nov. 8, 2007]

§ 25.613 Material strength properties and material design values.

(a) Material strength properties must be based on enough tests of material meeting approved specifications to establish design values on a statistical basis.

(b) Material design values must be chosen to minimize the probability of structural failures due to material variability. Except as provided in paragraphs (e) and (f) of this section, compliance must be shown by selecting material design values which assure material strength with the following probability:

(1) Where applied loads are eventually distributed through a single member within an assembly, the failure of which would result in loss of structural integrity of the component, 99 percent probability with 95 percent confidence.

(2) For redundant structure, in which the failure of individual elements would result in applied loads being safely distributed to other load carrying members, 90 percent probability with 95 percent confidence.

(c) The effects of environmental conditions, such as temperature and moisture, on material design values used in an essential component or structure must be considered where these effects are significant within the airplane operating envelope.

(d) [Reserved]

(e) Greater material design values may be used if a "premium selection" of the material is made in which a specimen of each individual item is tested before use to determine that the actual strength properties of that particular item will equal or exceed those used in design.

(f) Other material design values may be used if approved by the Administrator.

[Doc. No. 5066, 29 FR 18291, Dec. 24, 1964, as amended by Amdt. 25-46, 43 FR 50595, Oct. 30, 1978; Amdt. 25-72, 55 FR 29776, July 20, 1990; Amdt. 25-112, 68 FR 46431, Aug. 5, 2003]

§ 25.619 Special factors.

The factor of safety prescribed in § 25.303 must be multiplied by the highest pertinent special factor of safety prescribed in §§ 25.621 through 25.625 for each part of the structure whose strength is—

(a) Uncertain;

(b) Likely to deteriorate in service before normal replacement; or

(c) Subject to appreciable variability because of uncertainties in manufacturing processes or inspection methods.

[Doc. No. 5066, 29 FR 18291, Dec. 24, 1964, as amended by Amdt. 25-23, 35 FR 5674, Apr. 8, 1970]

§ 25.621 Casting factors.

(a) *General.* For castings used in structural applications, the factors, tests, and inspections specified in paragraphs (b) through (d) of this section must be applied in addition to those necessary to establish foundry quality control. The inspections must meet approved specifications. Paragraphs (c) and (d) of this section apply to any structural castings, except castings that are pressure tested as parts of hydraulic or other fluid systems and do not support structural loads.

(b) *Bearing stresses and surfaces.* The casting factors specified in paragraphs (c) and (d) of this section—

(1) Need not exceed 1.25 with respect to bearing stresses regardless of the method of inspection used; and

(2) Need not be used with respect to the bearing surfaces of a part whose bearing factor is larger than the applicable casting factor.

(c) *Critical castings.* Each casting whose failure could preclude continued safe flight and landing of the airplane or could result in serious injury to occupants is a critical casting. Each critical casting must have a factor associated with it for showing compliance with strength and deformation requirements of § 25.305, and must comply with the following criteria associated with that factor:

(1) A casting factor of 1.0 or greater may be used, provided that—

(i) It is demonstrated, in the form of process qualification, proof of product, and process monitoring that, for each casting design and part number, the castings produced by each foundry and process combination have coefficients of variation of the material properties that are equivalent to those of wrought alloy products of similar composition. Process monitoring must include testing of coupons cut from the prolongations of each casting (or each set of castings, if produced from a single pour into a single mold in a runner system) and, on a sampling basis, coupons cut from critical areas of production castings. The acceptance criteria for the process monitoring inspections and tests must be established and included in the process specifications to ensure the properties of the production castings are controlled to within levels used in design.

(ii) Each casting receives:

(A) Inspection of 100 percent of its surface, using visual inspection and liquid penetrant or equivalent inspection methods; and

(B) Inspection of structurally significant internal areas and areas where defects are likely to occur, using radiographic or equivalent inspection methods.

(iii) One casting undergoes a static test and is shown to meet the strength and deformation requirements of § 25.305(a) and (b).

(2) A casting factor of 1.25 or greater may be used, provided that—

(i) Each casting receives:

(A) Inspection of 100 percent of its surface, using visual inspection and liquid penetrant or equivalent inspection methods; and

(B) Inspection of structurally significant internal areas and areas where defects are likely to occur, using radiographic or equivalent inspection methods.

(ii) Three castings undergo static tests and are shown to meet:

(A) The strength requirements of § 25.305(b) at an ultimate load corresponding to a casting factor of 1.25; and

(B) The deformation requirements of § 25.305(a) at a load of 1.15 times the limit load.

(3) A casting factor of 1.50 or greater may be used, provided that—

(i) Each casting receives:

(A) Inspection of 100 percent of its surface, using visual inspection and liquid penetrant or equivalent inspection methods; and

(B) Inspection of structurally significant internal areas and areas where defects are likely to occur, using radiographic or equivalent inspection methods.

(ii) One casting undergoes a static test and is shown to meet:

(A) The strength requirements of § 25.305(b) at an ultimate load corresponding to a casting factor of 1.50; and

(B) The deformation requirements of § 25.305(a) at a load of 1.15 times the limit load.

(d) *Non-critical castings.* For each casting other than critical castings, as specified in paragraph (c) of this section, the following apply:

(1) A casting factor of 1.0 or greater may be used, provided that the requirements of (c)(1) of this section are met, or all of the following conditions are met:

(i) Castings are manufactured to approved specifications that specify the minimum mechanical properties of the material in the casting and provides for demonstration of these properties by testing of coupons cut from the castings on a sampling basis.

(ii) Each casting receives:

(A) Inspection of 100 percent of its surface, using visual inspection and liquid penetrant or equivalent inspection methods; and

(B) Inspection of structurally significant internal areas and areas where defects are likely to occur, using radiographic or equivalent inspection methods.

(iii) Three sample castings undergo static tests and are shown to meet the strength and deformation requirements of § 25.305(a) and (b).

(2) A casting factor of 1.25 or greater may be used, provided that each casting receives:

(i) Inspection of 100 percent of its surface, using visual inspection and liquid penetrant or equivalent inspection methods; and

(ii) Inspection of structurally significant internal areas and areas where defects are likely to occur, using radiographic or equivalent inspection methods.

(3) A casting factor of 1.5 or greater may be used, provided that each casting receives inspection of 100 percent of its surface using visual inspection and liquid penetrant or equivalent inspection methods.

(4) A casting factor of 2.0 or greater may be used, provided that each casting receives inspection of 100 percent of its surface using visual inspection methods.

(5) The number of castings per production batch to be inspected by non-visual methods in accordance with paragraphs (d)(2) and (3) of this section

may be reduced when an approved quality control procedure is established.

[Doc. No. 5066, 29 FR 18291, Dec. 24, 1964, as amended by Amdt. 25–139, 79 FR 59429, Oct. 2, 2014]

§ 25.623 Bearing factors.

(a) Except as provided in paragraph (b) of this section, each part that has clearance (free fit), and that is subject to pounding or vibration, must have a bearing factor large enough to provide for the effects of normal relative motion.

(b) No bearing factor need be used for a part for which any larger special factor is prescribed.

§ 25.625 Fitting factors.

For each fitting (a part or terminal used to join one structural member to another), the following apply:

(a) For each fitting whose strength is not proven by limit and ultimate load tests in which actual stress conditions are simulated in the fitting and surrounding structures, a fitting factor of at least 1.15 must be applied to each part of—

(1) The fitting;

(2) The means of attachment; and

(3) The bearing on the joined members.

(b) No fitting factor need be used—

(1) For joints made under approved practices and based on comprehensive test data (such as continuous joints in metal plating, welded joints, and scarf joints in wood); or

(2) With respect to any bearing surface for which a larger special factor is used.

(c) For each integral fitting, the part must be treated as a fitting up to the point at which the section properties become typical of the member.

(d) For each seat, berth, safety belt, and harness, the fitting factor specified in § 25.785(f)(3) applies.

[Doc. No. 5066, 29 FR 18291, Dec. 24, 1964, as amended by Amdt. 25–23, 35 FR 5674, Apr. 8, 1970; Amdt. 25–72, 55 FR 29776, July 20, 1990]

§ 25.629 Aeroelastic stability requirements.

(a) *General.* The aeroelastic stability evaluations required under this section include flutter, divergence, control reversal and any undue loss of stability and control as a result of structural deformation. The aeroelastic evaluation must include whirl modes associated with any propeller or rotating device that contributes significant dynamic forces. Compliance with this section must be shown by analyses, wind tunnel tests, ground vibration tests, flight tests, or other means found necessary by the Administrator.

(b) *Aeroelastic stability envelopes.* The airplane must be designed to be free from aeroelastic instability for all configurations and design conditions within the aeroelastic stability envelopes as follows:

(1) For normal conditions without failures, malfunctions, or adverse conditions, all combinations of altitudes and speeds encompassed by the V_D/M_D versus altitude envelope enlarged at all points by an increase of 15 percent in equivalent airspeed at both constant Mach number and constant altitude. In addition, a proper margin of stability must exist at all speeds up to V_D/M_D and, there must be no large and rapid reduction in stability as V_D/M_D is approached. The enlarged envelope may be limited to Mach 1.0 when M_D is less than 1.0 at all design altitudes, and

(2) For the conditions described in § 25.629(d) below, for all approved altitudes, any airspeed up to the greater airspeed defined by;

(i) The V_D/M_D envelope determined by § 25.335(b); or,

(ii) An altitude-airspeed envelope defined by a 15 percent increase in equivalent airspeed above V_C at constant altitude, from sea level to the altitude of the intersection of 1.15 V_C with the extension of the constant cruise Mach number line, M_C, then a linear variation in equivalent airspeed to M_C + .05 at the altitude of the lowest V_C/M_C intersection; then, at higher altitudes, up to the maximum flight altitude, the boundary defined by a .05 Mach increase in M_C at constant altitude.

(c) *Balance weights.* If concentrated balance weights are used, their effectiveness and strength, including supporting structure, must be substantiated.

(d) *Failures, malfunctions, and adverse conditions.* The failures, malfunctions, and adverse conditions which must be considered in showing compliance with this section are:

(1) Any critical fuel loading conditions, not shown to be extremely improbable, which may result from mismanagement of fuel.

(2) Any single failure in any flutter damper system.

(3) For airplanes not approved for operation in icing conditions, the maximum likely ice accumulation expected as a result of an inadvertent encounter.

(4) Failure of any single element of the structure supporting any engine, independently mounted propeller shaft, large auxiliary power unit, or large externally mounted aerodynamic body (such as an external fuel tank).

(5) For airplanes with engines that have propellers or large rotating devices capable of significant dynamic forces, any single failure of the engine structure that would reduce the rigidity of the rotational axis.

(6) The absence of aerodynamic or gyroscopic forces resulting from the most adverse combination of feathered propellers or other rotating devices capable of significant dynamic forces. In addition, the effect of a single feathered propeller or rotating device must be coupled with the failures of paragraphs (d)(4) and (d)(5) of this section.

(7) Any single propeller or rotating device capable of significant dynamic forces rotating at the highest likely overspeed.

(8) Any damage or failure condition, required or selected for investigation by § 25.571. The single structural failures described in paragraphs (d)(4) and (d)(5) of this section need not be considered in showing compliance with this section if;

(i) The structural element could not fail due to discrete source damage resulting from the conditions described in § 25.571(e), and

(ii) A damage tolerance investigation in accordance with § 25.571(b) shows that the maximum extent of damage assumed for the purpose of residual strength evaluation does not involve complete failure of the structural element.

(9) Any damage, failure, or malfunction considered under §§ 25.631, 25.671, 25.672, and 25.1309.

(10) Any other combination of failures, malfunctions, or adverse conditions not shown to be extremely improbable.

(e) *Flight flutter testing.* Full scale flight flutter tests at speeds up to V_{DF}/M_{DF} must be conducted for new type designs and for modifications to a type design unless the modifications have been shown to have an insignificant effect on the aeroelastic stability. These tests must demonstrate that the airplane has a proper margin of damping at all speeds up to V_{DF}/M_{DF}, and that there is no large and rapid reduction in damping as V_{DF}/M_{DF}, is approached. If a failure, malfunction, or adverse condition is simulated during flight test in showing compliance with paragraph (d) of this section, the maximum speed investigated need not exceed V_{FC}/M_{FC} if it is shown, by correlation of the flight test data with other test data or analyses, that the airplane is free from any aeroelastic instability at all speeds within the altitude-airspeed envelope described in paragraph (b)(2) of this section.

[Doc. No. 26007, 57 FR 28949, June 29, 1992]

§ 25.631 Bird strike damage.

The empennage structure must be designed to assure capability of continued safe flight and landing of the airplane after impact with an 8-pound bird when the velocity of the airplane (relative to the bird along the airplane's flight path) is equal to V_C at sea level, selected under § 25.335(a). Compliance with this section by provision of redundant structure and protected location of control system elements or protective devices such as splitter plates or energy absorbing material is acceptable. Where compliance is shown by analysis, tests, or both, use of data on airplanes having similar structural design is acceptable.

[Amdt. 25–23, 35 FR 5674, Apr. 8, 1970]

CONTROL SURFACES

§ 25.651 Proof of strength.

(a) Limit load tests of control surfaces are required. These tests must include the horn or fitting to which the control system is attached.

(b) Compliance with the special factors requirements of §§ 25.619 through 25.625 and 25.657 for control surface hinges must be shown by analysis or individual load tests.

§ 25.655 Installation.

(a) Movable tail surfaces must be installed so that there is no interference between any surfaces when one is held in its extreme position and the others are operated through their full angular movement.

(b) If an adjustable stabilizer is used, it must have stops that will limit its range of travel to the maximum for which the airplane is shown to meet the trim requirements of § 25.161.

§ 25.657 Hinges.

(a) For control surface hinges, including ball, roller, and self-lubricated bearing hinges, the approved rating of the bearing may not be exceeded. For nonstandard bearing hinge configurations, the rating must be established on the basis of experience or tests and, in the absence of a rational investigation, a factor of safety of not less than 6.67 must be used with respect to the ultimate bearing strength of the softest material used as a bearing.

(b) Hinges must have enough strength and rigidity for loads parallel to the hinge line.

[Amdt. 25–23, 35 FR 5674, Apr. 8, 1970]

CONTROL SYSTEMS

§ 25.671 General.

(a) Each control and control system must operate with the ease, smoothness, and positiveness appropriate to its function.

(b) Each element of each flight control system must be designed, or distinctively and permanently marked, to minimize the probability of incorrect assembly that could result in the malfunctioning of the system.

(c) The airplane must be shown by analysis, tests, or both, to be capable of continued safe flight and landing after any of the following failures or jamming in the flight control system and surfaces (including trim, lift, drag, and feel systems), within the normal flight envelope, without requiring exceptional piloting skill or strength.

Probable malfunctions must have only minor effects on control system operation and must be capable of being readily counteracted by the pilot.

(1) Any single failure, excluding jamming (for example, disconnection or failure of mechanical elements, or structural failure of hydraulic components, such as actuators, control spool housing, and valves).

(2) Any combination of failures not shown to be extremely improbable, excluding jamming (for example, dual electrical or hydraulic system failures, or any single failure in combination with any probable hydraulic or electrical failure).

(3) Any jam in a control position normally encountered during takeoff, climb, cruise, normal turns, descent, and landing unless the jam is shown to be extremely improbable, or can be alleviated. A runaway of a flight control to an adverse position and jam must be accounted for if such runaway and subsequent jamming is not extremely improbable.

(d) The airplane must be designed so that it is controllable if all engines fail. Compliance with this requirement may be shown by analysis where that method has been shown to be reliable.

[Doc. No. 5066, 29 FR 18291, Dec. 24, 1964, as amended by Amdt. 25–23, 35 FR 5674, Apr. 8, 1970]

§ 25.672 Stability augmentation and automatic and power-operated systems.

If the functioning of stability augmentation or other automatic or power-operated systems is necessary to show compliance with the flight characteristics requirements of this part, such systems must comply with § 25.671 and the following:

(a) A warning which is clearly distinguishable to the pilot under expected flight conditions without requiring his attention must be provided for any failure in the stability augmentation system or in any other automatic or power-operated system which could result in an unsafe condition if the pilot were not aware of the failure. Warning systems must not activate the control systems.

(b) The design of the stability augmentation system or of any other automatic or power-operated system must permit initial counteraction of failures of the type specified in § 25.671(c) without requiring exceptional pilot skill or strength, by either the deactivation of the system, or a failed portion thereof, or by overriding the failure by movement of the flight controls in the normal sense.

(c) It must be shown that after any single failure of the stability augmentation system or any other automatic or power-operated system—

(1) The airplane is safely controllable when the failure or malfunction occurs at any speed or altitude within the approved operating limitations that is critical for the type of failure being considered;

(2) The controllability and maneuverability requirements of this part are met within a practical operational flight envelope (for example, speed, altitude, normal acceleration, and airplane configurations) which is described in the Airplane Flight Manual; and

(3) The trim, stability, and stall characteristics are not impaired below a level needed to permit continued safe flight and landing.

[Amdt. 25–23, 35 FR 5675 Apr. 8, 1970]

§ 25.675 Stops.

(a) Each control system must have stops that positively limit the range of motion of each movable aerodynamic surface controlled by the system.

(b) Each stop must be located so that wear, slackness, or take-up adjustments will not adversely affect the control characteristics of the airplane because of a change in the range of surface travel.

(c) Each stop must be able to withstand any loads corresponding to the design conditions for the control system.

[Doc. No. 5066, 29 FR 18291, Dec. 24, 1964, as amended by Amdt. 25–38, 41 FR 55466, Dec. 20, 1976]

§ 25.677 Trim systems.

(a) Trim controls must be designed to prevent inadvertent or abrupt operation and to operate in the plane, and with the sense of motion, of the airplane.

(b) There must be means adjacent to the trim control to indicate the direction of the control movement relative to the airplane motion. In addition, there must be clearly visible means to indicate the position of the trim device with respect to the range of adjustment. The indicator must be clearly marked with the range within which it has been demonstrated that takeoff is safe for all center of gravity positions approved for takeoff.

(c) Trim control systems must be designed to prevent creeping in flight. Trim tab controls must be irreversible unless the tab is appropriately balanced and shown to be free from flutter.

(d) If an irreversible tab control system is used, the part from the tab to the attachment of the irreversible unit to the airplane structure must consist of a rigid connection.

[Doc. No. 5066, 29 FR 18291, Dec. 24, 1964, as amended by Amdt. 25–23, 35 FR 5675, Apr. 8, 1970; Amdt. 25–115, 69 FR 40527, July 2, 2004]

§ 25.679 Control system gust locks.

(a) There must be a device to prevent damage to the control surfaces (including tabs), and to the control system, from gusts striking the airplane while it is on the ground or water. If the device, when engaged, prevents normal operation of the control surfaces by the pilot, it must—

(1) Automatically disengage when the pilot operates the primary flight controls in a normal manner; or

(2) Limit the operation of the airplane so that the pilot receives unmistakable warning at the start of takeoff.

(b) The device must have means to preclude the possibility of it becoming inadvertently engaged in flight.

§ 25.681 Limit load static tests.

(a) Compliance with the limit load requirements of this Part must be shown by tests in which—

(1) The direction of the test loads produces the most severe loading in the control system; and

(2) Each fitting, pulley, and bracket used in attaching the system to the main structure is included.

(b) Compliance must be shown (by analyses or individual load tests) with the special factor requirements for control system joints subject to angular motion.

§ 25.683   Operation tests.

(a) It must be shown by operation tests that when portions of the control system subject to pilot effort loads are loaded to 80 percent of the limit load specified for the system and the powered portions of the control system are loaded to the maximum load expected in normal operation, the system is free from—

(1) Jamming;

(2) Excessive friction; and

(3) Excessive deflection.

(b) It must be shown by analysis and, where necessary, by tests, that in the presence of deflections of the airplane structure due to the separate application of pitch, roll, and yaw limit maneuver loads, the control system, when loaded to obtain these limit loads and operated within its operational range of deflections, can be exercised about all control axes and remain free from—

(1) Jamming;

(2) Excessive friction;

(3) Disconnection; and

(4) Any form of permanent damage.

(c) It must be shown that under vibration loads in the normal flight and ground operating conditions, no hazard can result from interference or contact with adjacent elements.

[Amdt. 25–139, 79 FR 59430, Oct. 2, 2014]

§ 25.685   Control system details.

(a) Each detail of each control system must be designed and installed to prevent jamming, chafing, and interference from cargo, passengers, loose objects, or the freezing of moisture.

(b) There must be means in the cockpit to prevent the entry of foreign objects into places where they would jam the system.

(c) There must be means to prevent the slapping of cables or tubes against other parts.

(d) Sections 25.689 and 25.693 apply to cable systems and joints.

[Doc. No. 5066, 29 FR 18291, Dec. 24, 1964, as amended by Amdt. 25–38, 41 FR 55466, Dec. 20, 1976]

§ 25.689   Cable systems.

(a) Each cable, cable fitting, turnbuckle, splice, and pulley must be approved. In addition—

(1) No cable smaller than $\frac{1}{8}$ inch in diameter may be used in the aileron, elevator, or rudder systems; and

(2) Each cable system must be designed so that there will be no hazardous change in cable tension throughout the range of travel under operating conditions and temperature variations.

(b) Each kind and size of pulley must correspond to the cable with which it is used. Pulleys and sprockets must have closely fitted guards to prevent the cables and chains from being displaced or fouled. Each pulley must lie in the plane passing through the cable so that the cable does not rub against the pulley flange.

(c) Fairleads must be installed so that they do not cause a change in cable direction of more than three degrees.

(d) Clevis pins subject to load or motion and retained only by cotter pins may not be used in the control system.

(e) Turnbuckles must be attached to parts having angular motion in a manner that will positively prevent binding throughout the range of travel.

(f) There must be provisions for visual inspection of fairleads, pulleys, terminals, and turnbuckles.

§ 25.693   Joints.

Control system joints (in push-pull systems) that are subject to angular motion, except those in ball and roller bearing systems, must have a special factor of safety of not less than 3.33 with respect to the ultimate bearing strength of the softest material used as a bearing. This factor may be reduced to 2.0 for joints in cable control systems. For ball or roller bearings, the approved ratings may not be exceeded.

[Amdt. 25–72, 55 FR 29777, July 20, 1990]

§ 25.697   Lift and drag devices, controls.

(a) Each lift device control must be designed so that the pilots can place the device in any takeoff, en route, approach, or landing position established under § 25.101(d). Lift and drag devices must maintain the selected positions, except for movement produced by an automatic positioning or load limiting device, without further attention by the pilots.

(b) Each lift and drag device control must be designed and located to make inadvertent operation improbable. Lift and drag devices intended for ground operation only must have means to prevent the inadvertent operation of their controls in flight if that operation could be hazardous.

(c) The rate of motion of the surfaces in response to the operation of the control and the characteristics of the automatic positioning or load limiting device must give satisfactory flight and performance characteristics under steady or changing conditions of airspeed, engine power, and airplane attitude.

(d) The lift device control must be designed to retract the surfaces from the fully extended position, during steady flight at maximum continuous engine power at any speed below $V_F + 9.0$ (knots).

[Amdt. 25–23, 35 FR 5675, Apr. 8, 1970, as amended by Amdt. 25–46, 43 FR 50595, Oct. 30, 1978; Amdt. 25–57, 49 FR 6848, Feb. 23, 1984]

§ 25.699 Lift and drag device indicator.

(a) There must be means to indicate to the pilots the position of each lift or drag device having a separate control in the cockpit to adjust its position. In addition, an indication of unsymmetrical operation or other malfunction in the lift or drag device systems must be provided when such indication is necessary to enable the pilots to prevent or counteract an unsafe flight or ground condition, considering the effects on flight characteristics and performance.

(b) There must be means to indicate to the pilots the takeoff, en route, approach, and landing lift device positions.

(c) If any extension of the lift and drag devices beyond the landing position is possible, the controls must be clearly marked to identify this range of extension.

[Amdt. 25–23, 35 FR 5675, Apr. 8, 1970]

§ 25.701 Flap and slat interconnection.

(a) Unless the airplane has safe flight characteristics with the flaps or slats retracted on one side and extended on the other, the motion of flaps or slats on opposite sides of the plane of symmetry must be synchronized by a mechanical interconnection or approved equivalent means.

(b) If a wing flap or slat interconnection or equivalent means is used, it must be designed to account for the applicable unsymmetrical loads, including those resulting from flight with the engines on one side of the plane of symmetry inoperative and the remaining engines at takeoff power.

(c) For airplanes with flaps or slats that are not subjected to slipstream conditions, the structure must be designed for the loads imposed when the wing flaps or slats on one side are carrying the most severe load occurring in the prescribed symmetrical conditions and those on the other side are carrying not more than 80 percent of that load.

(d) The interconnection must be designed for the loads resulting when interconnected flap or slat surfaces on one side of the plane of symmetry are jammed and immovable while the surfaces on the other side are free to move and the full power of the surface actuating system is applied.

[Amdt. 25–72, 55 FR 29777, July 20, 1990]

§ 25.703 Takeoff warning system.

A takeoff warning system must be installed and must meet the following requirements:

(a) The system must provide to the pilots an aural warning that is automatically activated during the initial portion of the takeoff roll if the airplane is in a configuration, including any of the following, that would not allow a safe takeoff:

(1) The wing flaps or leading edge devices are not within the approved range of takeoff positions.

(2) Wing spoilers (except lateral control spoilers meeting the requirements of § 25.671), speed brakes, or longitudinal trim devices are in a position that would not allow a safe takeoff.

(b) The warning required by paragraph (a) of this section must continue until—

(1) The configuration is changed to allow a safe takeoff;

(2) Action is taken by the pilot to terminate the takeoff roll;

(3) The airplane is rotated for takeoff; or

(4) The warning is manually deactivated by the pilot.

(c) The means used to activate the system must function properly throughout the ranges of takeoff weights, altitudes, and temperatures for which certification is requested.

[Amdt. 25–42, 43 FR 2323, Jan. 16, 1978]

LANDING GEAR

§ 25.721 General.

(a) The landing gear system must be designed so that when it fails due to overloads during takeoff and landing, the failure mode is not likely to cause spillage of enough fuel to constitute a fire hazard. The overloads must be assumed to act in the upward and aft directions in combination with side loads acting inboard and outboard. In the absence of a more rational analysis, the side loads must be assumed to be up to 20 percent of the vertical load or 20 percent of the drag load, whichever is greater.

(b) The airplane must be designed to avoid any rupture leading to the spillage of enough fuel to constitute a fire hazard as a result of a wheels-up landing on a paved runway, under the following minor crash landing conditions:

(1) Impact at 5 feet-per-second vertical velocity, with the airplane under control, at Maximum Design Landing Weight—

(i) With the landing gear fully retracted; and

(ii) With any one or more landing gear legs not extended.

(2) Sliding on the ground, with—

(i) The landing gear fully retracted and with up to a 20° yaw angle; and

(ii) Any one or more landing gear legs not extended and with 0° yaw angle.

(c) For configurations where the engine nacelle is likely to come into contact with the ground, the engine pylon or engine mounting must be designed so that when it fails due to overloads (assuming the overloads to act predominantly in the upward direction and separately, predominantly in the aft direction), the failure mode is not likely to cause the spillage of enough fuel to constitute a fire hazard.

[Amdt. 25–139, 79 FR 59430, Oct. 2, 2014]

§ 25.723 Shock absorption tests.

(a) The analytical representation of the landing gear dynamic characteristics that is used in determining the landing loads must be validated by energy absorption tests. A range of tests must be conducted to ensure that the analytical representation is valid for the design conditions specified in § 25.473.

(1) The configurations subjected to energy absorption tests at limit design conditions must include at least the design landing weight or the design takeoff weight, whichever produces the greater value of landing impact energy.

(2) The test attitude of the landing gear unit and the application of appropriate drag loads during the test must simulate the airplane landing conditions in a manner consistent with the development of rational or conservative limit loads.

(b) The landing gear may not fail in a test, demonstrating its reserve energy absorption capacity, simulating a descent velocity of 12 f.p.s. at design landing weight, assuming airplane lift not greater than airplane weight acting during the landing impact.

(c) In lieu of the tests prescribed in this section, changes in previously approved design weights and minor changes in design may be substantiated by analyses based on previous tests conducted on the same basic landing gear system that has similar energy absorption characteristics.

[Doc. No. 1999–5835, 66 FR 27394, May 16, 2001]

§§ 25.725–25.727 [Reserved]

§ 25.729 Retracting mechanism.

(a) *General.* For airplanes with retractable landing gear, the following apply:

(1) The landing gear retracting mechanism, wheel well doors, and supporting structure, must be designed for—

(i) The loads occurring in the flight conditions when the gear is in the retracted position,

(ii) The combination of friction loads, inertia loads, brake torque loads, air loads, and gyroscopic loads resulting from the wheels rotating at a peripheral speed equal to $1.23V_{SR}$ (with the wing-flaps in take-off position at design take-off weight), occurring during retraction and extension at any airspeed up to 1.5 V_{SR1} (with the wing-flaps in the approach position at design landing weight), and

(iii) Any load factor up to those specified in §25.345(a) for the wing-flaps extended condition.

(2) Unless there are other means to decelerate the airplane in flight at this speed, the landing gear, the retracting mechanism, and the airplane structure (including wheel well doors) must be designed to withstand the flight loads occurring with the landing gear in the extended position at any speed up to 0.67 V_C.

(3) Landing gear doors, their operating mechanism, and their supporting structures must be designed for the yawing maneuvers prescribed for the airplane in addition to the conditions of airspeed and load factor prescribed in paragraphs (a)(1) and (2) of this section.

(b) *Landing gear lock.* There must be positive means to keep the landing gear extended in flight and on the ground. There must be positive means to keep the landing gear and doors in the correct retracted position in flight, unless it can be shown that lowering of the landing gear or doors, or flight with the landing gear or doors extended, at any speed, is not hazardous.

(c) *Emergency operation.* There must be an emergency means for extending the landing gear in the event of—

(1) Any reasonably probable failure in the normal retraction system; or

(2) The failure of any single source of hydraulic, electric, or equivalent energy supply.

(d) *Operation test.* The proper functioning of the retracting mechanism must be shown by operation tests.

(e) *Position indicator and warning device.* If a retractable landing gear is used, there must be a landing gear position indicator easily visible to the pilot or to the appropriate crew members (as well as necessary devices to actuate the indicator) to indicate without ambiguity that the retractable units and their associated doors are secured in the extended (or retracted) position. The means must be designed as follows:

(1) If switches are used, they must be located and coupled to the landing gear mechanical systems in a manner that prevents an erroneous indication of "down and locked" if the landing gear is not in a fully extended position, or of "up and locked" if the landing gear is not in the fully retracted position. The switches may be located where they are operated by the actual landing gear locking latch or device.

(2) The flightcrew must be given an aural warning that functions continuously, or is periodically repeated, if a landing is attempted when the landing gear is not locked down.

(3) The warning must be given in sufficient time to allow the landing gear to be locked down or a go-around to be made.

(4) There must not be a manual shutoff means readily available to the flightcrew for the warning required by paragraph (e)(2) of this section such that it could be operated instinctively, inadvertently, or by habitual reflexive action.

(5) The system used to generate the aural warning must be designed to minimize false or inappropriate alerts.

(6) Failures of systems used to inhibit the landing gear aural warning, that would prevent the warning system from operating, must be improbable.

(7) A flightcrew alert must be provided whenever the landing gear position is not consistent with the landing gear selector lever position.

(f) *Protection of equipment on landing gear and in wheel wells.* Equipment that is essential to the safe operation of the airplane and that is located on the landing gear and in wheel wells must be protected from the damaging effects of—

(1) A bursting tire;

273

(2) A loose tire tread, unless it is shown that a loose tire tread cannot cause damage.

(3) Possible wheel brake temperatures.

[Doc. No. 5066, 29 FR 18291, Dec. 24, 1964, as amended by Amdt. 25–23, 35 FR 5676, Apr. 8, 1970; Amdt. 25–42, 43 FR 2323, Jan. 16, 1978; Amdt. 25–72, 55 FR 29777, July 20, 1990; Amdt. 25–75, 56 FR 63762, Dec. 5, 1991; Amdt. 25–136, 77 FR 1617, Jan. 11, 2012]

§ 25.731   Wheels.

(a) Each main and nose wheel must be approved.

(b) The maximum static load rating of each wheel may not be less than the corresponding static ground reaction with—

(1) Design maximum weight; and

(2) Critical center of gravity.

(c) The maximum limit load rating of each wheel must equal or exceed the maximum radial limit load determined under the applicable ground load requirements of this part.

(d) *Overpressure burst prevention.* Means must be provided in each wheel to prevent wheel failure and tire burst that may result from excessive pressurization of the wheel and tire assembly.

(e) *Braked wheels.* Each braked wheel must meet the applicable requirements of § 25.735.

[Doc. No. 5066, 29 FR 18291, Dec. 24, 1964, as amended by Amdt. 25–72, 55 FR 29777, July 20, 1990; Amdt. 25–107, 67 FR 20420, Apr. 24, 2002]

§ 25.733   Tires.

(a) When a landing gear axle is fitted with a single wheel and tire assembly, the wheel must be fitted with a suitable tire of proper fit with a speed rating approved by the Administrator that is not exceeded under critical conditions and with a load rating approved by the Administrator that is not exceeded under—

(1) The loads on the main wheel tire, corresponding to the most critical combination of airplane weight (up to maximum weight) and center of gravity position, and

(2) The loads corresponding to the ground reactions in paragraph (b) of this section, on the nose wheel tire, except as provided in paragraphs (b)(2) and (b)(3) of this section.

(b) The applicable ground reactions for nose wheel tires are as follows:

(1) The static ground reaction for the tire corresponding to the most critical combination of airplane weight (up to maximum ramp weight) and center of gravity position with a force of 1.0g acting downward at the center of gravity. This load may not exceed the load rating of the tire.

(2) The ground reaction of the tire corresponding to the most critical combination of airplane weight (up to maximum landing weight) and center of gravity position combined with forces of 1.0g downward and 0.31g forward acting at the center of gravity. The reactions in this case must be distributed to the nose and main wheels by the principles of statics with a drag reaction equal to 0.31 times the vertical load at each wheel with brakes capable of producing this ground reaction. This nose tire load may not exceed 1.5 times the load rating of the tire.

(3) The ground reaction of the tire corresponding to the most critical combination of airplane weight (up to maximum ramp weight) and center of gravity position combined with forces of 1.0g downward and 0.20g forward acting at the center of gravity. The reactions in this case must be distributed to the nose and main wheels by the principles of statics with a drag reaction equal to 0.20 times the vertical load at each wheel with brakes capable of producing this ground reaction. This nose tire load may not exceed 1.5 times the load rating of the tire.

(c) When a landing gear axle is fitted with more than one wheel and tire assembly, such as dual or dual-tandem, each wheel must be fitted with a suitable tire of proper fit with a speed rating approved by the Administrator that is not exceeded under critical conditions, and with a load rating approved by the Administrator that is not exceeded by—

(1) The loads on each main wheel tire, corresponding to the most critical combination of airplane weight (up to maximum weight) and center of gravity position, when multiplied by a factor of 1.07; and

(2) Loads specified in paragraphs (a)(2), (b)(1), (b)(2), and (b)(3) of this section on each nose wheel tire.

(d) Each tire installed on a retractable landing gear system must, at the maximum size of the tire type expected in service, have a clearance to surrounding structure and systems that is adequate to prevent unintended contact between the tire and any part of the structure or systems.

(e) For an airplane with a maximum certificated takeoff weight of more than 75,000 pounds, tires mounted on braked wheels must be inflated with dry nitrogen or other gases shown to be inert so that the gas mixture in the tire does not contain oxygen in excess of 5 percent by volume, unless it can be shown that the tire liner material will not produce a volatile gas when heated or that means are provided to prevent tire temperatures from reaching unsafe levels.

[Amdt. 25–48, 44 FR 68752, Nov. 29, 1979; Amdt. 25–72, 55 FR 29777, July 20, 1990, as amended by Amdt. 25–78, 58 FR 11781, Feb. 26, 1993]

§ 25.735 Brakes and braking systems.

(a) *Approval.* Each assembly consisting of a wheel(s) and brake(s) must be approved.

(b) *Brake system capability.* The brake system, associated systems and components must be designed and constructed so that:

(1) If any electrical, pneumatic, hydraulic, or mechanical connecting or transmitting element fails, or if any single source of hydraulic or other brake operating energy supply is lost, it is possible to bring the airplane to rest with a braked roll stopping distance of not more than two times that obtained in determining the landing distance as prescribed in § 25.125.

(2) Fluid lost from a brake hydraulic system following a failure in, or in the vicinity of, the brakes is insufficient to cause or support a hazardous fire on the ground or in flight.

(c) *Brake controls.* The brake controls must be designed and constructed so that:

(1) Excessive control force is not required for their operation.

(2) If an automatic braking system is installed, means are provided to:

(i) Arm and disarm the system, and

(ii) Allow the pilot(s) to override the system by use of manual braking.

(d) *Parking brake.* The airplane must have a parking brake control that, when selected on, will, without further attention, prevent the airplane from rolling on a dry and level paved runway when the most adverse combination of maximum thrust on one engine and up to maximum ground idle thrust on any, or all, other engine(s) is applied. The control must be suitably located or be adequately protected to prevent inadvertent operation. There must be indication in the cockpit when the parking brake is not fully released.

(e) *Antiskid system.* If an antiskid system is installed:

(1) It must operate satisfactorily over the range of expected runway conditions, without external adjustment.

(2) It must, at all times, have priority over the automatic braking system, if installed.

(f) *Kinetic energy capacity*—(1) *Design landing stop.* The design landing stop is an operational landing stop at maximum landing weight. The design landing stop brake kinetic energy absorption requirement of each wheel, brake, and tire assembly must be determined. It must be substantiated by dynamometer testing that the wheel, brake and tire assembly is capable of absorbing not less than this level of kinetic energy throughout the defined wear range of the brake. The energy absorption rate derived from the airplane manufacturer's braking requirements must be achieved. The mean deceleration must not be less than 10 fps^2.

(2) *Maximum kinetic energy accelerate-stop.* The maximum kinetic energy accelerate-stop is a rejected takeoff for the most critical combination of airplane takeoff weight and speed. The accelerate-stop brake kinetic energy absorption requirement of each wheel, brake, and tire assembly must be determined. It must be substantiated by dynamometer testing that the wheel, brake, and tire assembly is capable of absorbing not less than this level of kinetic energy throughout the defined wear range of the brake. The energy absorption rate derived from the airplane manufacturer's braking requirements must be achieved. The mean deceleration must not be less than 6 fps^2.

(3) *Most severe landing stop.* The most severe landing stop is a stop at the most critical combination of airplane landing weight and speed. The most severe landing stop brake kinetic energy absorption requirement of each wheel, brake, and tire assembly must be determined. It must be substantiated by dynamometer testing that, at the declared fully worn limit(s) of the brake heat sink, the wheel, brake and tire assembly is capable of absorbing not less than this level of kinetic energy. The most severe landing stop need not be considered for extremely improbable failure conditions or if the maximum kinetic energy accelerate-stop energy is more severe.

(g) *Brake condition after high kinetic energy dynamometer stop(s).* Following the high kinetic energy stop demonstration(s) required by paragraph (f) of this section, with the parking brake promptly and fully applied for at least 3 minutes, it must be demonstrated that for at least 5 minutes from application of the parking brake, no condition occurs (or has occurred during the stop), including fire associated with the tire or wheel and brake assembly, that could prejudice the safe and complete evacuation of the airplane.

(h) *Stored energy systems.* An indication to the flightcrew of the usable stored energy must be provided if a stored energy system is used to show compliance with paragraph (b)(1) of this section. The available stored energy must be sufficient for:

(1) At least 6 full applications of the brakes when an antiskid system is not operating; and

(2) Bringing the airplane to a complete stop when an antiskid system is operating, under all runway surface conditions for which the airplane is certificated.

(i) *Brake wear indicators.* Means must be provided for each brake assembly to indicate when the heat sink is worn to the permissible limit. The means must be reliable and readily visible.

(j) *Overtemperature burst prevention.* Means must be provided in each braked wheel to prevent a wheel failure, a tire burst, or both, that may result from elevated brake temperatures. Additionally, all wheels must meet the requirements of § 25.731(d).

(k) *Compatibility.* Compatibility of the wheel and brake assemblies with the airplane and its systems must be substantiated.

[Doc. No. FAA–1999–6063, 67 FR 20420, Apr. 24, 2002, as amended by Amdt. 25–108, 67 FR 70827, Nov. 26, 2002; 68 FR 1955, Jan. 15, 2003]

§ 25.737   Skis.

Each ski must be approved. The maximum limit load rating of each ski must equal or exceed the maximum limit load determined under the applicable ground load requirements of this part.

FLOATS AND HULLS

§ 25.751   Main float buoyancy.

Each main float must have—

(a) A buoyancy of 80 percent in excess of that required to support the maximum weight of the seaplane or amphibian in fresh water; and

(b) Not less than five watertight compartments approximately equal in volume.

§ 25.753   Main float design.

Each main float must be approved and must meet the requirements of § 25.521.

§ 25.755   Hulls.

(a) Each hull must have enough watertight compartments so that, with any two adjacent compartments flooded, the buoyancy of the hull and auxiliary floats (and wheel tires, if used) provides a margin of positive stability great enough to minimize the probability of capsizing in rough, fresh water.

(b) Bulkheads with watertight doors may be used for communication between compartments.

PERSONNEL AND CARGO
ACCOMMODATIONS

§ 25.771   Pilot compartment.

(a) Each pilot compartment and its equipment must allow the minimum flight crew (established under § 25.1523) to perform their duties without unreasonable concentration or fatigue.

(b) The primary controls listed in § 25.779(a), excluding cables and control rods, must be located with respect to the propellers so that no member of the minimum flight crew (established under §25.1523), or part of the controls, lies in the region between the plane of rotation of any inboard propeller and the surface generated by a line passing through the center of the propeller hub making an angle of five degrees forward or aft of the plane of rotation of the propeller.

(c) If provision is made for a second pilot, the airplane must be controllable with equal safety from either pilot seat.

(d) The pilot compartment must be constructed so that, when flying in rain or snow, it will not leak in a manner that will distract the crew or harm the structure.

(e) Vibration and noise characteristics of cockpit equipment may not interfere with safe operation of the airplane.

[Doc. No. 5066, 29 FR 18291, Dec. 24, 1964, as amended by Amdt. 25-4, 30 FR 6113, Apr. 30, 1965]

§25.772 Pilot compartment doors.

For an airplane that has a lockable door installed between the pilot compartment and the passenger compartment:

(a) For airplanes with a maximum passenger seating configuration of more than 20 seats, the emergency exit configuration must be designed so that neither crewmembers nor passengers require use of the flightdeck door in order to reach the emergency exits provided for them; and

(b) Means must be provided to enable flight crewmembers to directly enter the passenger compartment from the pilot compartment if the cockpit door becomes jammed.

(c) There must be an emergency means to enable a flight attendant to enter the pilot compartment in the event that the flightcrew becomes incapacitated.

[Doc. No. 24344, 55 FR 29777, July 20, 1990, as amended by Amdt. 25-106, 67 FR 2127, Jan. 15, 2002]

§25.773 Pilot compartment view.

(a) *Nonprecipitation conditions.* For nonprecipitation conditions, the following apply:

(1) Each pilot compartment must be arranged to give the pilots a sufficiently extensive, clear, and undistorted view, to enable them to safely perform any maneuvers within the operating limitations of the airplane, including taxiing takeoff, approach, and landing.

(2) Each pilot compartment must be free of glare and reflection that could interfere with the normal duties of the minimum flight crew (established under §25.1523). This must be shown in day and night flight tests under nonprecipitation conditions.

(b) *Precipitation conditions.* For precipitation conditions, the following apply:

(1) The airplane must have a means to maintain a clear portion of the windshield, during precipitation conditions, sufficient for both pilots to have a sufficiently extensive view along the flight path in normal flight attitudes of the airplane. This means must be designed to function, without continuous attention on the part of the crew, in—

(i) Heavy rain at speeds up to 1.5 V_{SR1} with lift and drag devices retracted; and

(ii) The icing conditions specified in Appendix C of this part and the following icing conditions specified in Appendix O of this part, if certification for flight in icing conditions is sought:

(A) For airplanes certificated in accordance with §25.1420(a)(1), the icing conditions that the airplane is certified to safely exit following detection.

(B) For airplanes certificated in accordance with §25.1420(a)(2), the icing conditions that the airplane is certified to safely operate in and the icing conditions that the airplane is certified to safely exit following detection.

(C) For airplanes certificated in accordance with §25.1420(a)(3) and for airplanes not subject to §25.1420, all icing conditions.

(2) No single failure of the systems used to provide the view required by paragraph (b)(1) of this section must cause the loss of that view by both pilots in the specified precipitation conditions.

(3) The first pilot must have a window that—

(i) Is openable under the conditions prescribed in paragraph (b)(1) of this section when the cabin is not pressurized;

(ii) Provides the view specified in paragraph (b)(1) of this section; and

(iii) Provides sufficient protection from the elements against impairment of the pilot's vision.

(4) The openable window specified in paragraph (b)(3) of this section need not be provided if it is shown that an area of the transparent surface will remain clear sufficient for at least one pilot to land the airplane safely in the event of—

(i) Any system failure or combination of failures which is not extremely improbable, in accordance with § 25.1309, under the precipitation conditions specified in paragraph (b)(1) of this section.

(ii) An encounter with severe hail, birds, or insects.

(c) *Internal windshield and window fogging.* The airplane must have a means to prevent fogging of the internal portions of the windshield and window panels over an area which would provide the visibility specified in paragraph (a) of this section under all internal and external ambient conditions, including precipitation conditions, in which the airplane is intended to be operated.

(d) Fixed markers or other guides must be installed at each pilot station to enable the pilots to position themselves in their seats for an optimum combination of outside visibility and instrument scan. If lighted markers or guides are used they must comply with the requirements specified in § 25.1381.

(e) *Vision systems with transparent displays.* A vision system with a transparent display surface located in the pilot's outside field of view, such as a head up-display, head mounted display, or other equivalent display, must meet the following requirements in nonprecipitation and precipitation conditions:

(1) While the vision system display is in operation, it must compensate for interference with the pilot's outside field of view such that the combination of what is visible in the display and what remains visible through and around it, enables the pilot to perform the maneuvers and normal duties of paragraph (a) of this section.

(2) The pilot's view of the external scene may not be distorted by the transparent display surface or by the vision system imagery. When the vision system displays imagery or any symbology that is referenced to the imagery and outside scene topography, including attitude symbology, flight path vector, and flight path angle reference cue, that imagery and symbology must be aligned with, and scaled to, the external scene.

(3) The vision system must provide a means to allow the pilot using the display to immediately deactivate and reactivate the vision system imagery, on demand, without removing the pilot's hands from the primary flight controls or thrust controls.

(4) When the vision system is not in operation it may not restrict the pilot from performing the maneuvers specified in paragraph (a)(1) of this section or the pilot compartment from meeting the provisions of paragraph (a)(2) of this section.

[Doc. No. 5066, 29 FR 18291, Dec. 24, 1964, as amended by Amdt. 25–23, 35 FR 5676, Apr. 8, 1970; Amdt. 25–46, 43 FR 50595, Oct. 30, 1978; Amdt. 25–72, 55 FR 29778, July 20, 1990; Amdt. 25–108, 67 FR 70827, Nov. 26, 2002; Amdt. 25–121, 72 FR 44669, Aug. 8, 2007; Amdt. 25–136, 77 FR 1618, Jan. 11, 2012; Amdt. 25–140, 79 FR 65525, Nov. 4, 2014; Docket FAA–2013–0485, Amdt. 25–144, 81 FR 90169, Dec. 13, 2016]

§ 25.775 Windshields and windows.

(a) Internal panes must be made of nonsplintering material.

(b) Windshield panes directly in front of the pilots in the normal conduct of their duties, and the supporting structures for these panes, must withstand, without penetration, the impact of a four-pound bird when the velocity of the airplane (relative to the bird along the airplane's flight path) is equal to the value of V_C at sea level, selected under § 25.335(a).

(c) Unless it can be shown by analysis or tests that the probability of occurrence of a critical windshield fragmentation condition is of a low order, the airplane must have a means to minimize the danger to the pilots from flying windshield fragments due to bird impact. This must be shown for each transparent pane in the cockpit that—

(1) Appears in the front view of the airplane;

(2) Is inclined 15 degrees or more to the longitudinal axis of the airplane; and

(3) Has any part of the pane located where its fragmentation will constitute a hazard to the pilots.

(d) The design of windshields and windows in pressurized airplanes must be based on factors peculiar to high altitude operation, including the effects of continuous and cyclic pressurization loadings, the inherent characteristics of the material used, and the effects of temperatures and temperature differentials. The windshield and window panels must be capable of withstanding the maximum cabin pressure differential loads combined with critical aerodynamic pressure and temperature effects after any single failure in the installation or associated systems. It may be assumed that, after a single failure that is obvious to the flight crew (established under § 25.1523), the cabin pressure differential is reduced from the maximum, in accordance with appropriate operating limitations, to allow continued safe flight of the airplane with a cabin pressure altitude of not more than 15,000 feet.

(e) The windshield panels in front of the pilots must be arranged so that, assuming the loss of vision through any one panel, one or more panels remain available for use by a pilot seated at a pilot station to permit continued safe flight and landing.

[Doc. No. 5066, 29 FR 18291, Dec. 24, 1964, as amended by Amdt. 25–23, 35 FR 5676, Apr. 8, 1970; Amdt. 25–38, 41 FR 55466, Dec. 20, 1976]

§ 25.777 Cockpit controls.

(a) Each cockpit control must be located to provide convenient operation and to prevent confusion and inadvertent operation.

(b) The direction of movement of cockpit controls must meet the requirements of § 25.779. Wherever practicable, the sense of motion involved in the operation of other controls must correspond to the sense of the effect of the operation upon the airplane or upon the part operated. Controls of a variable nature using a rotary motion must move clockwise from the off position, through an increasing range, to the full on position.

(c) The controls must be located and arranged, with respect to the pilots' seats, so that there is full and unrestricted movement of each control without interference from the cockpit structure or the clothing of the minimum flight crew (established under § 25.1523) when any member of this flight crew, from 5'2" to 6'3" in height, is seated with the seat belt and shoulder harness (if provided) fastened.

(d) Identical powerplant controls for each engine must be located to prevent confusion as to the engines they control.

(e) Wing flap controls and other auxiliary lift device controls must be located on top of the pedestal, aft of the throttles, centrally or to the right of the pedestal centerline, and not less than 10 inches aft of the landing gear control.

(f) The landing gear control must be located forward of the throttles and must be operable by each pilot when seated with seat belt and shoulder harness (if provided) fastened.

(g) Control knobs must be shaped in accordance with § 25.781. In addition, the knobs must be of the same color, and this color must contrast with the color of control knobs for other purposes and the surrounding cockpit.

(h) If a flight engineer is required as part of the minimum flight crew (established under § 25.1523), the airplane must have a flight engineer station located and arranged so that the flight crewmembers can perform their functions efficiently and without interfering with each other.

[Doc. No. 5066, 29 FR 18291, Dec. 24, 1964, as amended by Amdt. 25–46, 43 FR 50596, Oct. 30, 1978]

§ 25.779 Motion and effect of cockpit controls.

Cockpit controls must be designed so that they operate in accordance with the following movement and actuation:

(a) Aerodynamic controls:

(1) *Primary.*

Controls	Motion and effect
Aileron	Right (clockwise) for right wing down.
Elevator	Rearward for nose up.
Rudder	Right pedal forward for nose right.

(2) *Secondary.*

Controls	Motion and effect
Flaps (or auxiliary lift devices).	Forward for flaps up; rearward for flaps down.
Trim tabs (or equivalent).	Rotate to produce similar rotation of the airplane about an axis parallel to the axis of the control.

(b) Powerplant and auxiliary controls:
(1) *Powerplant.*

Controls	Motion and effect
Power or thrust	Forward to increase forward thrust and rearward to increase rearward thrust.
Propellers	Forward to increase rpm.
Mixture	Forward or upward for rich.

Controls	Motion and effect
Carburetor air heat	Forward or upward for cold.
Supercharger	Forward or upward for low blower. For turbosuperchargers, forward, upward, or clockwise, to increase pressure.

(2) *Auxiliary.*

Controls	Motion and effect
Landing gear	Down to extend.

[Doc. No. 5066, 29 FR 18291, Dec. 24, 1964, as amended by Amdt. 25–72, 55 FR 29778, July 20, 1990]

§ 25.781   Cockpit control knob shape.

Cockpit control knobs must conform to the general shapes (but not necessarily the exact sizes or specific proportions) in the following figure:

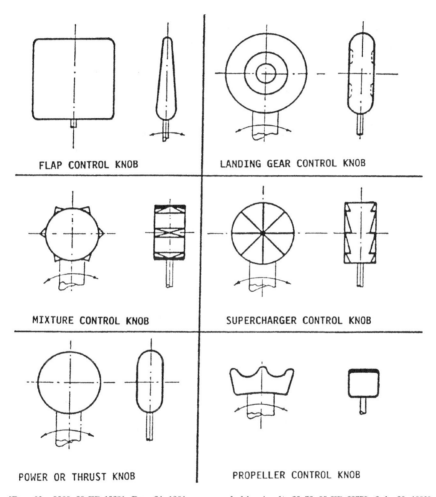

FLAP CONTROL KNOB

LANDING GEAR CONTROL KNOB

MIXTURE CONTROL KNOB

SUPERCHARGER CONTROL KNOB

POWER OR THRUST KNOB

PROPELLER CONTROL KNOB

[Doc. No. 5066, 29 FR 18291, Dec. 24, 1964, as amended by Amdt. 25–72, 55 FR 29779, July 20, 1990]

§ 25.783 Fuselage doors.

(a) *General.* This section applies to fuselage doors, which includes all doors, hatches, openable windows, access panels, covers, etc., on the exterior of the fuselage that do not require the use of tools to open or close. This also applies to each door or hatch through a pressure bulkhead, including any bulkhead that is specifically designed to function as a secondary bulkhead under the prescribed failure conditions of part 25. These doors must meet the requirements of this section, taking into account both pressurized and unpressurized flight, and must be designed as follows:

(1) Each door must have means to safeguard against opening in flight as a result of mechanical failure, or failure of any single structural element.

281

(2) Each door that could be a hazard if it unlatches must be designed so that unlatching during pressurized and unpressurized flight from the fully closed, latched, and locked condition is extremely improbable. This must be shown by safety analysis.

(3) Each element of each door operating system must be designed or, where impracticable, distinctively and permanently marked, to minimize the probability of incorrect assembly and adjustment that could result in a malfunction.

(4) All sources of power that could initiate unlocking or unlatching of any door must be automatically isolated from the latching and locking systems prior to flight and it must not be possible to restore power to the door during flight.

(5) Each removable bolt, screw, nut, pin, or other removable fastener must meet the locking requirements of § 25.607.

(6) Certain doors, as specified by § 25.807(h), must also meet the applicable requirements of §§ 25.809 through 25.812 for emergency exits.

(b) *Opening by persons.* There must be a means to safeguard each door against opening during flight due to inadvertent action by persons. In addition, design precautions must be taken to minimize the possibility for a person to open a door intentionally during flight. If these precautions include the use of auxiliary devices, those devices and their controlling systems must be designed so that—

(1) No single failure will prevent more than one exit from being opened; and

(2) Failures that would prevent opening of the exit after landing are improbable.

(c) *Pressurization prevention means.* There must be a provision to prevent pressurization of the airplane to an unsafe level if any door subject to pressurization is not fully closed, latched, and locked.

(1) The provision must be designed to function after any single failure, or after any combination of failures not shown to be extremely improbable.

(2) Doors that meet the conditions described in paragraph (h) of this section are not required to have a dedicated pressurization prevention means if, from every possible position of the door, it will remain open to the extent that it prevents pressurization or safely close and latch as pressurization takes place. This must also be shown with any single failure and malfunction, except that—

(i) With failures or malfunctions in the latching mechanism, it need not latch after closing; and

(ii) With jamming as a result of mechanical failure or blocking debris, the door need not close and latch if it can be shown that the pressurization loads on the jammed door or mechanism would not result in an unsafe condition.

(d) *Latching and locking.* The latching and locking mechanisms must be designed as follows:

(1) There must be a provision to latch each door.

(2) The latches and their operating mechanism must be designed so that, under all airplane flight and ground loading conditions, with the door latched, there is no force or torque tending to unlatch the latches. In addition, the latching system must include a means to secure the latches in the latched position. This means must be independent of the locking system.

(3) Each door subject to pressurization, and for which the initial opening movement is not inward, must—

(i) Have an individual lock for each latch;

(ii) Have the lock located as close as practicable to the latch; and

(iii) Be designed so that, during pressurized flight, no single failure in the locking system would prevent the locks from restraining the latches necessary to secure the door.

(4) Each door for which the initial opening movement is inward, and unlatching of the door could result in a hazard, must have a locking means to prevent the latches from becoming disengaged. The locking means must ensure sufficient latching to prevent opening of the door even with a single failure of the latching mechanism.

(5) It must not be possible to position the lock in the locked position if the latch and the latching mechanism are not in the latched position.

(6) It must not be possible to unlatch the latches with the locks in the locked position. Locks must be designed to withstand the limit loads resulting from—

(i) The maximum operator effort when the latches are operated manually;

(ii) The powered latch actuators, if installed; and

(iii) The relative motion between the latch and the structural counterpart.

(7) Each door for which unlatching would not result in a hazard is not required to have a locking mechanism meeting the requirements of paragraphs (d)(3) through (d)(6) of this section.

(e) *Warning, caution, and advisory indications.* Doors must be provided with the following indications:

(1) There must be a positive means to indicate at each door operator's station that all required operations to close, latch, and lock the door(s) have been completed.

(2) There must be a positive means clearly visible from each operator station for any door that could be a hazard if unlatched to indicate if the door is not fully closed, latched, and locked.

(3) There must be a visual means on the flight deck to signal the pilots if any door is not fully closed, latched, and locked. The means must be designed such that any failure or combination of failures that would result in an erroneous closed, latched, and locked indication is improbable for—

(i) Each door that is subject to pressurization and for which the initial opening movement is not inward; or

(ii) Each door that could be a hazard if unlatched.

(4) There must be an aural warning to the pilots prior to or during the initial portion of takeoff roll if any door is not fully closed, latched, and locked, and its opening would prevent a safe takeoff and return to landing.

(f) *Visual inspection provision.* Each door for which unlatching of the door could be a hazard must have a provision for direct visual inspection to determine, without ambiguity, if the door is fully closed, latched, and locked. The provision must be permanent and discernible under operational lighting conditions, or by means of a flashlight or equivalent light source.

(g) *Certain maintenance doors, removable emergency exits, and access panels.* Some doors not normally opened except for maintenance purposes or emergency evacuation and some access panels need not comply with certain paragraphs of this section as follows:

(1) Access panels that are not subject to cabin pressurization and would not be a hazard if open during flight need not comply with paragraphs (a) through (f) of this section, but must have a means to prevent inadvertent opening during flight.

(2) Inward-opening removable emergency exits that are not normally removed, except for maintenance purposes or emergency evacuation, and flight deck-openable windows need not comply with paragraphs (c) and (f) of this section.

(3) Maintenance doors that meet the conditions of paragraph (h) of this section, and for which a placard is provided limiting use to maintenance access, need not comply with paragraphs (c) and (f) of this section.

(h) *Doors that are not a hazard.* For the purposes of this section, a door is considered to be not a hazard in the unlatched condition during flight, provided it can be shown to meet all of the following conditions:

(1) Doors in pressurized compartments would remain in the fully closed position if not restrained by the latches when subject to a pressure greater than ½ psi. Opening by persons, either inadvertently or intentionally, need not be considered in making this determination.

(2) The door would remain inside the airplane or remain attached to the airplane if it opens either in pressurized or unpressurized portions of the flight. This determination must include the consideration of inadvertent and intentional opening by persons during either pressurized or unpressurized portions of the flight.

(3) The disengagement of the latches during flight would not allow depressurization of the cabin to an unsafe level. This safety assessment must include the physiological effects on the occupants.

283

(4) The open door during flight would not create aerodynamic interference that could preclude safe flight and landing.

(5) The airplane would meet the structural design requirements with the door open. This assessment must include the aeroelastic stability requirements of § 25.629, as well as the strength requirements of subpart C of this part.

(6) The unlatching or opening of the door must not preclude safe flight and landing as a result of interaction with other systems or structures.

[Doc. No. 2003-14193, 69 FR 24501, May 3, 2004]

§ 25.785 Seats, berths, safety belts, and harnesses.

(a) A seat (or berth for a nonambulant person) must be provided for each occupant who has reached his or her second birthday.

(b) Each seat, berth, safety belt, harness, and adjacent part of the airplane at each station designated as occupiable during takeoff and landing must be designed so that a person making proper use of these facilities will not suffer serious injury in an emergency landing as a result of the inertia forces specified in §§ 25.561 and 25.562.

(c) Each seat or berth must be approved.

(d) Each occupant of a seat that makes more than an 18-degree angle with the vertical plane containing the airplane centerline must be protected from head injury by a safety belt and an energy absorbing rest that will support the arms, shoulders, head, and spine, or by a safety belt and shoulder harness that will prevent the head from contacting any injurious object. Each occupant of any other seat must be protected from head injury by a safety belt and, as appropriate to the type, location, and angle of facing of each seat, by one or more of the following:

(1) A shoulder harness that will prevent the head from contacting any injurious object.

(2) The elimination of any injurious object within striking radius of the head.

(3) An energy absorbing rest that will support the arms, shoulders, head, and spine.

(e) Each berth must be designed so that the forward part has a padded end board, canvas diaphragm, or equivalent means, that can withstand the static load reaction of the occupant when subjected to the forward inertia force specified in § 25.561. Berths must be free from corners and protuberances likely to cause injury to a person occupying the berth during emergency conditions.

(f) Each seat or berth, and its supporting structure, and each safety belt or harness and its anchorage must be designed for an occupant weight of 170 pounds, considering the maximum load factors, inertia forces, and reactions among the occupant, seat, safety belt, and harness for each relevant flight and ground load condition (including the emergency landing conditions prescribed in § 25.561). In addition—

(1) The structural analysis and testing of the seats, berths, and their supporting structures may be determined by assuming that the critical load in the forward, sideward, downward, upward, and rearward directions (as determined from the prescribed flight, ground, and emergency landing conditions) acts separately or using selected combinations of loads if the required strength in each specified direction is substantiated. The forward load factor need not be applied to safety belts for berths.

(2) Each pilot seat must be designed for the reactions resulting from the application of the pilot forces prescribed in § 25.395.

(3) The inertia forces specified in § 25.561 must be multiplied by a factor of 1.33 (instead of the fitting factor prescribed in § 25.625) in determining the strength of the attachment of each seat to the structure and each belt or harness to the seat or structure.

(g) Each seat at a flight deck station must have a restraint system consisting of a combined safety belt and shoulder harness with a single-point release that permits the flight deck occupant, when seated with the restraint system fastened, to perform all of the occupant's necessary flight deck functions. There must be a means to secure each combined restraint system when not in use to prevent interference with the operation of the airplane and with rapid egress in an emergency.

(h) Each seat located in the passenger compartment and designated for use during takeoff and landing by a flight attendant required by the operating rules of this chapter must be:

(1) Near a required floor level emergency exit, except that another location is acceptable if the emergency egress of passengers would be enhanced with that location. A flight attendant seat must be located adjacent to each Type A or B emergency exit. Other flight attendant seats must be evenly distributed among the required floor-level emergency exits to the extent feasible.

(2) To the extent possible, without compromising proximity to a required floor level emergency exit, located to provide a direct view of the cabin area for which the flight attendant is responsible.

(3) Positioned so that the seat will not interfere with the use of a passageway or exit when the seat is not in use.

(4) Located to minimize the probability that occupants would suffer injury by being struck by items dislodged from service areas, stowage compartments, or service equipment.

(5) Either forward or rearward facing with an energy absorbing rest that is designed to support the arms, shoulders, head, and spine.

(6) Equipped with a restraint system consisting of a combined safety belt and shoulder harness unit with a single point release. There must be means to secure each restraint system when not in use to prevent interference with rapid egress in an emergency.

(i) Each safety belt must be equipped with a metal to metal latching device.

(j) If the seat backs do not provide a firm handhold, there must be a handgrip or rail along each aisle to enable persons to steady themselves while using the aisles in moderately rough air.

(k) Each projecting object that would injure persons seated or moving about the airplane in normal flight must be padded.

(l) Each forward observer's seat required by the operating rules must be shown to be suitable for use in conducting the necessary enroute inspection.

[Amdt. 25–72, 55 FR 29780, July 20, 1990, as amended by Amdt. 25–88, 61 FR 57956, Nov. 8, 1996]

§25.787 Stowage compartments.

(a) Each compartment for the stowage of cargo, baggage, carry-on articles, and equipment (such as life rafts), and any other stowage compartment, must be designed for its placarded maximum weight of contents and for the critical load distribution at the appropriate maximum load factors corresponding to the specified flight and ground load conditions, and to those emergency landing conditions of §25.561(b)(3) for which the breaking loose of the contents of such compartments in the specified direction could—

(1) Cause direct injury to occupants;

(2) Penetrate fuel tanks or lines or cause fire or explosion hazard by damage to adjacent systems; or

(3) Nullify any of the escape facilities provided for use after an emergency landing.

If the airplane has a passenger-seating configuration, excluding pilot seats, of 10 seats or more, each stowage compartment in the passenger cabin, except for under seat and overhead compartments for passenger convenience, must be completely enclosed.

(b) There must be a means to prevent the contents in the compartments from becoming a hazard by shifting, under the loads specified in paragraph (a) of this section. For stowage compartments in the passenger and crew cabin, if the means used is a latched door, the design must take into consideration the wear and deterioration expected in service.

(c) If cargo compartment lamps are installed, each lamp must be installed so as to prevent contact between lamp bulb and cargo.

[Doc. No. 5066, 29 FR 18291, Dec. 24, 1964, as amended by Amdt. 25–32, 37 FR 3969, Feb. 24, 1972; Amdt. 25–38, 41 FR 55466, Dec. 20, 1976; Amdt. 25–51, 45 FR 7755, Feb. 4, 1980; Amdt. 25–139, 79 FR 59430, Oct. 2, 2014]

§ 25.789 Retention of items of mass in passenger and crew compartments and galleys.

(a) Means must be provided to prevent each item of mass (that is part of the airplane type design) in a passenger or crew compartment or galley from becoming a hazard by shifting under the appropriate maximum load factors corresponding to the specified flight and ground load conditions, and to the emergency landing conditions of § 25.561(b).

(b) Each interphone restraint system must be designed so that when subjected to the load factors specified in § 25.561(b)(3), the interphone will remain in its stowed position.

[Amdt. 25–32, 37 FR 3969, Feb. 24, 1972, as amended by Amdt. 25–46, 43 FR 50596, Oct. 30, 1978]

§ 25.791 Passenger information signs and placards.

(a) If smoking is to be prohibited, there must be at least one placard so stating that is legible to each person seated in the cabin. If smoking is to be allowed, and if the crew compartment is separated from the passenger compartment, there must be at least one sign notifying when smoking is prohibited. Signs which notify when smoking is prohibited must be operable by a member of the flightcrew and, when illuminated, must be legible under all probable conditions of cabin illumination to each person seated in the cabin.

(b) Signs that notify when seat belts should be fastened and that are installed to comply with the operating rules of this chapter must be operable by a member of the flightcrew and, when illuminated, must be legible under all probable conditions of cabin illumination to each person seated in the cabin.

(c) A placard must be located on or adjacent to the door of each receptacle used for the disposal of flammable waste materials to indicate that use of the receptacle for disposal of cigarettes, etc., is prohibited.

(d) Lavatories must have "No Smoking" or "No Smoking in Lavatory" placards conspicuously located on or adjacent to each side of the entry door.

(e) Symbols that clearly express the intent of the sign or placard may be used in lieu of letters.

[Amdt. 25–72, 55 FR 29780, July 20, 1990]

§ 25.793 Floor surfaces.

The floor surface of all areas which are likely to become wet in service must have slip resistant properties.

[Amdt. 25–51, 45 FR 7755, Feb. 4, 1980]

§ 25.795 Security considerations.

(a) *Protection of flightcrew compartment.* If a flightdeck door is required by operating rules:

(1) The bulkhead, door, and any other accessible boundary separating the flightcrew compartment from occupied areas must be designed to resist forcible intrusion by unauthorized persons and be capable of withstanding impacts of 300 joules (221.3 foot pounds).

(2) The bulkhead, door, and any other accessible boundary separating the flightcrew compartment from occupied areas must be designed to resist a constant 250 pound (1,113 Newtons) tensile load on accessible handholds, including the doorknob or handle.

(3) The bulkhead, door, and any other boundary separating the flightcrew compartment from any occupied areas must be designed to resist penetration by small arms fire and fragmentation devices to a level equivalent to level IIIa of the National Institute of Justice (NIJ) Standard 0101.04.

(4) If required by the operating rules of this chapter, an installed physical secondary barrier (IPSB) must be installed to resist intrusion into the flightdeck whenever the flightdeck door is opened. When deployed, the IPSB must:

(i) Resist a 250 pound (1113 Newtons) static load in the direction of the passenger cabin applied at the most critical locations on the IPSB;

(ii) Resist a 600 pound (2669 Newtons) static load in the direction of the flightdeck applied at the most critical locations on the IPSB;

(iii) Delay a person attempting to access the flightdeck by at least the time required for a crewmember to open and reclose the flightdeck door, but no less than 5 seconds;

(iv) Prevent a person from reaching through and touching the flightdeck door;

(v) Allow for necessary crewmember activities; and

(vi) Provide line-of-sight visibility between the flightdeck door and the cabin.

(b) Airplanes with a maximum certificated passenger seating capacity of more than 60 persons or a maximum certificated takeoff gross weight of over 100,000 pounds (45,359 Kilograms) must be designed to limit the effects of an explosive or incendiary device as follows:

(1) *Flightdeck smoke protection.* Means must be provided to limit entry of smoke, fumes, and noxious gases into the flightdeck.

(2) *Passenger cabin smoke protection.* Means must be provided to prevent passenger incapacitation in the cabin resulting from smoke, fumes, and noxious gases as represented by the initial combined volumetric concentrations of 0.59% carbon monoxide and 1.23% carbon dioxide.

(3) *Cargo compartment fire suppression.* An extinguishing agent must be capable of suppressing a fire. All cargo-compartment fire suppression systems must be designed to withstand the following effects, including support structure displacements or adjacent materials displacing against the distribution system:

(i) Impact or damage from a 0.5-inch diameter aluminum sphere traveling at 430 feet per second (131.1 meters per second);

(ii) A 15-pound per square-inch (103.4 kPa) pressure load if the projected surface area of the component is greater than 4 square feet. Any single dimension greater than 4 feet (1.22 meters) may be assumed to be 4 feet (1.22 meters) in length; and

(iii) A 6-inch (0.152 meters) displacement, except where limited by the fuselage contour, from a single point force applied anywhere along the distribution system where relative movement between the system and its attachment can occur.

(iv) Paragraphs (b)(3)(i) through (iii) of this section do not apply to components that are redundant and separated in accordance with paragraph (c)(2) of

this section or are installed remotely from the cargo compartment.

(c) An airplane with a maximum certificated passenger seating capacity of more than 60 persons or a maximum certificated takeoff gross weight of over 100,000 pounds (45,359 Kilograms) must comply with the following:

(1) *Least risk bomb location.* An airplane must be designed with a designated location where a bomb or other explosive device could be placed to best protect flight-critical structures and systems from damage in the case of detonation.

(2) *Survivability of systems.* (i) Except where impracticable, redundant airplane systems necessary for continued safe flight and landing must be physically separated, at a minimum, by an amount equal to a sphere of diameter

$$D = 2\sqrt{\left(H_0/\pi\right)}$$

(where H_0 is defined under §25.365(e)(2) of this part and D need not exceed 5.05 feet (1.54 meters)). The sphere is applied everywhere within the fuselage— limited by the forward bulkhead and the aft bulkhead of the passenger cabin and cargo compartment beyond which only one-half the sphere is applied.

(ii) Where compliance with paragraph (c)(2)(i) of this section is impracticable, other design precautions must be taken to maximize the survivability of those systems.

(3) *Interior design to facilitate searches.* Design features must be incorporated that will deter concealment or promote discovery of weapons, explosives, or other objects from a simple inspection in the following areas of the airplane cabin:

(i) Areas above the overhead bins must be designed to prevent objects from being hidden from view in a simple search from the aisle. Designs that prevent concealment of objects with volumes 20 cubic inches and greater satisfy this requirement.

(ii) Toilets must be designed to prevent the passage of solid objects greater than 2.0 inches in diameter.

(iii) Life preservers or their storage locations must be designed so that tampering is evident.

(d) Each chemical oxygen generator or its installation must be designed to

287

be secure from deliberate manipulation by one of the following:

(1) By providing effective resistance to tampering,

(2) By providing an effective combination of resistance to tampering and active tamper-evident features,

(3) By installation in a location or manner whereby any attempt to access the generator would be immediately obvious, or

(4) By a combination of approaches specified in paragraphs (d)(1), (d)(2) and (d)(3) of this section that the Administrator finds provides a secure installation.

(e) *Exceptions.* Airplanes used solely to transport cargo only need to meet the requirements of paragraphs (b)(1), (b)(3), and (c)(2) of this section.

(f) *Material Incorporated by Reference.* You must use National Institute of Justice (NIJ) Standard 0101.04, Ballistic Resistance of Personal Body Armor, June 2001, Revision A, to establish ballistic resistance as required by paragraph (a)(3) of this section.

(1) The Director of the Federal Register approved the incorporation by reference of this document under 5 U.S.C. 552(a) and 1 CFR part 51.

(2) You may review copies of NIJ Standard 0101.04 at the:

(i) National Institute of Justice (NIJ), *http://www.ojp.usdoj.gov/nij*, telephone (202) 307–2942; or

(ii) National Archives and Records Administration (NARA). For information on the availability of this material at NARA, call (202) 741–6030, or go to *http://www.archives.gov/federal-register/cfr/ibr-locations.html.*

(3) You may obtain copies of NIJ Standard 0101.04 from the National Criminal Justice Reference Service, P.O. Box 6000, Rockville, MD 20849–6000, telephone (800) 851–3420.

[Amdt. 25–127; 121–341, 73 FR 63879, Oct. 28, 2008, as amended at 74 FR 22819, May 15, 2009; Amdt. 25–138, 79 FR 13519, Mar. 11, 2014; Doc. No. FAA–2018–0119, Amdt. 25–145, 83 FR 9169, Mar. 5, 2018; Amdt. 25–150, 88 FR 41308, June 26, 2023]

EMERGENCY PROVISIONS

§ 25.801 Ditching.

(a) If certification with ditching provisions is requested, the airplane must meet the requirements of this section and §§ 25.807(e), 25.1411, and 25.1415(a).

(b) Each practicable design measure, compatible with the general characteristics of the airplane, must be taken to minimize the probability that in an emergency landing on water, the behavior of the airplane would cause immediate injury to the occupants or would make it impossible for them to escape.

(c) The probable behavior of the airplane in a water landing must be investigated by model tests or by comparison with airplanes of similar configuration for which the ditching characteristics are known. Scoops, flaps, projections, and any other factor likely to affect the hydrodynamic characteristics of the airplane, must be considered.

(d) It must be shown that, under reasonably probable water conditions, the flotation time and trim of the airplane will allow the occupants to leave the airplane and enter the liferafts required by § 25.1415. If compliance with this provision is shown by buoyancy and trim computations, appropriate allowances must be made for probable structural damage and leakage. If the airplane has fuel tanks (with fuel jettisoning provisions) that can reasonably be expected to withstand a ditching without leakage, the jettisonable volume of fuel may be considered as buoyancy volume.

(e) Unless the effects of the collapse of external doors and windows are accounted for in the investigation of the probable behavior of the airplane in a water landing (as prescribed in paragraphs (c) and (d) of this section), the external doors and windows must be designed to withstand the probable maximum local pressures.

[Doc. No. 5066, 29 FR 18291, Dec. 24, 1964, as amended by Amdt. 25–72, 55 FR 29781, July 20, 1990]

§ 25.803 Emergency evacuation.

(a) Each crew and passenger area must have emergency means to allow rapid evacuation in crash landings, with the landing gear extended as well as with the landing gear retracted, considering the possibility of the airplane being on fire.

(b) [Reserved]

(c) For airplanes having a seating capacity of more than 44 passengers, it must be shown that the maximum seating capacity, including the number of crewmembers required by the operating rules for which certification is requested, can be evacuated from the airplane to the ground under simulated emergency conditions within 90 seconds. Compliance with this requirement must be shown by actual demonstration using the test criteria outlined in appendix J of this part unless the Administrator finds that a combination of analysis and testing will provide data equivalent to that which would be obtained by actual demonstration.

(d)–(e) [Reserved]

[Doc. No. 24344, 55 FR 29781, July 20, 1990]

§ 25.807 Emergency exits.

(a) *Type.* For the purpose of this part, the types of exits are defined as follows:

(1) *Type I.* This type is a floor-level exit with a rectangular opening of not less than 24 inches wide by 48 inches high, with corner radii not greater than eight inches.

(2) *Type II.* This type is a rectangular opening of not less than 20 inches wide by 44 inches high, with corner radii not greater than seven inches. Type II exits must be floor-level exits unless located over the wing, in which case they must not have a step-up inside the airplane of more than 10 inches nor a step-down outside the airplane of more than 17 inches.

(3) *Type III.* This type is a rectangular opening of not less than 20 inches wide by 36 inches high with corner radii not greater than seven inches, and with a step-up inside the airplane of not more than 20 inches. If the exit is located over the wing, the step-down outside the airplane may not exceed 27 inches.

(4) *Type IV.* This type is a rectangular opening of not less than 19 inches wide by 26 inches high, with corner radii not greater than 6.3 inches, located over the wing, with a step-up inside the airplane of not more than 29 inches and a step-down outside the airplane of not more than 36 inches.

(5) *Ventral.* This type is an exit from the passenger compartment through the pressure shell and the bottom fuselage skin. The dimensions and physical configuration of this type of exit must allow at least the same rate of egress as a Type I exit with the airplane in the normal ground attitude, with landing gear extended.

(6) *Tailcone.* This type is an aft exit from the passenger compartment through the pressure shell and through an openable cone of the fuselage aft of the pressure shell. The means of opening the tailcone must be simple and obvious and must employ a single operation.

(7) *Type A.* This type is a floor-level exit with a rectangular opening of not less than 42 inches wide by 72 inches high, with corner radii not greater than seven inches.

(8) *Type B.* This type is a floor-level exit with a rectangular opening of not less than 32 inches wide by 72 inches high, with corner radii not greater than six inches.

(9) *Type C.* This type is a floor-level exit with a rectangular opening of not less than 30 inches wide by 48 inches high, with corner radii not greater than 10 inches.

(b) *Step down distance.* Step down distance, as used in this section, means the actual distance between the bottom of the required opening and a usable foot hold, extending out from the fuselage, that is large enough to be effective without searching by sight or feel.

(c) *Over-sized exits.* Openings larger than those specified in this section, whether or not of rectangular shape, may be used if the specified rectangular opening can be inscribed within the opening and the base of the inscribed rectangular opening meets the specified step-up and step-down heights.

(d) *Asymmetry.* Exits of an exit pair need not be diametrically opposite each other nor of the same size; however, the number of passenger seats permitted under paragraph (g) of this section is based on the smaller of the two exits.

(e) *Uniformity.* Exits must be distributed as uniformly as practical, taking into account passenger seat distribution.

(f) *Location.* (1) Each required passenger emergency exit must be accessible to the passengers and located where it will afford the most effective means of passenger evacuation.

(2) If only one floor-level exit per side is prescribed, and the airplane does not have a tailcone or ventral emergency exit, the floor-level exits must be in the rearward part of the passenger compartment unless another location affords a more effective means of passenger evacuation.

(3) If more than one floor-level exit per side is prescribed, and the airplane does not have a combination cargo and passenger configuration, at least one floor-level exit must be located in each side near each end of the cabin.

(4) For an airplane that is required to have more than one passenger emergency exit for each side of the fuselage, no passenger emergency exit shall be more than 60 feet from any adjacent passenger emergency exit on the same side of the same deck of the fuselage, as measured parallel to the airplane's longitudinal axis between the nearest exit edges.

(g) *Type and number required.* The maximum number of passenger seats permitted depends on the type and number of exits installed in each side of the fuselage. Except as further restricted in paragraphs (g)(1) through (g)(9) of this section, the maximum number of passenger seats permitted for each exit of a specific type installed in each side of the fuselage is as follows:

Type A	110
Type B	75
Type C	55
Type I	45
Type II	40
Type III	35
Type IV	9

(1) For a passenger seating configuration of 1 to 9 seats, there must be at least one Type IV or larger overwing exit in each side of the fuselage or, if overwing exits are not provided, at least one exit in each side that meets the minimum dimensions of a Type III exit.

(2) For a passenger seating configuration of more than 9 seats, each exit must be a Type III or larger exit.

(3) For a passenger seating configuration of 10 to 19 seats, there must be at least one Type III or larger exit in each side of the fuselage.

(4) For a passenger seating configuration of 20 to 40 seats, there must be at least two exits, one of which must be a Type II or larger exit, in each side of the fuselage.

(5) For a passenger seating configuration of 41 to 110 seats, there must be at least two exits, one of which must be a Type I or larger exit, in each side of the fuselage.

(6) For a passenger seating configuration of more than 110 seats, the emergency exits in each side of the fuselage must include at least two Type I or larger exits.

(7) The combined maximum number of passenger seats permitted for all Type III exits is 70, and the combined maximum number of passenger seats permitted for two Type III exits in each side of the fuselage that are separated by fewer than three passenger seat rows is 65.

(8) If a Type A, Type B, or Type C exit is installed, there must be at least two Type C or larger exits in each side of the fuselage.

(9) If a passenger ventral or tailcone exit is installed and that exit provides at least the same rate of egress as a Type III exit with the airplane in the most adverse exit opening condition that would result from the collapse of one or more legs of the landing gear, an increase in the passenger seating configuration is permitted as follows:

(i) For a ventral exit, 12 additional passenger seats.

(ii) For a tailcone exit incorporating a floor level opening of not less than 20 inches wide by 60 inches high, with corner radii not greater than seven inches, in the pressure shell and incorporating an approved assist means in accordance with § 25.810(a), 25 additional passenger seats.

(iii) For a tailcone exit incorporating an opening in the pressure shell which is at least equivalent to a Type III emergency exit with respect to dimensions, step-up and step-down distance, and with the top of the opening not less than 56 inches from the passenger compartment floor, 15 additional passenger seats.

(h) *Other exits.* The following exits also must meet the applicable emergency exit requirements of §§ 25.809 through 25.812, and must be readily accessible:

(1) Each emergency exit in the passenger compartment in excess of the minimum number of required emergency exits.

(2) Any other floor-level door or exit that is accessible from the passenger compartment and is as large or larger than a Type II exit, but less than 46 inches wide.

(3) Any other ventral or tail cone passenger exit.

(i) *Ditching emergency exits for passengers.* Whether or not ditching certification is requested, ditching emergency exits must be provided in accordance with the following requirements, unless the emergency exits required by paragraph (g) of this section already meet them:

(1) For airplanes that have a passenger seating configuration of nine or fewer seats, excluding pilot seats, one exit above the waterline in each side of the airplane, meeting at least the dimensions of a Type IV exit.

(2) For airplanes that have a passenger seating configuration of 10 of more seats, excluding pilot seats, one exit above the waterline in a side of the airplane, meeting at least the dimensions of a Type III exit for each unit (or part of a unit) of 35 passenger seats, but no less than two such exits in the passenger cabin, with one on each side of the airplane. The passenger seat/exit ratio may be increased through the use of larger exits, or other means, provided it is shown that the evacuation capability during ditching has been improved accordingly.

(3) If it is impractical to locate side exits above the waterline, the side exits must be replaced by an equal number of readily accessible overhead hatches of not less than the dimensions of a Type III exit, except that for airplanes with a passenger configuration of 35 or fewer seats, excluding pilot seats, the two required Type III side exits need be replaced by only one overhead hatch.

(j) *Flightcrew emergency exits.* For airplanes in which the proximity of passenger emergency exits to the flightcrew area does not offer a convenient and readily accessible means of evacuation of the flightcrew, and for all airplanes having a passenger seating capacity greater than 20, flightcrew exits shall be located in the flightcrew area. Such exits shall be of sufficient size and so located as to permit rapid evacuation by the crew. One exit shall be provided on each side of the airplane; or, alternatively, a top hatch shall be provided. Each exit must encompass an unobstructed rectangular opening of at least 19 by 20 inches unless satisfactory exit utility can be demonstrated by a typical crewmember.

[Amdt. 25–72, 55 FR 29781, July 20, 1990, as amended by Amdt. 25–88, 61 FR 57956, Nov. 8, 1996; 62 FR 1817, Jan. 13, 1997; Amdt. 25–94, 63 FR 8848, Feb. 23, 1998; 63 FR 12862, Mar. 16, 1998; Amdt. 25–114, 69 FR 24502, May 3, 2004]

§ 25.809 Emergency exit arrangement.

(a) Each emergency exit, including each flightcrew emergency exit, must be a moveable door or hatch in the external walls of the fuselage, allowing an unobstructed opening to the outside. In addition, each emergency exit must have means to permit viewing of the conditions outside the exit when the exit is closed. The viewing means may be on or adjacent to the exit provided no obstructions exist between the exit and the viewing means. Means must also be provided to permit viewing of the likely areas of evacuee ground contact. The likely areas of evacuee ground contact must be viewable during all lighting conditions with the landing gear extended as well as in all conditions of landing gear collapse.

(b) Each emergency exit must be openable from the inside and the outside except that sliding window emergency exits in the flight crew area need not be openable from the outside if other approved exits are convenient and readily accessible to the flight crew area. Each emergency exit must be capable of being opened, when there is no fuselage deformation—

(1) With the airplane in the normal ground attitude and in each of the attitudes corresponding to collapse of one or more legs of the landing gear; and

(2) Within 10 seconds measured from the time when the opening means is actuated to the time when the exit is fully opened.

(3) Even though persons may be crowded against the door on the inside of the airplane.

(c) The means of opening emergency exits must be simple and obvious; may not require exceptional effort; and must be arranged and marked so that it can be readily located and operated, even in darkness. Internal exit-opening means involving sequence operations (such as operation of two handles or latches, or the release of safety catches) may be used for flightcrew emergency exits if it can be reasonably established that these means are simple and obvious to crewmembers trained in their use.

(d) If a single power-boost or single power-operated system is the primary system for operating more than one exit in an emergency, each exit must be capable of meeting the requirements of paragraph (b) of this section in the event of failure of the primary system. Manual operation of the exit (after failure of the primary system) is acceptable.

(e) Each emergency exit must be shown by tests, or by a combination of analysis and tests, to meet the requirements of paragraphs (b) and (c) of this section.

(f) Each door must be located where persons using them will not be endangered by the propellers when appropriate operating procedures are used.

(g) There must be provisions to minimize the probability of jamming of the emergency exits resulting from fuselage deformation in a minor crash landing.

(h) When required by the operating rules for any large passenger-carrying turbojet-powered airplane, each ventral exit and tailcone exit must be—

(1) Designed and constructed so that it cannot be opened during flight; and

(2) Marked with a placard readable from a distance of 30 inches and installed at a conspicuous location near the means of opening the exit, stating that the exit has been designed and constructed so that it cannot be opened during flight.

(i) Each emergency exit must have a means to retain the exit in the open position, once the exit is opened in an emergency. The means must not require separate action to engage when the exit is opened, and must require positive action to disengage.

[Doc. No. 5066, 29 FR 18291, Dec. 24, 1964, as amended by Amdt. 25–15, 32 FR 13264, Sept. 20, 1967; Amdt. 25–32, 37 FR 3970, Feb. 24, 1972; Amdt. 25–34, 37 FR 25355, Nov. 30, 1972; Amdt. 25–46, 43 FR 50597, Oct. 30, 1978; Amdt. 25–47, 44 FR 61325, Oct. 25, 1979; Amdt. 25–72, 55 FR 29782, July 20, 1990; Amdt. 25–114, 69 FR 24502, May 3, 2004; Amdt. 25–116, 69 FR 62788, Oct. 27, 2004]

§ 25.810 Emergency egress assist means and escape routes.

(a) Each non over-wing Type A, Type B or Type C exit, and any other non over-wing landplane emergency exit more than 6 feet from the ground with the airplane on the ground and the landing gear extended, must have an approved means to assist the occupants in descending to the ground.

(1) The assisting means for each passenger emergency exit must be a self-supporting slide or equivalent; and, in the case of Type A or Type B exits, it must be capable of carrying simultaneously two parallel lines of evacuees. In addition, the assisting means must be designed to meet the following requirements—

(i) It must be automatically deployed and deployment must begin during the interval between the time the exit opening means is actuated from inside the airplane and the time the exit is fully opened. However, each passenger emergency exit which is also a passenger entrance door or a service door must be provided with means to prevent deployment of the assisting means when it is opened from either the inside or the outside under non-emergency conditions for normal use.

(ii) Except for assisting means installed at Type C exits, it must be automatically erected within 6 seconds after deployment is begun. Assisting means installed at Type C exits must be automatically erected within 10 seconds from the time the opening means of the exit is actuated.

(iii) It must be of such length after full deployment that the lower end is self-supporting on the ground and provides safe evacuation of occupants to the ground after collapse of one or more legs of the landing gear.

(iv) It must have the capability, in 25-knot winds directed from the most critical angle, to deploy and, with the assistance of only one person, to remain usable after full deployment to evacuate occupants safely to the ground.

(v) For each system installation (mockup or airplane installed), five consecutive deployment and inflation tests must be conducted (per exit) without failure, and at least three tests of each such five-test series must be conducted using a single representative sample of the device. The sample devices must be deployed and inflated by the system's primary means after being subjected to the inertia forces specified in § 25.561(b). If any part of the system fails or does not function properly during the required tests, the cause of the failure or malfunction must be corrected by positive means and after that, the full series of five consecutive deployment and inflation tests must be conducted without failure.

(2) The assisting means for flightcrew emergency exits may be a rope or any other means demonstrated to be suitable for the purpose. If the assisting means is a rope, or an approved device equivalent to a rope, it must be—

(i) Attached to the fuselage structure at or above the top of the emergency exit opening, or, for a device at a pilot's emergency exit window, at another approved location if the stowed device, or its attachment, would reduce the pilot's view in flight;

(ii) Able (with its attachment) to withstand a 400-pound static load.

(b) Assist means from the cabin to the wing are required for each type A or Type B exit located above the wing and having a stepdown unless the exit without an assist-means can be shown to have a rate of passenger egress at least equal to that of the same type of non over-wing exit. If an assist means is required, it must be automatically deployed and automatically erected concurrent with the opening of the exit. In the case of assist means installed at Type C exits, it must be self-

supporting within 10 seconds from the time the opening means of the exits is actuated. For all other exit types, it must be self-supporting 6 seconds after deployment is begun.

(c) An escape route must be established from each overwing emergency exit, and (except for flap surfaces suitable as slides) covered with a slip resistant surface. Except where a means for channeling the flow of evacuees is provided—

(1) The escape route from each Type A or Type B passenger emergency exit, or any common escape route from two Type III passenger emergency exits, must be at least 42 inches wide; that from any other passenger emergency exit must be at least 24 inches wide; and

(2) The escape route surface must have a reflectance of at least 80 percent, and must be defined by markings with a surface-to-marking contrast ratio of at least 5:1.

(d) Means must be provided to assist evacuees to reach the ground for all Type C exits located over the wing and, if the place on the airplane structure at which the escape route required in paragraph (c) of this section terminates is more than 6 feet from the ground with the airplane on the ground and the landing gear extended, for all other exit types.

(1) If the escape route is over the flap, the height of the terminal edge must be measured with the flap in the takeoff or landing position, whichever is higher from the ground.

(2) The assisting means must be usable and self-supporting with one or more landing gear legs collapsed and under a 25-knot wind directed from the most critical angle.

(3) The assisting means provided for each escape route leading from a Type A or B emergency exit must be capable of carrying simultaneously two parallel lines of evacuees; and, the assisting means leading from any other exit type must be capable of carrying as many parallel lines of evacuees as there are required escape routes.

(4) The assisting means provided for each escape route leading from a Type C exit must be automatically erected within 10 seconds from the time the opening means of the exit is actuated,

and that provided for the escape route leading from any other exit type must be automatically erected within 10 seconds after actuation of the erection system.

(e) If an integral stair is installed in a passenger entry door that is qualified as a passenger emergency exit, the stair must be designed so that, under the following conditions, the effectiveness of passenger emergency egress will not be impaired:

(1) The door, integral stair, and operating mechanism have been subjected to the inertia forces specified in § 25.561(b)(3), acting separately relative to the surrounding structure.

(2) The airplane is in the normal ground attitude and in each of the attitudes corresponding to collapse of one or more legs of the landing gear.

[Amdt. 25–72, 55 FR 29782, July 20, 1990, as amended by Amdt. 25–88, 61 FR 57958, Nov. 8, 1996; 62 FR 1817, Jan. 13, 1997; Amdt. 25–114, 69 FR 24502, May 3, 2004]

§ 25.811 Emergency exit marking.

(a) Each passenger emergency exit, its means of access, and its means of opening must be conspicuously marked.

(b) The identity and location of each passenger emergency exit must be recognizable from a distance equal to the width of the cabin.

(c) Means must be provided to assist the occupants in locating the exits in conditions of dense smoke.

(d) The location of each passenger emergency exit must be indicated by a sign visible to occupants approaching along the main passenger aisle (or aisles). There must be—

(1) A passenger emergency exit locator sign above the aisle (or aisles) near each passenger emergency exit, or at another overhead location if it is more practical because of low headroom, except that one sign may serve more than one exit if each exit can be seen readily from the sign;

(2) A passenger emergency exit marking sign next to each passenger emergency exit, except that one sign may serve two such exits if they both can be seen readily from the sign; and

(3) A sign on each bulkhead or divider that prevents fore and aft vision along the passenger cabin to indicate emergency exits beyond and obscured by the bulkhead or divider, except that if this is not possible the sign may be placed at another appropriate location.

(e) The location of the operating handle and instructions for opening exits from the inside of the airplane must be shown in the following manner:

(1) Each passenger emergency exit must have, on or near the exit, a marking that is readable from a distance of 30 inches.

(2) Each Type A, Type B, Type C or Type I passenger emergency exit operating handle must—

(i) Be self-illuminated with an initial brightness of at least 160 microlamberts; or

(ii) Be conspicuously located and well illuminated by the emergency lighting even in conditions of occupant crowding at the exit.

(3) [Reserved]

(4) Each Type A, Type B, Type C, Type I, or Type II passenger emergency exit with a locking mechanism released by rotary motion of the handle must be marked—

(i) With a red arrow, with a shaft at least three-fourths of an inch wide and a head twice the width of the shaft, extending along at least 70 degrees of arc at a radius approximately equal to three-fourths of the handle length.

(ii) So that the centerline of the exit handle is within ±1 inch of the projected point of the arrow when the handle has reached full travel and has released the locking mechanism, and

(iii) With the word "open" in red letters 1 inch high, placed horizontally near the head of the arrow.

(f) Each emergency exit that is required to be openable from the outside, and its means of opening, must be marked on the outside of the airplane. In addition, the following apply:

(1) The outside marking for each passenger emergency exit in the side of the fuselage must include a 2-inch colored band outlining the exit.

(2) Each outside marking including the band, must have color contrast to be readily distinguishable from the surrounding fuselage surface. The contrast must be such that if the reflectance of the darker color is 15 percent or less, the reflectance of the lighter color

must be at least 45 percent. "Reflectance" is the ratio of the luminous flux reflected by a body to the luminous flux it receives. When the reflectance of the darker color is greater than 15 percent, at least a 30-percent difference between its reflectance and the reflectance of the lighter color must be provided.

(3) In the case of exists other than those in the side of the fuselage, such as ventral or tailcone exists, the external means of opening, including instructions if applicable, must be conspicuously marked in red, or bright chrome yellow if the background color is such that red is inconspicuous. When the opening means is located on only one side of the fuselage, a conspicuous marking to that effect must be provided on the other side.

(g) Each sign required by paragraph (d) of this section may use the word "exit" in its legend in place of the term "emergency exit".

[Amdt. 25–15, 32 FR 13264, Sept. 20, 1967, as amended by Amdt. 25–32, 37 FR 3970, Feb. 24, 1972; Amdt. 25–46, 43 FR 50597, Oct. 30, 1978; 43 FR 52495, Nov. 13, 1978; Amdt. 25–79, 58 FR 45229, Aug. 26, 1993; Amdt. 25–88, 61 FR 57958, Nov. 8, 1996]

§ 25.812 Emergency lighting.

(a) An emergency lighting system, independent of the main lighting system, must be installed. However, the sources of general cabin illumination may be common to both the emergency and the main lighting systems if the power supply to the emergency lighting system is independent of the power supply to the main lighting system. The emergency lighting system must include:

(1) Illuminated emergency exit marking and locating signs, sources of general cabin illumination, interior lighting in emergency exit areas, and floor proximity escape path marking.

(2) Exterior emergency lighting.

(b) Emergency exit signs—

(1) For airplanes that have a passenger seating configuration, excluding pilot seats, of 10 seats or more must meet the following requirements:

(i) Each passenger emergency exit locator sign required by § 25.811(d)(1) and each passenger emergency exit marking sign required by § 25.811(d)(2) must

have red letters at least 1½ inches high on an illuminated white background, and must have an area of at least 21 square inches excluding the letters. The lighted background-to-letter contrast must be at least 10:1. The letter height to stroke-width ratio may not be more than 7:1 nor less than 6:1. These signs must be internally electrically illuminated with a background brightness of at least 25 foot-lamberts and a high-to-low background contrast no greater than 3:1.

(ii) Each passenger emergency exit sign required by § 25.811(d)(3) must have red letters at least 1½ inches high on a white background having an area of at least 21 square inches excluding the letters. These signs must be internally electrically illuminated or self-illuminated by other than electrical means and must have an initial brightness of at least 400 microlamberts. The colors may be reversed in the case of a sign that is self-illuminated by other than electrical means.

(2) For airplanes that have a passenger seating configuration, excluding pilot seats, of nine seats or less, that are required by § 25.811(d)(1), (2), and (3) must have red letters at least 1 inch high on a white background at least 2 inches high. These signs may be internally electrically illuminated, or self-illuminated by other than electrical means, with an initial brightness of at least 160 microlamberts. The colors may be reversed in the case of a sign that is self-illuminated by other than electrical means.

(c) General illumination in the passenger cabin must be provided so that when measured along the centerline of main passenger aisle(s), and cross aisle(s) between main aisles, at seat arm-rest height and at 40-inch intervals, the average illumination is not less than 0.05 foot-candle and the illumination at each 40-inch interval is not less than 0.01 foot-candle. A main passenger aisle(s) is considered to extend along the fuselage from the most forward passenger emergency exit or cabin occupant seat, whichever is farther forward, to the most rearward passenger emergency exit or cabin occupant seat, whichever is farther aft.

(d) The floor of the passageway leading to each floor-level passenger emergency exit, between the main aisles and the exit openings, must be provided with illumination that is not less than 0.02 foot-candle measured along a line that is within 6 inches of and parallel to the floor and is centered on the passenger evacuation path.

(e) Floor proximity emergency escape path marking must provide emergency evacuation guidance for passengers when all sources of illumination more than 4 feet above the cabin aisle floor are totally obscured. In the dark of the night, the floor proximity emergency escape path marking must enable each passenger to—

(1) After leaving the passenger seat, visually identify the emergency escape path along the cabin aisle floor to the first exits or pair of exits forward and aft of the seat; and

(2) Readily identify each exit from the emergency escape path by reference only to markings and visual features not more than 4 feet above the cabin floor.

(f) Except for subsystems provided in accordance with paragraph (h) of this section that serve no more than one assist means, are independent of the airplane's main emergency lighting system, and are automatically activated when the assist means is erected, the emergency lighting system must be designed as follows.

(1) The lights must be operable manually from the flight crew station and from a point in the passenger compartment that is readily accessible to a normal flight attendant seat.

(2) There must be a flight crew warning light which illuminates when power is on in the airplane and the emergency lighting control device is not armed.

(3) The cockpit control device must have an "on," "off," and "armed" position so that when armed in the cockpit or turned on at either the cockpit or flight attendant station the lights will either light or remain lighted upon interruption (except an interruption caused by a transverse vertical separation of the fuselage during crash landing) of the airplane's normal electric power. There must be a means to safeguard against inadvertent operation of

the control device from the "armed" or "on" positions.

(g) Exterior emergency lighting must be provided as follows:

(1) At each overwing emergency exit the illumination must be—

(i) Not less than 0.03 foot-candle (measured normal to the direction of the incident light) on a 2-square-foot area where an evacuee is likely to make his first step outside the cabin;

(ii) Not less than 0.05 foot-candle (measured normal to the direction of the incident light) for a minimum width of 42 inches for a Type A overwing emergency exit and two feet for all other overwing emergency exits along the 30 percent of the slip-resistant portion of the escape route required in § 25.810(c) that is farthest from the exit; and

(iii) Not less than 0.03 foot-candle on the ground surface with the landing gear extended (measured normal to the direction of the incident light) where an evacuee using the established escape route would normally make first contact with the ground.

(2) At each non-overwing emergency exit not required by § 25.810(a) to have descent assist means the illumination must be not less than 0.03 foot-candle (measured normal to the direction of the incident light) on the ground surface with the landing gear extended where an evacuee is likely to make first contact with the ground outside the cabin.

(h) The means required in §§ 25.810(a)(1) and (d) to assist the occupants in descending to the ground must be illuminated so that the erected assist means is visible from the airplane.

(1) If the assist means is illuminated by exterior emergency lighting, it must provide illumination of not less than 0.03 foot-candle (measured normal to the direction of the incident light) at the ground end of the erected assist means where an evacuee using the established escape route would normally make first contact with the ground, with the airplane in each of the attitudes corresponding to the collapse of one or more legs of the landing gear.

(2) If the emergency lighting subsystem illuminating the assist means

serves no other assist means, is independent of the airplane's main emergency lighting system, and is automatically activated when the assist means is erected, the lighting provisions—

(i) May not be adversely affected by stowage; and

(ii) Must provide illumination of not less than 0.03 foot-candle (measured normal to the direction of incident light) at the ground and of the erected assist means where an evacuee would normally make first contact with the ground, with the airplane in each of the attitudes corresponding to the collapse of one or more legs of the landing gear.

(i) The energy supply to each emergency lighting unit must provide the required level of illumination for at least 10 minutes at the critical ambient conditions after emergency landing.

(j) If storage batteries are used as the energy supply for the emergency lighting system, they may be recharged from the airplane's main electric power system: *Provided,* That, the charging circuit is designed to preclude inadvertent battery discharge into charging circuit faults.

(k) Components of the emergency lighting system, including batteries, wiring relays, lamps, and switches must be capable of normal operation after having been subjected to the inertia forces listed in §25.561(b).

(l) The emergency lighting system must be designed so that after any single transverse vertical separation of the fuselage during crash landing—

(1) Not more than 25 percent of all electrically illuminated emergency lights required by this section are rendered inoperative, in addition to the lights that are directly damaged by the separation;

(2) Each electrically illuminated exit sign required under §25.811(d)(2) remains operative exclusive of those that are directly damaged by the separation; and

(3) At least one required exterior emergency light for each side of the airplane remains operative exclusive of those that are directly damaged by the separation.

[Amdt. 25–15, 32 FR 13265, Sept. 20, 1967, as amended by Amdt. 25–28, 36 FR 16899, Aug. 26, 1971; Amdt. 25–32, 37 FR 3971, Feb. 24, 1972; Amdt. 25–46, 43 FR 50597, Oct. 30, 1978; Amdt. 25–58, 49 FR 43186, Oct. 26, 1984; Amdt. 25–88, 61 FR 57958, Nov. 8, 1996; Amdt. 25–116, 69 FR 62788, Oct. 27, 2004; Amdt. 25–128, 74 FR 25645, May 29, 2009]

§25.813 Emergency exit access.

Each required emergency exit must be accessible to the passengers and located where it will afford an effective means of evacuation. Emergency exit distribution must be as uniform as practical, taking passenger distribution into account; however, the size and location of exits on both sides of the cabin need not be symmetrical. If only one floor level exit per side is prescribed, and the airplane does not have a tailcone or ventral emergency exit, the floor level exit must be in the rearward part of the passenger compartment, unless another location affords a more effective means of passenger evacuation. Where more than one floor level exit per side is prescribed, at least one floor level exit per side must be located near each end of the cabin, except that this provision does not apply to combination cargo/passenger configurations. In addition—

(a) There must be a passageway leading from the nearest main aisle to each Type A, Type B, Type C, Type I, or Type II emergency exit and between individual passenger areas. Each passageway leading to a Type A or Type B exit must be unobstructed and at least 36 inches wide. Passageways between individual passenger areas and those leading to Type I, Type II, or Type C emergency exits must be unobstructed and at least 20 inches wide. Unless there are two or more main aisles, each Type A or B exit must be located so that there is passenger flow along the main aisle to that exit from both the forward and aft directions. If two or more main aisles are provided, there must be unobstructed cross-aisles at least 20 inches wide between main aisles. There must be—

(1) A cross-aisle which leads directly to each passageway between the nearest main aisle and a Type A or B exit; and

(2) A cross-aisle which leads to the immediate vicinity of each passageway between the nearest main aisle and a Type 1, Type II, or Type III exit; except that when two Type III exits are located within three passenger rows of each other, a single cross-aisle may be used if it leads to the vicinity between the passageways from the nearest main aisle to each exit.

(b) Adequate space to allow crewmember(s) to assist in the evacuation of passengers must be provided as follows:

(1) Each assist space must be a rectangle on the floor, of sufficient size to enable a crewmember, standing erect, to effectively assist evacuees. The assist space must not reduce the unobstructed width of the passageway below that required for the exit.

(2) For each Type A or B exit, assist space must be provided at each side of the exit regardless of whether an assist means is required by § 25.810(a).

(3) For each Type C, I or II exit installed in an airplane with seating for more than 80 passengers, an assist space must be provided at one side of the passageway regardless of whether an assist means is required by § 25.810(a).

(4) For each Type C, I or II exit, an assist space must be provided at one side of the passageway if an assist means is required by § 25.810(a).

(5) For any tailcone exit that qualifies for 25 additional passenger seats under the provisions of § 25.807(g)(9)(ii), an assist space must be provided, if an assist means is required by § 25.810(a).

(6) There must be a handle, or handles, at each assist space, located to enable the crewmember to steady himself or herself:

(i) While manually activating the assist means (where applicable) and,

(ii) While assisting passengers during an evacuation.

(c) The following must be provided for each Type III or Type IV exit—(1) There must be access from the nearest aisle to each exit. In addition, for each Type III exit in an airplane that has a passenger seating configuration of 60 or more—

(i) Except as provided in paragraph (c)(1)(ii), the access must be provided by an unobstructed passageway that is at least 10 inches in width for interior arrangements in which the adjacent seat rows on the exit side of the aisle contain no more than two seats, or 20 inches in width for interior arrangements in which those rows contain three seats. The width of the passageway must be measured with adjacent seats adjusted to their most adverse position. The centerline of the required passageway width must not be displaced more than 5 inches horizontally from that of the exit.

(ii) In lieu of one 10- or 20-inch passageway, there may be two passageways, between seat rows only, that must be at least 6 inches in width and lead to an unobstructed space adjacent to each exit. (Adjacent exits must not share a common passageway.) The width of the passageways must be measured with adjacent seats adjusted to their most adverse position. The unobstructed space adjacent to the exit must extend vertically from the floor to the ceiling (or bottom of sidewall stowage bins), inboard from the exit for a distance not less than the width of the narrowest passenger seat installed on the airplane, and from the forward edge of the forward passageway to the aft edge of the aft passageway. The exit opening must be totally within the fore and aft bounds of the unobstructed space.

(2) In addition to the access—

(i) For airplanes that have a passenger seating configuration of 20 or more, the projected opening of the exit provided must not be obstructed and there must be no interference in opening the exit by seats, berths, or other protrusions (including any seatback in the most adverse position) for a distance from that exit not less than the width of the narrowest passenger seat installed on the airplane.

(ii) For airplanes that have a passenger seating configuration of 19 or fewer, there may be minor obstructions in this region, if there are compensating factors to maintain the effectiveness of the exit.

(3) For each Type III exit, regardless of the passenger capacity of the airplane in which it is installed, there must be placards that—

(i) Are readable by all persons seated adjacent to and facing a passageway to the exit;

(ii) Accurately state or illustrate the proper method of opening the exit, including the use of handholds; and

(iii) If the exit is a removable hatch, state the weight of the hatch and indicate an appropriate location to place the hatch after removal.

(d) If it is necessary to pass through a passageway between passenger compartments to reach any required emergency exit from any seat in the passenger cabin, the passageway must be unobstructed. However, curtains may be used if they allow free entry through the passageway.

(e) No door may be installed between any passenger seat that is occupiable for takeoff and landing and any passenger emergency exit, such that the door crosses any egress path (including aisles, crossaisles and passageways).

(f) If it is necessary to pass through a doorway separating any crewmember seat (except those seats on the flightdeck), occupiable for takeoff and landing, from any emergency exit, the door must have a means to latch it in the open position. The latching means must be able to withstand the loads imposed upon it when the door is subjected to the ultimate inertia forces, relative to the surrounding structure, listed in § 25.561(b).

[Amdt. 25–1, 30 FR 3204, Mar. 9, 1965, as amended by Amdt. 25–15, 32 FR 13265, Sept. 20, 1967; Amdt. 25–32, 37 FR 3971, Feb. 24, 1972; Amdt. 25–46, 43 FR 50597, Oct. 30, 1978; Amdt. 25–72, 55 FR 29783, July 20, 1990; Amdt. 25–76, 57 FR 19244, May 4, 1992; Amdt. 25–76, 57 FR 29120, June 30, 1992; Amdt. 25–88, 61 FR 57958, Nov. 8, 1996; Amdt. 25–116, 69 FR 62788, Oct. 27, 2004; Amdt. 25–128, 74 FR 25645, May 29, 2009]

§ 25.815 Width of aisle.

The passenger aisle width at any point between seats must equal or exceed the values in the following table:

Passenger seating capacity	Minimum passenger aisle width (inches)	
	Less than 25 in. from floor	25 in. and more from floor
10 or less	[1] 12	15
11 through 19	12	20
20 or more	15	20

[1] A narrower width not less than 9 inches may be approved when substantiated by tests found necessary by the Administrator.

[Amdt. 25–15, 32 FR 13265, Sept. 20, 1967, as amended by Amdt. 25–38, 41 FR 55466, Dec. 20, 1976]

§ 25.817 Maximum number of seats abreast.

On airplanes having only one passenger aisle, no more than three seats abreast may be placed on each side of the aisle in any one row.

[Amdt. 25–15, 32 FR 13265, Sept. 20, 1967]

§ 25.819 Lower deck service compartments (including galleys).

For airplanes with a service compartment located below the main deck, which may be occupied during taxi or flight but not during takeoff or landing, the following apply:

(a) There must be at least two emergency evacuation routes, one at each end of each lower deck service compartment or two having sufficient separation within each compartment, which could be used by each occupant of the lower deck service compartment to rapidly evacuate to the main deck under normal and emergency lighting conditions. The routes must provide for the evacuation of incapacitated persons, with assistance. The use of the evacuation routes may not be dependent on any powered device. The routes must be designed to minimize the possibility of blockage which might result from fire, mechanical or structural failure, or persons standing on top of or against the escape routes. In the event the airplane's main power system or compartment main lighting system should fail, emergency illumination for each lower deck service compartment must be automatically provided.

(b) There must be a means for two-way voice communication between the flight deck and each lower deck service compartment, which remains available following loss of normal electrical power generating system.

(c) There must be an aural emergency alarm system, audible during normal and emergency conditions, to enable crewmembers on the flight deck and at each required floor level emergency exit to alert occupants of each lower deck service compartment of an emergency situation.

(d) There must be a means, readily detectable by occupants of each lower deck service compartment, that indicates when seat belts should be fastened.

(e) If a public address system is installed in the airplane, speakers must be provided in each lower deck service compartment.

(f) For each occupant permitted in a lower deck service compartment, there must be a forward or aft facing seat which meets the requirements of § 25.785(d), and must be able to withstand maximum flight loads when occupied.

(g) For each powered lift system installed between a lower deck service compartment and the main deck for the carriage of persons or equipment, or both, the system must meet the following requirements:

(1) Each lift control switch outside the lift, except emergency stop buttons, must be designed to prevent the activation of the life if the lift door, or the hatch required by paragraph (g)(3) of this section, or both are open.

(2) An emergency stop button, that when activated will immediately stop the lift, must be installed within the lift and at each entrance to the lift.

(3) There must be a hatch capable of being used for evacuating persons from the lift that is openable from inside and outside the lift without tools, with the lift in any position.

[Amdt. 25–53, 45 FR 41593, June 19, 1980; 45 FR 43154, June 26, 1980; Amdt. 25–110; 68 FR 36883, June 19, 2003]

§ 25.820   Lavatory doors.

All lavatory doors must be designed to preclude anyone from becoming trapped inside the lavatory. If a locking mechanism is installed, it must be capable of being unlocked from the outside without the aid of special tools.

[Doc. No. 2003–14193, 69 FR 24502, May 3, 2004]

VENTILATION AND HEATING

§ 25.831   Ventilation.

(a) Under normal operating conditions and in the event of any probable failure conditions of any system which would adversely affect the ventilating air, the ventilation system must be designed to provide a sufficient amount of uncontaminated air to enable the crewmembers to perform their duties without undue discomfort or fatigue and to provide reasonable passenger comfort. For normal operating conditions, the ventilation system must be designed to provide each occupant with an airflow containing at least 0.55 pounds of fresh air per minute.

(b) Crew and passenger compartment air must be free from harmful or hazardous concentrations of gases or vapors. In meeting this requirement, the following apply:

(1) Carbon monoxide concentrations in excess of 1 part in 20,000 parts of air are considered hazardous. For test purposes, any acceptable carbon monoxide detection method may be used.

(2) Carbon dioxide concentration during flight must be shown not to exceed 0.5 percent by volume (sea level equivalent) in compartments normally occupied by passengers or crewmembers.

(c) There must be provisions made to ensure that the conditions prescribed in paragraph (b) of this section are met after reasonably probable failures or malfunctioning of the ventilating, heating, pressurization, or other systems and equipment.

(d) If accumulation of hazardous quantities of smoke in the cockpit area is reasonably probable, smoke evacuation must be readily accomplished, starting with full pressurization and without depressurizing beyond safe limits.

(e) Except as provided in paragraph (f) of this section, means must be provided to enable the occupants of the following compartments and areas to control the temperature and quantity of ventilating air supplied to their compartment or area independently of the temperature and quantity of air supplied to other compartments and areas:

(1) The flight crew compartment.

(2) Crewmember compartments and areas other than the flight crew compartment unless the crewmember compartment or area is ventilated by air interchange with other compartments or areas under all operating conditions.

(f) Means to enable the flight crew to control the temperature and quantity of ventilating air supplied to the flight crew compartment independently of the temperature and quantity of ventilating air supplied to other compartments are not required if all of the following conditions are met:

(1) The total volume of the flight crew and passenger compartments is 800 cubic feet or less.

(2) The air inlets and passages for air to flow between flight crew and passenger compartments are arranged to provide compartment temperatures within 5 degrees F. of each other and adequate ventilation to occupants in both compartments.

(3) The temperature and ventilation controls are accessible to the flight crew.

(g) The exposure time at any given temperature must not exceed the values shown in the following graph after any improbable failure condition.

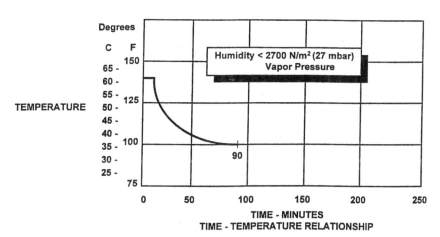

TIME - TEMPERATURE RELATIONSHIP

[Doc. No. 5066, 29 FR 18291, Dec. 24, 1964, as amended by Amdt. 25–41, 42 FR 36970, July 18, 1977; Amdt. 25–87, 61 FR 28695, June 5, 1996; Amdt. 25–89, 61 FR 63956, Dec. 2, 1996]

§ 25.832 Cabin ozone concentration.

(a) The airplane cabin ozone concentration during flight must be shown not to exceed—

(1) 0.25 parts per million by volume, sea level equivalent, at any time above flight level 320; and

(2) 0.1 parts per million by volume, sea level equivalent, time-weighted average during any 3-hour interval above flight level 270.

(b) For the purpose of this section, "sea level equivalent" refers to conditions of 25 °C and 760 millimeters of mercury pressure.

(c) Compliance with this section must be shown by analysis or tests based on airplane operational procedures and performance limitations, that demonstrate that either—

(1) The airplane cannot be operated at an altitude which would result in cabin ozone concentrations exceeding the limits prescribed by paragraph (a) of this section; or

(2) The airplane ventilation system, including any ozone control equipment,

will maintain cabin ozone concentrations at or below the limits prescribed by paragraph (a) of this section.

[Amdt. 25–50, 45 FR 3883, Jan. 1, 1980, as amended by Amdt. 25–56, 47 FR 58489, Dec. 30, 1982; Amdt. 25–94, 63 FR 8848, Feb. 23, 1998]

§ 25.833 Combustion heating systems.

Combustion heaters must be approved.

[Amdt. 25–72, 55 FR 29783, July 20, 1990]

PRESSURIZATION

§ 25.841 Pressurized cabins.

(a) Except as provided in paragraph (c) of this section, pressurized cabins and compartments to be occupied must be equipped to provide a cabin pressure altitude of not more than 8,000 feet under normal operating conditions.

(1) If certification for operation above 25,000 feet is requested, the airplane must be designed so that occupants will not be exposed to cabin pressure altitudes in excess of 15,000 feet after any probable failure condition in the pressurization system except as provided in paragraph (c) of this section.

(2) The airplane must be designed so that occupants will not be exposed to a cabin pressure altitude that exceeds the following after decompression from any failure condition not shown to be extremely improbable:

(i) Twenty-five thousand (25,000) feet for more than 2 minutes; or

(ii) Forty thousand (40,000) feet for any duration.

(3) Fuselage structure, engine and system failures are to be considered in evaluating the cabin decompression.

(b) Pressurized cabins must have at least the following valves, controls, and indicators for controlling cabin pressure:

(1) Two pressure relief valves to automatically limit the positive pressure differential to a predetermined value at the maximum rate of flow delivered by the pressure source. The combined capacity of the relief valves must be large enough so that the failure of any one valve would not cause an appreciable rise in the pressure differential. The pressure differential is positive when the internal pressure is greater than the external.

(2) Two reverse pressure differential relief valves (or their equivalents) to automatically prevent a negative pressure differential that would damage the structure. One valve is enough, however, if it is of a design that reasonably precludes its malfunctioning.

(3) A means by which the pressure differential can be rapidly equalized.

(4) An automatic or manual regulator for controlling the intake or exhaust airflow, or both, for maintaining the required internal pressures and airflow rates.

(5) Instruments at the pilot or flight engineer station to show the pressure differential, the cabin pressure altitude, and the rate of change of the cabin pressure altitude.

(6) Warning indication to the flightcrew when the safe or preset pressure differential or cabin pressure altitude limit is exceeded. Appropriate warning markings on the cabin pressure differential indicator meet the warning requirement for pressure differential limits. An alert meets the warning requirement for cabin pressure altitude limits if it warns the flightcrew when the cabin pressure altitude exceeds 10,000 feet, except as provided in paragraph (d) of this section.

(7) A warning placard at the pilot or flight engineer station if the structure is not designed for pressure differentials up to the maximum relief valve setting in combination with landing loads.

(8) The pressure sensors necessary to meet the requirements of paragraphs (b)(5) and (b)(6) of this section and § 25.1447(c), must be located and the sensing system designed so that, in the event of loss of cabin pressure in any passenger or crew compartment (including upper and lower lobe galleys), the warning and automatic presentation devices, required by those provisions, will be actuated without any delay that would significantly increase the hazards resulting from decompression.

(c) When operating into or out of airports with elevations at or above 8,000 feet, the cabin pressure altitude in pressurized cabins and occupied compartments may be up to, or greater than, the airport elevation by 2,000 feet, provided—

(1) In the event of probable failure conditions of the cabin pressurization system, the cabin pressure altitude must not exceed 15,000 feet, or 2,000 feet above the airport elevation, whichever is higher; and

(2) The cabin pressurization system is designed to minimize the time in flight that occupants may be exposed to cabin pressure altitudes exceeding 8,000 feet.

(d) When operating into or out of airports with elevations at or above 8,000 feet, the cabin pressure high altitude warning alert may be provided at up to 15,000 feet, or 2,000 feet above the airplane's maximum takeoff and landing altitude, whichever is greater, provided:

(1) During landing, the change in cabin pressure high altitude warning alert may not occur before the start of descent into the high elevation airport and, following takeoff, the cabin pressure high altitude warning alert must be reset to 10,000 feet before beginning cruise operation;

(2) Indication is provided to the flightcrew that the cabin pressure high altitude warning alert has shifted above 10,000 feet cabin pressure altitude; and

(3) Either an alerting system is installed that notifies the flightcrew members on flight deck duty when to don oxygen in accordance with the applicable operating regulations, or a limitation is provided in the airplane flight manual that requires the pilot flying the airplane to don oxygen when the cabin pressure altitude warning has shifted above 10,000 feet, and requires other flightcrew members on flight deck duty to monitor the cabin pressure and utilize oxygen in accordance with the applicable operating regulations.

[Doc. No. 5066, 29 FR 18291, Dec. 24, 1964, as amended by Amdt. 25-38, 41 FR 55466, Dec. 20, 1976; Amdt. 25-87, 61 FR 28696, June 5, 1996; Amdt. No. 25-151, 88 FR 39160, June 15, 2023; 88 FR 44032, July 11, 2023]

§25.843 Tests for pressurized cabins.

(a) *Strength test.* The complete pressurized cabin, including doors, windows, and valves, must be tested as a pressure vessel for the pressure differential specified in §25.365(d).

(b) *Functional tests.* The following functional tests must be performed:

(1) Tests of the functioning and capacity of the positive and negative pressure differential valves, and of the emergency release valve, to stimulate the effects of closed regulator valves.

(2) Tests of the pressurization system to show proper functioning under each possible condition of pressure, temperature, and moisture, up to the maximum altitude for which certification is requested.

(3) Flight tests, to show the performance of the pressure supply, pressure and flow regulators, indicators, and warning signals, in steady and stepped climbs and descents at rates corresponding to the maximum attainable within the operating limitations of the airplane, up to the maximum altitude for which certification is requested.

(4) Tests of each door and emergency exit, to show that they operate properly after being subjected to the flight tests prescribed in paragraph (b)(3) of this section.

FIRE PROTECTION

§25.851 Fire extinguishers.

(a) *Hand fire extinguishers.* (1) The following minimum number of hand fire extinguishers must be conveniently located and evenly distributed in passenger compartments:

Passenger capacity	No. of extinguishers
7 through 30	1
31 through 60	2
61 through 200	3
201 through 300	4
301 through 400	5
401 through 500	6
501 through 600	7
601 through 700	8

(2) At least one hand fire extinguisher must be conveniently located in the pilot compartment.

(3) At least one readily accessible hand fire extinguisher must be available for use in each Class A or Class B cargo or baggage compartment and in each Class E or Class F cargo or baggage compartment that is accessible to crewmembers in flight.

(4) At least one hand fire extinguisher must be located in, or readily

accessible for use in, each galley located above or below the passenger compartment.

(5) Each hand fire extinguisher must be approved.

(6) At least one of the required fire extinguishers located in the passenger compartment of an airplane with a passenger capacity of at least 31 and not more than 60, and at least two of the fire extinguishers located in the passenger compartment of an airplane with a passenger capacity of 61 or more must contain Halon 1211 (bromochlorodifluoromethane CBrC$_1$F$_2$), or equivalent, as the extinguishing agent. The type of extinguishing agent used in any other extinguisher required by this section must be appropriate for the kinds of fires likely to occur where used.

(7) The quantity of extinguishing agent used in each extinguisher required by this section must be appropriate for the kinds of fires likely to occur where used.

(8) Each extinguisher intended for use in a personnel compartment must be designed to minimize the hazard of toxic gas concentration.

(b) Built-in fire extinguishers. If a built-in fire extinguisher is provided—

(1) Each built-in fire extinguishing system must be installed so that—

(i) No extinguishing agent likely to enter personnel compartments will be hazardous to the occupants; and

(ii) No discharge of the extinguisher can cause structural damage.

(2) The capacity of each required built-in fire extinguishing system must be adequate for any fire likely to occur in the compartment where used, considering the volume of the compartment and the ventilation rate. The capacity of each system is adequate if there is sufficient quantity of agent to extinguish the fire or suppress the fire anywhere baggage or cargo is placed within the cargo compartment for the duration required to land and evacuate the airplane.

[Amdt. 25–74, 56 FR 15456, Apr. 16, 1991, as amended by Doc. No. Docket FAA–2014–0001, Amdt. 25–142, 81 FR 7703, Feb. 16, 2016]

§ 25.853 Compartment interiors.

For each compartment occupied by the crew or passengers, the following apply:

(a) Materials (including finishes or decorative surfaces applied to the materials) must meet the applicable test criteria prescribed in part I of appendix F of this part, or other approved equivalent methods, regardless of the passenger capacity of the airplane.

(b) [Reserved]

(c) In addition to meeting the requirements of paragraph (a) of this section, seat cushions, except those on flight crewmember seats, must meet the test requirements of part II of appendix F of this part, or other equivalent methods, regardless of the passenger capacity of the airplane.

(d) Except as provided in paragraph (e) of this section, the following interior components of airplanes with passenger capacities of 20 or more must also meet the test requirements of parts IV and V of appendix F of this part, or other approved equivalent method, in addition to the flammability requirements prescribed in paragraph (a) of this section:

(1) Interior ceiling and wall panels, other than lighting lenses and windows;

(2) Partitions, other than transparent panels needed to enhance cabin safety;

(3) Galley structure, including exposed surfaces of stowed carts and standard containers and the cavity walls that are exposed when a full complement of such carts or containers is not carried; and

(4) Large cabinets and cabin stowage compartments, other than underseat stowage compartments for stowing small items such as magazines and maps.

(e) The interiors of compartments, such as pilot compartments, galleys, lavatories, crew rest quarters, cabinets and stowage compartments, need not meet the standards of paragraph (d) of this section, provided the interiors of such compartments are isolated from the main passenger cabin by doors or equivalent means that would normally be closed during an emergency landing condition.

(f) Smoking is not allowed in lavatories. If smoking is allowed in any area occupied by the crew or passengers, an adequate number of self-contained, removable ashtrays must be provided in designated smoking sections for all seated occupants.

(g) Regardless of whether smoking is allowed in any other part of the airplane, lavatories must have self-contained, removable ashtrays located conspicuously on or near the entry side of each lavatory door, except that one ashtray may serve more than one lavatory door if the ashtray can be seen readily from the cabin side of each lavatory served.

(h) Each receptacle used for the disposal of flammable waste material must be fully enclosed, constructed of at least fire resistant materials, and must contain fires likely to occur in it under normal use. The capability of the receptacle to contain those fires under all probable conditions of wear, misalignment, and ventilation expected in service must be demonstrated by test.

[Amdt. 25–83, 60 FR 6623, Feb. 2, 1995, as amended by Amdt. 25–116, 69 FR 62788, Oct. 27, 2004]

§ 25.854   Lavatory fire protection.

For airplanes with a passenger capacity of 20 or more:

(a) Each lavatory must be equipped with a smoke detector system or equivalent that provides a warning light in the cockpit, or provides a warning light or audible warning in the passenger cabin that would be readily detected by a flight attendant; and

(b) Each lavatory must be equipped with a built-in fire extinguisher for each disposal receptacle for towels, paper, or waste, located within the lavatory. The extinguisher must be designed to discharge automatically into each disposal receptacle upon occurrence of a fire in that receptacle.

[Amdt. 25–74, 56 FR 15456, Apr. 16, 1991]

§ 25.855   Cargo or baggage compartments.

For each cargo or baggage compartment, the following apply:

(a) The compartment must meet one of the class requirements of § 25.857.

(b) Each of the following cargo or baggage compartments, as defined in § 25.857, must have a liner that is separate from, but may be attached to, the airplane structure:

(1) Any Class B through Class E cargo or baggage compartment, and

(2) Any Class F cargo or baggage compartment, unless other means of containing a fire and protecting critical systems and structure are provided.

(c) Ceiling and sidewall liner panels of Class C cargo or baggage compartments, and ceiling and sidewall liner panels in Class F cargo or baggage compartments, if installed to meet the requirements of paragraph (b)(2) of this section, must meet the test requirements of part III of appendix F of this part or other approved equivalent methods.

(d) All other materials used in the construction of the cargo or baggage compartment must meet the applicable test criteria prescribed in part I of appendix F of this part or other approved equivalent methods.

(e) No compartment may contain any controls, lines, equipment, or accessories whose damage or failure would affect safe operation, unless those items are protected so that—

(1) They cannot be damaged by the movement of cargo in the compartment, and

(2) Their breakage or failure will not create a fire hazard.

(f) There must be means to prevent cargo or baggage from interfering with the functioning of the fire protective features of the compartment.

(g) Sources of heat within the compartment must be shielded and insulated to prevent igniting the cargo or baggage.

(h) Flight tests must be conducted to show compliance with the provisions of § 25.857 concerning—

(1) Compartment accessibility,

(2) The entries of hazardous quantities of smoke or extinguishing agent into compartments occupied by the crew or passengers, and

(3) The dissipation of the extinguishing agent in all Class C compartments and, if applicable, in any Class F compartments.

(i) During the above tests, it must be shown that no inadvertent operation of smoke or fire detectors in any compartment would occur as a result of fire contained in any other compartment, either during or after extinguishment, unless the extinguishing system floods each such compartment simultaneously.

(j) Cargo or baggage compartment electrical wiring interconnection system components must meet the requirements of § 25.1721.

[Amdt. 25–72, 55 FR 29784, July 20, 1990, as amended by Amdt. 25–93, 63 FR 8048, Feb. 17, 1998; Amdt. 25–116, 69 FR 62788, Oct. 27, 2004; Amdt. 25–123, 72 FR 63405, Nov. 8, 2007; Doc. No. Docket FAA–2014–0001, Amdt. 25–142, 81 FR 7704, Feb. 16, 2016]

§ 25.856 Thermal/Acoustic insulation materials.

(a) Thermal/acoustic insulation material installed in the fuselage must meet the flame propagation test requirements of part VI of Appendix F to this part, or other approved equivalent test requirements. This requirement does not apply to "small parts," as defined in part I of Appendix F of this part.

(b) For airplanes with a passenger capacity of 20 or greater, thermal/acoustic insulation materials (including the means of fastening the materials to the fuselage) installed in the lower half of the airplane fuselage must meet the flame penetration resistance test requirements of part VII of Appendix F to this part, or other approved equivalent test requirements. This requirement does not apply to thermal/acoustic insulation installations that the FAA finds would not contribute to fire penetration resistance.

[Amdt. 25–111, 68 FR 45059, July 31, 2003]

§ 25.857 Cargo compartment classification.

(a) *Class A;* A Class A cargo or baggage compartment is one in which—

(1) The presence of a fire would be easily discovered by a crewmember while at his station; and

(2) Each part of the compartment is easily accessible in flight.

(b) *Class B.* A Class B cargo or baggage compartment is one in which—

(1) There is sufficient access in flight to enable a crewmember, standing at any one access point and without stepping into the compartment, to extinguish a fire occurring in any part of the compartment using a hand fire extinguisher;

(2) When the access provisions are being used, no hazardous quantity of smoke, flames, or extinguishing agent, will enter any compartment occupied by the crew or passengers;

(3) There is a separate approved smoke detector or fire detector system to give warning at the pilot or flight engineer station.

(c) *Class C.* A Class C cargo or baggage compartment is one not meeting the requirements for either a Class A or B compartment but in which—

(1) There is a separate approved smoke detector or fire detector system to give warning at the pilot or flight engineer station;

(2) There is an approved built-in fire extinguishing or suppression system controllable from the cockpit.

(3) There are means to exclude hazardous quantities of smoke, flames, or extinguishing agent, from any compartment occupied by the crew or passengers;

(4) There are means to control ventilation and drafts within the compartment so that the extinguishing agent used can control any fire that may start within the compartment.

(d) [Reserved]

(e) *Class E.* A Class E cargo compartment is one on airplanes used only for the carriage of cargo and in which—

(1) [Reserved]

(2) There is a separate approved smoke or fire detector system to give warning at the pilot or flight engineer station;

(3) There are means to shut off the ventilating airflow to, or within, the compartment, and the controls for these means are accessible to the flight crew in the crew compartment;

(4) There are means to exclude hazardous quantities of smoke, flames, or noxious gases, from the flight crew compartment; and

(5) The required crew emergency exits are accessible under any cargo loading condition.

(f) Class F. A Class F cargo or baggage compartment must be located on the main deck and is one in which—

(1) There is a separate approved smoke detector or fire detector system

to give warning at the pilot or flight engineer station;

(2) There are means to extinguish or control a fire without requiring a crewmember to enter the compartment; and

(3) There are means to exclude hazardous quantities of smoke, flames, or extinguishing agent from any compartment occupied by the crew or passengers.

[Doc. No. 5066, 29 FR 18291, Dec. 24, 1964, as amended by Amdt. 25–32, 37 FR 3972, Feb. 24, 1972; Amdt. 25–60, 51 FR 18243, May 16, 1986; Amdt. 25–93, 63 FR 8048, Feb. 17, 1998; Doc. No. Docket FAA–2014–0001, Amdt. 25–142, 81 FR 7704, Feb. 16, 2016]

§ 25.858 Cargo or baggage compartment smoke or fire detection systems.

If certification with cargo or baggage compartment smoke or fire detection provisions is requested, the following must be met for each cargo or baggage compartment with those provisions:

(a) The detection system must provide a visual indication to the flight crew within one minute after the start of a fire.

(b) The system must be capable of detecting a fire at a temperature significantly below that at which the structural integrity of the airplane is substantially decreased.

(c) There must be means to allow the crew to check in flight, the functioning of each fire detector circuit.

(d) The effectiveness of the detection system must be shown for all approved operating configurations and conditions.

[Amdt. 25–54, 45 FR 60173, Sept. 11, 1980, as amended by Amdt. 25–93, 63 FR 8048, Feb. 17, 1998]

§ 25.859 Combustion heater fire protection.

(a) *Combustion heater fire zones.* The following combustion heater fire zones must be protected from fire in accordance with the applicable provisions of §§ 25.1181 through 25.1191 and §§ 25.1195 through 25.1203;

(1) The region surrounding the heater, if this region contains any flammable fluid system components (excluding the heater fuel system), that could—

(i) Be damaged by heater malfunctioning; or

(ii) Allow flammable fluids or vapors to reach the heater in case of leakage.

(2) The region surrounding the heater, if the heater fuel system has fittings that, if they leaked, would allow fuel or vapors to enter this region.

(3) The part of the ventilating air passage that surrounds the combustion chamber. However, no fire extinguishment is required in cabin ventilating air passages.

(b) *Ventilating air ducts.* Each ventilating air duct passing through any fire zone must be fireproof. In addition—

(1) Unless isolation is provided by fireproof valves or by equally effective means, the ventilating air duct downstream of each heater must be fireproof for a distance great enough to ensure that any fire originating in the heater can be contained in the duct; and

(2) Each part of any ventilating duct passing through any region having a flammable fluid system must be constructed or isolated from that system so that the malfunctioning of any component of that system cannot introduce flammable fluids or vapors into the ventilating airstream.

(c) *Combustion air ducts.* Each combustion air duct must be fireproof for a distance great enough to prevent damage from backfiring or reverse flame propagation. In addition—

(1) No combustion air duct may have a common opening with the ventilating airstream unless flames from backfires or reverse burning cannot enter the ventilating airstream under any operating condition, including reverse flow or malfunctioning of the heater or its associated components; and

(2) No combustion air duct may restrict the prompt relief of any backfire that, if so restricted, could cause heater failure.

(d) *Heater controls; general.* Provision must be made to prevent the hazardous accumulation of water or ice on or in any heater control component, control system tubing, or safety control.

(e) *Heater safety controls.* For each combustion heater there must be the following safety control means:

(1) *Means* independent of the components provided for the normal continuous control of air temperature, airflow, and fuel flow must be provided, for each heater, to automatically shut off the ignition and fuel supply to that heater at a point remote from that heater when any of the following occurs:

(i) The heat exchanger temperature exceeds safe limits.

(ii) The ventilating air temperature exceeds safe limits.

(iii) The combustion airflow becomes inadequate for safe operation.

(iv) The ventilating airflow becomes inadequate for safe operation.

(2) The means of complying with paragraph (e)(1) of this section for any individual heater must—

(i) Be independent of components serving any other heater whose heat output is essential for safe operation; and

(ii) Keep the heater off until restarted by the crew.

(3) There must be means to warn the crew when any heater whose heat output is essential for safe operation has been shut off by the automatic means prescribed in paragraph (e)(1) of this section.

(f) *Air intakes.* Each combustion and ventilating air intake must be located so that no flammable fluids or vapors can enter the heater system under any operating condition—

(1) During normal operation; or

(2) As a result of the malfunctioning of any other component.

(g) *Heater exhaust.* Heater exhaust systems must meet the provisions of §§ 25.1121 and 25.1123. In addition, there must be provisions in the design of the heater exhaust system to safely expel the products of combustion to prevent the occurrence of—

(1) Fuel leakage from the exhaust to surrounding compartments;

(2) Exhaust gas impingement on surrounding equipment or structure;

(3) Ignition of flammable fluids by the exhaust, if the exhaust is in a compartment containing flammable fluid lines; and

(4) Restriction by the exhaust of the prompt relief of backfires that, if so restricted, could cause heater failure.

(h) *Heater fuel systems.* Each heater fuel system must meet each powerplant fuel system requirement affecting safe heater operation. Each heater fuel system component within the ventilating airstream must be protected by shrouds so that no leakage from those components can enter the ventilating airstream.

(i) *Drains.* There must be means to safely drain fuel that might accumulate within the combustion chamber or the heat exchanger. In addition—

(1) Each part of any drain that operates at high temperatures must be protected in the same manner as heater exhausts; and

(2) Each drain must be protected from hazardous ice accumulation under any operating condition.

[Doc. No. 5066, 29 FR 18291, Dec. 24, 1964, as amended by Amdt. 25–11, 32 FR 6912, May 5, 1967; Amdt. 25–23, 35 FR 5676, Apr. 8, 1970]

§ 25.863 Flammable fluid fire protection.

(a) In each area where flammable fluids or vapors might escape by leakage of a fluid system, there must be means to minimize the probability of ignition of the fluids and vapors, and the resultant hazards if ignition does occur.

(b) Compliance with paragraph (a) of this section must be shown by analysis or tests, and the following factors must be considered:

(1) Possible sources and paths of fluid leakage, and means of detecting leakage.

(2) Flammability characteristics of fluids, including effects of any combustible or absorbing materials.

(3) Possible ignition sources, including electrical faults, overheating of equipment, and malfunctioning of protective devices.

(4) Means available for controlling or extinguishing a fire, such as stopping flow of fluids, shutting down equipment, fireproof containment, or use of extinguishing agents.

(5) Ability of airplane components that are critical to safety of flight to withstand fire and heat.

(c) If action by the flight crew is required to prevent or counteract a fluid fire (e.g., equipment shutdown or actuation of a fire extinguisher) quick acting means must be provided to alert the crew.

(d) Each area where flammable fluids or vapors might escape by leakage of a fluid system must be identified and defined.

[Amdt. 25–23, 35 FR 5676, Apr. 8, 1970, as amended by Amdt. 25–46, 43 FR 50597, Oct. 30, 1978]

§25.865 Fire protection of flight controls, engine mounts, and other flight structure.

Essential flight controls, engine mounts, and other flight structures located in designated fire zones or in adjacent areas which would be subjected to the effects of fire in the fire zone must be constructed of fireproof material or shielded so that they are capable of withstanding the effects of fire.

[Amdt. 25–23, 35 FR 5676, Apr. 8, 1970]

§25.867 Fire protection: other components.

(a) Surfaces to the rear of the nacelles, within one nacelle diameter of the nacelle centerline, must be at least fire-resistant.

(b) Paragraph (a) of this section does not apply to tail surfaces to the rear of the nacelles that could not be readily affected by heat, flames, or sparks coming from a designated fire zone or engine compartment of any nacelle.

[Amdt. 25–23, 35 FR 5676, Apr. 8, 1970]

§25.869 Fire protection: systems.

(a) Electrical system components:

(1) Components of the electrical system must meet the applicable fire and smoke protection requirements of §§25.831(c) and 25.863.

(2) Equipment that is located in designated fire zones and is used during emergency procedures must be at least fire resistant.

(3) EWIS components must meet the requirements of §25.1713.

(b) Each vacuum air system line and fitting on the discharge side of the pump that might contain flammable vapors or fluids must meet the requirements of §25.1183 if the line or fitting is in a designated fire zone. Other vacuum air systems components in designated fire zones must be at least fire resistant.

(c) Oxygen equipment and lines must—

(1) Not be located in any designated fire zone,

(2) Be protected from heat that may be generated in, or escape from, any designated fire zone, and

(3) Be installed so that escaping oxygen cannot cause ignition of grease, fluid, or vapor accumulations that are present in normal operation or as a result of failure or malfunction of any system.

[Amdt. 25–72, 55 FR 29784, July 20, 1990, as amended by Amdt. 25–113, 69 FR 12530, Mar. 16, 2004; Amdt. 25–123, 72 FR 63405, Nov. 8, 2007]

MISCELLANEOUS

§25.871 Leveling means.

There must be means for determining when the airplane is in a level position on the ground.

[Amdt. 25–23, 35 FR 5676, Apr. 8, 1970]

§25.875 Reinforcement near propellers.

(a) Each part of the airplane near the propeller tips must be strong and stiff enough to withstand the effects of the induced vibration and of ice thrown from the propeller.

(b) No window may be near the propeller tips unless it can withstand the most severe ice impact likely to occur.

§25.899 Electrical bonding and protection against static electricity.

(a) Electrical bonding and protection against static electricity must be designed to minimize accumulation of electrostatic charge that would cause—

(1) Human injury from electrical shock,

(2) Ignition of flammable vapors, or

(3) Interference with installed electrical/electronic equipment.

(b) Compliance with paragraph (a) of this section may be shown by—

(1) Bonding the components properly to the airframe; or

(2) Incorporating other acceptable means to dissipate the static charge so

as not to endanger the airplane, personnel, or operation of the installed electrical/electronic systems.

[Amdt. 25–123, 72 FR 63405, Nov. 8, 2007]

Subpart E—Powerplant

GENERAL

§ 25.901 Installation.

(a) For the purpose of this part, the airplane powerplant installation includes each component that—

(1) Is necessary for propulsion;

(2) Affects the control of the major propulsive units; or

(3) Affects the safety of the major propulsive units between normal inspections or overhauls.

(b) For each powerplant—

(1) The installation must comply with—

(i) The installation instructions provided under §§ 33.5 and 35.3 of this chapter; and

(ii) The applicable provisions of this subpart;

(2) The components of the installation must be constructed, arranged, and installed so as to ensure their continued safe operation between normal inspections or overhauls;

(3) The installation must be accessible for necessary inspections and maintenance; and

(4) The major components of the installation must be electrically bonded to the other parts of the airplane.

(c) For each powerplant and auxiliary power unit installation, it must be established that no single failure or malfunction or probable combination of failures will jeopardize the safe operation of the airplane except that the failure of structural elements need not be considered if the probability of such failure is extremely remote.

(d) Each auxiliary power unit installation must meet the applicable provisions of this subpart.

[Doc. No. 5066, 29 FR 18291, Dec. 24, 1964, as amended by Amdt. 25–23, 35 FR 5676, Apr. 8, 1970; Amdt. 25–40, 42 FR 15042, Mar. 17, 1977; Amdt. 25–46, 43 FR 50597, Oct. 30, 1978; Amdt. 25–126, 73 FR 63345, Oct. 24, 2008]

§ 25.903 Engines.

(a) *Engine type certificate.* (1) Each engine must have a type certificate and must meet the applicable requirements of part 34 of this chapter.

(2) Each turbine engine must comply with one of the following:

(i) Sections 33.76, 33.77 and 33.78 of this chapter in effect on December 13, 2000, or as subsequently amended; or

(ii) Sections 33.77 and 33.78 of this chapter in effect on April 30, 1998, or as subsequently amended before December 13, 2000; or

(iii) Comply with § 33.77 of this chapter in effect on October 31, 1974, or as subsequently amended prior to April 30, 1998, unless that engine's foreign object ingestion service history has resulted in an unsafe condition; or

(iv) Be shown to have a foreign object ingestion service history in similar installation locations which has not resulted in any unsafe condition.

NOTE: § 33.77 of this chapter in effect on October 31, 1974, was published in 14 CFR parts 1 to 59, Revised as of January 1, 1975. See 39 FR 35467, October 1, 1974.

(3) Each turbine engine must comply with one of the following paragraphs:

(i) Section 33.68 of this chapter in effect on January 5, 2015, or as subsequently amended; or

(ii) Section 33.68 of this chapter in effect on March 26, 1984, or as subsequently amended before January 5, 2015, unless that engine's ice accumulation service history has resulted in an unsafe condition; or

(iii) Section 33.68 of this chapter in effect on October 31, 1974, or as subsequently amended prior to February 23, 1984, unless that engine's ice accumulation service history has resulted in an unsafe condition; or

(iv) Be shown to have an ice accumulation service history in similar installation locations which has not resulted in any unsafe conditions.

(b) *Engine isolation.* The powerplants must be arranged and isolated from each other to allow operation, in at least one configuration, so that the failure or malfunction of any engine, or of any system that can affect the engine, will not—

(1) Prevent the continued safe operation of the remaining engines; or

(2) Require immediate action by any crewmember for continued safe operation.

(c) *Control of engine rotation.* There must be means for stopping the rotation of any engine individually in flight, except that, for turbine engine installations, the means for stopping the rotation of any engine need be provided only where continued rotation could jeopardize the safety of the airplane. Each component of the stopping system on the engine side of the firewall that might be exposed to fire must be at least fire-resistant. If hydraulic propeller feathering systems are used for this purpose, the feathering lines must be at least fire resistant under the operating conditions that may be expected to exist during feathering.

(d) *Turbine engine installations.* For turbine engine installations—

(1) Design precautions must be taken to minimize the hazards to the airplane in the event of an engine rotor failure or of a fire originating within the engine which burns through the engine case.

(2) The powerplant systems associated with engine control devices, systems, and instrumentation, must be designed to give reasonable assurance that those engine operating limitations that adversely affect turbine rotor structural integrity will not be exceeded in service.

(e) *Restart capability.* (1) Means to restart any engine in flight must be provided.

(2) An altitude and airspeed envelope must be established for in-flight engine restarting, and each engine must have a restart capability within that envelope.

(3) For turbine engine powered airplanes, if the minimum windmilling speed of the engines, following the inflight shutdown of all engines, is insufficient to provide the necessary electrical power for engine ignition, a power source independent of the engine-driven electrical power generating system must be provided to permit inflight engine ignition for restarting.

(f) *Auxiliary Power Unit.* Each auxiliary power unit must be approved or meet the requirements of the category for its intended use.

[Doc. No. 5066, 29 FR 18291, Dec. 24, 1964, as amended by Amdt. 25–23, 35 FR 5676, Apr. 8, 1970; Amdt. 25–40, 42 FR 15042, Mar. 17, 1977; Amdt. 25–57, 49 FR 6848, Feb. 23, 1984; Amdt. 25–72, 55 FR 29784, July 20, 1990; Amdt. 25–73, 55 FR 32861, Aug. 10, 1990; Amdt. 25–94, 63 FR 8848, Feb. 23, 1998; Amdt. 25–95, 63 FR 14798, Mar. 26, 1998; Amdt. 25–100, 65 FR 55854, Sept. 14, 2000; Amdt. 25–140, 79 FR 65525, Nov. 4, 2014; Amdt. No. 25–148, 87 FR 75710, Dec. 9, 2022; 88 FR 2813, Jan. 18, 2023]

§ 25.904 **Automatic takeoff thrust control system (ATTCS).**

Each applicant seeking approval for installation of an engine power control system that automatically resets the power or thrust on the operating engine(s) when any engine fails during the takeoff must comply with the requirements of appendix I of this part.

[Amdt. 25–62, 52 FR 43156, Nov. 9, 1987]

§ 25.905 **Propellers.**

(a) Each propeller must have a type certificate.

(b) Engine power and propeller shaft rotational speed may not exceed the limits for which the propeller is certificated.

(c) The propeller blade pitch control system must meet the requirements of §§ 35.21, 35.23, 35.42 and 35.43 of this chapter.

(d) Design precautions must be taken to minimize the hazards to the airplane in the event a propeller blade fails or is released by a hub failure. The hazards which must be considered include damage to structure and vital systems due to impact of a failed or released blade and the unbalance created by such failure or release.

[Doc. No. 5066, 29 FR 18291, Dec. 24, 1964, as amended by Amdt. 25–54, 45 FR 60173, Sept. 11, 1980; Amdt. 25–57, 49 FR 6848, Feb. 23, 1984; Amdt. 25–72, 55 FR 29784, July 20, 1990; Amdt. 25–126, 73 FR 63345, Oct. 24, 2008]

§ 25.907 **Propeller vibration and fatigue.**

This section does not apply to fixed-pitch wood propellers of conventional design.

(a) The applicant must determine the magnitude of the propeller vibration stresses or loads, including any stress peaks and resonant conditions, throughout the operational envelope of the airplane by either:

(1) Measurement of stresses or loads through direct testing or analysis based on direct testing of the propeller on the airplane and engine installation for which approval is sought; or

(2) Comparison of the propeller to similar propellers installed on similar airplane installations for which these measurements have been made.

(b) The applicant must demonstrate by tests, analysis based on tests, or previous experience on similar designs that the propeller does not experience harmful effects of flutter throughout the operational envelope of the airplane.

(c) The applicant must perform an evaluation of the propeller to show that failure due to fatigue will be avoided throughout the operational life of the propeller using the fatigue and structural data obtained in accordance with part 35 of this chapter and the vibration data obtained from compliance with paragraph (a) of this section. For the purpose of this paragraph, the propeller includes the hub, blades, blade retention component and any other propeller component whose failure due to fatigue could be catastrophic to the airplane. This evaluation must include:

(1) The intended loading spectra including all reasonably foreseeable propeller vibration and cyclic load patterns, identified emergency conditions, allowable overspeeds and overtorques, and the effects of temperatures and humidity expected in service.

(2) The effects of airplane and propeller operating and airworthiness limitations.

[Amdt. 25-126, 73 FR 63345, Oct. 24, 2008]

§ 25.925 Propeller clearance.

Unless smaller clearances are substantiated, propeller clearances with the airplane at maximum weight, with the most adverse center of gravity, and with the propeller in the most adverse pitch position, may not be less than the following:

(a) *Ground clearance.* There must be a clearance of at least seven inches (for each airplane with nose wheel landing gear) or nine inches (for each airplane with tail wheel landing gear) between

each propeller and the ground with the landing gear statically deflected and in the level takeoff, or taxiing attitude, whichever is most critical. In addition, there must be positive clearance between the propeller and the ground when in the level takeoff attitude with the critical tire(s) completely deflated and the corresponding landing gear strut bottomed.

(b) *Water clearance.* There must be a clearance of at least 18 inches between each propeller and the water, unless compliance with § 25.239(a) can be shown with a lesser clearance.

(c) *Structural clearance.* There must be—

(1) At least one inch radial clearance between the blade tips and the airplane structure, plus any additional radial clearance necessary to prevent harmful vibration;

(2) At least one-half inch longitudinal clearance between the propeller blades or cuffs and stationary parts of the airplane; and

(3) Positive clearance between other rotating parts of the propeller or spinner and stationary parts of the airplane.

[Doc. No. 5066, 29 FR 18291, Dec. 24, 1964, as amended by Amdt. 25-72, 55 FR 29784, July 20, 1990]

§ 25.929 Propeller deicing.

(a) If certification for flight in icing is sought there must be a means to prevent or remove hazardous ice accumulations that could form in the icing conditions defined in Appendix C of this part and in the portions of Appendix O of this part for which the airplane is approved for flight on propellers or on accessories where ice accumulation would jeopardize engine performance.

(b) If combustible fluid is used for propeller deicing, §§ 25.1181 through 25.1185 and 25.1189 apply.

[Doc. No. 5066, 29 FR 18291, Dec. 24, 1964, as amended by Amdt. 25-140, 79 FR 65525, Nov. 4, 2014]

§ 25.933 Reversing systems.

(a) For turbojet reversing systems—

(1) Each system intended for ground operation only must be designed so that during any reversal in flight the engine will produce no more than flight

idle thrust. In addition, it must be shown by analysis or test, or both, that—

(i) Each operable reverser can be restored to the forward thrust position; and

(ii) The airplane is capable of continued safe flight and landing under any possible position of the thrust reverser.

(2) Each system intended for inflight use must be designed so that no unsafe condition will result during normal operation of the system, or from any failure (or reasonably likely combination of failures) of the reversing system, under any anticipated condition of operation of the airplane including ground operation. Failure of structural elements need not be considered if the probability of this kind of failure is extremely remote.

(3) Each system must have means to prevent the engine from producing more than idle thrust when the reversing system malfunctions, except that it may produce any greater forward thrust that is shown to allow directional control to be maintained, with aerodynamic means alone, under the most critical reversing condition expected in operation.

(b) For propeller reversing systems—

(1) Each system intended for ground operation only must be designed so that no single failure (or reasonably likely combination of failures) or malfunction of the system will result in unwanted reverse thrust under any expected operating condition. Failure of structural elements need not be considered if this kind of failure is extremely remote.

(2) Compliance with this section may be shown by failure analysis or testing, or both, for propeller systems that allow propeller blades to move from the flight low-pitch position to a position that is substantially less than that at the normal flight low-pitch position. The analysis may include or be supported by the analysis made to show compliance with the requirements of §35.21 of this chapter for the propeller and associated installation components.

[Amdt. 25–72, 55 FR 29784, July 20, 1990]

§25.934 Turbojet engine thrust reverser system tests.

Thrust reversers installed on turbojet engines must meet the requirements of §33.97 of this chapter.

[Amdt. 25–23, 35 FR 5677, Apr. 8, 1970]

§25.937 Turbopropeller-drag limiting systems.

Turbopropeller power airplane propeller-drag limiting systems must be designed so that no single failure or malfunction of any of the systems during normal or emergency operation results in propeller drag in excess of that for which the airplane was designed under §25.367. Failure of structural elements of the drag limiting systems need not be considered if the probability of this kind of failure is extremely remote.

§25.939 Turbine engine operating characteristics.

(a) Turbine engine operating characteristics must be investigated in flight to determine that no adverse characteristics (such as stall, surge, or flameout) are present, to a hazardous degree, during normal and emergency operation within the range of operating limitations of the airplane and of the engine.

(b) [Reserved]

(c) The turbine engine air inlet system may not, as a result of air flow distortion during normal operation, cause vibration harmful to the engine.

[Amdt. 25–11, 32 FR 6912, May 5, 1967, as amended by Amdt. 25–40, 42 FR 15043, Mar. 17, 1977]

§25.941 Inlet, engine, and exhaust compatibility.

For airplanes using variable inlet or exhaust system geometry, or both—

(a) The system comprised of the inlet, engine (including thrust augmentation systems, if incorporated), and exhaust must be shown to function properly under all operating conditions for which approval is sought, including all engine rotating speeds and power settings, and engine inlet and exhaust configurations;

(b) The dynamic effects of the operation of these (including consideration of probable malfunctions) upon the aerodynamic control of the airplane may not result in any condition that would require exceptional skill, alertness, or strength on the part of the pilot to avoid exceeding an operational or structural limitation of the airplane; and

(c) In showing compliance with paragraph (b) of this section, the pilot strength required may not exceed the limits set forth in § 25.143(d), subject to the conditions set forth in paragraphs (e) and (f) of § 25.143.

[Amdt. 25–38, 41 FR 55467, Dec. 20, 1976, as amended by Amdt. 25–121, 72 FR 44669, Aug. 8, 2007]

§ 25.943 Negative acceleration.

No hazardous malfunction of an engine, an auxiliary power unit approved for use in flight, or any component or system associated with the powerplant or auxiliary power unit may occur when the airplane is operated at the negative accelerations within the flight envelopes prescribed in § 25.333. This must be shown for the greatest duration expected for the acceleration.

[Amdt. 25–40, 42 FR 15043, Mar. 17, 1977]

§ 25.945 Thrust or power augmentation system.

(a) *General.* Each fluid injection system must provide a flow of fluid at the rate and pressure established for proper engine functioning under each intended operating condition. If the fluid can freeze, fluid freezing may not damage the airplane or adversely affect airplane performance.

(b) *Fluid tanks.* Each augmentation system fluid tank must meet the following requirements:

(1) Each tank must be able to withstand without failure the vibration, inertia, fluid, and structural loads that it may be subject to in operation.

(2) The tanks as mounted in the airplane must be able to withstand without failure or leakage an internal pressure 1.5 times the maximum operating pressure.

(3) If a vent is provided, the venting must be effective under all normal flight conditions.

(4) [Reserved]

(5) Each tank must have an expansion space of not less than 2 percent of the tank capacity. It must be impossible to fill the expansion space inadvertently with the airplane in the normal ground attitude.

(c) Augmentation system drains must be designed and located in accordance with § 25.1455 if—

(1) The augmentation system fluid is subject to freezing; and

(2) The fluid may be drained in flight or during ground operation.

(d) The augmentation liquid tank capacity available for the use of each engine must be large enough to allow operation of the airplane under the approved procedures for the use of liquid-augmented power. The computation of liquid consumption must be based on the maximum approved rate appropriate for the desired engine output and must include the effect of temperature on engine performance as well as any other factors that might vary the amount of liquid required.

(e) This section does not apply to fuel injection systems.

[Amdt. 25–40, 42 FR 15043, Mar. 17, 1977, as amended by Amdt. 25–72, 55 FR 29785, July 20, 1990; Amdt. 25–115, 69 FR 40527, July 2, 2004]

FUEL SYSTEM

§ 25.951 General.

(a) Each fuel system must be constructed and arranged to ensure a flow of fuel at a rate and pressure established for proper engine and auxiliary power unit functioning under each likely operating condition, including any maneuver for which certification is requested and during which the engine or auxiliary power unit is permitted to be in operation.

(b) Each fuel system must be arranged so that any air which is introduced into the system will not result in—

(1) Power interruption for more than 20 seconds for reciprocating engines; or

(2) Flameout for turbine engines.

(c) Each fuel system for a turbine engine must be capable of sustained operation throughout its flow and pressure range with fuel initially saturated with water at 80 °F and having 0.75cc of free water per gallon added and cooled to the most critical condition for icing likely to be encountered in operation.

(d) Each fuel system for a turbine engine powered airplane must meet the applicable fuel venting requirements of part 34 of this chapter.

[Doc. No. 5066, 29 FR 18291, Dec. 24, 1964, as amended by Amdt. 25–23, 35 FR 5677, Apr. 8, 1970; Amdt. 25–36, 39 FR 35460, Oct. 1, 1974; Amdt. 25–38, 41 FR 55467, Dec. 20, 1976; Amdt. 25–73, 55 FR 32861, Aug. 10, 1990]

§ 25.952 Fuel system analysis and test.

(a) Proper fuel system functioning under all probable operating conditions must be shown by analysis and those tests found necessary by the Administrator. Tests, if required, must be made using the airplane fuel system or a test article that reproduces the operating characteristics of the portion of the fuel system to be tested.

(b) The likely failure of any heat exchanger using fuel as one of its fluids may not result in a hazardous condition.

[Amdt. 25–40, 42 FR 15043, Mar. 17, 1977]

§ 25.953 Fuel system independence.

Each fuel system must meet the requirements of § 25.903(b) by—

(a) Allowing the supply of fuel to each engine through a system independent of each part of the system supplying fuel to any other engine; or

(b) Any other acceptable method.

§ 25.954 Fuel system lightning protection.

(a) For purposes of this section—

(1) A critical lightning strike is a lightning strike that attaches to the airplane in a location that, when combined with the failure of any design feature or structure, could create an ignition source.

(2) A fuel system includes any component within either the fuel tank structure or the fuel tank systems, and any airplane structure or system components that penetrate, connect to, or are located within a fuel tank.

(b) The design and installation of a fuel system must prevent catastrophic fuel vapor ignition due to lightning and its effects, including:

(1) Direct lightning strikes to areas having a high probability of stroke attachment;

(2) Swept lightning strokes to areas where swept strokes are highly probable; and

(3) Lightning-induced or conducted electrical transients.

(c) To comply with paragraph (b) of this section, catastrophic fuel vapor ignition must be extremely improbable, taking into account flammability, critical lightning strikes, and failures within the fuel system.

(d) To protect design features that prevent catastrophic fuel vapor ignition caused by lightning, the type design must include critical design configuration control limitations (CDCCLs) identifying those features and providing information to protect them. To ensure the continued effectiveness of those design features, the type design must also include inspection and test procedures, intervals between repetitive inspections and tests, and mandatory replacement times for those design features used in demonstrating compliance to paragraph (b) of this section. The applicant must include the information required by this paragraph in the Airworthiness Limitations section of the Instructions for Continued Airworthiness required by § 25.1529.

[Doc. No. FAA–2014–1027, Amdt. 25–146, 83 FR 47556, Sept. 20, 2018]

§ 25.955 Fuel flow.

(a) Each fuel system must provide at least 100 percent of the fuel flow required under each intended operating condition and maneuver. Compliance must be shown as follows:

(1) Fuel must be delivered to each engine at a pressure within the limits specified in the engine type certificate.

(2) The quantity of fuel in the tank may not exceed the amount established as the unusable fuel supply for that tank under the requirements of § 25.959 plus that necessary to show compliance with this section.

(3) Each main pump must be used that is necessary for each operating condition and attitude for which compliance with this section is shown, and the appropriate emergency pump must be substituted for each main pump so used.

(4) If there is a fuel flowmeter, it must be blocked and the fuel must flow through the meter or its bypass.

(b) If an engine can be supplied with fuel from more than one tank, the fuel system must—

(1) For each reciprocating engine, supply the full fuel pressure to that engine in not more than 20 seconds after switching to any other fuel tank containing usable fuel when engine malfunctioning becomes apparent due to the depletion of the fuel supply in any tank from which the engine can be fed; and

(2) For each turbine engine, in addition to having appropriate manual switching capability, be designed to prevent interruption of fuel flow to that engine, without attention by the flight crew, when any tank supplying fuel to that engine is depleted of usable fuel during normal operation, and any other tank, that normally supplies fuel to that engine alone, contains usable fuel.

[Doc. No. 5066, 29 FR 18291, Dec. 24, 1964, as amended by Amdt. 25–11, 32 FR 6912, May 5, 1967]

§ 25.957 Flow between interconnected tanks.

If fuel can be pumped from one tank to another in flight, the fuel tank vents and the fuel transfer system must be designed so that no structural damage to the tanks can occur because of overfilling.

§ 25.959 Unusable fuel supply.

The unusable fuel quantity for each fuel tank and its fuel system components must be established at not less than the quantity at which the first evidence of engine malfunction occurs under the most adverse fuel feed condition for all intended operations and flight maneuvers involving fuel feeding from that tank. Fuel system component failures need not be considered.

[Amdt. 25–23, 35 FR 5677, Apr. 8, 1970, as amended by Amdt. 25–40, 42 FR 15043, Mar. 17, 1977]

§ 25.961 Fuel system hot weather operation.

(a) The fuel system must perform satisfactorily in hot weather operation. This must be shown by showing that the fuel system from the tank outlets to each engine is pressurized, under all intended operations, so as to prevent vapor formation, or must be shown by climbing from the altitude of the airport elected by the applicant to the maximum altitude established as an operating limitation under § 25.1527. If a climb test is elected, there may be no evidence of vapor lock or other malfunctioning during the climb test conducted under the following conditions:

(1) For reciprocating engine powered airplanes, the engines must operate at maximum continuous power, except that takeoff power must be used for the altitudes from 1,000 feet below the critical altitude through the critical altitude. The time interval during which takeoff power is used may not be less than the takeoff time limitation.

(2) For turbine engine powered airplanes, the engines must operate at takeoff power for the time interval selected for showing the takeoff flight path, and at maximum continuous power for the rest of the climb.

(3) The weight of the airplane must be the weight with full fuel tanks, minimum crew, and the ballast necessary to maintain the center of gravity within allowable limits.

(4) The climb airspeed may not exceed—

(i) For reciprocating engine powered airplanes, the maximum airspeed established for climbing from takeoff to the maximum operating altitude with the airplane in the following configuration:

(A) Landing gear retracted.

(B) Wing flaps in the most favorable position.

(C) Cowl flaps (or other means of controlling the engine cooling supply) in the position that provides adequate cooling in the hot-day condition.

(D) Engine operating within the maximum continuous power limitations.

(E) Maximum takeoff weight; and

(ii) For turbine engine powered airplanes, the maximum airspeed established for climbing from takeoff to the maximum operating altitude.

(5) The fuel temperature must be at least 110 °F.

(b) The test prescribed in paragraph (a) of this section may be performed in flight or on the ground under closely

316

simulated flight conditions. If a flight test is performed in weather cold enough to interfere with the proper conduct of the test, the fuel tank surfaces, fuel lines, and other fuel system parts subject to cold air must be insulated to simulate, insofar as practicable, flight in hot weather.

[Amdt. 25–11, 32 FR 6912, May 5, 1967, as amended by Amdt. 25–57, 49 FR 6848, Feb. 23, 1984]

§25.963 Fuel tanks: general.

(a) Each fuel tank must be able to withstand, without failure, the vibration, inertia, fluid, and structural loads that it may be subjected to in operation.

(b) Flexible fuel tank liners must be approved or must be shown to be suitable for the particular application.

(c) Integral fuel tanks must have facilities for interior inspection and repair.

(d) Fuel tanks must, so far as it is practicable, be designed, located, and installed so that no fuel is released in or near the fuselage, or near the engines, in quantities that would constitute a fire hazard in otherwise survivable emergency landing conditions, and—

(1) Fuel tanks must be able to resist rupture and retain fuel under ultimate hydrostatic design conditions in which the pressure P within the tank varies in accordance with the formula:

$$P = K\rho gL$$

Where—

P = fuel pressure at each point within the tank
ρ = typical fuel density
g = acceleration due to gravity
L = a reference distance between the point of pressure and the tank farthest boundary in the direction of loading
K = 4.5 for the forward loading condition for those parts of fuel tanks outside the fuselage pressure boundary
K = 9 for the forward loading condition for those parts of fuel tanks within the fuselage pressure boundary, or that form part of the fuselage pressure boundary
K = 1.5 for the aft loading condition
K = 3.0 for the inboard and outboard loading conditions for those parts of fuel tanks within the fuselage pressure boundary, or that form part of the fuselage pressure boundary

K = 1.5 for the inboard and outboard loading conditions for those parts of fuel tanks outside the fuselage pressure boundary
K = 6 for the downward loading condition
K = 3 for the upward loading condition

(2) For those parts of wing fuel tanks near the fuselage or near the engines, the greater of the fuel pressures resulting from paragraphs (d)(2)(i) or (d)(2)(ii) of this section must be used:

(i) The fuel pressures resulting from paragraph (d)(1) of this section, and

(ii) The lesser of the two following conditions:

(A) Fuel pressures resulting from the accelerations specified in §25.561(b)(3) considering the fuel tank full of fuel at maximum fuel density. Fuel pressures based on the 9.0g forward acceleration may be calculated using the fuel static head equal to the streamwise local chord of the tank. For inboard and outboard conditions, an acceleration of 1.5g may be used in lieu of 3.0g as specified in §25.561(b)(3).

(B) Fuel pressures resulting from the accelerations as specified in §25.561(b)(3) considering a fuel volume beyond 85 percent of the maximum permissible volume in each tank using the static head associated with the 85 percent fuel level. A typical density of the appropriate fuel may be used. For inboard and outboard conditions, an acceleration of 1.5g may be used in lieu of 3.0g as specified in §25.561(b)(3).

(3) Fuel tank internal barriers and baffles may be considered as solid boundaries if shown to be effective in limiting fuel flow.

(4) For each fuel tank and surrounding airframe structure, the effects of crushing and scraping actions with the ground must not cause the spillage of enough fuel, or generate temperatures that would constitute a fire hazard under the conditions specified in §25.721(b).

(5) Fuel tank installations must be such that the tanks will not rupture as a result of the landing gear or an engine pylon or engine mount tearing away as specified in §25.721(a) and (c).

(e) Fuel tank access covers must comply with the following criteria in order to avoid loss of hazardous quantities of fuel:

(1) All covers located in an area where experience or analysis indicates

a strike is likely must be shown by analysis or tests to minimize penetration and deformation by tire fragments, low energy engine debris, or other likely debris.

(2) All covers must be fire resistant as defined in part 1 of this chapter.

(f) For pressurized fuel tanks, a means with fail-safe features must be provided to prevent the buildup of an excessive pressure difference between the inside and the outside of the tank.

[Doc. No. 5066, 29 FR 18291, Dec. 24, 1964, as amended by Amdt. 25–40, 42 FR 15043, Mar. 17, 1977; Amdt. 25–69, 54 FR 40354, Sept. 29, 1989; Amdt. 25–139, 79 FR 59430, Oct. 2, 2014]

§ 25.965 Fuel tank tests.

(a) It must be shown by tests that the fuel tanks, as mounted in the airplane, can withstand, without failure or leakage, the more critical of the pressures resulting from the conditions specified in paragraphs (a)(1) and (2) of this section. In addition, it must be shown by either analysis or tests, that tank surfaces subjected to more critical pressures resulting from the condition of paragraphs (a)(3) and (4) of this section, are able to withstand the following pressures:

(1) An internal pressure of 3.5 psi.

(2) 125 percent of the maximum air pressure developed in the tank from ram effect.

(3) Fluid pressures developed during maximum limit accelerations, and deflections, of the airplane with a full tank.

(4) Fluid pressures developed during the most adverse combination of airplane roll and fuel load.

(b) Each metallic tank with large unsupported or unstiffened flat surfaces, whose failure or deformation could cause fuel leakage, must be able to withstand the following test, or its equivalent, without leakage or excessive deformation of the tank walls:

(1) Each complete tank assembly and its supports must be vibration tested while mounted to simulate the actual installation.

(2) Except as specified in paragraph (b)(4) of this section, the tank assembly must be vibrated for 25 hours at an amplitude of not less than 1/32 of an inch (unless another amplitude is substan-

tiated) while 2/3 filled with water or other suitable test fluid.

(3) The test frequency of vibration must be as follows:

(i) If no frequency of vibration resulting from any r.p.m. within the normal operating range of engine speeds is critical, the test frequency of vibration must be 2,000 cycles per minute.

(ii) If only one frequency of vibration resulting from any r.p.m. within the normal operating range of engine speeds is critical, that frequency of vibration must be the test frequency.

(iii) If more than one frequency of vibration resulting from any r.p.m. within the normal operating range of engine speeds is critical, the most critical of these frequencies must be the test frequency.

(4) Under paragraphs (b)(3)(ii) and (iii) of this section, the time of test must be adjusted to accomplish the same number of vibration cycles that would be accomplished in 25 hours at the frequency specified in paragraph (b)(3)(i) of this section.

(5) During the test, the tank assembly must be rocked at the rate of 16 to 20 complete cycles per minute, through an angle of 15° on both sides of the horizontal (30° total), about the most critical axis, for 25 hours. If motion about more than one axis is likely to be critical, the tank must be rocked about each critical axis for 12½ hours.

(c) Except where satisfactory operating experience with a similar tank in a similar installation is shown, nonmetallic tanks must withstand the test specified in paragraph (b)(5) of this section, with fuel at a temperature of 110 °F. During this test, a representative specimen of the tank must be installed in a supporting structure simulating the installation in the airplane.

(d) For pressurized fuel tanks, it must be shown by analysis or tests that the fuel tanks can withstand the maximum pressure likely to occur on the ground or in flight.

[Doc. No. 5066, 29 FR 18291, Dec. 24, 1964, as amended by Amdt. 25–11, 32 FR 6913, May 5, 1967; Amdt. 25–40, 42 FR 15043, Mar. 17, 1977]

§ 25.967 Fuel tank installations.

(a) Each fuel tank must be supported so that tank loads (resulting from the weight of the fuel in the tanks) are not concentrated on unsupported tank surfaces. In addition—

(1) There must be pads, if necessary, to prevent chafing between the tank and its supports;

(2) Padding must be nonabsorbent or treated to prevent the absorption of fluids;

(3) If a flexible tank liner is used, it must be supported so that it is not required to withstand fluid loads; and

(4) Each interior surface of the tank compartment must be smooth and free of projections that could cause wear of the liner unless—

(i) Provisions are made for protection of the liner at these points; or

(ii) The construction of the liner itself provides that protection.

(b) Spaces adjacent to tank surfaces must be ventilated to avoid fume accumulation due to minor leakage. If the tank is in a sealed compartment, ventilation may be limited to drain holes large enough to prevent excessive pressure resulting from altitude changes.

(c) The location of each tank must meet the requirements of § 25.1185(a).

(d) No engine nacelle skin immediately behind a major air outlet from the engine compartment may act as the wall of an integral tank.

(e) Each fuel tank must be isolated from personnel compartments by a fumeproof and fuelproof enclosure.

§ 25.969 Fuel tank expansion space.

Each fuel tank must have an expansion space of not less than 2 percent of the tank capacity. It must be impossible to fill the expansion space inadvertently with the airplane in the normal ground attitude. For pressure fueling systems, compliance with this section may be shown with the means provided to comply with § 25.979(b).

[Amdt. 25–11, 32 FR 6913, May 5, 1967]

§ 25.971 Fuel tank sump.

(a) Each fuel tank must have a sump with an effective capacity, in the normal ground attitude, of not less than the greater of 0.10 percent of the tank capacity or one-sixteenth of a gallon unless operating limitations are established to ensure that the accumulation of water in service will not exceed the sump capacity.

(b) Each fuel tank must allow drainage of any hazardous quantity of water from any part of the tank to its sump with the airplane in the ground attitude.

(c) Each fuel tank sump must have an accessible drain that—

(1) Allows complete drainage of the sump on the ground;

(2) Discharges clear of each part of the airplane; and

(3) Has manual or automatic means for positive locking in the closed position.

§ 25.973 Fuel tank filler connection.

Each fuel tank filler connection must prevent the entrance of fuel into any part of the airplane other than the tank itself. In addition—

(a) [Reserved]

(b) Each recessed filler connection that can retain any appreciable quantity of fuel must have a drain that discharges clear of each part of the airplane;

(c) Each filler cap must provide a fuel-tight seal; and

(d) Each fuel filling point must have a provision for electrically bonding the airplane to ground fueling equipment.

[Doc. No. 5066, 29 FR 18291, Dec. 24, 1964, as amended by Amdt. 25–40, 42 FR 15043, Mar. 17, 1977; Amdt. 25–72, 55 FR 29785, July 20, 1990; Amdt. 25–115, 69 FR 40527, July 2, 2004]

§ 25.975 Fuel tank vents and carburetor vapor vents.

(a) *Fuel tank vents.* Each fuel tank must be vented from the top part of the expansion space so that venting is effective under any normal flight condition. In addition—

(1) Each vent must be arranged to avoid stoppage by dirt or ice formation;

(2) The vent arrangement must prevent siphoning of fuel during normal operation;

(3) The venting capacity and vent pressure levels must maintain acceptable differences of pressure between the interior and exterior of the tank, during—

(i) Normal flight operation;

(ii) Maximum rate of ascent and descent; and

(iii) Refueling and defueling (where applicable);

(4) Airspaces of tanks with interconnected outlets must be interconnected;

(5) There may be no point in any vent line where moisture can accumulate with the airplane in the ground attitude or the level flight attitude, unless drainage is provided;

(6) No vent or drainage provision may end at any point—

(i) Where the discharge of fuel from the vent outlet would constitute a fire hazard; or

(ii) From which fumes could enter personnel compartments; and

(7) Each fuel tank vent system must prevent explosions, for a minimum of 2 minutes and 30 seconds, caused by propagation of flames from outside the tank through the fuel tank vents into fuel tank vapor spaces when any fuel tank vent is continuously exposed to flame.

(b) *Carburetor vapor vents.* Each carburetor with vapor elimination connections must have a vent line to lead vapors back to one of the fuel tanks. In addition—

(1) Each vent system must have means to avoid stoppage by ice; and

(2) If there is more than one fuel tank, and it is necessary to use the tanks in a definite sequence, each vapor vent return line must lead back to the fuel tank used for takeoff and landing.

[Doc. No. 5066, 29 FR 18291, Dec. 24, 1964, as amended by Docket No. FAA-2014-0500, Amdt. No. 25-143, 81 FR 41207, June 24, 2016]

§ 25.977 Fuel tank outlet.

(a) There must be a fuel strainer for the fuel tank outlet or for the booster pump. This strainer must—

(1) For reciprocating engine powered airplanes, have 8 to 16 meshes per inch; and

(2) For turbine engine powered airplanes, prevent the passage of any object that could restrict fuel flow or damage any fuel system component.

(b) [Reserved]

(c) The clear area of each fuel tank outlet strainer must be at least five times the area of the outlet line.

(d) The diameter of each strainer must be at least that of the fuel tank outlet.

(e) Each finger strainer must be accessible for inspection and cleaning.

[Amdt. 25-11, 32 FR 6913, May 5, 1967, as amended by Amdt. 25-36, 39 FR 35460, Oct. 1, 1974]

§ 25.979 Pressure fueling system.

For pressure fueling systems, the following apply:

(a) Each pressure fueling system fuel manifold connection must have means to prevent the escape of hazardous quantities of fuel from the system if the fuel entry valve fails.

(b) An automatic shutoff means must be provided to prevent the quantity of fuel in each tank from exceeding the maximum quantity approved for that tank. This means must—

(1) Allow checking for proper shutoff operation before each fueling of the tank; and

(2) Provide indication at each fueling station of failure of the shutoff means to stop the fuel flow at the maximum quantity approved for that tank.

(c) A means must be provided to prevent damage to the fuel system in the event of failure of the automatic shutoff means prescribed in paragraph (b) of this section.

(d) The airplane pressure fueling system (not including fuel tanks and fuel tank vents) must withstand an ultimate load that is 2.0 times the load arising from the maximum pressures, including surge, that is likely to occur during fueling. The maximum surge pressure must be established with any combination of tank valves being either intentionally or inadvertently closed.

(e) The airplane defueling system (not including fuel tanks and fuel tank vents) must withstand an ultimate load that is 2.0 times the load arising from the maximum permissible defueling pressure (positive or negative) at the airplane fueling connection.

[Amdt. 25-11, 32 FR 6913, May 5, 1967, as amended by Amdt. 25-38, 41 FR 55467, Dec. 20, 1976; Amdt. 25-72, 55 FR 29785, July 20, 1990]

§ 25.981 Fuel tank explosion prevention.

(a) No ignition source may be present at each point in the fuel tank or fuel tank system where catastrophic failure could occur due to ignition of fuel or vapors. This must be shown by:

(1) Determining the highest temperature allowing a safe margin below the lowest expected autoignition temperature of the fuel in the fuel tanks.

(2) Demonstrating that no temperature at each place inside each fuel tank where fuel ignition is possible will exceed the temperature determined under paragraph (a)(1) of this section. This must be verified under all probable operating, failure, and malfunction conditions of each component whose operation, failure, or malfunction could increase the temperature inside the tank.

(3) Except for ignition sources due to lightning addressed by § 25.954, demonstrating that an ignition source could not result from each single failure, from each single failure in combination with each latent failure condition not shown to be extremely remote, and from all combinations of failures not shown to be extremely improbable, taking into account the effects of manufacturing variability, aging, wear, corrosion, and likely damage.

(b) Except as provided in paragraphs (b)(2) and (c) of this section, no fuel tank Fleet Average Flammability Exposure on an airplane may exceed three percent of the Flammability Exposure Evaluation Time (FEET) as defined in Appendix N of this part, or that of a fuel tank within the wing of the airplane model being evaluated, whichever is greater. If the wing is not a conventional unheated aluminum wing, the analysis must be based on an assumed Equivalent Conventional Unheated Aluminum Wing Tank.

(1) Fleet Average Flammability Exposure is determined in accordance with Appendix N of this part. The assessment must be done in accordance with the methods and procedures set forth in the Fuel Tank Flammability Assessment Method User's Manual, dated May 2008, document number DOT/FAA/AR–05/8 (incorporated by reference, see § 25.5).

(2) Any fuel tank other than a main fuel tank on an airplane must meet the flammability exposure criteria of Appendix M to this part if any portion of the tank is located within the fuselage contour.

(3) As used in this paragraph,

(i) *Equivalent Conventional Unheated Aluminum Wing Tank* is an integral tank in an unheated semi-monocoque aluminum wing of a subsonic airplane that is equivalent in aerodynamic performance, structural capability, fuel tank capacity and tank configuration to the designed wing.

(ii) *Fleet Average Flammability Exposure* is defined in Appendix N to this part and means the percentage of time each fuel tank ullage is flammable for a fleet of an airplane type operating over the range of flight lengths.

(iii) *Main Fuel Tank* means a fuel tank that feeds fuel directly into one or more engines and holds required fuel reserves continually throughout each flight.

(c) Paragraph (b) of this section does not apply to a fuel tank if means are provided to mitigate the effects of an ignition of fuel vapors within that fuel tank such that no damage caused by an ignition will prevent continued safe flight and landing.

(d) To protect design features that prevent catastrophic ignition sources within the fuel tank or fuel tank system according to paragraph (a) of this section, and to prevent increasing the flammability exposure of the tanks above that permitted in paragraph (b) of this section, the type design must include critical design configuration control limitations (CDCCLs) identifying those features and providing instructions on how to protect them. To ensure the continued effectiveness of those features, and prevent degradation of the performance and reliability of any means provided according to paragraphs (a), (b), or (c) of this section, the type design must also include necessary inspection and test procedures, intervals between repetitive inspections and tests, and mandatory replacement times for those features. The applicant must include information required by this paragraph in the Airworthiness Limitations section of the Instructions for Continued Airworthiness required by § 25.1529. The type design must also include visible means of identifying critical features of the design in areas of the airplane where foreseeable maintenance actions, repairs, or alterations may compromise the CDCCLs.

[Doc. No. 1999–6411, 66 FR 23129, May 7, 2001, as amended by Doc. No. FAA–2005–22997, 73 FR 42494, July 21, 2008; Doc. No. FAA– 2014–1027, Amdt. No. 25–146, 83 FR 47556, Sept. 20, 2018]

FUEL SYSTEM COMPONENTS

§ 25.991 Fuel pumps.

(a) *Main pumps.* Each fuel pump required for proper engine operation, or required to meet the fuel system requirements of this subpart (other than those in paragraph (b) of this section, is a main pump. For each main pump, provision must be made to allow the bypass of each positive displacement fuel pump other than a fuel injection pump (a pump that supplies the proper flow and pressure for fuel injection when the injection is not accomplished in a carburetor) approved as part of the engine.

(b) *Emergency pumps.* There must be emergency pumps or another main pump to feed each engine immediately after failure of any main pump (other than a fuel injection pump approved as part of the engine).

§ 25.993 Fuel system lines and fittings.

(a) Each fuel line must be installed and supported to prevent excessive vibration and to withstand loads due to fuel pressure and accelerated flight conditions.

(b) Each fuel line connected to components of the airplane between which relative motion could exist must have provisions for flexibility.

(c) Each flexible connection in fuel lines that may be under pressure and subjected to axial loading must use flexible hose assemblies.

(d) Flexible hose must be approved or must be shown to be suitable for the particular application.

(e) No flexible hose that might be adversely affected by exposure to high temperatures may be used where excessive temperatures will exist during operation or after engine shut-down.

(f) Each fuel line within the fuselage must be designed and installed to allow

a reasonable degree of deformation and stretching without leakage.

[Doc. No. 5066, 29 FR 18291, Dec. 24, 1964, as amended by Amdt. 25–15, 32 FR 13266, Sept. 20, 1967]

§ 25.994 Fuel system components.

Fuel system components in an engine nacelle or in the fuselage must be protected from damage that could result in spillage of enough fuel to constitute a fire hazard as a result of a wheels-up landing on a paved runway under each of the conditions prescribed in § 25.721(b).

[Amdt. 25–139, 79 FR 59430, Oct. 2, 2014]

§ 25.995 Fuel valves.

In addition to the requirements of § 25.1189 for shutoff means, each fuel valve must—

(a) [Reserved]

(b) Be supported so that no loads resulting from their operation or from accelerated flight conditions are transmitted to the lines attached to the valve.

[Doc. No. 5066, 29 FR 18291, Dec. 24, 1964, as amended by Amdt. 25–40, 42 FR 15043, Mar. 17, 1977]

§ 25.997 Fuel strainer or filter.

There must be a fuel strainer or filter between the fuel tank outlet and the inlet of either the fuel metering device or an engine driven positive displacement pump, whichever is nearer the fuel tank outlet. This fuel strainer or filter must—

(a) Be accessible for draining and cleaning and must incorporate a screen or element which is easily removable;

(b) Have a sediment trap and drain except that it need not have a drain if the strainer or filter is easily removable for drain purposes;

(c) Be mounted so that its weight is not supported by the connecting lines or by the inlet or outlet connections of the strainer or filter itself, unless adequate strength margins under all loading conditions are provided in the lines and connections; and

(d) Have the capacity (with respect to operating limitations established for the engine) to ensure that engine fuel system functioning is not impaired, with the fuel contaminated to a degree (with respect to particle size and density) that is greater than that established for the engine in Part 33 of this chapter.

[Amdt. 25–36, 39 FR 35460, Oct. 1, 1974, as amended by Amdt. 25–57, 49 FR 6848, Feb. 23, 1984]

§25.999 Fuel system drains.

(a) Drainage of the fuel system must be accomplished by the use of fuel strainer and fuel tank sump drains.

(b) Each drain required by paragraph (a) of this section must—

(1) Discharge clear of all parts of the airplane;

(2) Have manual or automatic means for positive locking in the closed position; and

(3) Have a drain valve—

(i) That is readily accessible and which can be easily opened and closed; and

(ii) That is either located or protected to prevent fuel spillage in the event of a landing with landing gear retracted.

[Doc. No. 5066, 29 FR 18291, Dec. 24, 1964, as amended by Amdt. 25–38, 41 FR 55467, Dec. 20, 1976]

§25.1001 Fuel jettisoning system.

(a) A fuel jettisoning system must be installed on each airplane unless it is shown that the airplane meets the climb requirements of §§25.119 and 25.121(d) at maximum takeoff weight, less the actual or computed weight of fuel necessary for a 15-minute flight comprised of a takeoff, go-around, and landing at the airport of departure with the airplane configuration, speed, power, and thrust the same as that used in meeting the applicable takeoff, approach, and landing climb performance requirements of this part.

(b) If a fuel jettisoning system is required it must be capable of jettisoning enough fuel within 15 minutes, starting with the weight given in paragraph (a) of this section, to enable the airplane to meet the climb requirements of §§25.119 and 25.121(d), assuming that the fuel is jettisoned under the conditions, except weight, found least favorable during the flight tests prescribed in paragraph (c) of this section.

(c) Fuel jettisoning must be demonstrated beginning at maximum take-off weight with flaps and landing gear up and in—

(1) A power-off glide at 1.3 V_{SR1};

(2) A climb at the one-engine inoperative best rate-of-climb speed, with the critical engine inoperative and the remaining engines at maximum continuous power; and

(3) Level flight at 1.3 V_{SR1}; if the results of the tests in the conditions specified in paragraphs (c)(1) and (2) of this section show that this condition could be critical.

(d) During the flight tests prescribed in paragraph (c) of this section, it must be shown that—

(1) The fuel jettisoning system and its operation are free from fire hazard;

(2) The fuel discharges clear of any part of the airplane;

(3) Fuel or fumes do not enter any parts of the airplane; and

(4) The jettisoning operation does not adversely affect the controllability of the airplane.

(e) For reciprocating engine powered airplanes, means must be provided to prevent jettisoning the fuel in the tanks used for takeoff and landing below the level allowing 45 minutes flight at 75 percent maximum continuous power. However, if there is an auxiliary control independent of the main jettisoning control, the system may be designed to jettison the remaining fuel by means of the auxiliary jettisoning control.

(f) For turbine engine powered airplanes, means must be provided to prevent jettisoning the fuel in the tanks used for takeoff and landing below the level allowing climb from sea level to 10,000 feet and thereafter allowing 45 minutes cruise at a speed for maximum range. However, if there is an auxiliary control independent of the main jettisoning control, the system may be designed to jettison the remaining fuel by means of the auxiliary jettisoning control.

(g) The fuel jettisoning valve must be designed to allow flight personnel to close the valve during any part of the jettisoning operation.

(h) Unless it is shown that using any means (including flaps, slots, and slats) for changing the airflow across or around the wings does not adversely affect fuel jettisoning, there must be a placard, adjacent to the jettisoning control, to warn flight crewmembers against jettisoning fuel while the means that change the airflow are being used.

(i) The fuel jettisoning system must be designed so that any reasonably probable single malfunction in the system will not result in a hazardous condition due to unsymmetrical jettisoning of, or inability to jettison, fuel.

[Doc. No. 5066, 29 FR 18291, Dec. 24, 1964, as amended by Amdt. 25–18, 33 FR 12226, Aug. 30, 1968; Amdt. 25–57, 49 FR 6848, Feb. 23, 1984; Amdt. 25–108, 67 FR 70827, Nov. 26, 2002]

OIL SYSTEM

§ 25.1011 General.

(a) Each engine must have an independent oil system that can supply it with an appropriate quantity of oil at a temperature not above that safe for continuous operation.

(b) The usable oil capacity may not be less than the product of the endurance of the airplane under critical operating conditions and the approved maximum allowable oil consumption of the engine under the same conditions, plus a suitable margin to ensure system circulation. Instead of a rational analysis of airplane range for the purpose of computing oil requirements for reciprocating engine powered airplanes, the following fuel/oil ratios may be used:

(1) For airplanes without a reserve oil or oil transfer system, a fuel/oil ratio of 30:1 by volume.

(2) For airplanes with either a reserve oil or oil transfer system, a fuel/oil ratio of 40:1 by volume.

(c) Fuel/oil ratios higher than those prescribed in paragraphs (b)(1) and (2) of this section may be used if substantiated by data on actual engine oil consumption.

§ 25.1013 Oil tanks.

(a) *Installation.* Each oil tank installation must meet the requirements of § 25.967.

(b) *Expansion space.* Oil tank expansion space must be provided as follows:

(1) Each oil tank used with a reciprocating engine must have an expansion space of not less than the greater of 10 percent of the tank capacity or 0.5 gal-

lon, and each oil tank used with a turbine engine must have an expansion space of not less than 10 percent of the tank capacity.

(2) Each reserve oil tank not directly connected to any engine may have an expansion space of not less than two percent of the tank capacity.

(3) It must be impossible to fill the expansion space inadvertently with the airplane in the normal ground attitude.

(c) *Filler connection.* Each recessed oil tank filler connection that can retain any appreciable quantity of oil must have a drain that discharges clear of each part of the airplane. In addition, each oil tank filler cap must provide an oil-tight seal.

(d) *Vent.* Oil tanks must be vented as follows:

(1) Each oil tank must be vented from the top part of the expansion space so that venting is effective under any normal flight condition.

(2) Oil tank vents must be arranged so that condensed water vapor that might freeze and obstruct the line cannot accumulate at any point.

(e) *Outlet.* There must be means to prevent entrance into the tank itself, or into the tank outlet, of any object that might obstruct the flow of oil through the system. No oil tank outlet may be enclosed by any screen or guard that would reduce the flow of oil below a safe value at any operating temperature. There must be a shutoff valve at the outlet of each oil tank used with a turbine engine, unless the external portion of the oil system (including the oil tank supports) is fireproof.

(f) *Flexible oil tank liners.* Each flexible oil tank liner must be approved or must be shown to be suitable for the particular application.

[Doc. No. 5066, 29 FR 18291, Dec. 24, 1964, as amended by Amdt. 25–19, 33 FR 15410, Oct. 17, 1968; Amdt. 25–23, 35 FR 5677, Apr. 8, 1970; Amdt. 25–36, 39 FR 35460, Oct. 1, 1974; Amdt. 25–57, 49 FR 6848, Feb. 23, 1984; Amdt. 25–72, 55 FR 29785, July 20, 1990]

§ 25.1015 Oil tank tests.

Each oil tank must be designed and installed so that—

(a) It can withstand, without failure, each vibration, inertia, and fluid load that it may be subjected to in operation; and

(b) It meets the provisions of §25.965, except—

(1) The test pressure—

(i) For pressurized tanks used with a turbine engine, may not be less than 5 p.s.i. plus the maximum operating pressure of the tank instead of the pressure specified in §25.965(a); and

(ii) For all other tanks may not be less than 5 p.s.i. instead of the pressure specified in §25.965(a); and

(2) The test fluid must be oil at 250 °F. instead of the fluid specified in §25.965(c).

[Doc. No. 5066, 29 FR 18291, Dec. 24, 1964, as amended by Amdt. 25–36, 39 FR 35461, Oct. 1, 1974]

§25.1017 Oil lines and fittings.

(a) Each oil line must meet the requirements of §25.993 and each oil line and fitting in any designated fire zone must meet the requirements of §25.1183.

(b) Breather lines must be arranged so that—

(1) Condensed water vapor that might freeze and obstruct the line cannot accumulate at any point;

(2) The breather discharge does not constitute a fire hazard if foaming occurs or causes emitted oil to strike the pilot's windshield; and

(3) The breather does not discharge into the engine air induction system.

§25.1019 Oil strainer or filter.

(a) Each turbine engine installation must incorporate an oil strainer or filter through which all of the engine oil flows and which meets the following requirements:

(1) Each oil strainer or filter that has a bypass must be constructed and installed so that oil will flow at the normal rate through the rest of the system with the strainer or filter completely blocked.

(2) The oil strainer or filter must have the capacity (with respect to operating limitations established for the engine) to ensure that engine oil system functioning is not impaired when the oil is contaminated to a degree (with respect to particle size and density) that is greater than that established for the engine under Part 33 of this chapter.

(3) The oil strainer or filter, unless it is installed at an oil tank outlet, must incorporate an indicator that will indicate contamination before it reaches the capacity established in accordance with paragraph (a)(2) of this section.

(4) The bypass of a strainer or filter must be constructed and installed so that the release of collected contaminants is minimized by appropriate location of the bypass to ensure that collected contaminants are not in the bypass flow path.

(5) An oil strainer or filter that has no bypass, except one that is installed at an oil tank outlet, must have a means to connect it to the warning system required in §25.1305(c)(7).

(b) Each oil strainer or filter in a powerplant installation using reciprocating engines must be constructed and installed so that oil will flow at the normal rate through the rest of the system with the strainer or filter element completely blocked.

[Amdt. 25–36, 39 FR 35461, Oct. 1, 1974, as amended by Amdt. 25–57, 49 FR 6848, Feb. 23, 1984]

§25.1021 Oil system drains.

A drain (or drains) must be provided to allow safe drainage of the oil system. Each drain must—

(a) Be accessible; and

(b) Have manual or automatic means for positive locking in the closed position.

[Amdt. 25–57, 49 FR 6848, Feb. 23, 1984]

§25.1023 Oil radiators.

(a) Each oil radiator must be able to withstand, without failure, any vibration, inertia, and oil pressure load to which it would be subjected in operation.

(b) Each oil radiator air duct must be located so that, in case of fire, flames coming from normal openings of the engine nacelle cannot impinge directly upon the radiator.

§25.1025 Oil valves.

(a) Each oil shutoff must meet the requirements of §25.1189.

(b) The closing of oil shutoff means may not prevent propeller feathering.

(c) Each oil valve must have positive stops or suitable index provisions in

325

the "on" and "off" positions and must be supported so that no loads resulting from its operation or from accelerated flight conditions are transmitted to the lines attached to the valve.

§ 25.1027 Propeller feathering system.

(a) If the propeller feathering system depends on engine oil, there must be means to trap an amount of oil in the tank if the supply becomes depleted due to failure of any part of the lubricating system other than the tank itself.

(b) The amount of trapped oil must be enough to accomplish the feathering operation and must be available only to the feathering pump.

(c) The ability of the system to accomplish feathering with the trapped oil must be shown. This may be done on the ground using an auxiliary source of oil for lubricating the engine during operation.

(d) Provision must be made to prevent sludge or other foreign matter from affecting the safe operation of the propeller feathering system.

[Doc. No. 5066, 29 FR 18291, Dec. 24, 1964, as amended by Amdt. 25–38, 41 FR 55467, Dec. 20, 1976]

COOLING

§ 25.1041 General.

The powerplant and auxiliary power unit cooling provisions must be able to maintain the temperatures of powerplant components, engine fluids, and auxiliary power unit components and fluids within the temperature limits established for these components and fluids, under ground, water, and flight operating conditions, and after normal engine or auxiliary power unit shutdown, or both.

[Amdt. 25–38, 41 FR 55467, Dec. 20, 1976]

§ 25.1043 Cooling tests.

(a) *General.* Compliance with § 25.1041 must be shown by tests, under critical ground, water, and flight operating conditions. For these tests, the following apply:

(1) If the tests are conducted under conditions deviating from the maximum ambient atmospheric temperature, the recorded powerplant tempera-

tures must be corrected under paragraphs (c) and (d) of this section.

(2) No corrected temperatures determined under paragraph (a)(1) of this section may exceed established limits.

(3) For reciprocating engines, the fuel used during the cooling tests must be the minimum grade approved for the engines, and the mixture settings must be those normally used in the flight stages for which the cooling tests are conducted. The test procedures must be as prescribed in § 25.1045.

(b) *Maximum ambient atmospheric temperature.* A maximum ambient atmospheric temperature corresponding to sea level conditions of at least 100 degrees F must be established. The assumed temperature lapse rate is 3.6 degrees F per thousand feet of altitude above sea level until a temperature of −69.7 degrees F is reached, above which altitude the temperature is considered constant at −69.7 degrees F. However, for winterization installations, the applicant may select a maximum ambient atmospheric temperature corresponding to sea level conditions of less than 100 degrees F.

(c) *Correction factor (except cylinder barrels).* Unless a more rational correction applies, temperatures of engine fluids and powerplant components (except cylinder barrels) for which temperature limits are established, must be corrected by adding to them the difference between the maximum ambient atmospheric temperature and the temperature of the ambient air at the time of the first occurrence of the maximum component or fluid temperature recorded during the cooling test.

(d) *Correction factor for cylinder barrel temperatures.* Unless a more rational correction applies, cylinder barrel temperatures must be corrected by adding to them 0.7 times the difference between the maximum ambient atmospheric temperature and the temperature of the ambient air at the time of the first occurrence of the maximum cylinder barrel temperature recorded during the cooling test.

[Doc. No. 5066, 29 FR 18291, Dec. 24, 1964, as amended by Amdt. 25–42, 43 FR 2323, Jan. 16, 1978]

§ 25.1045 Cooling test procedures.

(a) Compliance with § 25.1041 must be shown for the takeoff, climb, en route, and landing stages of flight that correspond to the applicable performance requirements. The cooling tests must be conducted with the airplane in the configuration, and operating under the conditions, that are critical relative to cooling during each stage of flight. For the cooling tests, a temperature is "stabilized" when its rate of change is less than two degrees F. per minute.

(b) Temperatures must be stabilized under the conditions from which entry is made into each stage of flight being investigated, unless the entry condition normally is not one during which component and the engine fluid temperatures would stabilize (in which case, operation through the full entry condition must be conducted before entry into the stage of flight being investigated in order to allow temperatures to reach their natural levels at the time of entry). The takeoff cooling test must be preceded by a period during which the powerplant component and engine fluid temperatures are stabilized with the engines at ground idle.

(c) Cooling tests for each stage of flight must be continued until—

(1) The component and engine fluid temperatures stabilize;

(2) The stage of flight is completed; or

(3) An operating limitation is reached.

(d) For reciprocating engine powered airplanes, it may be assumed, for cooling test purposes, that the takeoff stage of flight is complete when the airplane reaches an altitude of 1,500 feet above the takeoff surface or reaches a point in the takeoff where the transition from the takeoff to the en route configuration is completed and a speed is reached at which compliance with § 25.121(c) is shown, whichever point is at a higher altitude. The airplane must be in the following configuration:

(1) Landing gear retracted.

(2) Wing flaps in the most favorable position.

(3) Cowl flaps (or other means of controlling the engine cooling supply) in the position that provides adequate cooling in the hot-day condition.

(4) Critical engine inoperative and its propeller stopped.

(5) Remaining engines at the maximum continuous power available for the altitude.

(e) For hull seaplanes and amphibians, cooling must be shown during taxiing downwind for 10 minutes, at five knots above step speed.

[Doc. No. 5066, 29 FR 18291, Dec. 24, 1964, as amended by Amdt. 25–57, 49 FR 6848, Feb. 23, 1984]

INDUCTION SYSTEM

§ 25.1091 Air induction.

(a) The air induction system for each engine and auxiliary power unit must supply—

(1) The air required by that engine and auxiliary power unit under each operating condition for which certification is requested; and

(2) The air for proper fuel metering and mixture distribution with the induction system valves in any position.

(b) Each reciprocating engine must have an alternate air source that prevents the entry of rain, ice, or any other foreign matter.

(c) Air intakes may not open within the cowling, unless—

(1) That part of the cowling is isolated from the engine accessory section by means of a fireproof diaphragm; or

(2) For reciprocating engines, there are means to prevent the emergence of backfire flames.

(d) For turbine engine powered airplanes and airplanes incorporating auxiliary power units—

(1) There must be means to prevent hazardous quantities of fuel leakage or overflow from drains, vents, or other components of flammable fluid systems from entering the engine or auxiliary power unit intake system; and

(2) The airplane must be designed to prevent water or slush on the runway, taxiway, or other airport operating surfaces from being directed into the engine or auxiliary power unit air inlet ducts in hazardous quantities, and the air inlet ducts must be located or protected so as to minimize the ingestion of foreign matter during takeoff, landing, and taxiing.

(e) If the engine induction system contains parts or components that could be damaged by foreign objects entering the air inlet, it must be shown by tests or, if appropriate, by analysis that the induction system design can withstand the foreign object ingestion test conditions of §§ 33.76, 33.77 and 33.78(a)(1) of this chapter without failure of parts or components that could create a hazard.

[Doc. No. 5066, 29 FR 18291, Dec. 24, 1964, as amended by Amdt. 25–38, 41 FR 55467, Dec. 20, 1976; Amdt. 25–40, 42 FR 15043, Mar. 17, 1977; Amdt. 25–57, 49 FR 6849, Feb. 23, 1984; Amdt. 25–100, 65 FR 55854, Sept. 14, 2000]

§ 25.1093 Induction system icing protection.

(a) *Reciprocating engines.* Each reciprocating engine air induction system must have means to prevent and eliminate icing. Unless this is done by other means, it must be shown that, in air free of visible moisture at a temperature of 30 F., each airplane with altitude engines using—

(1) Conventional venturi carburetors have a preheater that can provide a heat rise of 120 F. with the engine at 60 percent of maximum continuous power; or

(2) Carburetors tending to reduce the probability of ice formation has a preheater that can provide a heat rise of 100 °F. with the engine at 60 percent of maximum continuous power.

(b) *Turbine engines.* Except as provided in paragraph (b)(3) of this section, each engine, with all icing protection systems operating, must:

(1) Operate throughout its flight power range, including the minimum descent idling speeds, in the icing conditions defined in Appendices C and O of this part, and Appendix D of part 33 of this chapter, and in falling and blowing snow within the limitations established for the airplane for such operation, without the accumulation of ice on the engine, inlet system components, or airframe components that would do any of the following:

(i) Adversely affect installed engine operation or cause a sustained loss of power or thrust; or an unacceptable increase in gas path operating temperature; or an airframe/engine incompatibility; or

(ii) Result in unacceptable temporary power loss or engine damage; or

(iii) Cause a stall, surge, or flameout or loss of engine controllability (for example, rollback).

(2) Operate at ground idle speed for a minimum of 30 minutes on the ground in the following icing conditions shown in Table 1 of this section, unless replaced by similar test conditions that are more critical. These conditions must be demonstrated with the available air bleed for icing protection at its critical condition, without adverse effect, followed by an acceleration to takeoff power or thrust in accordance with the procedures defined in the airplane flight manual. During the idle operation, the engine may be run up periodically to a moderate power or thrust setting in a manner acceptable to the Administrator. Analysis may be used to show ambient temperatures below the tested temperature are less critical. The applicant must document the engine run-up procedure (including the maximum time interval between run-ups from idle, run-up power setting, and duration at power), the associated minimum ambient temperature, and the maximum time interval. These conditions must be used in the analysis that establishes the airplane operating limitations in accordance with § 25.1521.

(3) For the purposes of this section, the icing conditions defined in appendix O of this part, including the conditions specified in Condition 3 of Table 1 of this section, are not applicable to airplanes with a maximum takeoff weight equal to or greater than 60,000 pounds.

TABLE 1—ICING CONDITIONS FOR GROUND TESTS

Condition	Total air temperature	Water concentration (minimum)	Mean effective particle diameter	Demonstration
1. Rime ice condition	0 to 15 °F (18 to −9 °C)	Liquid—0.3 g/m³	15–25 microns	By test, analysis or combination of the two.
2. Glaze ice condition	20 to 30 °F (−7 to −1 °C).	Liquid—0.3 g/m³	15–25 microns	By test, analysis or combination of the two.

TABLE 1—ICING CONDITIONS FOR GROUND TESTS—Continued

Condition	Total air temperature	Water concentration (minimum)	Mean effective particle diameter	Demonstration
3. Large drop condition.	15 to 30 °F (−9 to −1 °C).	Liquid—0.3 g/m³	100 microns (minimum).	By test, analysis or combination of the two.

(c) *Supercharged reciprocating engines.* For each engine having a supercharger to pressurize the air before it enters the carburetor, the heat rise in the air caused by that supercharging at any altitude may be utilized in determining compliance with paragraph (a) of this section if the heat rise utilized is that which will be available, automatically, for the applicable altitude and operating condition because of supercharging.

[Doc. No. 5066, 29 FR 18291, Dec. 24, 1964, as amended by Amdt. 25–38, 41 FR 55467, Dec. 20, 1976; Amdt. 25–40, 42 FR 15043, Mar. 17, 1977; Amdt. 25–57, 49 FR 6849, Feb. 23, 1984; Amdt. 25–72, 55 FR 29785, July 20, 1990; Amdt. 25–140, 79 FR 65526, Nov. 4, 2014]

§25.1101 Carburetor air preheater design.

Each carburetor air preheater must be designed and constructed to—

(a) Ensure ventilation of the preheater when the engine is operated in cold air;

(b) Allow inspection of the exhaust manifold parts that it surrounds; and

(c) Allow inspection of critical parts of the preheater itself.

§25.1103 Induction system ducts and air duct systems.

(a) Each induction system duct upstream of the first stage of the engine supercharger and of the auxiliary power unit compressor must have a drain to prevent the hazardous accumulation of fuel and moisture in the ground attitude. No drain may discharge where it might cause a fire hazard.

(b) Each induction system duct must be—

(1) Strong enough to prevent induction system failures resulting from normal backfire conditions; and

(2) Fire-resistant if it is in any fire zone for which a fire-extinguishing system is required, except that ducts for auxiliary power units must be fireproof within the auxiliary power unit fire zone.

(c) Each duct connected to components between which relative motion could exist must have means for flexibility.

(d) For turbine engine and auxiliary power unit bleed air duct systems, no hazard may result if a duct failure occurs at any point between the air duct source and the airplane unit served by the air.

(e) Each auxiliary power unit induction system duct must be fireproof for a sufficient distance upstream of the auxiliary power unit compartment to prevent hot gas reverse flow from burning through auxiliary power unit ducts and entering any other compartment or area of the airplane in which a hazard would be created resulting from the entry of hot gases. The materials used to form the remainder of the induction system duct and plenum chamber of the auxiliary power unit must be capable of resisting the maximum heat conditions likely to occur.

(f) Each auxiliary power unit induction system duct must be constructed of materials that will not absorb or trap hazardous quantities of flammable fluids that could be ignited in the event of a surge or reverse flow condition.

[Doc. No. 5066, 29 FR 18291, Dec. 24, 1964, as amended by Amdt. 25–46, 43 FR 50597, Oct. 30, 1978]

§25.1105 Induction system screens.

If induction system screens are used—

(a) Each screen must be upstream of the carburetor;

(b) No screen may be in any part of the induction system that is the only passage through which air can reach the engine, unless it can be deiced by heated air;

(c) No screen may be deiced by alcohol alone; and

(d) It must be impossible for fuel to strike any screen.

§ 25.1107 Inter-coolers and after-coolers.

Each inter-cooler and after-cooler must be able to withstand any vibration, inertia, and air pressure load to which it would be subjected in operation.

EXHAUST SYSTEM

§ 25.1121 General.

For powerplant and auxiliary power unit installations the following apply:

(a) Each exhaust system must ensure safe disposal of exhaust gases without fire hazard or carbon monoxide contamination in any personnel compartment. For test purposes, any acceptable carbon monoxide detection method may be used to show the absence of carbon monoxide.

(b) Each exhaust system part with a surface hot enough to ignite flammable fluids or vapors must be located or shielded so that leakage from any system carrying flammable fluids or vapors will not result in a fire caused by impingement of the fluids or vapors on any part of the exhaust system including shields for the exhaust system.

(c) Each component that hot exhaust gases could strike, or that could be subjected to high temperatures from exhaust system parts, must be fireproof. All exhaust system components must be separated by fireproof shields from adjacent parts of the airplane that are outside the engine and auxiliary power unit compartments.

(d) No exhaust gases may discharge so as to cause a fire hazard with respect to any flammable fluid vent or drain.

(e) No exhaust gases may discharge where they will cause a glare seriously affecting pilot vision at night.

(f) Each exhaust system component must be ventilated to prevent points of excessively high temperature.

(g) Each exhaust shroud must be ventilated or insulated to avoid, during normal operation, a temperature high enough to ignite any flammable fluids or vapors external to the shroud.

[Doc. No. 5066, 29 FR 18291, Dec. 24, 1964, as amended by Amdt. 25–40, 42 FR 15043, Mar. 17, 1977]

§ 25.1123 Exhaust piping.

For powerplant and auxiliary power unit installations, the following apply:

(a) Exhaust piping must be heat and corrosion resistant, and must have provisions to prevent failure due to expansion by operating temperatures.

(b) Piping must be supported to withstand any vibration and inertia loads to which it would be subjected in operation; and

(c) Piping connected to components between which relative motion could exist must have means for flexibility.

[Doc. No. 5066, 29 FR 18291, Dec. 24, 1964, as amended by Amdt. 25–40, 42 FR 15044, Mar. 17, 1977]

§ 25.1125 Exhaust heat exchangers.

For reciprocating engine powered airplanes, the following apply:

(a) Each exhaust heat exchanger must be constructed and installed to withstand each vibration, inertia, and other load to which it would be subjected in operation. In addition—

(1) Each exchanger must be suitable for continued operation at high temperatures and resistant to corrosion from exhaust gases;

(2) There must be means for the inspection of the critical parts of each exchanger;

(3) Each exchanger must have cooling provisions wherever it is subject to contact with exhaust gases; and

(4) No exhaust heat exchanger or muff may have any stagnant areas or liquid traps that would increase the probability of ignition of flammable fluids or vapors that might be present in case of the failure or malfunction of components carrying flammable fluids.

(b) If an exhaust heat exchanger is used for heating ventilating air—

(1) There must be a secondary heat exchanger between the primary exhaust gas heat exchanger and the ventilating air system; or

(2) Other means must be used to preclude the harmful contamination of the ventilating air.

[Doc. No. 5066, 29 FR 18291, Dec. 24, 1964, as amended by Amdt. 25–38, 41 FR 55467, Dec. 20, 1976]

§25.1127 Exhaust driven turbo-superchargers.

(a) Each exhaust driven turbo-supercharger must be approved or shown to be suitable for the particular application. It must be installed and supported to ensure safe operation between normal inspections and overhauls. In addition, there must be provisions for expansion and flexibility between exhaust conduits and the turbine.

(b) There must be provisions for lubricating the turbine and for cooling turbine parts where temperatures are critical.

(c) If the normal turbo-supercharger control system malfunctions, the turbine speed may not exceed its maximum allowable value. Except for the waste gate operating components, the components provided for meeting this requirement must be independent of the normal turbo-supercharger controls.

POWERPLANT CONTROLS AND ACCESSORIES

§25.1141 Powerplant controls: general.

Each powerplant control must be located, arranged, and designed under §§25.777 through 25.781 and marked under §25.1555. In addition, it must meet the following requirements:

(a) Each control must be located so that it cannot be inadvertently operated by persons entering, leaving, or moving normally in, the cockpit.

(b) Each flexible control must be approved or must be shown to be suitable for the particular application.

(c) Each control must have sufficient strength and rigidity to withstand operating loads without failure and without excessive deflection.

(d) Each control must be able to maintain any set position without constant attention by flight crewmembers and without creep due to control loads or vibration.

(e) The portion of each powerplant control located in a designated fire zone that is required to be operated in the event of fire must be at least fire resistant.

(f) For powerplant valve controls located in the flight deck there must be a means:

(1) For the flightcrew to select each intended position or function of the valve; and

(2) To indicate to the flightcrew:

(i) The selected position or function of the valve; and

(ii) When the valve has not responded as intended to the selected position or function.

[Doc. No. 5066, 29 FR 18291, Dec. 24, 1964, as amended by Amdt. 25–40, 42 FR 15044, Mar. 17, 1977; Amdt. 25–72, 55 FR 29785, July 20, 1990; Amdt. 25–115, 69 FR 40527, July 2, 2004]

§25.1142 Auxiliary power unit controls.

Means must be provided on the flight deck for starting, stopping, and emergency shutdown of each installed auxiliary power unit.

[Amdt. 25–46, 43 FR 50598, Oct. 30, 1978]

§25.1143 Engine controls.

(a) There must be a separate power or thrust control for each engine.

(b) Power and thrust controls must be arranged to allow—

(1) Separate control of each engine; and

(2) Simultaneous control of all engines.

(c) Each power and thrust control must provide a positive and immediately responsive means of controlling its engine.

(d) For each fluid injection (other than fuel) system and its controls not provided and approved as part of the engine, the applicant must show that the flow of the injection fluid is adequately controlled.

(e) If a power or thrust control incorporates a fuel shutoff feature, the control must have a means to prevent the inadvertent movement of the control into the shutoff position. The means must—

(1) Have a positive lock or stop at the idle position; and

(2) Require a separate and distinct operation to place the control in the shutoff position.

[Amdt. 25–23, 35 FR 5677, Apr. 8, 1970, as amended by Amdt. 25–38, 41 FR 55467, Dec. 20, 1976; Amdt. 25–57, 49 FR 6849, Feb. 23, 1984]

§ 25.1145 Ignition switches.

(a) Ignition switches must control each engine ignition circuit on each engine.

(b) There must be means to quickly shut off all ignition by the grouping of switches or by a master ignition control.

(c) Each group of ignition switches, except ignition switches for turbine engines for which continuous ignition is not required, and each master ignition control must have a means to prevent its inadvertent operation.

[Doc. No. 5066, 29 FR 18291, Dec. 24, 1964, as amended by Amdt. 25–40, 42 FR 15044 Mar. 17, 1977]

§ 25.1147 Mixture controls.

(a) If there are mixture controls, each engine must have a separate control. The controls must be grouped and arranged to allow—

(1) Separate control of each engine; and

(2) Simultaneous control of all engines.

(b) Each intermediate position of the mixture controls that corresponds to a normal operating setting must be identifiable by feel and sight.

(c) The mixture controls must be accessible to both pilots. However, if there is a separate flight engineer station with a control panel, the controls need be accessible only to the flight engineer.

§ 25.1149 Propeller speed and pitch controls.

(a) There must be a separate propeller speed and pitch control for each propeller.

(b) The controls must be grouped and arranged to allow—

(1) Separate control of each propeller; and

(2) Simultaneous control of all propellers.

(c) The controls must allow synchronization of all propellers.

(d) The propeller speed and pitch controls must be to the right of, and at least one inch below, the pilot's throttle controls.

§ 25.1153 Propeller feathering controls.

(a) There must be a separate propeller feathering control for each propeller. The control must have means to prevent its inadvertent operation.

(b) If feathering is accomplished by movement of the propeller pitch or speed control lever, there must be means to prevent the inadvertent movement of this lever to the feathering position during normal operation.

[Doc. No. 5066, 29 FR 18291, Dec. 24, 1964, as amended by Amdt. 25–11, 32 FR 6913, May 5, 1967]

§ 25.1155 Reverse thrust and propeller pitch settings below the flight regime.

Each control for reverse thrust and for propeller pitch settings below the flight regime must have means to prevent its inadvertent operation. The means must have a positive lock or stop at the flight idle position and must require a separate and distinct operation by the crew to displace the control from the flight regime (forward thrust regime for turbojet powered airplanes).

[Amdt. 25–11, 32 FR 6913, May 5, 1967]

§ 25.1157 Carburetor air temperature controls.

There must be a separate carburetor air temperature control for each engine.

§ 25.1159 Supercharger controls.

Each supercharger control must be accessible to the pilots or, if there is a separate flight engineer station with a control panel, to the flight engineer.

§ 25.1161 Fuel jettisoning system controls.

Each fuel jettisoning system control must have guards to prevent inadvertent operation. No control may be near any fire extinguisher control or other control used to combat fire.

§ 25.1163 Powerplant accessories.

(a) Each engine mounted accessory must—

(1) Be approved for mounting on the engine involved;

(2) Use the provisions on the engine for mounting; and

(3) Be sealed to prevent contamination of the engine oil system and the accessory system.

(b) Electrical equipment subject to arcing or sparking must be installed to minimize the probability of contact with any flammable fluids or vapors that might be present in a free state.

(c) If continued rotation of an engine-driven cabin supercharger or of any remote accessory driven by the engine is hazardous if malfunctioning occurs, there must be means to prevent rotation without interfering with the continued operation of the engine.

[Doc. No. 5066, 29 FR 18291, Dec. 24, 1964, as amended by Amdt. 25–57, 49 FR 6849, Feb. 23, 1984]

§ 25.1165 Engine ignition systems.

(a) Each battery ignition system must be supplemented by a generator that is automatically available as an alternate source of electrical energy to allow continued engine operation if any battery becomes depleted.

(b) The capacity of batteries and generators must be large enough to meet the simultaneous demands of the engine ignition system and the greatest demands of any electrical system components that draw electrical energy from the same source.

(c) The design of the engine ignition system must account for—

(1) The condition of an inoperative generator;

(2) The condition of a completely depleted battery with the generator running at its normal operating speed; and

(3) The condition of a completely depleted battery with the generator operating at idling speed, if there is only one battery.

(d) Magneto ground wiring (for separate ignition circuits) that lies on the engine side of the fire wall, must be installed, located, or protected, to minimize the probability of simultaneous failure of two or more wires as a result of mechanical damage, electrical faults, or other cause.

(e) No ground wire for any engine may be routed through a fire zone of another engine unless each part of that wire within that zone is fireproof.

(f) Each ignition system must be independent of any electrical circuit, not used for assisting, controlling, or analyzing the operation of that system.

(g) There must be means to warn appropriate flight crewmembers if the malfunctioning of any part of the electrical system is causing the continuous discharge of any battery necessary for engine ignition.

(h) Each engine ignition system of a turbine powered airplane must be considered an essential electrical load.

[Doc. No. 5066, 29 FR 18291, Dec. 24, 1964, as amended by Amdt. 25–23, 35 FR 5677, Apr. 8, 1970; Amdt. 25–72, 55 FR 29785, July 20, 1990]

§ 25.1167 Accessory gearboxes.

For airplanes equipped with an accessory gearbox that is not certificated as part of an engine—

(a) The engine with gearbox and connecting transmissions and shafts attached must be subjected to the tests specified in § 33.49 or § 33.87 of this chapter, as applicable;

(b) The accessory gearbox must meet the requirements of §§ 33.25 and 33.53 or 33.91 of this chapter, as applicable; and

(c) Possible misalignments and torsional loadings of the gearbox, transmission, and shaft system, expected to result under normal operating conditions must be evaluated.

[Amdt. 25–38, 41 FR 55467, Dec. 20, 1976]

POWERPLANT FIRE PROTECTION

§ 25.1181 Designated fire zones; regions included.

(a) Designated fire zones are—

(1) The engine power section;

(2) The engine accessory section;

(3) Except for reciprocating engines, any complete powerplant compartment in which no isolation is provided between the engine power section and the engine accessory section;

(4) Any auxiliary power unit compartment;

(5) Any fuel-burning heater and other combustion equipment installation described in § 25.859;

(6) The compressor and accessory sections of turbine engines; and

(7) Combustor, turbine, and tailpipe sections of turbine engine installations that contain lines or components carrying flammable fluids or gases.

(b) Each designated fire zone must meet the requirements of §§ 25.863, 25.865, 25.867, 25.869, and 25.1185 through 25.1203.

[Doc. No. 5066, 29 FR 18291, Dec. 24, 1964, as amended by Amdt. 25–11, 32 FR 6913, May 5, 1967; Amdt. 25–23, 35 FR 5677, Apr. 8, 1970; Amdt. 25–72, 55 FR 29785, July 20, 1990; Amdt. 25–115, 69 FR 40527, July 2, 2004]

§ 25.1182 Nacelle areas behind fire-walls, and engine pod attaching structures containing flammable fluid lines.

(a) Each nacelle area immediately behind the firewall, and each portion of any engine pod attaching structure containing flammable fluid lines, must meet each requirement of §§ 25.1103(b), 25.1165 (d) and (e), 25.1183, 25.1185(c), 25.1187, 25.1189, and 25.1195 through 25.1203, including those concerning designated fire zones. However, engine pod attaching structures need not contain fire detection or extinguishing means.

(b) For each area covered by paragraph (a) of this section that contains a retractable landing gear, compliance with that paragraph need only be shown with the landing gear retracted.

[Amdt. 25–11, 32 FR 6913, May 5, 1967]

§ 25.1183 Flammable fluid-carrying components.

(a) Except as provided in paragraph (b) of this section, each line, fitting, and other component carrying flammable fluid in any area subject to engine fire conditions, and each component which conveys or contains flammable fluid in a designated fire zone must be fire resistant, except that flammable fluid tanks and supports in a designated fire zone must be fireproof or be enclosed by a fireproof shield unless damage by fire to any non-fireproof part will not cause leakage or spillage of flammable fluid. Components must be shielded or located to safeguard against the ignition of leaking flammable fluid. An integral oil sump of less than 25-quart capacity on a reciprocating engine need not be fireproof nor be enclosed by a fireproof shield.

(b) Paragraph (a) of this section does not apply to—

(1) Lines, fittings, and components which are already approved as part of a type certificated engine; and

(2) Vent and drain lines, and their fittings, whose failure will not result in, or add to, a fire hazard.

(c) All components, including ducts, within a designated fire zone must be fireproof if, when exposed to or damaged by fire, they could—

(1) Result in fire spreading to other regions of the airplane; or

(2) Cause unintentional operation of, or inability to operate, essential services or equipment.

[Doc. No. 5066, 29 FR 18291, Dec. 24, 1964, as amended by Amdt. 25–11, 32 FR 6913, May 5, 1967; Amdt. 25–36, 39 FR 35461, Oct. 1, 1974; Amdt. 25–57, 49 FR 6849, Feb. 23, 1984; Amdt. 25–101, 65 FR 79710, Dec. 19, 2000]

§ 25.1185 Flammable fluids.

(a) Except for the integral oil sumps specified in § 25.1183(a), no tank or reservoir that is a part of a system containing flammable fluids or gases may be in a designated fire zone unless the fluid contained, the design of the system, the materials used in the tank, the shut-off means, and all connections, lines, and control provide a degree of safety equal to that which would exist if the tank or reservoir were outside such a zone.

(b) There must be at least one-half inch of clear airspace between each tank or reservoir and each firewall or shroud isolating a designated fire zone.

(c) Absorbent materials close to flammable fluid system components that might leak must be covered or treated to prevent the absorption of hazardous quantities of fluids.

[Doc. No. 5066, 29 FR 18291, Dec. 24, 1964, as amended by Amdt. 25–19, 33 FR 15410, Oct. 17, 1968; Amdt. 25–94, 63 FR 8848, Feb. 23, 1998]

§ 25.1187 Drainage and ventilation of fire zones.

(a) There must be complete drainage of each part of each designated fire zone to minimize the hazards resulting from failure or malfunctioning of any component containing flammable fluids. The drainage means must be—

(1) Effective under conditions expected to prevail when drainage is needed; and

(2) Arranged so that no discharged fluid will cause an additional fire hazard.

(b) Each designated fire zone must be ventilated to prevent the accumulation of flammable vapors.

(c) No ventilation opening may be where it would allow the entry of flammable fluids, vapors, or flame from other zones.

(d) Each ventilation means must be arranged so that no discharged vapors will cause an additional fire hazard.

(e) Unless the extinguishing agent capacity and rate of discharge are based on maximum air flow through a zone, there must be means to allow the crew to shut off sources of forced ventilation to any fire zone except the engine power section of the nacelle and the combustion heater ventilating air ducts.

§25.1189 Shutoff means.

(a) Each engine installation and each fire zone specified in §25.1181(a)(4) and (5) must have a means to shut off or otherwise prevent hazardous quantities of fuel, oil, deicer, and other flammable fluids, from flowing into, within, or through any designated fire zone, except that shutoff means are not required for—

(1) Lines, fittings, and components forming an integral part of an engine; and

(2) Oil systems for turbine engine installations in which all components of the system in a designated fire zone, including oil tanks, are fireproof or located in areas not subject to engine fire conditions.

(b) The closing of any fuel shutoff valve for any engine may not make fuel unavailable to the remaining engines.

(c) Operation of any shutoff may not interfere with the later emergency operation of other equipment, such as the means for feathering the propeller.

(d) Each flammable fluid shutoff means and control must be fireproof or must be located and protected so that any fire in a fire zone will not affect its operation.

(e) No hazardous quantity of flammable fluid may drain into any designated fire zone after shutoff.

(f) There must be means to guard against inadvertent operation of the shutoff means and to make it possible for the crew to reopen the shutoff means in flight after it has been closed.

(g) Each tank-to-engine shutoff valve must be located so that the operation of the valve will not be affected by powerplant or engine mount structural failure.

(h) Each shutoff valve must have a means to relieve excessive pressure accumulation unless a means for pressure relief is otherwise provided in the system.

[Doc. No. 5066, 29 FR 18291, Dec. 24, 1964, as amended by Amdt. 25–23, 35 FR 5677, Apr. 8, 1970; Amdt. 25–57, 49 FR 6849, Feb. 23, 1984]

§25.1191 Firewalls.

(a) Each engine, auxiliary power unit, fuel-burning heater, other combustion equipment intended for operation in flight, and the combustion, turbine, and tailpipe sections of turbine engines, must be isolated from the rest of the airplane by firewalls, shrouds, or equivalent means.

(b) Each firewall and shroud must be—

(1) Fireproof;

(2) Constructed so that no hazardous quantity of air, fluid, or flame can pass from the compartment to other parts of the airplane;

(3) Constructed so that each opening is sealed with close fitting fireproof grommets, bushings, or firewall fittings; and

(4) Protected against corrosion.

§25.1192 Engine accessory section diaphragm.

For reciprocating engines, the engine power section and all portions of the exhaust system must be isolated from the engine accessory compartment by a diaphragm that complies with the firewall requirements of §25.1191.

[Amdt. 25–23, 35 FR 5678, Apr. 8, 1970]

§25.1193 Cowling and nacelle skin.

(a) Each cowling must be constructed and supported so that it can resist any vibration, inertia, and air load to which it may be subjected in operation.

(b) Cowling must meet the drainage and ventilation requirements of § 25.1187.

(c) On airplanes with a diaphragm isolating the engine power section from the engine accessory section, each part of the accessory section cowling subject to flame in case of fire in the engine power section of the powerplant must—

(1) Be fireproof; and

(2) Meet the requirements of § 25.1191.

(d) Each part of the cowling subject to high temperatures due to its nearness to exhaust system parts or exhaust gas impingement must be fireproof.

(e) Each airplane must—

(1) Be designed and constructed so that no fire originating in any fire zone can enter, either through openings or by burning through external skin, any other zone or region where it would create additional hazards;

(2) Meet paragraph (e)(1) of this section with the landing gear retracted (if applicable); and

(3) Have fireproof skin in areas subject to flame if a fire starts in the engine power or accessory sections.

§ 25.1195 Fire extinguishing systems.

(a) Except for combustor, turbine, and tail pipe sections of turbine engine installations that contain lines or components carrying flammable fluids or gases for which it is shown that a fire originating in these sections can be controlled, there must be a fire extinguisher system serving each designated fire zone.

(b) The fire extinguishing system, the quantity of the extinguishing agent, the rate of discharge, and the discharge distribution must be adequate to extinguish fires. It must be shown by either actual or simulated flights tests that under critical airflow conditions in flight the discharge of the extinguishing agent in each designated fire zone specified in paragraph (a) of this section will provide an agent concentration capable of extinguishing fires in that zone and of minimizing the probability of reignition. An individual "one-shot" system may be used for auxiliary power units, fuel burning

heaters, and other combustion equipment. For each other designated fire zone, two discharges must be provided each of which produces adequate agent concentration.

(c) The fire extinguishing system for a nacelle must be able to simultaneously protect each zone of the nacelle for which protection is provided.

[Doc. No. 5066, 29 FR 18291, Dec. 24, 1964, as amended by Amdt. 25–46, 43 FR 50598, Oct. 30, 1978]

§ 25.1197 Fire extinguishing agents.

(a) Fire extinguishing agents must—

(1) Be capable of extinguishing flames emanating from any burning of fluids or other combustible materials in the area protected by the fire extinguishing system; and

(2) Have thermal stability over the temperature range likely to be experienced in the compartment in which they are stored.

(b) If any toxic extinguishing agent is used, provisions must be made to prevent harmful concentrations of fluid or fluid vapors (from leakage during normal operation of the airplane or as a result of discharging the fire extinguisher on the ground or in flight) from entering any personnel compartment, even though a defect may exist in the extinguishing system. This must be shown by test except for built-in carbon dioxide fuselage compartment fire extinguishing systems for which—

(1) Five pounds or less of carbon dioxide will be discharged, under established fire control procedures, into any fuselage compartment; or

(2) There is protective breathing equipment for each flight crewmember on flight deck duty.

[Doc. No. 5066, 29 FR 18291, Dec. 24, 1964, as amended by Amdt. 25–38, 41 FR 55467, Dec. 20, 1976; Amdt. 25–40, 42 FR 15044, Mar. 17, 1977]

§ 25.1199 Extinguishing agent containers.

(a) Each extinguishing agent container must have a pressure relief to prevent bursting of the container by excessive internal pressures.

(b) The discharge end of each discharge line from a pressure relief connection must be located so that discharge of the fire extinguishing agent would not damage the airplane. The line must also be located or protected to prevent clogging caused by ice or other foreign matter.

(c) There must be a means for each fire extinguishing agent container to indicate that the container has discharged or that the charging pressure is below the established minimum necessary for proper functioning.

(d) The temperature of each container must be maintained, under intended operating conditions, to prevent the pressure in the container from—

(1) Falling below that necessary to provide an adequate rate of discharge; or

(2) Rising high enough to cause premature discharge.

(e) If a pyrotechnic capsule is used to discharge the extinguishing agent, each container must be installed so that temperature conditions will not cause hazardous deterioration of the pyrotechnic capsule.

[Doc. No. 5066, 29 FR 18291, Dec. 24, 1964, as amended by Amdt. 25–23, 35 FR 5678, Apr. 8, 1970; Amdt. 25–40, 42 FR 15044, Mar. 17, 1977]

§25.1201 Fire extinguishing system materials.

(a) No material in any fire extinguishing system may react chemically with any extinguishing agent so as to create a hazard.

(b) Each system component in an engine compartment must be fireproof.

§25.1203 Fire detector system.

(a) There must be approved, quick acting fire or overheat detectors in each designated fire zone, and in the combustion, turbine, and tailpipe sections of turbine engine installations, in numbers and locations ensuring prompt detection of fire in those zones.

(b) Each fire detector system must be constructed and installed so that—

(1) It will withstand the vibration, inertia, and other loads to which it may be subjected in operation;

(2) There is a means to warn the crew in the event that the sensor or associated wiring within a designated fire zone is severed at one point, unless the system continues to function as a satisfactory detection system after the severing; and

(3) There is a means to warn the crew in the event of a short circuit in the sensor or associated wiring within a designated fire zone, unless the system continues to function as a satisfactory detection system after the short circuit.

(c) No fire or overheat detector may be affected by any oil, water, other fluids or fumes that might be present.

(d) There must be means to allow the crew to check, in flight, the functioning of each fire or overheat detector electric circuit.

(e) Components of each fire or overheat detector system in a fire zone must be fire-resistant.

(f) No fire or overheat detector system component for any fire zone may pass through another fire zone, unless—

(1) It is protected against the possibility of false warnings resulting from fires in zones through which it passes; or

(2) Each zone involved is simultaneously protected by the same detector and extinguishing system.

(g) Each fire detector system must be constructed so that when it is in the configuration for installation it will not exceed the alarm activation time approved for the detectors using the response time criteria specified in the appropriate Technical Standard Order for the detector.

(h) EWIS for each fire or overheat detector system in a fire zone must meet the requirements of §25.1731.

[Doc. No. 5066, 29 FR 18291, Dec. 24, 1964, as amended by Amdt. 25–23, 35 FR 5678, Apr. 8, 1970; Amdt. 25–26, 36 FR 5493, Mar. 24, 1971; Amdt. 25–123, 72 FR 63405, Nov. 8, 2007]

§25.1207 Compliance.

Unless otherwise specified, compliance with the requirements of §§25.1181 through 25.1203 must be shown by a full scale fire test or by one or more of the following methods:

(a) Tests of similar powerplant configurations;

(b) Tests of components;

(c) Service experience of aircraft with similar powerplant configurations;

(d) Analysis.

[Amdt. 25–46, 43 FR 50598, Oct. 30, 1978]

Subpart F—Equipment

GENERAL

§ 25.1301 Function and installation.

(a) Each item of installed equipment must—

(1) Be of a kind and design appropriate to its intended function;

(2) Be labeled as to its identification, function, or operating limitations, or any applicable combination of these factors;

(3) Be installed according to limitations specified for that equipment; and

(4) Function properly when installed.

(b) EWIS must meet the requirements of subpart H of this part.

[Doc. No. 5066, 29 FR 18333, Dec. 24, 1964, as amended by Amdt. 25–123, 72 FR 63405, Nov. 8, 2007]

§ 25.1302 Installed systems and equipment for use by the flightcrew.

This section applies to installed systems and equipment intended for flightcrew members' use in operating the airplane from their normally seated positions on the flight deck. The applicant must show that these systems and installed equipment, individually and in combination with other such systems and equipment, are designed so that qualified flightcrew members trained in their use can safely perform all of the tasks associated with the systems' and equipment's intended functions. Such installed equipment and systems must meet the following requirements:

(a) Flight deck controls must be installed to allow accomplishment of all the tasks required to safely perform the equipment's intended function, and information must be provided to the flightcrew that is necessary to accomplish the defined tasks.

(b) Flight deck controls and information intended for the flightcrew's use must:

(1) Be provided in a clear and unambiguous manner at a resolution and precision appropriate to the task;

(2) Be accessible and usable by the flightcrew in a manner consistent with the urgency, frequency, and duration of their tasks; and

(3) Enable flightcrew awareness, if awareness is required for safe operation, of the effects on the airplane or systems resulting from flightcrew actions.

(c) Operationally-relevant behavior of the installed equipment must be:

(1) Predictable and unambiguous; and

(2) Designed to enable the flightcrew to intervene in a manner appropriate to the task.

(d) To the extent practicable, installed equipment must incorporate means to enable the flightcrew to manage errors resulting from the kinds of flightcrew interactions with the equipment that can be reasonably expected in service. This paragraph does not apply to any of the following:

(1) Skill-related errors associated with manual control of the airplane;

(2) Errors that result from decisions, actions, or omissions committed with malicious intent;

(3) Errors arising from a crewmember's reckless decisions, actions, or omissions reflecting a substantial disregard for safety; and

(4) Errors resulting from acts or threats of violence, including actions taken under duress.

[Doc. No. FAA–2010–1175, 78 FR 25846, May 3, 2013]

§ 25.1303 Flight and navigation instruments.

(a) The following flight and navigation instruments must be installed so that the instrument is visible from each pilot station:

(1) A free air temperature indicator or an air-temperature indicator which provides indications that are convertible to free-air temperature.

(2) A clock displaying hours, minutes, and seconds with a sweep-second pointer or digital presentation.

(3) A direction indicator (nonstabilized magnetic compass).

(b) The following flight and navigation instruments must be installed at each pilot station:

(1) An airspeed indicator. If airspeed limitations vary with altitude, the indicator must have a maximum allowable airspeed indicator showing the variation of V_{MO} with altitude.

(2) An altimeter (sensitive).

(3) A rate-of-climb indicator (vertical speed).

(4) A gyroscopic rate-of-turn indicator combined with an integral slipskid indicator (turn-and-bank indicator) except that only a slip-skid indicator is required on large airplanes with a third attitude instrument system useable through flight attitudes of 360° of pitch and roll and installed in accordance with §121.305(k) of this title.

(5) A bank and pitch indicator (gyroscopically stabilized).

(6) A direction indicator (gyroscopically stabilized, magnetic or nonmagnetic).

(c) The following flight and navigation instruments are required as prescribed in this paragraph:

(1) A speed warning device is required for turbine engine powered airplanes and for airplanes with V_{MO}/M_{MO} greater than 0.8 V_{DF}/M_{DF} or 0.8 V_{D}/M_D. The speed warning device must give effective aural warning (differing distinctively from aural warnings used for other purposes) to the pilots, whenever the speed exceeds V_{MO} plus 6 knots or M_{MO} + 0.01. The upper limit of the production tolerance for the warning device may not exceed the prescribed warning speed.

(2) A machmeter is required at each pilot station for airplanes with compressibility limitations not otherwise indicated to the pilot by the airspeed indicating system required under paragraph (b)(1) of this section.

[Amdt. 25–23, 35 FR 5678, Apr. 8, 1970, as amended by Amdt. 25–24, 35 FR 7108, May 6, 1970; Amdt. 25–38, 41 FR 55467, Dec. 20, 1976; Amdt. 25–90, 62 FR 13253, Mar. 19, 1997]

§ 25.1305 Powerplant instruments.

The following are required powerplant instruments:

(a) *For all airplanes.* (1) A fuel pressure warning means for each engine, or a master warning means for all engines with provision for isolating the individual warning means from the master warning means.

(2) A fuel quantity indicator for each fuel tank.

(3) An oil quantity indicator for each oil tank.

(4) An oil pressure indicator for each independent pressure oil system of each engine.

(5) An oil pressure warning means for each engine, or a master warning means for all engines with provision for isolating the individual warning means from the master warning means.

(6) An oil temperature indicator for each engine.

(7) Fire-warning devices that provide visual and audible warning.

(8) An augmentation liquid quantity indicator (appropriate for the manner in which the liquid is to be used in operation) for each tank.

(b) *For reciprocating engine-powered airplanes.* In addition to the powerplant instruments required by paragraph (a) of this section, the following powerplant instruments are required:

(1) A carburetor air temperature indicator for each engine.

(2) A cylinder head temperature indicator for each air-cooled engine.

(3) A manifold pressure indicator for each engine.

(4) A fuel pressure indicator (to indicate the pressure at which the fuel is supplied) for each engine.

(5) A fuel flowmeter, or fuel mixture indicator, for each engine without an automatic altitude mixture control.

(6) A tachometer for each engine.

(7) A device that indicates, to the flight crew (during flight), any change in the power output, for each engine with—

(i) An automatic propeller feathering system, whose operation is initiated by a power output measuring system; or

(ii) A total engine piston displacement of 2,000 cubic inches or more.

(8) A means to indicate to the pilot when the propeller is in reverse pitch, for each reversing propeller.

(c) *For turbine engine-powered airplanes.* In addition to the powerplant instruments required by paragraph (a) of this section, the following powerplant instruments are required:

(1) A gas temperature indicator for each engine.

(2) A fuel flowmeter indicator for each engine.

(3) A tachometer (to indicate the speed of the rotors with established limiting speeds) for each engine.

(4) A means to indicate, to the flight crew, the operation of each engine starter that can be operated continuously but that is neither designed for

continuous operation nor designed to prevent hazard if it failed.

(5) An indicator to indicate the functioning of the powerplant ice protection system for each engine.

(6) An indicator for the fuel strainer or filter required by § 25.997 to indicate the occurrence of contamination of the strainer or filter before it reaches the capacity established in accordance with § 25.997(d).

(7) A warning means for the oil strainer or filter required by § 25.1019, if it has no bypass, to warn the pilot of the occurrence of contamination of the strainer or filter screen before it reaches the capacity established in accordance with § 25.1019(a)(2).

(8) An indicator to indicate the proper functioning of any heater used to prevent ice clogging of fuel system components.

(d) *For turbojet engine powered airplanes.* In addition to the powerplant instruments required by paragraphs (a) and (c) of this section, the following powerplant instruments are required:

(1) An indicator to indicate thrust, or a parameter that is directly related to thrust, to the pilot. The indication must be based on the direct measurement of thrust or of parameters that are directly related to thrust. The indicator must indicate a change in thrust resulting from any engine malfunction, damage, or deterioration.

(2) A position indicating means to indicate to the flightcrew when the thrust reversing device—

(i) Is not in the selected position, and

(ii) Is in the reverse thrust position, for each engine using a thrust reversing device.

(3) An indicator to indicate rotor system unbalance.

(e) *For turbopropeller-powered airplanes.* In addition to the powerplant instruments required by paragraphs (a) and (c) of this section, the following powerplant instruments are required:

(1) A torque indicator for each engine.

(2) Position indicating means to indicate to the flight crew when the propeller blade angle is below the flight low pitch position, for each propeller.

(f) For airplanes equipped with fluid systems (other than fuel) for thrust or power augmentation, an approved means must be provided to indicate the proper functioning of that system to the flight crew.

[Amdt. 25–23, 35 FR 5678, Apr. 8, 1970, as amended by Amdt. 25–35, 39 FR 1831, Jan. 15, 1974; Amdt. 25–36, 39 FR 35461, Oct. 1, 1974; Amdt. 25–38, 41 FR 55467, Dec. 20, 1976; Amdt. 25–54, 45 FR 60173, Sept. 11, 1980; Amdt. 25–72, 55 FR 29785, July 20, 1990; Amdt. 25–115, 69 FR 40527, July 2, 2004]

§ 25.1307 **Miscellaneous equipment.**

The following is required miscellaneous equipment:

(a) [Reserved]

(b) Two or more independent sources of electrical energy.

(c) Electrical protective devices, as prescribed in this part.

(d) Two systems for two-way radio communications, with controls for each accessible from each pilot station, designed and installed so that failure of one system will not preclude operation of the other system. The use of a common antenna system is acceptable if adequate reliability is shown.

(e) Two systems for radio navigation, with controls for each accessible from each pilot station, designed and installed so that failure of one system will not preclude operation of the other system. The use of a common antenna system is acceptable if adequate reliability is shown.

[Amdt. 25–23, 35 FR 5678, Apr. 8, 1970, as amended by Amdt. 25–46, 43 FR 50598, Oct. 30, 1978; Amdt. 25–54, 45 FR 60173, Sept. 11, 1980; Amdt. 25–72, 55 FR 29785, July 20, 1990]

§ 25.1309 **Equipment, systems, and installations.**

(a) The equipment, systems, and installations whose functioning is required by this subchapter, must be designed to ensure that they perform their intended functions under any foreseeable operating condition.

(b) The airplane systems and associated components, considered separately and in relation to other systems, must be designed so that—

(1) The occurrence of any failure condition which would prevent the continued safe flight and landing of the airplane is extremely improbable, and

(2) The occurrence of any other failure conditions which would reduce the capability of the airplane or the ability

of the crew to cope with adverse operating conditions is improbable.

(c) Warning information must be provided to alert the crew to unsafe system operating conditions, and to enable them to take appropriate corrective action. Systems, controls, and associated monitoring and warning means must be designed to minimize crew errors which could create additional hazards.

(d) Compliance with the requirements of paragraph (b) of this section must be shown by analysis, and where necessary, by appropriate ground, flight, or simulator tests. The analysis must consider—

(1) Possible modes of failure, including malfunctions and damage from external sources.

(2) The probability of multiple failures and undetected failures.

(3) The resulting effects on the airplane and occupants, considering the stage of flight and operating conditions, and

(4) The crew warning cues, corrective action required, and the capability of detecting faults.

(e) In showing compliance with paragraphs (a) and (b) of this section with regard to the electrical system and equipment design and installation, critical environmental conditions must be considered. For electrical generation, distribution, and utilization equipment required by or used in complying with this chapter, except equipment covered by Technical Standard Orders containing environmental test procedures, the ability to provide continuous, safe service under foreseeable environmental conditions may be shown by environmental tests, design analysis, or reference to previous comparable service experience on other aircraft.

(f) EWIS must be assessed in accordance with the requirements of § 25.1709.

[Amdt. 25–23, 35 FR 5679, Apr. 8, 1970, as amended by Amdt. 25–38, 41 FR 55467, Dec. 20, 1976; Amdt. 25–41, 42 FR 36970, July 18, 1977; Amdt. 25–123, 72 FR 63405, Nov. 8, 2007]

§ 25.1310 Power source capacity and distribution.

(a) Each installation whose functioning is required for type certification or under operating rules and that requires a power supply is an "essential load" on the power supply. The power sources and the system must be able to supply the following power loads in probable operating combinations and for probable durations:

(1) Loads connected to the system with the system functioning normally.

(2) Essential loads, after failure of any one prime mover, power converter, or energy storage device.

(3) Essential loads after failure of—

(i) Any one engine on two-engine airplanes; and

(ii) Any two engines on airplanes with three or more engines.

(4) Essential loads for which an alternate source of power is required, after any failure or malfunction in any one power supply system, distribution system, or other utilization system.

(b) In determining compliance with paragraphs (a)(2) and (3) of this section, the power loads may be assumed to be reduced under a monitoring procedure consistent with safety in the kinds of operation authorized. Loads not required in controlled flight need not be considered for the two-engine-inoperative condition on airplanes with three or more engines.

[Amdt. 25–123, 72 FR 63405, Nov. 8, 2007]

§ 25.1316 Electrical and electronic system lightning protection.

(a) Each electrical and electronic system that performs a function, for which failure would prevent the continued safe flight and landing of the airplane, must be designed and installed so that—

(1) The function is not adversely affected during and after the time the airplane is exposed to lightning; and

(2) The system automatically recovers normal operation of that function in a timely manner after the airplane is exposed to lightning.

(b) Each electrical and electronic system that performs a function, for which failure would reduce the capability of the airplane or the ability of the flightcrew to respond to an adverse operating condition, must be designed and installed so that the function recovers normal operation in a timely manner after the airplane is exposed to lightning.

[Doc. No. FAA–2010–0224, Amdt. 25–134, 76 FR 33135, June 8, 2011]

§ 25.1317 High-intensity Radiated Fields (HIRF) Protection.

(a) Except as provided in paragraph (d) of this section, each electrical and electronic system that performs a function whose failure would prevent the continued safe flight and landing of the airplane must be designed and installed so that—

(1) The function is not adversely affected during and after the time the airplane is exposed to HIRF environment I, as described in appendix L to this part;

(2) The system automatically recovers normal operation of that function, in a timely manner, after the airplane is exposed to HIRF environment I, as described in appendix L to this part, unless the system's recovery conflicts with other operational or functional requirements of the system; and

(3) The system is not adversely affected during and after the time the airplane is exposed to HIRF environment II, as described in appendix L to this part.

(b) Each electrical and electronic system that performs a function whose failure would significantly reduce the capability of the airplane or the ability of the flightcrew to respond to an adverse operating condition must be designed and installed so the system is not adversely affected when the equipment providing these functions is exposed to equipment HIRF test level 1 or 2, as described in appendix L to this part.

(c) Each electrical and electronic system that performs a function whose failure would reduce the capability of the airplane or the ability of the flightcrew to respond to an adverse operating condition must be designed and installed so the system is not adversely affected when the equipment providing the function is exposed to equipment HIRF test level 3, as described in appendix L to this part.

(d) Before December 1, 2012, an electrical or electronic system that performs a function whose failure would prevent the continued safe flight and landing of an airplane may be designed and installed without meeting the provisions of paragraph (a) provided—

(1) The system has previously been shown to comply with special conditions for HIRF, prescribed under § 21.16, issued before December 1, 2007;

(2) The HIRF immunity characteristics of the system have not changed since compliance with the special conditions was demonstrated; and

(3) The data used to demonstrate compliance with the special conditions is provided.

[Doc. No. FAA–2006–23657, 72 FR 44025, Aug. 6, 2007]

INSTRUMENTS: INSTALLATION

§ 25.1321 Arrangement and visibility.

(a) Each flight, navigation, and powerplant instrument for use by any pilot must be plainly visible to him from his station with the minimum practicable deviation from his normal position and line of vision when he is looking forward along the flight path.

(b) The flight instruments required by § 25.1303 must be grouped on the instrument panel and centered as nearly as practicable about the vertical plane of the pilot's forward vision. In addition—

(1) The instrument that most effectively indicates attitude must be on the panel in the top center position;

(2) The instrument that most effectively indicates airspeed must be adjacent to and directly to the left of the instrument in the top center position:

(3) The instrument that most effectively indicates altitude must be adjacent to and directly to the right of the instrument in the top center position; and

(4) The instrument that most effectively indicates direction of flight must be adjacent to and directly below the instrument in the top center position.

(c) Required powerplant instruments must be closely grouped on the instrument panel. In addition—

(1) The location of identical powerplant instruments for the engines must prevent confusion as to which engine each instrument relates; and

(2) Powerplant instruments vital to the safe operation of the airplane must

be plainly visible to the appropriate crewmembers.

(d) Instrument panel vibration may not damage or impair the accuracy of any instrument.

(e) If a visual indicator is provided to indicate malfunction of an instrument, it must be effective under all probable cockpit lighting conditions.

[Amdt. 25–23, 35 FR 5679, Apr. 8, 1970, as amended by Amdt. 25–41, 42 FR 36970, July 18, 1977]

§ 25.1322 Flightcrew alerting.

(a) Flightcrew alerts must:

(1) Provide the flightcrew with the information needed to:

(i) Identify non-normal operation or airplane system conditions, and

(ii) Determine the appropriate actions, if any.

(2) Be readily and easily detectable and intelligible by the flightcrew under all foreseeable operating conditions, including conditions where multiple alerts are provided.

(3) Be removed when the alerting condition no longer exists.

(b) Alerts must conform to the following prioritization hierarchy based on the urgency of flightcrew awareness and response.

(1) Warning: For conditions that require immediate flightcrew awareness and immediate flightcrew response.

(2) Caution: For conditions that require immediate flightcrew awareness and subsequent flightcrew response.

(3) Advisory: For conditions that require flightcrew awareness and may require subsequent flightcrew response.

(c) Warning and caution alerts must:

(1) Be prioritized within each category, when necessary.

(2) Provide timely attention-getting cues through at least two different senses by a combination of aural, visual, or tactile indications.

(3) Permit each occurrence of the attention-getting cues required by paragraph (c)(2) of this section to be acknowledged and suppressed, unless they are required to be continuous.

(d) The alert function must be designed to minimize the effects of false and nuisance alerts. In particular, it must be designed to:

(1) Prevent the presentation of an alert that is inappropriate or unnecessary.

(2) Provide a means to suppress an attention-getting component of an alert caused by a failure of the alerting function that interferes with the flightcrew's ability to safely operate the airplane. This means must not be readily available to the flightcrew so that it could be operated inadvertently or by habitual reflexive action. When an alert is suppressed, there must be a clear and unmistakable annunciation to the flightcrew that the alert has been suppressed.

(e) Visual alert indications must:

(1) Conform to the following color convention:

(i) Red for warning alert indications.

(ii) Amber or yellow for caution alert indications.

(iii) Any color except red or green for advisory alert indications.

(2) Use visual coding techniques, together with other alerting function elements on the flight deck, to distinguish between warning, caution, and advisory alert indications, if they are presented on monochromatic displays that are not capable of conforming to the color convention in paragraph (e)(1) of this section.

(f) Use of the colors red, amber, and yellow on the flight deck for functions other than flightcrew alerting must be limited and must not adversely affect flightcrew alerting.

[Amdt. 25–131, 75 FR 67209, Nov. 2, 2010]

§ 25.1323 Airspeed indicating system.

For each airspeed indicating system, the following apply:

(a) Each airspeed indicating instrument must be approved and must be calibrated to indicate true airspeed (at sea level with a standard atmosphere) with a minimum practicable instrument calibration error when the corresponding pitot and static pressures are applied.

(b) Each system must be calibrated to determine the system error (that is, the relation between IAS and CAS) in flight and during the accelerated take-off ground run. The ground run calibration must be determined—

(1) From 0.8 of the minimum value of V_1 to the maximum value of V_2, considering the approved ranges of altitude and weight; and

(2) With the flaps and power settings corresponding to the values determined in the establishment of the takeoff path under § 25.111 assuming that the critical engine fails at the minimum value of V_1.

(c) The airspeed error of the installation, excluding the airspeed indicator instrument calibration error, may not exceed three percent or five knots, whichever is greater, throughout the speed range, from—

(1) V_{MO} to 1.23 V_{SR1}, with flaps retracted; and

(2) 1.23 V_{SR0} to V_{FE} with flaps in the landing position.

(d) From 1.23 V_{SR} to the speed at which stall warning begins, the IAS must change perceptibly with CAS and in the same sense, and at speeds below stall warning speed the IAS must not change in an incorrect sense.

(e) From V_{MO} to V_{MO} + ⅔ (V_{DF} − V_{MO}), the IAS must change perceptibly with CAS and in the same sense, and at higher speeds up to V_{DF} the IAS must not change in an incorrect sense.

(f) There must be no indication of airspeed that would cause undue difficulty to the pilot during the takeoff between the initiation of rotation and the achievement of a steady climbing condition.

(g) The effects of airspeed indicating system lag may not introduce significant takeoff indicated airspeed bias, or significant errors in takeoff or accelerate-stop distances.

(h) Each system must be arranged, so far as practicable, to prevent malfunction or serious error due to the entry of moisture, dirt, or other substances.

(i) Each system must have a heated pitot tube or an equivalent means of preventing malfunction in the heavy rain conditions defined in Table 1 of this section; mixed phase and ice crystal conditions as defined in part 33, Appendix D, of this chapter; the icing conditions defined in Appendix C of this part; and the following icing conditions specified in Appendix O of this part:

(1) For airplanes certificated in accordance with § 25.1420(a)(1), the icing conditions that the airplane is certified to safely exit following detection.

(2) For airplanes certificated in accordance with § 25.1420(a)(2), the icing conditions that the airplane is certified to safely operate in and the icing conditions that the airplane is certified to safely exit following detection.

(3) For airplanes certificated in accordance with § 25.1420(a)(3) and for airplanes not subject to § 25.1420, all icing conditions.

TABLE 1—HEAVY RAIN CONDITIONS FOR AIRSPEED INDICATING SYSTEM TESTS

Altitude range		Liquid water content	Horizontal extent		Droplet MVD
(ft)	(m)	(g/m3)	(km)	(nmiles)	(µm)
0 to 10 000	0 to 3000	1	100	50	1000
		6	5	3	2000
		15	1	0.5	2000

(j) Where duplicate airspeed indicators are required, their respective pitot tubes must be far enough apart to avoid damage to both tubes in a collision with a bird.

[Doc. No. 5066, 29 FR 18291, Dec. 24, 1964, as amended by Amdt. 25–57, 49 FR 6849, Feb. 23, 1984; Amdt. 25–108, 67 FR 70828, Nov. 26, 2002; Amdt. 25–109, 67 FR 76656, Dec. 12, 2002; Amdt. 25–140, 79 FR 65526, Nov. 4, 2014]

§ 25.1324 Angle of attack system.

Each angle of attack system sensor must be heated or have an equivalent means of preventing malfunction in the heavy rain conditions defined in Table 1 of § 25.1323, the mixed phase and ice crystal conditions as defined in part 33, Appendix D, of this chapter, the icing conditions defined in Appendix C of this part, and the following icing conditions specified in Appendix O of this part:

(a) For airplanes certificated in accordance with § 25.1420(a)(1), the icing conditions that the airplane is certified to safely exit following detection.

(b) For airplanes certificated in accordance with § 25.1420(a)(2), the icing conditions that the airplane is certified to safely operate in and the icing conditions that the airplane is certified to safely exit following detection.

(c) For airplanes certificated in accordance with § 25.1420(a)(3) and for airplanes not subject to § 25.1420, all icing conditions.

[Amdt. 25–140, 79 FR 65527, Nov. 4, 2014]

§ 25.1325 Static pressure systems.

(a) Each instrument with static air case connections must be vented to the outside atmosphere through an appropriate piping system.

(b) Each static port must be designed and located so that:

(1) The static pressure system performance is least affected by airflow variation, or by moisture or other foreign matter; and

(2) The correlation between air pressure in the static pressure system and true ambient atmospheric static pressure is not changed when the airplane is exposed to the icing conditions defined in Appendix C of this part, and the following icing conditions specified in Appendix O of this part:

(i) For airplanes certificated in accordance with § 25.1420(a)(1), the icing conditions that the airplane is certified to safely exit following detection.

(ii) For airplanes certificated in accordance with § 25.1420(a)(2), the icing conditions that the airplane is certified to safely operate in and the icing conditions that the airplane is certified to safely exit following detection.

(iii) For airplanes certificated in accordance with § 25.1420(a)(3) and for airplanes not subject to § 25.1420, all icing conditions.

(c) The design and installation of the static pressure system must be such that—

(1) Positive drainage of moisture is provided; chafing of the tubing and excessive distortion or restriction at bends in the tubing is avoided; and the materials used are durable, suitable for the purpose intended, and protected against corrosion; and

(2) It is airtight except for the port into the atmosphere. A proof test must be conducted to demonstrate the integrity of the static pressure system in the following manner:

(i) *Unpressurized airplanes.* Evacuate the static pressure system to a pressure differential of approximately 1 inch of mercury or to a reading on the altimeter, 1,000 feet above the airplane elevation at the time of the test. Without additional pumping for a period of 1 minute, the loss of indicated altitude must not exceed 100 feet on the altimeter.

(ii) *Pressurized airplanes.* Evacuate the static pressure system until a pressure differential equivalent to the maximum cabin pressure differential for which the airplane is type certificated is achieved. Without additional pumping for a period of 1 minute, the loss of indicated altitude must not exceed 2 percent of the equivalent altitude of the maximum cabin differential pressure or 100 feet, whichever is greater.

(d) Each pressure altimeter must be approved and must be calibrated to indicate pressure altitude in a standard atmosphere, with a minimum practicable calibration error when the corresponding static pressures are applied.

(e) Each system must be designed and installed so that the error in indicated pressure altitude, at sea level, with a standard atmosphere, excluding instrument calibration error, does not result in an error of more than ±30 feet per 100 knots speed for the appropriate configuration in the speed range between 1.23 V_{SR0} with flaps extended and 1.7 V_{SR1} with flaps retracted. However, the error need not be less than ±30 feet.

(f) If an altimeter system is fitted with a device that provides corrections to the altimeter indication, the device must be designed and installed in such manner that it can be bypassed when it malfunctions, unless an alternate altimeter system is provided. Each correction device must be fitted with a means for indicating the occurrence of reasonably probable malfunctions, including power failure, to the flight crew. The indicating means must be effective for any cockpit lighting condition likely to occur.

(g) Except as provided in paragraph (h) of this section, if the static pressure system incorporates both a primary and an alternate static pressure source, the means for selecting one or the other source must be designed so that—

(1) When either source is selected, the other is blocked off; and

(2) Both sources cannot be blocked off simultaneously.

(h) For unpressurized airplanes, paragraph (g)(1) of this section does not apply if it can be demonstrated that the static pressure system calibration, when either static pressure source is selected, is not changed by the other static pressure source being open or blocked.

[Doc. No. 5066, 29 FR 18291, Dec. 24, 1964, as amended by Amdt. 25–5, 30 FR 8261, June 29, 1965; Amdt. 25–12, 32 FR 7587, May 24, 1967; Amdt. 25–41, 42 FR 36970, July 18, 1977; Amdt. 25–108, 67 FR 70828, Nov. 26, 2002; Amdt. 25–140, 79 FR 65527, Nov. 4, 2014]

§ 25.1326　Pitot heat indication systems.

If a flight instrument pitot heating system is installed, an indication system must be provided to indicate to the flight crew when that pitot heating system is not operating. The indication system must comply with the following requirements:

(a) The indication provided must incorporate an amber light that is in clear view of a flight crewmember.

(b) The indication provided must be designed to alert the flight crew if either of the following conditions exist:

(1) The pitot heating system is switched "off".

(2) The pitot heating system is switched "on" and any pitot tube heating element is inoperative.

[Amdt. 25–43, 43 FR 10339, Mar. 13, 1978]

§ 25.1327　Magnetic direction indicator.

(a) Each magnetic direction indicator must be installed so that its accuracy is not excessively affected by the airplane's vibration or magnetic fields.

(b) The compensated installation may not have a deviation, in level flight, greater than 10 degrees on any heading.

§ 25.1329　Flight guidance system.

(a) Quick disengagement controls for the autopilot and autothrust functions must be provided for each pilot. The autopilot quick disengagement controls must be located on both control wheels (or equivalent). The autothrust quick disengagement controls must be located on the thrust control levers. Quick disengagement controls must be readily accessible to each pilot while operating the control wheel (or equivalent) and thrust control levers.

(b) The effects of a failure of the system to disengage the autopilot or autothrust functions when manually commanded by the pilot must be assessed in accordance with the requirements of § 25.1309.

(c) Engagement or switching of the flight guidance system, a mode, or a sensor may not cause a transient response of the airplane's flight path any greater than a minor transient, as defined in paragraph (n)(1) of this section.

(d) Under normal conditions, the disengagement of any automatic control function of a flight guidance system may not cause a transient response of the airplane's flight path any greater than a minor transient.

(e) Under rare normal and non-normal conditions, disengagement of any automatic control function of a flight guidance system may not result in a transient any greater than a significant transient, as defined in paragraph (n)(2) of this section.

(f) The function and direction of motion of each command reference control, such as heading select or vertical speed, must be plainly indicated on, or adjacent to, each control if necessary to prevent inappropriate use or confusion.

(g) Under any condition of flight appropriate to its use, the flight guidance system may not produce hazardous loads on the airplane, nor create hazardous deviations in the flight path. This applies to both fault-free operation and in the event of a malfunction, and assumes that the pilot begins corrective action within a reasonable period of time.

(h) When the flight guidance system is in use, a means must be provided to avoid excursions beyond an acceptable margin from the speed range of the normal flight envelope. If the airplane experiences an excursion outside this range, a means must be provided to prevent the flight guidance system from providing guidance or control to an unsafe speed.

(i) The flight guidance system functions, controls, indications, and alerts must be designed to minimize flightcrew errors and confusion concerning the behavior and operation of the flight guidance system. Means must be provided to indicate the current mode of operation, including any armed modes, transitions, and reversions. Selector switch position is not an acceptable means of indication. The controls and indications must be grouped and presented in a logical and consistent manner. The indications must be visible to each pilot under all expected lighting conditions.

(j) Following disengagement of the autopilot, a warning (visual and auditory) must be provided to each pilot and be timely and distinct from all other cockpit warnings.

(k) Following disengagement of the autothrust function, a caution must be provided to each pilot.

(l) The autopilot may not create a potential hazard when the flightcrew applies an override force to the flight controls.

(m) During autothrust operation, it must be possible for the flightcrew to move the thrust levers without requiring excessive force. The autothrust may not create a potential hazard when the flightcrew applies an override force to the thrust levers.

(n) For purposes of this section, a transient is a disturbance in the control or flight path of the airplane that is not consistent with response to flightcrew inputs or environmental conditions.

(1) A minor transient would not significantly reduce safety margins and would involve flightcrew actions that are well within their capabilities. A minor transient may involve a slight increase in flightcrew workload or some physical discomfort to passengers or cabin crew.

(2) A significant transient may lead to a significant reduction in safety margins, an increase in flightcrew workload, discomfort to the flightcrew, or physical distress to the passengers or cabin crew, possibly including nonfatal injuries. Significant transients do not require, in order to remain within

or recover to the normal flight envelope, any of the following:

(i) Exceptional piloting skill, alertness, or strength.

(ii) Forces applied by the pilot which are greater than those specified in §25.143(c).

(iii) Accelerations or attitudes in the airplane that might result in further hazard to secured or non-secured occupants.

[Doc. No. FAA–2004–18775, 71 FR 18191, Apr. 11, 2006]

§25.1331 Instruments using a power supply.

(a) For each instrument required by §25.1303(b) that uses a power supply, the following apply:

(1) Each instrument must have a visual means integral with, the instrument, to indicate when power adequate to sustain proper instrument performance is not being supplied. The power must be measured at or near the point where it enters the instruments. For electric instruments, the power is considered to be adequate when the voltage is within approved limits.

(2) Each instrument must, in the event of the failure of one power source, be supplied by another power source. This may be accomplished automatically or by manual means.

(3) If an instrument presenting navigation data receives information from sources external to that instrument and loss of that information would render the presented data unreliable, the instrument must incorporate a visual means to warn the crew, when such loss of information occurs, that the presented data should not be relied upon.

(b) As used in this section, "instrument" includes devices that are physically contained in one unit, and devices that are composed of two or more physically separate units or components connected together (such as a remote indicating gyroscopic direction indicator that includes a magnetic sensing element, a gyroscopic unit, an amplifier and an indicator connected together).

[Doc. No. 5066, 29 FR 18291, Dec. 24, 1964, as amended by Amdt. 25–41, 42 FR 36970, July 18, 1977]

347

§ 25.1333   Instrument systems.

For systems that operate the instruments required by § 25.1303(b) which are located at each pilot's station—

(a) Means must be provided to connect the required instruments at the first pilot's station to operating systems which are independent of the operating systems at other flight crew stations, or other equipment;

(b) The equipment, systems, and installations must be designed so that one display of the information essential to the safety of flight which is provided by the instruments, including attitude, direction, airspeed, and altitude will remain available to the pilots, without additional crewmember action, after any single failure or combination of failures that is not shown to be extremely improbable; and

(c) Additional instruments, systems, or equipment may not be connected to the operating systems for the required instruments, unless provisions are made to ensure the continued normal functioning of the required instruments in the event of any malfunction of the additional instruments, systems, or equipment which is not shown to be extremely improbable.

[Amdt. 25–23, 35 FR 5679, Apr. 8, 1970, as amended by Amdt. 25–41, 42 FR 36970, July 18, 1977]

§ 25.1337   Powerplant instruments.

(a) *Instruments and instrument lines.* (1) Each powerplant and auxiliary power unit instrument must meet the requirements of §§ 25.993 and 25.1183.

(2) Each line carrying flammable fluids under pressure must—

(i) Have restricting orifices or other safety devices at the source of pressure to prevent the escape of excessive fluid if the line fails; and

(ii) Be installed and located so that the escape of fluids would not create a hazard.

(3) Each powerplant and auxiliary power unit instrument that utilizes flammable fluids must be installed and located so that the escape of fluid would not create a hazard.

(b) *Fuel quantity indicator.* There must be means to indicate to the flight crewmembers, the quantity, in gallons or equivalent units, of usable fuel in each tank during flight. In addition—

(1) Each fuel quantity indicator must be calibrated to read "zero" during level flight when the quantity of fuel remaining in the tank is equal to the unusable fuel supply determined under § 25.959;

(2) Tanks with interconnected outlets and airspaces may be treated as one tank and need not have separate indicators; and

(3) Each exposed sight gauge, used as a fuel quantity indicator, must be protected against damage.

(c) *Fuel flowmeter system.* If a fuel flowmeter system is installed, each metering component must have a means for bypassing the fuel supply if malfunction of that component severely restricts fuel flow.

(d) *Oil quantity indicator.* There must be a stick gauge or equivalent means to indicate the quantity of oil in each tank. If an oil transfer or reserve oil supply system is installed, there must be a means to indicate to the flight crew, in flight, the quantity of oil in each tank.

(e) *Turbopropeller blade position indicator.* Required turbopropeller blade position indicators must begin indicating before the blade moves more than eight degrees below the flight low pitch stop. The source of indication must directly sense the blade position.

(f) *Fuel pressure indicator.* There must be means to measure fuel pressure, in each system supplying reciprocating engines, at a point downstream of any fuel pump except fuel injection pumps. In addition—

(1) If necessary for the maintenance of proper fuel delivery pressure, there must be a connection to transmit the carburetor air intake static pressure to the proper pump relief valve connection; and

(2) If a connection is required under paragraph (f)(1) of this section, the gauge balance lines must be independently connected to the carburetor inlet pressure to avoid erroneous readings.

[Doc. No. 5066, 29 FR 18291, Dec. 24, 1964, as amended by Amdt. 25–40, 42 FR 15044, Mar. 17, 1977]

ELECTRICAL SYSTEMS AND EQUIPMENT

§ 25.1351 General.

(a) *Electrical system capacity.* The required generating capacity, and number and kinds of power sources must—

(1) Be determined by an electrical load analysis; and

(2) Meet the requirements of § 25.1309.

(b) *Generating system.* The generating system includes electrical power sources, main power busses, transmission cables, and associated control, regulation, and protective devices. It must be designed so that—

(1) Power sources function properly when independent and when connected in combination;

(2) No failure or malfunction of any power source can create a hazard or impair the ability of remaining sources to supply essential loads;

(3) The system voltage and frequency (as applicable) at the terminals of all essential load equipment can be maintained within the limits for which the equipment is designed, during any probable operating condition; and

(4) System transients due to switching, fault clearing, or other causes do not make essential loads inoperative, and do not cause a smoke or fire hazard.

(5) There are means accessible, in flight, to appropriate crewmembers for the individual and collective disconnection of the electrical power sources from the system.

(6) There are means to indicate to appropriate crewmembers the generating system quantities essential for the safe operation of the system, such as the voltage and current supplied by each generator.

(c) *External power.* If provisions are made for connecting external power to the airplane, and that external power can be electrically connected to equipment other than that used for engine starting, means must be provided to ensure that no external power supply having a reverse polarity, or a reverse phase sequence, can supply power to the airplane's electrical system.

(d) *Operation without normal electrical power.* It must be shown by analysis, tests, or both, that the airplane can be operated safely in VFR conditions, for a period of not less than five minutes, with the normal electrical power (electrical power sources excluding the battery) inoperative, with critical type fuel (from the standpoint of flameout and restart capability), and with the airplane initially at the maximum certificated altitude. Parts of the electrical system may remain on if—

(1) A single malfunction, including a wire bundle or junction box fire, cannot result in loss of both the part turned off and the part turned on; and

(2) The parts turned on are electrically and mechanically isolated from the parts turned off.

[Doc. No. 5066, 29 FR 18291, Dec. 24, 1964, as amended by Amdt. 25–41, 42 FR 36970, July 18, 1977; Amdt. 25–72, 55 FR 29785, July 20, 1990]

§ 25.1353 Electrical equipment and installations.

(a) Electrical equipment and controls must be installed so that operation of any one unit or system of units will not adversely affect the simultaneous operation of any other electrical unit or system essential to safe operation. Any electrical interference likely to be present in the airplane must not result in hazardous effects on the airplane or its systems.

(b) Storage batteries must be designed and installed as follows:

(1) Safe cell temperatures and pressures must be maintained during any probable charging or discharging condition. No uncontrolled increase in cell temperature may result when the battery is recharged (after previous complete discharge)—

(i) At maximum regulated voltage or power;

(ii) During a flight of maximum duration; and

(iii) Under the most adverse cooling condition likely to occur in service.

(2) Compliance with paragraph (b)(1) of this section must be shown by test unless experience with similar batteries and installations has shown that maintaining safe cell temperatures and pressures presents no problem.

(3) No explosive or toxic gases emitted by any battery in normal operation, or as the result of any probable malfunction in the charging system or battery installation, may accumulate in hazardous quantities within the airplane.

349

(4) No corrosive fluids or gases that may escape from the battery may damage surrounding airplane structures or adjacent essential equipment.

(5) Each nickel cadmium battery installation must have provisions to prevent any hazardous effect on structure or essential systems that may be caused by the maximum amount of heat the battery can generate during a short circuit of the battery or of individual cells.

(6) Nickel cadmium battery installations must have—

(i) A system to control the charging rate of the battery automatically so as to prevent battery overheating;

(ii) A battery temperature sensing and over-temperature warning system with a means for disconnecting the battery from its charging source in the event of an over-temperature condition; or

(iii) A battery failure sensing and warning system with a means for disconnecting the battery from its charging source in the event of battery failure.

(c) Electrical bonding must provide an adequate electrical return path under both normal and fault conditions, on airplanes having grounded electrical systems.

[Amdt. 25–123, 72 FR 63405, Nov. 8, 2007]

§ 25.1355 Distribution system.

(a) The distribution system includes the distribution busses, their associated feeders, and each control and protective device.

(b) [Reserved]

(c) If two independent sources of electrical power for particular equipment or systems are required by this chapter, in the event of the failure of one power source for such equipment or system, another power source (including its separate feeder) must be automatically provided or be manually selectable to maintain equipment or system operation.

[Doc. No. 5066, 29 FR 18291, Dec. 24, 1964, as amended by Amdt. 25–23, 35 FR 5679, Apr. 8, 1970; Amdt. 25–38, 41 FR 55468, Dec. 20, 1976]

§ 25.1357 Circuit protective devices.

(a) Automatic protective devices must be used to minimize distress to the electrical system and hazard to the airplane in the event of wiring faults or serious malfunction of the system or connected equipment.

(b) The protective and control devices in the generating system must be designed to de-energize and disconnect faulty power sources and power transmission equipment from their associated busses with sufficient rapidity to provide protection from hazardous over-voltage and other malfunctioning.

(c) Each resettable circuit protective device must be designed so that, when an overload or circuit fault exists, it will open the circuit irrespective of the position of the operating control.

(d) If the ability to reset a circuit breaker or replace a fuse is essential to safety in flight, that circuit breaker or fuse must be located and identified so that it can be readily reset or replaced in flight. Where fuses are used, there must be spare fuses for use in flight equal to at least 50% of the number of fuses of each rating required for complete circuit protection.

(e) Each circuit for essential loads must have individual circuit protection. However, individual protection for each circuit in an essential load system (such as each position light circuit in a system) is not required.

(f) For airplane systems for which the ability to remove or reset power during normal operations is necessary, the system must be designed so that circuit breakers are not the primary means to remove or reset system power unless specifically designed for use as a switch.

(g) Automatic reset circuit breakers may be used as integral protectors for electrical equipment (such as thermal cut-outs) if there is circuit protection to protect the cable to the equipment.

[Doc. No. 5066, 29 FR 18291, Dec. 24, 1964, as amended by Amdt. 25–123, 72 FR 63405, Nov. 8, 2007]

§ 25.1360 Precautions against injury.

(a) Shock. The electrical system must be designed to minimize risk of electric shock to crew, passengers, and servicing personnel and to maintenance personnel using normal precautions.

(b) *Burns.* The temperature of any part that may be handled by a crewmember during normal operations must not cause dangerous inadvertent movement by the crewmember or injury to the crewmember.

[Amdt. 25–123, 72 FR 63406, Nov. 8, 2007]

§25.1362 Electrical supplies for emergency conditions.

A suitable electrical supply must be provided to those services required for emergency procedures after an emergency landing or ditching. The circuits for these services must be designed, protected, and installed so that the risk of the services being rendered ineffective under these emergency conditions is minimized.

[Amdt. 25–123, 72 FR 63406, Nov. 8, 2007]

§25.1363 Electrical system tests.

(a) When laboratory tests of the electrical system are conducted—

(1) The tests must be performed on a mock-up using the same generating equipment used in the airplane;

(2) The equipment must simulate the electrical characteristics of the distribution wiring and connected loads to the extent necessary for valid test results; and

(3) Laboratory generator drives must simulate the actual prime movers on the airplane with respect to their reaction to generator loading, including loading due to faults.

(b) For each flight condition that cannot be simulated adequately in the laboratory or by ground tests on the airplane, flight tests must be made.

§25.1365 Electrical appliances, motors, and transformers.

(a) Domestic appliances must be designed and installed so that in the event of failures of the electrical supply or control system, the requirements of §25.1309(b), (c), and (d) will be satisfied. Domestic appliances are items such as cooktops, ovens, coffee makers, water heaters, refrigerators, and toilet flush systems that are placed on the airplane to provide service amenities to passengers.

(b) Galleys and cooking appliances must be installed in a way that minimizes risk of overheat or fire.

(c) Domestic appliances, particularly those in galley areas, must be installed or protected so as to prevent damage or contamination of other equipment or systems from fluids or vapors which may be present during normal operation or as a result of spillage, if such damage or contamination could create a hazardous condition.

(d) Unless compliance with §25.1309(b) is provided by the circuit protective device required by §25.1357(a), electric motors and transformers, including those installed in domestic systems, must have a suitable thermal protection device to prevent overheating under normal operation and failure conditions, if overheating could create a smoke or fire hazard.

[Amdt. 25–123, 72 FR 63406, Nov. 8, 2007]

LIGHTS

§25.1381 Instrument lights.

(a) The instrument lights must—

(1) Provide sufficient illumination to make each instrument, switch and other device necessary for safe operation easily readable unless sufficient illumination is available from another source; and

(2) Be installed so that—

(i) Their direct rays are shielded from the pilot's eyes; and

(ii) No objectionable reflections are visible to the pilot.

(b) Unless undimmed instrument lights are satisfactory under each expected flight condition, there must be a means to control the intensity of illumination.

[Doc. No. 5066, 29 FR 18291, Dec. 24, 1964, as amended by Amdt. 25–72, 55 FR 29785, July 20, 1990]

§25.1383 Landing lights.

(a) Each landing light must be approved, and must be installed so that—

(1) No objectionable glare is visible to the pilot;

(2) The pilot is not adversely affected by halation; and

(3) It provides enough light for night landing.

(b) Except when one switch is used for the lights of a multiple light installation at one location, there must be a separate switch for each light.

(c) There must be a means to indicate to the pilots when the landing lights are extended.

§ 25.1385 Position light system installation.

(a) *General.* Each part of each position light system must meet the applicable requirements of this section and each system as a whole must meet the requirements of §§ 25.1387 through 25.1397.

(b) *Forward position lights.* Forward position lights must consist of a red and a green light spaced laterally as far apart as practicable and installed forward on the airplane so that, with the airplane in the normal flying position, the red light is on the left side and the green light is on the right side. Each light must be approved.

(c) *Rear position light.* The rear position light must be a white light mounted as far aft as practicable on the tail or on each wing tip, and must be approved.

(d) *Light covers and color filters.* Each light cover or color filter must be at least flame resistant and may not change color or shape or lose any appreciable light transmission during normal use.

[Doc. No. 5066, 29 FR 18291, Dec. 24, 1964, as amended by Amdt. 25–38, 41 FR 55468, Dec. 20, 1976]

§ 25.1387 Position light system dihedral angles.

(a) Except as provided in paragraph (e) of this section, each forward and rear position light must, as installed, show unbroken light within the dihedral angles described in this section.

(b) Dihedral angle L (left) is formed by two intersecting vertical planes, the first parallel to the longitudinal axis of the airplane, and the other at 110 degrees to the left of the first, as viewed when looking forward along the longitudinal axis.

(c) Dihedral angle R (right) is formed by two intersecting vertical planes, the first parallel to the longitudinal axis of the airplane, and the other at 110 degrees to the right of the first, as viewed when looking forward along the longitudinal axis.

(d) Dihedral angle A (aft) is formed by two intersecting vertical planes making angles of 70 degrees to the right and to the left, respectively, to a vertical plane passing through the longitudinal axis, as viewed when looking aft along the longitudinal axis.

(e) If the rear position light, when mounted as far aft as practicable in accordance with § 25.1385(c), cannot show unbroken light within dihedral angle A (as defined in paragraph (d) of this section), a solid angle or angles of obstructed visibility totaling not more than 0.04 steradians is allowable within that dihedral angle, if such solid angle is within a cone whose apex is at the rear position light and whose elements make an angle of 30° with a vertical line passing through the rear position light.

[Doc. No. 5066, 29 FR 18291, Dec. 24, 1964, as amended by Amdt. 25–30, 36 FR 21278, Nov. 5, 1971]

§ 25.1389 Position light distribution and intensities.

(a) *General.* The intensities prescribed in this section must be provided by new equipment with light covers and color filters in place. Intensities must be determined with the light source operating at a steady value equal to the average luminous output of the source at the normal operating voltage of the airplane. The light distribution and intensity of each position light must meet the requirements of paragraph (b) of this section.

(b) *Forward and rear position lights.* The light distribution and intensities of forward and rear position lights must be expressed in terms of minimum intensities in the horizontal plane, minimum intensities in any vertical plane, and maximum intensities in overlapping beams, within dihedral angles L, R, and A, and must meet the following requirements:

(1) *Intensities in the horizontal plane.* Each intensity in the horizontal plane (the plane containing the longitudinal axis of the airplane and perpendicular to the plane of symmetry of the airplane) must equal or exceed the values in § 25.1391.

(2) *Intensities in any vertical plane.* Each intensity in any vertical plane (the plane perpendicular to the horizontal plane) must equal or exceed the appropriate value in § 25.1393, where I is

the minimum intensity prescribed in §25.1391 for the corresponding angles in the horizontal plane.

(3) *Intensities in overlaps between adjacent signals.* No intensity in any overlap between adjacent signals may exceed the values given in §25.1395, except that higher intensities in overlaps may be used with main beam intensities substantially greater than the minima specified in §§25.1391 and 25.1393 if the overlap intensities in relation to the main beam intensities do not adversely affect signal clarity. When the peak intensity of the forward position lights is more than 100 candles, the maximum overlap intensities between them may exceed the values given in §25.1395 if the overlap intensity in Area A is not more than 10 percent of peak position light intensity and the overlap intensity in Area B is not greater than 2.5 percent of peak position light intensity.

§25.1391 Minimum intensities in the horizontal plane of forward and rear position lights.

Each position light intensity must equal or exceed the applicable values in the following table:

Dihedral angle (light included)	Angle from right or left of longitudinal axis, measured from dead ahead	Intensity (candles)
L and R (forward red and green).	0° to 10°	40
	10° to 20°	30
	20° to 110°	5
A (rear white)	110° to 180°	20

§25.1393 Minimum intensities in any vertical plane of forward and rear position lights.

Each position light intensity must equal or exceed the applicable values in the following table:

Angle above or below the horizontal plane	Intensity, I
0°	1.00
0° to 5°	0.90
5° to 10°	0.80
10° to 15°	0.70
15° to 20°	0.50
20° to 30°	0.30
30° to 40°	0.10
40° to 90°	0.05

§25.1395 Maximum intensities in overlapping beams of forward and rear position lights.

No position light intensity may exceed the applicable values in the following table, except as provided in §25.1389(b)(3).

Overlaps	Maximum intensity	
	Area A (candles)	Area B (candles)
Green in dihedral angle L	10	1
Red in dihedral angle R	10	1
Green in dihedral angle A	5	1
Red in dihedral angle A	5	1
Rear white in dihedral angle L	5	1
Rear white in dihedral angle R	5	1

Where—

(a) Area A includes all directions in the adjacent dihedral angle that pass through the light source and intersect the common boundary plane at more than 10 degrees but less than 20 degrees; and

(b) Area B includes all directions in the adjacent dihedral angle that pass through the light source and intersect the common boundary plane at more than 20 degrees.

§25.1397 Color specifications.

Each position light color must have the applicable International Commission on Illumination chromaticity coordinates as follows:

(a) *Aviation red—*

y is not greater than 0.335; and
z is not greater than 0.002.

(b) *Aviation green—*

x is not greater than $0.440 - 0.320y$;
x is not greater than $y - 0.170$; and
y is not less than $0.390 - 0.170x$.

(c) *Aviation white—*

x is not less than 0.300 and not greater than 0.540;
y is not less than $x - 0.040$; or $y_0 - 0.010$, whichever is the smaller; and
y is not greater than $x + 0.020$ nor $0.636 - 0.400x$;
Where y_0 is the y coordinate of the Planckian radiator for the value of x considered.

[Doc. No. 5066, 29 FR 18291, Dec. 24, 1964, as amended by Amdt. 25–27, 36 FR 12972, July 10, 1971]

§ 25.1399 Riding light.

(a) Each riding (anchor) light required for a seaplane or amphibian must be installed so that it can—

(1) Show a white light for at least 2 nautical miles at night under clear atmospheric conditions; and

(2) Show the maximum unbroken light practicable when the airplane is moored or drifting on the water.

(b) Externally hung lights may be used.

§ 25.1401 Anticollision light system.

(a) *General.* The airplane must have an anticollision light system that—

(1) Consists of one or more approved anticollision lights located so that their light will not impair the crew's vision or detract from the conspicuity of the position lights; and

(2) Meets the requirements of paragraphs (b) through (f) of this section.

(b) *Field of coverage.* The system must consist of enough lights to illuminate the vital areas around the airplane considering the physical configuration and flight characteristics of the airplane. The field of coverage must extend in each direction within at least 75 degrees above and 75 degrees below the horizontal plane of the airplane, except that a solid angle or angles of obstructed visibility totaling not more than 0.03 steradians is allowable within a solid angle equal to 0.15 steradians centered about the longitudinal axis in the rearward direction.

(c) *Flashing characteristics.* The arrangement of the system, that is, the number of light sources, beam width, speed of rotation, and other characteristics, must give an effective flash frequency of not less than 40, nor more than 100 cycles per minute. The effective flash frequency is the frequency at which the airplane's complete anticollision light system is observed from a distance, and applies to each sector of light including any overlaps that exist when the system consists of more than one light source. In overlaps, flash frequencies may exceed 100, but not 180 cycles per minute.

(d) *Color.* Each anticollision light must be either aviation red or aviation white and must meet the applicable requirements of § 25.1397.

(e) *Light intensity.* The minimum light intensities in all vertical planes, measured with the red filter (if used) and expressed in terms of "effective" intensities, must meet the requirements of paragraph (f) of this section. The following relation must be assumed:

$$ I_e = \frac{\int_{t_1}^{t_2} I(t)dt}{0.2 + (t_2 - t_1)} $$

where:

I_e = effective intensity (candles).
$I(t)$ = instantaneous intensity as a function of time.
$t_2 - t_1$ = flash time interval (seconds).

Normally, the maximum value of effective intensity is obtained when t_2 and t_1 are chosen so that the effective intensity is equal to the instantaneous intensity at t_2 and t_1.

(f) *Minimum effective intensities for anticollision lights.* Each anticollision light effective intensity must equal or exceed the applicable values in the following table.

Angle above or below the horizontal plane	Effective intensity (candles)
0° to 5°	400
5° to 10°	240
10° to 20°	80
20° to 30°	40
30° to 75°	20

[Doc. No. 5066, 29 FR 18291, Dec. 24, 1964, as amended by Amdt. 25-27, 36 FR 12972, July 10, 1971; Amdt. 25-41, 42 FR 36970, July 18, 1977]

§ 25.1403 Wing icing detection lights.

Unless operations at night in known or forecast icing conditions are prohibited by an operating limitation, a means must be provided for illuminating or otherwise determining the formation of ice on the parts of the wings that are critical from the standpoint of ice accumulation. Any illumination that is used must be of a type that will not cause glare or reflection that would handicap crewmembers in the performance of their duties.

[Amdt. 25-38, 41 FR 55468, Dec. 20, 1976]

SAFETY EQUIPMENT

§25.1411 General.

(a) *Accessibility.* Required safety equipment to be used by the crew in an emergency must be readily accessible.

(b) *Stowage provisions.* Stowage provisions for required emergency equipment must be furnished and must—

(1) Be arranged so that the equipment is directly accessible and its location is obvious; and

(2) Protect the safety equipment from inadvertent damage.

(c) *Emergency exit descent device.* The stowage provisions for the emergency exit descent devices required by §25.810(a) must be at each exit for which they are intended.

(d) *Liferafts.* (1) The stowage provisions for the liferafts described in §25.1415 must accommodate enough rafts for the maximum number of occupants for which certification for ditching is requested.

(2) Liferafts must be stowed near exits through which the rafts can be launched during an unplanned ditching.

(3) Rafts automatically or remotely released outside the airplane must be attached to the airplane by means of the static line prescribed in §25.1415.

(4) The stowage provisions for each portable liferaft must allow rapid detachment and removal of the raft for use at other than the intended exits.

(e) *Long-range signaling device.* The stowage provisions for the long-range signaling device required by §25.1415 must be near an exit available during an unplanned ditching.

(f) *Life preserver stowage provisions.* The stowage provisions for life preservers described in §25.1415 must accommodate one life preserver for each occupant for which certification for ditching is requested. Each life preserver must be within easy reach of each seated occupant.

(g) *Life line stowage provisions.* If certification for ditching under §25.801 is requested, there must be provisions to store life lines. These provisions must—

(1) Allow one life line to be attached to each side of the fuselage; and

(2) Be arranged to allow the life lines to be used to enable the occupants to stay on the wing after ditching.

[Doc. No. 5066, 29 FR 18291, Dec. 24, 1964, as amended by Amdt. 25–32, 37 FR 3972, Feb. 24, 1972; Amdt. 25–46, 43 FR 50598, Oct. 30, 1978; Amdt. 25–53, 45 FR 41593, June 19, 1980; Amdt. 25–70, 54 FR 43925, Oct. 27, 1989; Amdt. 25–79, 58 FR 45229, Aug. 26, 1993; Amdt. 25–116, 69 FR 62789, Oct. 27, 2004]

§25.1415 Ditching equipment.

(a) Ditching equipment used in airplanes to be certificated for ditching under §25.801, and required by the operating rules of this chapter, must meet the requirements of this section.

(b) Each liferaft and each life preserver must be approved. In addition—

(1) Unless excess rafts of enough capacity are provided, the buoyancy and seating capacity beyond the rated capacity of the rafts must accommodate all occupants of the airplane in the event of a loss of one raft of the largest rated capacity; and

(2) Each raft must have a trailing line, and must have a static line designed to hold the raft near the airplane but to release it if the airplane becomes totally submerged.

(c) Approved survival equipment must be attached to each liferaft.

(d) There must be an approved survival type emergency locator transmitter for use in one life raft.

(e) For airplanes not certificated for ditching under §25.801 and not having approved life preservers, there must be an approved flotation means for each occupant. This means must be within easy reach of each seated occupant and must be readily removable from the airplane.

[Doc. No. 5066, 29 FR 18291, Dec. 24, 1964, as amended by Amdt. 25–29, 36 FR 18722, Sept. 21, 1971; Amdt. 25–50, 45 FR 38348, June 9, 1980; Amdt. 25–72, 55 FR 29785, July 20, 1990; Amdt. 25–82, 59 FR 32057, June 21, 1994]

§25.1419 Ice protection.

If the applicant seeks certification for flight in icing conditions, the airplane must be able to safely operate in the continuous maximum and intermittent maximum icing conditions of appendix C. To establish this—

(a) An analysis must be performed to establish that the ice protection for the various components of the airplane is adequate, taking into account the various airplane operational configurations; and

(b) To verify the ice protection analysis, to check for icing anomalies, and to demonstrate that the ice protection system and its components are effective, the airplane or its components must be flight tested in the various operational configurations, in measured natural atmospheric icing conditions and, as found necessary, by one or more of the following means:

(1) Laboratory dry air or simulated icing tests, or a combination of both, of the components or models of the components.

(2) Flight dry air tests of the ice protection system as a whole, or of its individual components.

(3) Flight tests of the airplane or its components in measured simulated icing conditions.

(c) Caution information, such as an amber caution light or equivalent, must be provided to alert the flightcrew when the anti-ice or de-ice system is not functioning normally.

(d) For turbine engine powered airplanes, the ice protection provisions of this section are considered to be applicable primarily to the airframe. For the powerplant installation, certain additional provisions of subpart E of this part may be found applicable.

(e) One of the following methods of icing detection and activation of the airframe ice protection system must be provided:

(1) A primary ice detection system that automatically activates or alerts the flightcrew to activate the airframe ice protection system;

(2) A definition of visual cues for recognition of the first sign of ice accretion on a specified surface combined with an advisory ice detection system that alerts the flightcrew to activate the airframe ice protection system; or

(3) Identification of conditions conducive to airframe icing as defined by an appropriate static or total air temperature and visible moisture for use by the flightcrew to activate the airframe ice protection system.

(f) Unless the applicant shows that the airframe ice protection system need not be operated during specific phases of flight, the requirements of paragraph (e) of this section are applicable to all phases of flight.

(g) After the initial activation of the airframe ice protection system—

(1) The ice protection system must be designed to operate continuously;

(2) The airplane must be equipped with a system that automatically cycles the ice protection system; or

(3) An ice detection system must be provided to alert the flightcrew each time the ice protection system must be cycled.

(h) Procedures for operation of the ice protection system, including activation and deactivation, must be established and documented in the Airplane Flight Manual.

[Amdt. 25–72, 55 FR 29785, July 20, 1990, as amended by Amdt. 25–121, 72 FR 44669, Aug. 8, 2007; Amdt. 25–129, 74 FR 38339, Aug. 3, 2009]

§ 25.1420 Supercooled large drop icing conditions.

(a) If certification for flight in icing conditions is sought, in addition to the requirements of § 25.1419, an airplane with a maximum takeoff weight less than 60,000 pounds or with reversible flight controls must be capable of operating in accordance with paragraphs (a)(1), (2), or (3), of this section.

(1) Operating safely after encountering the icing conditions defined in Appendix O of this part:

(i) The airplane must have a means to detect that it is operating in Appendix O icing conditions; and

(ii) Following detection of Appendix O icing conditions, the airplane must be capable of operating safely while exiting all icing conditions.

(2) Operating safely in a portion of the icing conditions defined in Appendix O of this part as selected by the applicant:

(i) The airplane must have a means to detect that it is operating in conditions that exceed the selected portion of Appendix O icing conditions; and

(ii) Following detection, the airplane must be capable of operating safely while exiting all icing conditions.

(3) Operating safely in the icing conditions defined in Appendix O of this part.

(b) To establish that the airplane can operate safely as required in paragraph

(a) of this section, an applicant must show through analysis that the ice protection for the various components of the airplane is adequate, taking into account the various airplane operational configurations. To verify the analysis, one, or more as found necessary, of the following methods must be used:

(1) Laboratory dry air or simulated icing tests, or a combination of both, of the components or models of the components.

(2) Laboratory dry air or simulated icing tests, or a combination of both, of models of the airplane.

(3) Flight tests of the airplane or its components in simulated icing conditions, measured as necessary to support the analysis.

(4) Flight tests of the airplane with simulated ice shapes.

(5) Flight tests of the airplane in natural icing conditions, measured as necessary to support the analysis.

(c) For an airplane certified in accordance with paragraph (a)(2) or (3) of this section, the requirements of §25.1419(e), (f), (g), and (h) must be met for the icing conditions defined in Appendix O of this part in which the airplane is certified to operate.

(d) For the purposes of this section, the following definitions apply:

(1) *Reversible Flight Controls.* Flight controls in the normal operating configuration that have force or motion originating at the airplane's control surface (for example, through aerodynamic loads, static imbalance, or trim or servo tab inputs) that is transmitted back to flight deck controls. This term refers to flight deck controls connected to the pitch, roll, or yaw control surfaces by direct mechanical linkages, cables, or push-pull rods in such a way that pilot effort produces motion or force about the hinge line.

(2) *Simulated Icing Test.* Testing conducted in simulated icing conditions, such as in an icing tunnel or behind an icing tanker.

(3) *Simulated Ice Shape.* Ice shape fabricated from wood, epoxy, or other materials by any construction technique.

[Amdt. 25–140, 79 FR 65528, Nov. 4, 2014]

§25.1421 Megaphones.

If a megaphone is installed, a restraining means must be provided that is capable of restraining the megaphone when it is subjected to the ultimate inertia forces specified in §25.561(b)(3).

[Amdt. 25–41, 42 FR 36970, July 18, 1977]

§25.1423 Public address system.

A public address system required by this chapter must—

(a) Be powerable when the aircraft is in flight or stopped on the ground, after the shutdown or failure of all engines and auxiliary power units, or the disconnection or failure of all power sources dependent on their continued operation, for—

(1) A time duration of at least 10 minutes, including an aggregate time duration of at least 5 minutes of announcements made by flight and cabin crewmembers, considering all other loads which may remain powered by the same source when all other power sources are inoperative; and

(2) An additional time duration in its standby state appropriate or required for any other loads that are powered by the same source and that are essential to safety of flight or required during emergency conditions.

(b) Be capable of operation within 3 seconds from the time a microphone is removed from its stowage.

(c) Be intelligible at all passenger seats, lavatories, and flight attendant seats and work stations.

(d) Be designed so that no unused, unstowed microphone will render the system inoperative.

(e) Be capable of functioning independently of any required crewmember interphone system.

(f) Be accessible for immediate use from each of two flight crewmember stations in the pilot compartment.

(g) For each required floor-level passenger emergency exit which has an adjacent flight attendant seat, have a microphone which is readily accessible to the seated flight attendant, except that one microphone may serve more than one exit, provided the proximity of the exits allows unassisted verbal communication between seated flight attendants.

[Doc. No. 26003, 58 FR 45229, Aug. 26, 1993, as amended by Amdt. 25–115, 69 FR 40527, July 2, 2004]

MISCELLANEOUS EQUIPMENT

§ 25.1431 Electronic equipment.

(a) In showing compliance with § 25.1309 (a) and (b) with respect to radio and electronic equipment and their installations, critical environmental conditions must be considered.

(b) Radio and electronic equipment must be supplied with power under the requirements of § 25.1355(c).

(c) Radio and electronic equipment, controls, and wiring must be installed so that operation of any one unit or system of units will not adversely affect the simultaneous operation of any other radio unit, or electronic unit, or system of units, required by this chapter.

(d) Electronic equipment must be designed and installed such that it does not cause essential loads to become inoperative as a result of electrical power supply transients or transients from other causes.

[Doc. No. 5066, 29 FR 18291, Dec. 24, 1964, as amended by Amdt. 25–113, 69 FR 12530, Mar. 16, 2004]

§ 25.1433 Vacuum systems.

There must be means, in addition to the normal pressure relief, to automatically relieve the pressure in the discharge lines from the vacuum air pump when the delivery temperature of the air becomes unsafe.

[Doc. No. 5066, 29 FR 18291, Dec. 24, 1964, as amended by Amdt. 25–72, 55 FR 29785, July 20, 1990]

§ 25.1435 Hydraulic systems.

(a) *Element design.* Each element of the hydraulic system must be designed to:

(1) Withstand the proof pressure without permanent deformation that would prevent it from performing its intended functions, and the ultimate pressure without rupture. The proof and ultimate pressures are defined in terms of the design operating pressure (DOP) as follows:

Element	Proof (xDOP)	Ultimate (xDOP)
1. Tubes and fittings.	1.5	3.0
2. Pressure vessels containing gas:		
High pressure (e.g., accumulators)	3.0	4.0
Low pressure (e.g., reservoirs) ..	1.5	3.0
3. Hoses	2.0	4.0
4. All other elements	1.5	2.0

(2) Withstand, without deformation that would prevent it from performing its intended function, the design operating pressure in combination with limit structural loads that may be imposed;

(3) Withstand, without rupture, the design operating pressure multiplied by a factor of 1.5 in combination with ultimate structural load that can reasonably occur simultaneously;

(4) Withstand the fatigue effects of all cyclic pressures, including transients, and associated externally induced loads, taking into account the consequences of element failure; and

(5) Perform as intended under all environmental conditions for which the airplane is certificated.

(b) *System design.* Each hydraulic system must:

(1) Have means located at a flightcrew station to indicate appropriate system parameters, if

(i) It performs a function necessary for continued safe flight and landing; or

(ii) In the event of hydraulic system malfunction, corrective action by the crew to ensure continued safe flight and landing is necessary;

(2) Have means to ensure that system pressures, including transient pressures and pressures from fluid volumetric changes in elements that are likely to remain closed long enough for such changes to occur, are within the design capabilities of each element, such that they meet the requirements defined in § 25.1435(a)(1) through (a)(5);

(3) Have means to minimize the release of harmful or hazardous concentrations of hydraulic fluid or vapors into the crew and passenger compartments during flight;

(4) Meet the applicable requirements of §§ 25.863, 25.1183, 25.1185, and 25.1189 if a flammable hydraulic fluid is used; and

(5) Be designed to use any suitable hydraulic fluid specified by the airplane manufacturer, which must be identified by appropriate markings as required by §25.1541.

(c) *Tests*. Tests must be conducted on the hydraulic system(s), and/or subsystem(s) and elements, except that analysis may be used in place of or to supplement testing, where the analysis is shown to be reliable and appropriate. All internal and external influences must be taken into account to an extent necessary to evaluate their effects, and to assure reliable system and element functioning and integration. Failure or unacceptable deficiency of an element or system must be corrected and be sufficiently retested, where necessary.

(1) The system(s), subsystem(s), or element(s) must be subjected to performance, fatigue, and endurance tests representative of airplane ground and flight operations.

(2) The complete system must be tested to determine proper functional performance and relation to the other systems, including simulation of relevant failure conditions, and to support or validate element design.

(3) The complete hydraulic system(s) must be functionally tested on the airplane in normal operation over the range of motion of all associated user systems. The test must be conducted at the system relief pressure or 1.25 times the DOP if a system pressure relief device is not part of the system design. Clearances between hydraulic system elements and other systems or structural elements must remain adequate and there must be no detrimental effects.

[Doc. No. 28617, 66 FR 27402, May 16, 2001]

§25.1438 Pressurization and pneumatic systems.

(a) Pressurization system elements must be burst pressure tested to 2.0 times, and proof pressure tested to 1.5 times, the maximum normal operating pressure.

(b) Pneumatic system elements must be burst pressure tested to 3.0 times, and proof pressure tested to 1.5 times, the maximum normal operating pressure.

(c) An analysis, or a combination of analysis and test, may be substituted for any test required by paragraph (a) or (b) of this section if the Administrator finds it equivalent to the required test.

[Amdt. 25–41, 42 FR 36971, July 18, 1977]

§25.1439 Protective breathing equipment.

(a) Fixed (stationary, or built in) protective breathing equipment must be installed for the use of the flightcrew, and at least one portable protective breathing equipment shall be located at or near the flight deck for use by a flight crewmember. In addition, portable protective breathing equipment must be installed for the use of appropriate crewmembers for fighting fires in compartments accessible in flight other than the flight deck. This includes isolated compartments and upper and lower lobe galleys, in which crewmember occupancy is permitted during flight. Equipment must be installed for the maximum number of crewmembers expected to be in the area during any operation.

(b) For protective breathing equipment required by paragraph (a) of this section or by the applicable Operating Regulations:

(1) The equipment must be designed to protect the appropriate crewmember from smoke, carbon dioxide, and other harmful gases while on flight deck duty or while combating fires.

(2) The equipment must include—

(i) Masks covering the eyes, nose and mouth, or

(ii) Masks covering the nose and mouth, plus accessory equipment to cover the eyes.

(3) Equipment, including portable equipment, must allow communication with other crewmembers while in use. Equipment available at flightcrew assigned duty stations must also enable the flightcrew to use radio equipment.

(4) The part of the equipment protecting the eyes shall not cause any appreciable adverse effect on vision and must allow corrective glasses to be worn.

(5) The equipment must supply protective oxygen of 15 minutes duration per crewmember at a pressure altitude of 8,000 feet with a respiratory minute

volume of 30 liters per minute BTPD. The equipment and system must be designed to prevent any inward leakage to the inside of the device and prevent any outward leakage causing significant increase in the oxygen content of the local ambient atmosphere. If a demand oxygen system is used, a supply of 300 liters of free oxygen at 70 °F. and 760 mm. Hg. pressure is considered to be of 15-minute duration at the prescribed altitude and minute volume. If a continuous flow open circuit protective breathing system is used, a flow rate of 60 liters per minute at 8,000 feet (45 liters per minute at sea level) and a supply of 600 liters of free oxygen at 70 °F. and 760 mm. Hg. pressure is considered to be of 15-minute duration at the prescribed altitude and minute volume. Continuous flow systems must not increase the ambient oxygen content of the local atmosphere above that of demand systems. BTPD refers to body temperature conditions (that is, 37 °C., at ambient pressure, dry).

(6) The equipment must meet the requirements of § 25.1441.

[Doc. No. FAA–2002–13859, 69 FR 40528, July 2, 2004]

§ 25.1441 Oxygen equipment and supply.

(a) If certification with supplemental oxygen equipment is requested, the equipment must meet the requirements of this section and §§ 25.1443 through 25.1453.

(b) The oxygen system must be free from hazards in itself, in its method of operation, and in its effect upon other components.

(c) There must be a means to allow the crew to readily determine, during flight, the quantity of oxygen available in each source of supply.

(d) The oxygen flow rate and the oxygen equipment for airplanes for which certification for operation above 40,000 feet is requested must be approved.

§ 25.1443 Minimum mass flow of supplemental oxygen.

(a) If continuous flow equipment is installed for use by flight crewmembers, the minimum mass flow of supplemental oxygen required for each crewmember may not be less than the flow required to maintain, during inspiration, a mean tracheal oxygen partial pressure of 149 mm. Hg. when breathing 15 liters per minute, BTPS, and with a maximum tidal volume of 700 cc. with a constant time interval between respirations.

(b) If demand equipment is installed for use by flight crewmembers, the minimum mass flow of supplemental oxygen required for each crewmember may not be less than the flow required to maintain, during inspiration, a mean tracheal oxygen partial pressure of 122 mm. Hg., up to and including a cabin pressure altitude of 35,000 feet, and 95 percent oxygen between cabin pressure altitudes of 35,000 and 40,000 feet, when breathing 20 liters per minute BTPS. In addition, there must be means to allow the crew to use undiluted oxygen at their discretion.

(c) For passengers and cabin attendants, the minimum mass flow of supplemental oxygen required for each person at various cabin pressure altitudes may not be less than the flow required to maintain, during inspiration and while using the oxygen equipment (including masks) provided, the following mean tracheal oxygen partial pressures:

(1) At cabin pressure altitudes above 10,000 feet up to and including 18,500 feet, a mean tracheal oxygen partial pressure of 100 mm. Hg. when breathing 15 liters per minute, BTPS, and with a tidal volume of 700 cc. with a constant time interval between respirations.

(2) At cabin pressure altitudes above 18,500 feet up to and including 40,000 feet, a mean tracheal oxygen partial pressure of 83.8 mm. Hg. when breathing 30 liters per minute, BTPS, and with a tidal volume of 1,100 cc. with a constant time interval between respirations.

(d) If first-aid oxygen equipment is installed, the minimum mass flow of oxygen to each user may not be less than four liters per minute, STPD. However, there may be a means to decrease this flow to not less than two liters per minute, STPD, at any cabin altitude. The quantity of oxygen required is based upon an average flow rate of three liters per minute per person for whom first-aid oxygen is required.

(e) If portable oxygen equipment is installed for use by crewmembers, the minimum mass flow of supplemental oxygen is the same as specified in paragraph (a) or (b) of this section, whichever is applicable.

§ 25.1445 Equipment standards for the oxygen distributing system.

(a) When oxygen is supplied to both crew and passengers, the distribution system must be designed for either—

(1) A source of supply for the flight crew on duty and a separate source for the passengers and other crewmembers; or

(2) A common source of supply with means to separately reserve the minimum supply required by the flight crew on duty.

(b) Portable walk-around oxygen units of the continuous flow, diluter-demand, and straight demand kinds may be used to meet the crew or passenger breathing requirements.

§ 25.1447 Equipment standards for oxygen dispensing units.

If oxygen dispensing units are installed, the following apply:

(a) There must be an individual dispensing unit for each occupant for whom supplemental oxygen is to be supplied. Units must be designed to cover the nose and mouth and must be equipped with a suitable means to retain the unit in position on the face. Flight crew masks for supplemental oxygen must have provisions for the use of communication equipment.

(b) If certification for operation up to and including 25,000 feet is requested, an oxygen supply terminal and unit of oxygen dispensing equipment for the immediate use of oxygen by each crewmember must be within easy reach of that crewmember. For any other occupants, the supply terminals and dispensing equipment must be located to allow the use of oxygen as required by the operating rules in this chapter.

(c) If certification for operation above 25,000 feet is requested, there must be oxygen dispensing equipment meeting the following requirements:

(1) There must be an oxygen dispensing unit connected to oxygen supply terminals immediately available to each occupant wherever seated, and at least two oxygen dispensing units connected to oxygen terminals in each lavatory. The total number of dispensing units and outlets in the cabin must exceed the number of seats by at least 10 percent. The extra units must be as uniformly distributed throughout the cabin as practicable. Except as provided in paragraph (c)(5) of this section, if certification for operation above 30,000 feet is requested, the dispensing units providing the required oxygen flow must be automatically presented to the occupants before the cabin pressure altitude exceeds 15,000 feet. The crewmembers must be provided with a manual means of making the dispensing units immediately available in the event of failure of the automatic system.

(2) Each flight crewmember on flight deck duty must be provided with a quick-donning type oxygen dispensing unit connected to an oxygen supply terminal. This dispensing unit must be immediately available to the flight crewmember when seated at his station, and installed so that it:

(i) Can be placed on the face from its ready position, properly secured, sealed, and supplying oxygen upon demand, with one hand, within five seconds and without disturbing eyeglasses or causing delay in proceeding with emergency duties; and

(ii) Allows, while in place, the performance of normal communication functions.

(3) The oxygen dispensing equipment for the flight crewmembers must be:

(i) The diluter demand or pressure demand (pressure demand mask with a diluter demand pressure breathing regulator) type, or other approved oxygen equipment shown to provide the same degree of protection, for airplanes to be operated above 25,000 feet.

(ii) The pressure demand (pressure demand mask with a diluter demand pressure breathing regulator) type with mask-mounted regulator, or other approved oxygen equipment shown to provide the same degree of protection, for airplanes operated at altitudes where decompressions that are not extremely improbable may expose the flightcrew to cabin pressure altitudes in excess of 34,000 feet.

(4) Portable oxygen equipment must be immediately available for each cabin attendant. The portable oxygen equipment must have the oxygen dispensing unit connected to the portable oxygen supply.

(5) When operating into or out of airports with elevations above 13,000 feet, the dispensing units providing the required oxygen flow must be automatically presented to the occupants at cabin pressure altitudes no higher than 2,000 feet above the airplane's maximum takeoff and landing altitude.

[Doc. No. 5066, 29 FR 18291, Dec. 24, 1964, as amended by Amdt. 25–41, 42 FR 36971, July 18, 1977; Amdt. 25–87, 61 FR 28696, June 5, 1996; Amdt. 25–116, 69 FR 62789, Oct. 27, 2004; Amdt. No. 25–151, 88 FR 39161, June 15, 2023; 88 FR 44032, July 11, 2023]

§ 25.1449 Means for determining use of oxygen.

There must be a means to allow the crew to determine whether oxygen is being delivered to the dispensing equipment.

§ 25.1450 Chemical oxygen generators.

(a) For the purpose of this section, a chemical oxygen generator is defined as a device which produces oxygen by chemical reaction.

(b) Each chemical oxygen generator must be designed and installed in accordance with the following requirements:

(1) Surface temperature developed by the generator during operation may not create a hazard to the airplane or to its occupants.

(2) Means must be provided to relieve any internal pressure that may be hazardous.

(3) Except as provided in SFAR 109, each chemical oxygen generator installation must meet the requirements of § 25.795(d).

(c) In addition to meeting the requirements in paragraph (b) of this section, each portable chemical oxygen generator that is capable of sustained operation by successive replacement of a generator element must be placarded to show—

(1) The rate of oxygen flow, in liters per minute;

(2) The duration of oxygen flow, in minutes, for the replaceable generator element; and

(3) A warning that the replaceable generator element may be hot, unless the element construction is such that the surface temperature cannot exceed 100 degrees F.

[Amdt. 25–41, 42 FR 36971, July 18, 1977, as amended at 79 FR 13519, Mar. 11, 2014]

§ 25.1453 Protection of oxygen equipment from rupture.

Oxygen pressure tanks, and lines between tanks and the shutoff means, must be—

(a) Protected from unsafe temperatures; and

(b) Located where the probability and hazards of rupture in a crash landing are minimized.

§ 25.1455 Draining of fluids subject to freezing.

If fluids subject to freezing may be drained overboard in flight or during ground operation, the drains must be designed and located to prevent the formation of hazardous quantities of ice on the airplane as a result of the drainage.

[Amdt. 25–23, 35 FR 5680, Apr. 8, 1970]

§ 25.1457 Cockpit voice recorders.

(a) Each cockpit voice recorder required by the operating rules of this chapter must be approved and must be installed so that it will record the following:

(1) Voice communications transmitted from or received in the airplane by radio.

(2) Voice communications of flight crewmembers on the flight deck.

(3) Voice communications of flight crewmembers on the flight deck, using the airplane's interphone system.

(4) Voice or audio signals identifying navigation or approach aids introduced into a headset or speaker.

(5) Voice communications of flight crewmembers using the passenger loudspeaker system, if there is such a system and if the fourth channel is available in accordance with the requirements of paragraph (c)(4)(ii) of this section.

(6) If datalink communication equipment is installed, all datalink communications, using an approved data message set. Datalink messages must be recorded as the output signal from the communications unit that translates the signal into usable data.

(b) The recording requirements of paragraph (a)(2) of this section must be met by installing a cockpit-mounted area microphone, located in the best position for recording voice communications originating at the first and second pilot stations and voice communications of other crewmembers on the flight deck when directed to those stations. The microphone must be so located and, if necessary, the pre-amplifiers and filters of the recorder must be so adjusted or supplemented, that the intelligibility of the recorded communications is as high as practicable when recorded under flight cockpit noise conditions and played back. Repeated aural or visual playback of the record may be used in evaluating intelligibility.

(c) Each cockpit voice recorder must be installed so that the part of the communication or audio signals specified in paragraph (a) of this section obtained from each of the following sources is recorded on a separate channel:

(1) For the first channel, from each boom, mask, or hand-held microphone, headset, or speaker used at the first pilot station.

(2) For the second channel from each boom, mask, or hand-held microphone, headset, or speaker used at the second pilot station.

(3) For the third channel—from the cockpit-mounted area microphone.

(4) For the fourth channel, from—

(i) Each boom, mask, or hand-held microphone, headset, or speaker used at the station for the third and fourth crew members; or

(ii) If the stations specified in paragraph (c)(4)(i) of this section are not required or if the signal at such a station is picked up by another channel, each microphone on the flight deck that is used with the passenger loudspeaker system, if its signals are not picked up by another channel.

(5) As far as is practicable all sounds received by the microphone listed in paragraphs (c)(1), (2), and (4) of this section must be recorded without interruption irrespective of the position of the interphone-transmitter key switch. The design shall ensure that sidetone for the flight crew is produced only when the interphone, public address system, or radio transmitters are in use.

(d) Each cockpit voice recorder must be installed so that—

(1)(i) It receives its electrical power from the bus that provides the maximum reliability for operation of the cockpit voice recorder without jeopardizing service to essential or emergency loads.

(ii) It remains powered for as long as possible without jeopardizing emergency operation of the airplane.

(2) There is an automatic means to simultaneously stop the recorder and prevent each erasure feature from functioning, within 10 minutes after crash impact;

(3) There is an aural or visual means for preflight checking of the recorder for proper operation;

(4) Any single electrical failure external to the recorder does not disable both the cockpit voice recorder and the flight data recorder;

(5) It has an independent power source—

(i) That provides 10 ±1 minutes of electrical power to operate both the cockpit voice recorder and cockpit-mounted area microphone;

(ii) That is located as close as practicable to the cockpit voice recorder; and

(iii) To which the cockpit voice recorder and cockpit-mounted area microphone are switched automatically in the event that all other power to the cockpit voice recorder is interrupted either by normal shutdown or by any other loss of power to the electrical power bus; and

(6) It is in a separate container from the flight data recorder when both are required. If used to comply with only the cockpit voice recorder requirements, a combination unit may be installed.

(e) The recorder container must be located and mounted to minimize the probability of rupture of the container

363

as a result of crash impact and consequent heat damage to the recorder from fire.

(1) Except as provided in paragraph (e)(2) of this section, the recorder container must be located as far aft as practicable, but need not be outside of the pressurized compartment, and may not be located where aft-mounted engines may crush the container during impact.

(2) If two separate combination digital flight data recorder and cockpit voice recorder units are installed instead of one cockpit voice recorder and one digital flight data recorder, the combination unit that is installed to comply with the cockpit voice recorder requirements may be located near the cockpit.

(f) If the cockpit voice recorder has a bulk erasure device, the installation must be designed to minimize the probability of inadvertent operation and actuation of the device during crash impact.

(g) Each recorder container must—

(1) Be either bright orange or bright yellow;

(2) Have reflective tape affixed to its external surface to facilitate its location under water; and

(3) Have an underwater locating device, when required by the operating rules of this chapter, on or adjacent to the container which is secured in such manner that they are not likely to be separated during crash impact.

[Doc. No. 5066, 29 FR 18291, Dec. 24, 1964, as amended by Amdt. 25-2, 30 FR 3932, Mar. 26, 1965; Amdt. 25-16, 32 FR 13914, Oct. 6, 1967; Amdt. 25-41, 42 FR 36971, July 18, 1977; Amdt. 25-65, 53 FR 26143, July 11, 1988; Amdt. 25-124, 73 FR 12563, Mar. 7, 2008; 74 FR 32800, July 9, 2009]

§ 25.1459　Flight data recorders.

(a) Each flight recorder required by the operating rules of this chapter must be installed so that—

(1) It is supplied with airspeed, altitude, and directional data obtained from sources that meet the accuracy requirements of §§ 25.1323, 25.1325, and 25.1327, as appropriate;

(2) The vertical acceleration sensor is rigidly attached, and located longitudinally either within the approved center of gravity limits of the airplane, or at a distance forward or aft of these limits that does not exceed 25 percent of the airplane's mean aerodynamic chord;

(3)(i) It receives its electrical power from the bus that provides the maximum reliability for operation of the flight data recorder without jeopardizing service to essential or emergency loads.

(ii) It remains powered for as long as possible without jeopardizing emergency operation of the airplane.

(4) There is an aural or visual means for preflight checking of the recorder for proper recording of data in the storage medium;

(5) Except for recorders powered solely by the engine-driven electrical generator system, there is an automatic means to simultaneously stop a recorder that has a data erasure feature and prevent each erasure feature from functioning, within 10 minutes after crash impact;

(6) There is a means to record data from which the time of each radio transmission either to or from ATC can be determined;

(7) Any single electrical failure external to the recorder does not disable both the cockpit voice recorder and the flight data recorder; and

(8) It is in a separate container from the cockpit voice recorder when both are required. If used to comply with only the flight data recorder requirements, a combination unit may be installed. If a combination unit is installed as a cockpit voice recorder to comply with § 25.1457(e)(2), a combination unit must be used to comply with this flight data recorder requirement.

(b) Each nonejectable record container must be located and mounted so as to minimize the probability of container rupture resulting from crash impact and subsequent damage to the record from fire. In meeting this requirement the record container must be located as far aft as practicable, but need not be aft of the pressurized compartment, and may not be where aft-mounted engines may crush the container upon impact.

(c) A correlation must be established between the flight recorder readings of airspeed, altitude, and heading and the corresponding readings (taking into account correction factors) of the first pilot's instruments. The correlation must cover the airspeed range over which the airplane is to be operated, the range of altitude to which the airplane is limited, and 360 degrees of heading. Correlation may be established on the ground as appropriate.

(d) Each recorder container must—

(1) Be either bright orange or bright yellow;

(2) Have reflective tape affixed to its external surface to facilitate its location under water; and

(3) Have an underwater locating device, when required by the operating rules of this chapter, on or adjacent to the container which is secured in such a manner that they are not likely to be separated during crash impact.

(e) Any novel or unique design or operational characteristics of the aircraft shall be evaluated to determine if any dedicated parameters must be recorded on flight recorders in addition to or in place of existing requirements.

[Amdt. 25–8, 31 FR 127, Jan. 6, 1966, as amended by Amdt. 25–25, 35 FR 13192, Aug. 19, 1970; Amdt. 25–37, 40 FR 2577, Jan. 14, 1975; Amdt. 25–41, 42 FR 36971, July 18, 1977; Amdt. 25–65, 53 FR 26144, July 11, 1988; Amdt. 25–124, 73 FR 12563, Mar. 7, 2008; 74 FR 32800, July 9, 2009]

§ 25.1461 Equipment containing high energy rotors.

(a) Equipment containing high energy rotors must meet paragraph (b), (c), or (d) of this section.

(b) High energy rotors contained in equipment must be able to withstand damage caused by malfunctions, vibration, abnormal speeds, and abnormal temperatures. In addition—

(1) Auxiliary rotor cases must be able to contain damage caused by the failure of high energy rotor blades; and

(2) Equipment control devices, systems, and instrumentation must reasonably ensure that no operating limitations affecting the integrity of high energy rotors will be exceeded in service.

(c) It must be shown by test that equipment containing high energy rotors can contain any failure of a high energy rotor that occurs at the highest speed obtainable with the normal speed control devices inoperative.

(d) Equipment containing high energy rotors must be located where rotor failure will neither endanger the occupants nor adversely affect continued safe flight.

[Amdt. 25–41, 42 FR 36971, July 18, 1977]

Subpart G—Operating Limitations and Information

§ 25.1501 General.

(a) Each operating limitation specified in §§ 25.1503 through 25.1533 and other limitations and information necessary for safe operation must be established.

(b) The operating limitations and other information necessary for safe operation must be made available to the crewmembers as prescribed in §§ 25.1541 through 25.1587.

[Amdt. 25–42, 43 FR 2323, Jan. 16, 1978]

OPERATING LIMITATIONS

§ 25.1503 Airspeed limitations: general.

When airspeed limitations are a function of weight, weight distribution, altitude, or Mach number, limitations corresponding to each critical combination of these factors must be established.

§ 25.1505 Maximum operating limit speed.

The maximum operating limit speed (V_{MO}/M_{MO} airspeed or Mach Number, whichever is critical at a particular altitude) is a speed that may not be deliberately exceeded in any regime of flight (climb, cruise, or descent), unless a higher speed is authorized for flight test or pilot training operations. V_{MO}/M_{MO} must be established so that it is not greater than the design cruising speed V_C and so that it is sufficiently below V_D/M_D or V_{DF}/M_{DF}, to make it highly improbable that the latter speeds will be inadvertently exceeded in operations. The speed margin between V_{MO}/M_{MO} and V_D/M_D or $V_{DF}M/_{DF}$ may not be less than that determined under § 25.335(b) or found necessary during the flight tests conducted under § 25.253.

[Amdt. 25–23, 35 FR 5680, Apr. 8, 1970]

§ 25.1507 Maneuvering speed.

The maneuvering speed must be established so that it does not exceed the design maneuvering speed V_A determined under § 25.335(c).

§ 25.1511 Flap extended speed.

The established flap extended speed V_{FE} must be established so that it does not exceed the design flap speed V_F chosen under §§ 25.335(e) and 25.345, for the corresponding flap positions and engine powers.

§ 25.1513 Minimum control speed.

The minimum control speed V_{MC} determined under § 25.149 must be established as an operating limitation.

§ 25.1515 Landing gear speeds.

(a) The established landing gear operating speed or speeds, V_{LO}, may not exceed the speed at which it is safe both to extend and to retract the landing gear, as determined under § 25.729 or by flight characteristics. If the extension speed is not the same as the retraction speed, the two speeds must be designated as $V_{LO(EXT)}$ and $V_{LO(RET)}$, respectively.

(b) The established landing gear extended speed V_{LE} may not exceed the speed at which it is safe to fly with the landing gear secured in the fully extended position, and that determined under § 25.729.

[Doc. No. 5066, 29 FR 18291, Dec. 24, 1964, as amended by Amdt. 25–38, 41 FR 55468, Dec. 20, 1976]

§ 25.1516 Other speed limitations.

Any other limitation associated with speed must be established.

[Doc. No. 2000–8511, 66 FR 34024, June 26, 2001]

§ 25.1517 Rough air speed, V_{RA}.

(a) A rough air speed, V_{RA}, for use as the recommended turbulence penetration airspeed, and a rough air Mach number, M_{RA}, for use as the recommended turbulence penetration Mach number, must be established. V_{RA}/M_{RA} must be sufficiently less than V_{MO}/M_{MO} to ensure that likely speed variation during rough air encounters will not cause the overspeed warning to operate too frequently.

(b) At altitudes where V_{MO} is not limited by Mach number, in the absence of a rational investigation substantiating the use of other values, V_{RA} must be less than V_{MO} minus 35 KTAS.

(c) At altitudes where V_{MO} is limited by Mach number, M_{RA} may be chosen to provide an optimum margin between low and high speed buffet boundaries.

[Amdt. 25–141, 79 FR 73469, Dec. 11, 2014, as amended by FAA–2022–1355; Amdt. No. 25–148, 87 FR 75710, Dec. 9, 2022; 88 FR 2813, Jan. 18, 2023]

§ 25.1519 Weight, center of gravity, and weight distribution.

The airplane weight, center of gravity, and weight distribution limitations determined under §§ 25.23 through 25.27 must be established as operating limitations.

§ 25.1521 Powerplant limitations.

(a) *General.* The powerplant limitations prescribed in this section must be established so that they do not exceed the corresponding limits for which the engines or propellers are type certificated and do not exceed the values on which compliance with any other requirement of this part is based.

(b) *Reciprocating engine installations.* Operating limitations relating to the following must be established for reciprocating engine installations:

(1) Horsepower or torque, r.p.m., manifold pressure, and time at critical pressure altitude and sea level pressure altitude for—

(i) Maximum continuous power (relating to unsupercharged operation or to operation in each supercharger mode as applicable); and

(ii) Takeoff power (relating to unsupercharged operation or to operation in each supercharger mode as applicable).

(2) Fuel grade or specification.

(3) Cylinder head and oil temperatures.

(4) Any other parameter for which a limitation has been established as part of the engine type certificate except that a limitation need not be established for a parameter that cannot be exceeded during normal operation due to the design of the installation or to another established limitation.

(c) *Turbine engine installations.* Operating limitations relating to the following must be established for turbine engine installations:

(1) Horsepower, torque or thrust, r.p.m., gas temperature, and time for—

(i) Maximum continuous power or thrust (relating to augmented or unaugmented operation as applicable).

(ii) Takeoff power or thrust (relating to augmented or unaugmented operation as applicable).

(2) Fuel designation or specification.

(3) Maximum time interval between engine run-ups from idle, run-up power setting and duration at power for ground operation in icing conditions, as defined in §25.1093(b)(2).

(4) Any other parameter for which a limitation has been established as part of the engine type certificate except that a limitation need not be established for a parameter that cannot be exceeded during normal operation due to the design of the installation or to another established limitation.

(d) *Ambient temperature.* An ambient temperature limitation (including limitations for winterization installations, if applicable) must be established as the maximum ambient atmospheric temperature established in accordance with §25.1043(b).

[Amdt. 25–72, 55 FR 29786, July 20, 1990, as amended by Amdt. 25–140, 79 FR 65528, Nov. 4, 2014]

§25.1522 Auxiliary power unit limitations.

If an auxiliary power unit is installed in the airplane, limitations established for the auxiliary power unit, including categories of operation, must be specified as operating limitations for the airplane.

[Amdt. 25–72, 55 FR 29786, July 20, 1990]

§25.1523 Minimum flight crew.

The minimum flight crew must be established so that it is sufficient for safe operation, considering—

(a) The workload on individual crewmembers;

(b) The accessibility and ease of operation of necessary controls by the appropriate crewmember; and

(c) The kind of operation authorized under §25.1525.

The criteria used in making the determinations required by this section are set forth in appendix D.

[Doc. No. 5066, 29 FR 18291, Dec. 24, 1964, as amended by Amdt. 25–3, 30 FR 6067, Apr. 29, 1965]

§25.1525 Kinds of operation.

The kinds of operation to which the airplane is limited are established by the category in which it is eligible for certification and by the installed equipment.

§25.1527 Ambient air temperature and operating altitude.

The extremes of the ambient air temperature and operating altitude for which operation is allowed, as limited by flight, structural, powerplant, functional, or equipment characteristics, must be established.

[Doc. No. 2000–8511, 66 FR 34024, June 26, 2001]

§25.1529 Instructions for Continued Airworthiness.

The applicant must prepare Instructions for Continued Airworthiness in accordance with appendix H to this part that are acceptable to the Administrator. The instructions may be incomplete at type certification if a program exists to ensure their completion prior to delivery of the first airplane or issuance of a standard certificate of airworthiness, whichever occurs later.

[Amdt. 25–54, 45 FR 60173, Sept. 11, 1980]

§25.1531 Maneuvering flight load factors.

Load factor limitations, not exceeding the positive limit load factors determined from the maneuvering diagram in §25.333(b), must be established.

§25.1533 Additional operating limitations.

(a) Additional operating limitations must be established as follows:

(1) The maximum takeoff weights must be established as the weights at which compliance is shown with the applicable provisions of this part (including the takeoff climb provisions of §25.121(a) through (c), for altitudes and ambient temperatures).

(2) The maximum landing weights must be established as the weights at

which compliance is shown with the applicable provisions of this part (including the landing and approach climb provisions of §§ 25.119 and 25.121(d) for altitudes and ambient temperatures).

(3) The minimum takeoff distances must be established as the distances at which compliance is shown with the applicable provisions of this part (including the provisions of §§ 25.109 and 25.113, for weights, altitudes, temperatures, wind components, runway surface conditions (dry and wet), and runway gradients) for smooth, hard-surfaced runways. Additionally, at the option of the applicant, wet runway takeoff distances may be established for runway surfaces that have been grooved or treated with a porous friction course, and may be approved for use on runways where such surfaces have been designed constructed, and maintained in a manner acceptable to the Administrator.

(b) The extremes for variable factors (such as altitude, temperature, wind, and runway gradients) are those at which compliance with the applicable provisions of this part is shown.

(c) For airplanes certified in accordance with § 25.1420(a)(1) or (2), an operating limitation must be established to:

(1) Prohibit intentional flight, including takeoff and landing, into icing conditions defined in Appendix O of this part for which the airplane has not been certified to safely operate; and

(2) Require exiting all icing conditions if icing conditions defined in Appendix O of this part are encountered for which the airplane has not been certified to safely operate.

[Doc. No. 5066, 29 FR 18291, Dec. 24, 1964, as amended by Amdt. 25–38, 41 FR 55468, Dec. 20, 1976; Amdt. 25–72, 55 FR 29786, July 20, 1990; Amdt. 25–92, 63 FR 8321, Feb. 18, 1998; Amdt. 25–140, 79 FR 65528, Nov. 4, 2014]

§ 25.1535 ETOPS approval.

Except as provided in § 25.3, each applicant seeking ETOPS type design approval must comply with the provisions of Appendix K of this part.

[Doc. No. FAA–2002–6717, 72 FR 1873, Jan. 16, 2007]

MARKINGS AND PLACARDS

§ 25.1541 General.

(a) The airplane must contain—

(1) The specified markings and placards; and

(2) Any additional information, instrument markings, and placards required for the safe operation if there are unusual design, operating, or handling characteristics.

(b) Each marking and placard prescribed in paragraph (a) of this section—

(1) Must be displayed in a conspicuous place; and

(2) May not be easily erased, disfigured, or obscured.

§ 25.1543 Instrument markings: general.

For each instrument—

(a) When markings are on the cover glass of the instrument, there must be means to maintain the correct alignment of the glass cover with the face of the dial; and

(b) Each instrument marking must be clearly visible to the appropriate crewmember.

[Doc. No. 5066, 29 FR 18291, Dec. 24, 1964, as amended by Amdt. 25–72, 55 FR 29786, July 20, 1990]

§ 25.1545 Airspeed limitation information.

The airspeed limitations required by § 25.1583 (a) must be easily read and understood by the flight crew.

§ 25.1547 Magnetic direction indicator.

(a) A placard meeting the requirements of this section must be installed on, or near, the magnetic direction indicator.

(b) The placard must show the calibration of the instrument in level flight with the engines operating.

(c) The placard must state whether the calibration was made with radio receivers on or off.

(d) Each calibration reading must be in terms of magnetic heading in not more than 45 degree increments.

§ 25.1549 Powerplant and auxiliary power unit instruments.

For each required powerplant and auxiliary power unit instrument, as appropriate to the type of instrument—

(a) Each maximum and, if applicable, minimum safe operating limit must be marked with a red radial or a red line;

(b) Each normal operating range must be marked with a green arc or green line, not extending beyond the maximum and minimum safe limits;

(c) Each takeoff and precautionary range must be marked with a yellow arc or a yellow line; and

(d) Each engine, auxiliary power unit, or propeller speed range that is restricted because of excessive vibration stresses must be marked with red arcs or red lines.

[Amdt. 25–40, 42 FR 15044, Mar. 17, 1977]

§ 25.1551 Oil quantity indication.

Each oil quantity indicating means must be marked to indicate the quantity of oil readily and accurately.

[Amdt. 25–72, 55 FR 29786, July 20, 1990]

§ 25.1553 Fuel quantity indicator.

If the unusable fuel supply for any tank exceeds one gallon, or five percent of the tank capacity, whichever is greater, a red arc must be marked on its indicator extending from the calibrated zero reading to the lowest reading obtainable in level flight.

§ 25.1555 Control markings.

(a) Each cockpit control, other than primary flight controls and controls whose function is obvious, must be plainly marked as to its function and method of operation.

(b) Each aerodynamic control must be marked under the requirements of §§ 25.677 and 25.699.

(c) For powerplant fuel controls—

(1) Each fuel tank selector control must be marked to indicate the position corresponding to each tank and to each existing cross feed position;

(2) If safe operation requires the use of any tanks in a specific sequence, that sequence must be marked on, or adjacent to, the selector for those tanks; and

(3) Each valve control for each engine must be marked to indicate the position corresponding to each engine controlled.

(d) For accessory, auxiliary, and emergency controls—

(1) Each emergency control (including each fuel jettisoning and fluid shutoff must be colored red; and

(2) Each visual indicator required by § 25.729(e) must be marked so that the pilot can determine at any time when the wheels are locked in either extreme position, if retractable landing gear is used.

§ 25.1557 Miscellaneous markings and placards.

(a) *Baggage and cargo compartments and ballast location.* Each baggage and cargo compartment, and each ballast location must have a placard stating any limitations on contents, including weight, that are necessary under the loading requirements. However, underseat compartments designed for the storage of carry-on articles weighing not more than 20 pounds need not have a loading limitation placard.

(b) *Powerplant fluid filler openings.* The following apply:

(1) Fuel filler openings must be marked at or near the filler cover with—

(i) The word "fuel";

(ii) For reciprocating engine powered airplanes, the minimum fuel grade;

(iii) For turbine engine powered airplanes, the permissible fuel designations; and

(iv) For pressure fueling systems, the maximum permissible fueling supply pressure and the maximum permissible defueling pressure.

(2) Oil filler openings must be marked at or near the filler cover with the word "oil".

(3) Augmentation fluid filler openings must be marked at or near the filler cover to identify the required fluid.

(c) *Emergency exit placards.* Each emergency exit placard must meet the requirements of § 25.811.

(d) *Doors.* Each door that must be used in order to reach any required emergency exit must have a suitable placard stating that the door is to be latched in the open position during takeoff and landing.

[Doc. No. 5066, 29 FR 18291, Dec. 24, 1964, as amended by Amdt. 25–32, 37 FR 3972, Feb. 24, 1972; Amdt. 25–38, 41 FR 55468, Dec. 20, 1976; Amdt. 25–72, 55 FR 29786, July 20, 1990]

§ 25.1561 Safety equipment.

(a) Each safety equipment control to be operated by the crew in emergency, such as controls for automatic liferaft releases, must be plainly marked as to its method of operation.

(b) Each location, such as a locker or compartment, that carries any fire extinguishing, signaling, or other life saving equipment must be marked accordingly.

(c) Stowage provisions for required emergency equipment must be conspicuously marked to identify the contents and facilitate the easy removal of the equipment.

(d) Each liferaft must have obviously marked operating instructions.

(e) Approved survival equipment must be marked for identification and method of operation.

[Doc. No. 5066, 29 FR 18291, Dec. 24, 1964, as amended by Amdt. 25–46, 43 FR 50598, Oct. 30, 1978]

§ 25.1563 Airspeed placard.

A placard showing the maximum airspeeds for flap extension for the take-off, approach, and landing positions must be installed in clear view of each pilot.

AIRPLANE FLIGHT MANUAL

§ 25.1581 General.

(a) *Furnishing information.* An Airplane Flight Manual must be furnished with each airplane, and it must contain the following:

(1) Information required by §§ 25.1583 through 25.1587.

(2) Other information that is necessary for safe operation because of design, operating, or handling characteristics.

(3) Any limitation, procedure, or other information established as a condition of compliance with the applicable noise standards of part 36 of this chapter.

(b) *Approved information.* Each part of the manual listed in §§ 25.1583 through 25.1587, that is appropriate to the airplane, must be furnished, verified, and approved, and must be segregated, identified, and clearly distinguished from each unapproved part of that manual.

(c) [Reserved]

(d) Each Airplane Flight Manual must include a table of contents if the complexity of the manual indicates a need for it.

[Amdt. 25–42, 43 FR 2323, Jan. 16, 1978, as amended by Amdt. 25–72, 55 FR 29786, July 20, 1990]

§ 25.1583 Operating limitations.

(a) *Airspeed limitations.* The following airspeed limitations and any other airspeed limitations necessary for safe operation must be furnished:

(1) The maximum operating limit speed V_{MO}/M_{MO} and a statement that this speed limit may not be deliberately exceeded in any regime of flight (climb, cruise, or descent) unless a higher speed is authorized for flight test or pilot training.

(2) If an airspeed limitation is based upon compressibility effects, a statement to this effect and information as to any symptoms, the probable behavior of the airplane, and the recommended recovery procedures.

(3) The maneuvering speed established under § 25.1507 and statements, as applicable to the particular design, explaining that:

(i) Full application of pitch, roll, or yaw controls should be confined to speeds below the maneuvering speed; and

(ii) Rapid and large alternating control inputs, especially in combination with large changes in pitch, roll, or yaw, and full control inputs in more than one axis at the same time, should be avoided as they may result in structural failures at any speed, including below the maneuvering speed.

(4) The flap extended speed V_{FE} and the pertinent flap positions and engine powers.

(5) The landing gear operating speed or speeds, and a statement explaining the speeds as defined in § 25.1515(a).

(6) The landing gear extended speed V_{LE} if greater than V_{LO} and a statement that this is the maximum speed

at which the airplane can be safely flown with the landing gear extended.

(b) *Powerplant limitations.* The following information must be furnished:

(1) Limitations required by § 25.1521 and § 25.1522.

(2) Explanation of the limitations, when appropriate.

(3) Information necessary for marking the instruments required by §§ 25.1549 through 25.1553.

(c) *Weight and loading distribution.* The weight and center of gravity limitations established under § 25.1519 must be furnished in the Airplane Flight Manual. All of the following information, including the weight distribution limitations established under § 25.1519, must be presented either in the Airplane Flight Manual or in a separate weight and balance control and loading document that is incorporated by reference in the Airplane Flight Manual:

(1) The condition of the airplane and the items included in the empty weight as defined in accordance with § 25.29.

(2) Loading instructions necessary to ensure loading of the airplane within the weight and center of gravity limits, and to maintain the loading within these limits in flight.

(3) If certification for more than one center of gravity range is requested, the appropriate limitations, with regard to weight and loading procedures, for each separate center of gravity range.

(d) *Flight crew.* The number and functions of the minimum flight crew determined under § 25.1523 must be furnished.

(e) *Kinds of operation.* The kinds of operation approved under § 25.1525 must be furnished.

(f) *Ambient air temperatures and operating altitudes.* The extremes of the ambient air temperatures and operating altitudes established under § 25.1527 must be furnished.

(g) [Reserved]

(h) *Additional operating limitations.* The operating limitations established under § 25.1533 must be furnished.

(i) *Maneuvering flight load factors.* The positive maneuvering limit load factors for which the structure is proven,

described in terms of accelerations, must be furnished.

[Doc. No. 5066, 29 FR 1891, Dec. 24, 1964, as amended by Amdt. 25–38, 41 FR 55468, Dec. 20, 1976; Amdt. 25–42, 43 FR 2323, Jan. 16, 1978; Amdt. 25–46, 43 FR 50598, Oct. 30, 1978; Amdt. 25–72, 55 FR 29787, July 20, 1990; Amdt. 25–105, 66 FR 34024, June 26, 2001; 75 FR 49818, Aug. 16, 2010]

§ 25.1585 Operating procedures.

(a) Operating procedures must be furnished for—

(1) Normal procedures peculiar to the particular type or model encountered in connection with routine operations;

(2) Non-normal procedures for malfunction cases and failure conditions involving the use of special systems or the alternative use of regular systems; and

(3) Emergency procedures for foreseeable but unusual situations in which immediate and precise action by the crew may be expected to substantially reduce the risk of catastrophe.

(b) Information or procedures not directly related to airworthiness or not under the control of the crew, must not be included, nor must any procedure that is accepted as basic airmanship.

(c) Information identifying each operating condition in which the fuel system independence prescribed in § 25.953 is necessary for safety must be furnished, together with instructions for placing the fuel system in a configuration used to show compliance with that section.

(d) The buffet onset envelopes, determined under § 25.251 must be furnished. The buffet onset envelopes presented may reflect the center of gravity at which the airplane is normally loaded during cruise if corrections for the effect of different center of gravity locations are furnished.

(e) Information must be furnished that indicates that when the fuel quantity indicator reads "zero" in level flight, any fuel remaining in the fuel tank cannot be used safely in flight.

(f) Information on the total quantity of usable fuel for each fuel tank must be furnished.

[Doc. No. 2000–8511, 66 FR 34024, June 26, 2001]

§ 25.1587 Performance information.

(a) Each Airplane Flight Manual must contain information to permit conversion of the indicated temperature to free air temperature if other than a free air temperature indicator is used to comply with the requirements of § 25.1303(a)(1).

(b) Each Airplane Flight Manual must contain the performance information computed under the applicable provisions of this part (including §§ 25.115, 25.123, and 25.125 for the weights, altitudes, temperatures, wind components, and runway gradients, as applicable) within the operational limits of the airplane, and must contain the following:

(1) In each case, the conditions of power, configuration, and speeds, and the procedures for handling the airplane and any system having a significant effect on the performance information.

(2) V_{SR} determined in accordance with § 25.103.

(3) The following performance information (determined by extrapolation and computed for the range of weights between the maximum landing weight and the maximum takeoff weight):

(i) Climb in the landing configuration.

(ii) Climb in the approach configuration.

(iii) Landing distance.

(4) Procedures established under § 25.101(f) and (g) that are related to the limitations and information required by § 25.1533 and by this paragraph (b) in the form of guidance material, including any relevant limitations or information.

(5) An explanation of significant or unusual flight or ground handling characteristics of the airplane.

(6) Corrections to indicated values of airspeed, altitude, and outside air temperature.

(7) An explanation of operational landing runway length factors included in the presentation of the landing distance, if appropriate.

[Doc. No. 2000–8511, 66 FR 34024, June 26, 2001, as amended by Amdt. 25–108, 67 FR 70828, Nov. 26, 2002]

Subpart H—Electrical Wiring Interconnection Systems (EWIS)

SOURCE: Docket No. FAA–2004–18379, 72 FR 63406, Nov. 8, 2007, unless otherwise noted.

§ 25.1701 Definition.

(a) As used in this chapter, electrical wiring interconnection system (EWIS) means any wire, wiring device, or combination of these, including termination devices, installed in any area of the airplane for the purpose of transmitting electrical energy, including data and signals, between two or more intended termination points. This includes:

(1) Wires and cables.

(2) Bus bars.

(3) The termination point on electrical devices, including those on relays, interrupters, switches, contactors, terminal blocks and circuit breakers, and other circuit protection devices.

(4) Connectors, including feed-through connectors.

(5) Connector accessories.

(6) Electrical grounding and bonding devices and their associated connections.

(7) Electrical splices.

(8) Materials used to provide additional protection for wires, including wire insulation, wire sleeving, and conduits that have electrical termination for the purpose of bonding.

(9) Shields or braids.

(10) Clamps and other devices used to route and support the wire bundle.

(11) Cable tie devices.

(12) Labels or other means of identification.

(13) Pressure seals.

(14) EWIS components inside shelves, panels, racks, junction boxes, distribution panels, and back-planes of equipment racks, including, but not limited to, circuit board back-planes, wire integration units, and external wiring of equipment.

(b) Except for the equipment indicated in paragraph (a)(14) of this section, EWIS components inside the following equipment, and the external connectors that are part of that equipment, are excluded from the definition in paragraph (a) of this section:

(1) Electrical equipment or avionics that are qualified to environmental conditions and testing procedures when those conditions and procedures are—

(i) Appropriate for the intended function and operating environment, and

(ii) Acceptable to the FAA.

(2) Portable electrical devices that are not part of the type design of the airplane. This includes personal entertainment devices and laptop computers.

(3) Fiber optics.

§ 25.1703 Function and installation: EWIS.

(a) Each EWIS component installed in any area of the aircraft must:

(1) Be of a kind and design appropriate to its intended function.

(2) Be installed according to limitations specified for the EWIS components.

(3) Perform the function for which it was intended without degrading the airworthiness of the airplane.

(4) Be designed and installed in a way that will minimize mechanical strain.

(b) Selection of wires must take into account known characteristics of the wire in relation to each installation and application to minimize the risk of wire damage, including any arc tracking phenomena.

(c) The design and installation of the main power cables (including generator cables) in the fuselage must allow for a reasonable degree of deformation and stretching without failure.

(d) EWIS components located in areas of known moisture accumulation must be protected to minimize any hazardous effects due to moisture.

§ 25.1705 Systems and functions: EWIS.

(a) EWIS associated with any system required for type certification or by operating rules must be considered an integral part of that system and must be considered in showing compliance with the applicable requirements for that system.

(b) For systems to which the following rules apply, the components of EWIS associated with those systems must be considered an integral part of that system or systems and must be considered in showing compliance with

the applicable requirements for that system.

(1) § 25.773(b)(2) Pilot compartment view.

(2) § 25.981 Fuel tank ignition prevention.

(3) § 25.1165 Engine ignition systems.

(4) § 25.1310 Power source capacity and distribution.

(5) § 25.1316 System lightning protection.

(6) § 25.1331(a)(2) Instruments using a power supply.

(7) § 25.1351 General.

(8) § 25.1355 Distribution system.

(9) § 25.1360 Precautions against injury.

(10) § 25.1362 Electrical supplies for emergency conditions.

(11) § 25.1365 Electrical appliances, motors, and transformers.

(12) § 25.1431(c) and (d) Electronic equipment.

§ 25.1707 System separation: EWIS.

(a) Each EWIS must be designed and installed with adequate physical separation from other EWIS and airplane systems so that an EWIS component failure will not create a hazardous condition. Unless otherwise stated, for the purposes of this section, adequate physical separation must be achieved by separation distance or by a barrier that provides protection equivalent to that separation distance.

(b) Each EWIS must be designed and installed so that any electrical interference likely to be present in the airplane will not result in hazardous effects upon the airplane or its systems.

(c) Wires and cables carrying heavy current, and their associated EWIS components, must be designed and installed to ensure adequate physical separation and electrical isolation so that damage to circuits associated with essential functions will be minimized under fault conditions.

(d) Each EWIS associated with independent airplane power sources or power sources connected in combination must be designed and installed to ensure adequate physical separation and electrical isolation so that a fault in any one airplane power source EWIS will not adversely affect any other independent power sources. In addition:

(1) Airplane independent electrical power sources must not share a common ground terminating location.

(2) Airplane system static grounds must not share a common ground terminating location with any of the airplane's independent electrical power sources.

(e) Except to the extent necessary to provide electrical connection to the fuel systems components, the EWIS must be designed and installed with adequate physical separation from fuel lines and other fuel system components, so that:

(1) An EWIS component failure will not create a hazardous condition.

(2) Any fuel leakage onto EWIS components will not create a hazardous condition.

(f) Except to the extent necessary to provide electrical connection to the hydraulic systems components, EWIS must be designed and installed with adequate physical separation from hydraulic lines and other hydraulic system components, so that:

(1) An EWIS component failure will not create a hazardous condition.

(2) Any hydraulic fluid leakage onto EWIS components will not create a hazardous condition.

(g) Except to the extent necessary to provide electrical connection to the oxygen systems components, EWIS must be designed and installed with adequate physical separation from oxygen lines and other oxygen system components, so that an EWIS component failure will not create a hazardous condition.

(h) Except to the extent necessary to provide electrical connection to the water/waste systems components, EWIS must be designed and installed with adequate physical separation from water/waste lines and other water/waste system components, so that:

(1) An EWIS component failure will not create a hazardous condition.

(2) Any water/waste leakage onto EWIS components will not create a hazardous condition.

(i) EWIS must be designed and installed with adequate physical separation between the EWIS and flight or other mechanical control systems cables and associated system components, so that:

(1) Chafing, jamming, or other interference are prevented.

(2) An EWIS component failure will not create a hazardous condition.

(3) Failure of any flight or other mechanical control systems cables or systems components will not damage the EWIS and create a hazardous condition.

(j) EWIS must be designed and installed with adequate physical separation between the EWIS components and heated equipment, hot air ducts, and lines, so that:

(1) An EWIS component failure will not create a hazardous condition.

(2) Any hot air leakage or heat generated onto EWIS components will not create a hazardous condition.

(k) For systems for which redundancy is required, by certification rules, by operating rules, or as a result of the assessment required by § 25.1709, EWIS components associated with those systems must be designed and installed with adequate physical separation.

(l) Each EWIS must be designed and installed so there is adequate physical separation between it and other aircraft components and aircraft structure, and so that the EWIS is protected from sharp edges and corners, to minimize potential for abrasion/chafing, vibration damage, and other types of mechanical damage.

§ 25.1709 System safety: EWIS.

Each EWIS must be designed and installed so that:

(a) Each catastrophic failure condition—

(1) Is extremely improbable; and

(2) Does not result from a single failure.

(b) Each hazardous failure condition is extremely remote.

§ 25.1711 Component identification: EWIS.

(a) EWIS components must be labeled or otherwise identified using a consistent method that facilitates identification of the EWIS component, its function, and its design limitations, if any.

(b) For systems for which redundancy is required, by certification rules, by operating rules, or as a result of the assessment required by §25.1709, EWIS components associated with those systems must be specifically identified with component part number, function, and separation requirement for bundles.

(1) The identification must be placed along the wire, cable, or wire bundle at appropriate intervals and in areas of the airplane where it is readily visible to maintenance, repair, or alteration personnel.

(2) If an EWIS component cannot be marked physically, then other means of identification must be provided.

(c) The identifying markings required by paragraphs (a) and (b) of this section must remain legible throughout the expected service life of the EWIS component.

(d) The means used for identifying each EWIS component as required by this section must not have an adverse effect on the performance of that component throughout its expected service life.

(e) Identification for EWIS modifications to the type design must be consistent with the identification scheme of the original type design.

§25.1713 Fire protection: EWIS.

(a) All EWIS components must meet the applicable fire and smoke protection requirements of §25.831(c) of this part.

(b) EWIS components that are located in designated fire zones and are used during emergency procedures must be fire resistant.

(c) Insulation on electrical wire and electrical cable, and materials used to provide additional protection for the wire and cable, installed in any area of the airplane, must be self-extinguishing when tested in accordance with the applicable portions of Appendix F, part I, of 14 CFR part 25.

§25.1715 Electrical bonding and protection against static electricity: EWIS.

(a) EWIS components used for electrical bonding and protection against static electricity must meet the requirements of §25.899.

(b) On airplanes having grounded electrical systems, electrical bonding provided by EWIS components must provide an electrical return path capable of carrying both normal and fault currents without creating a shock hazard or damage to the EWIS components, other airplane system components, or airplane structure.

§25.1717 Circuit protective devices: EWIS.

Electrical wires and cables must be designed and installed so they are compatible with the circuit protection devices required by §25.1357, so that a fire or smoke hazard cannot be created under temporary or continuous fault conditions.

§25.1719 Accessibility provisions: EWIS.

Access must be provided to allow inspection and replacement of any EWIS component as necessary for continued airworthiness.

§25.1721 Protection of EWIS.

(a) No cargo or baggage compartment may contain any EWIS whose damage or failure may affect safe operation, unless the EWIS is protected so that:

(1) It cannot be damaged by movement of cargo or baggage in the compartment.

(2) Its breakage or failure will not create a fire hazard.

(b) EWIS must be designed and installed to minimize damage and risk of damage to EWIS by movement of people in the airplane during all phases of flight, maintenance, and servicing.

(c) EWIS must be designed and installed to minimize damage and risk of damage to EWIS by items carried onto the aircraft by passengers or cabin crew.

§25.1723 Flammable fluid fire protection: EWIS.

EWIS components located in each area where flammable fluid or vapors might escape by leakage of a fluid system must be considered a potential ignition source and must meet the requirements of §25.863.

§25.1725 Powerplants: EWIS.

(a) EWIS associated with any powerplant must be designed and installed so that the failure of an EWIS component will not prevent the continued safe operation of the remaining powerplants or require immediate action by any crewmember for continued safe operation, in accordance with the requirements of § 25.903(b).

(b) Design precautions must be taken to minimize hazards to the airplane due to EWIS damage in the event of a powerplant rotor failure or a fire originating within the powerplant that burns through the powerplant case, in accordance with the requirements of § 25.903(d)(1).

§ 25.1727 Flammable fluid shutoff means: EWIS.

EWIS associated with each flammable fluid shutoff means and control must be fireproof or must be located and protected so that any fire in a fire zone will not affect operation of the flammable fluid shutoff means, in accordance with the requirements of § 25.1189.

§ 25.1729 Instructions for Continued Airworthiness: EWIS.

The applicant must prepare Instructions for Continued Airworthiness applicable to EWIS in accordance with Appendix H sections H25.4 and H25.5 to this part that are approved by the FAA.

§ 25.1731 Powerplant and APU fire detector system: EWIS.

(a) EWIS that are part of each fire or overheat detector system in a fire zone must be fire-resistant.

(b) No EWIS component of any fire or overheat detector system for any fire zone may pass through another fire zone, unless:

(1) It is protected against the possibility of false warnings resulting from fires in zones through which it passes; or

(2) Each zone involved is simultaneously protected by the same detector and extinguishing system.

(c) EWIS that are part of each fire or overheat detector system in a fire zone must meet the requirements of § 25.1203.

§ 25.1733 Fire detector systems, general: EWIS.

EWIS associated with any installed fire protection system, including those required by §§ 25.854 and 25.858, must be considered an integral part of the system in showing compliance with the applicable requirements for that system.

Subpart I—Special Federal Aviation Regulations

SOURCE: Docket No. FAA–2011–0186, Amdt. 25–133, 76 FR 12555, Mar. 8, 2011, unless otherwise noted.

§ 25.1801 SFAR No. 111—Lavatory Oxygen Systems.

The requirements of § 121.1500 of this chapter also apply to this part.

APPENDIX A TO PART 25

Appendix A

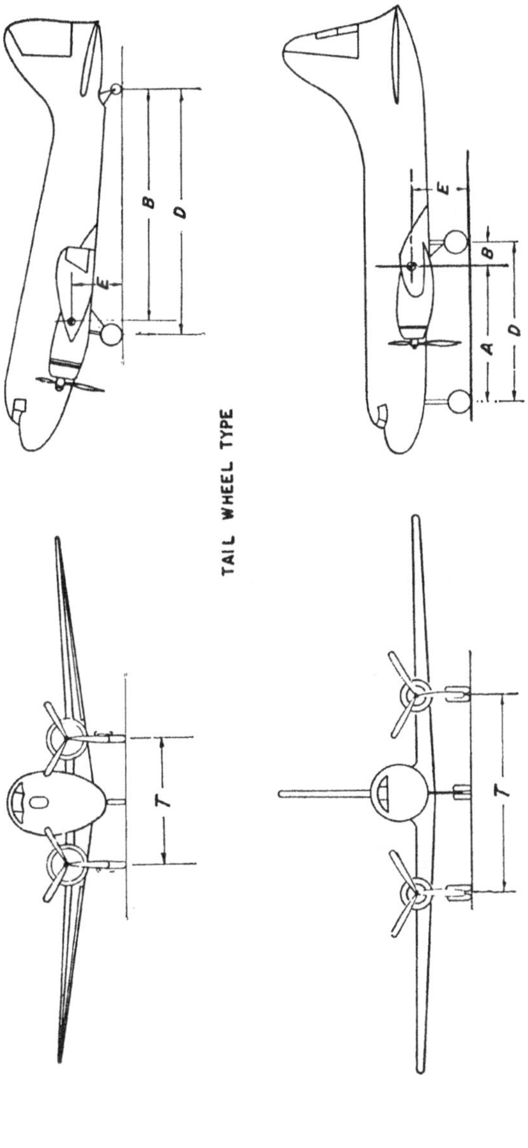

FIGURE 1—Basic landing gear dimension data.

TAIL WHEEL TYPE

NOSE WHEEL TYPE

377

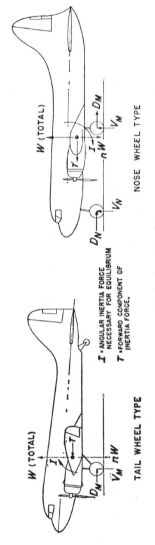

FIGURE 2—Level landing.

I = ANGULAR INERTIA FORCE
NECESSARY FOR EQUILIBRIUM

T = FORWARD COMPONENT OF
INERTIA FORCE.

FIGURE 3—Tail-down landing.

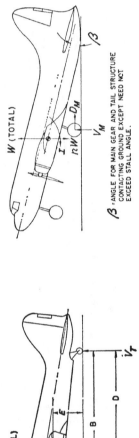

β = ANGLE FOR MAIN GEAR AND TAIL STRUCTURE
CONTACTING GROUND EXCEPT NEED NOT
EXCEED STALL ANGLE.

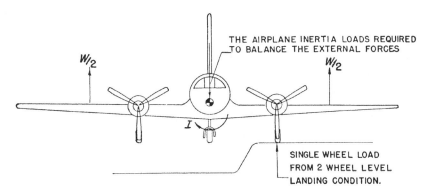

THE AIRPLANE INERTIA LOADS REQUIRED
TO BALANCE THE EXTERNAL FORCES

SINGLE WHEEL LOAD
FROM 2 WHEEL LEVEL
LANDING CONDITION.

NOSE OR TAIL WHEEL TYPE

FIGURE 5—Lateral drift landing.

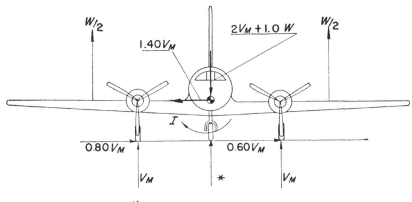

V_M = ONE-HALF THE MAXIMUM VERTICAL GROUND REACTION
OBTAINED AT EACH MAIN GEAR IN THE LEVEL LANDING CONDITIONS.

* NOSE GEAR GROUND REACTION = 0

NOSE OR TAIL WHEEL TYPE AIRPLANE IN LEVEL ALTITUDE

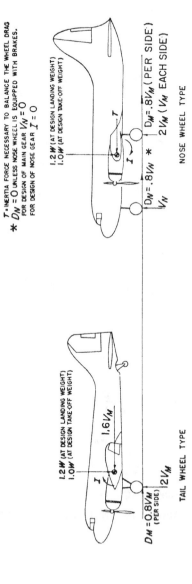

FIGURE 6—Braked roll.

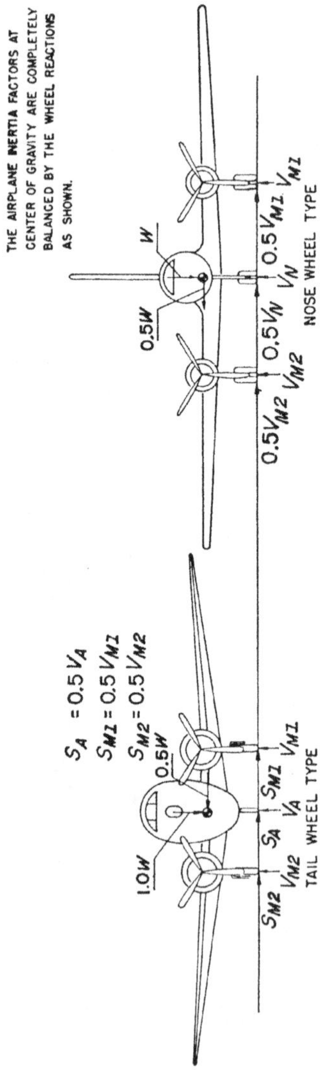

FIGURE 7—Ground turning.

FIGURE 8—Pivoting, nose or tail wheel type.

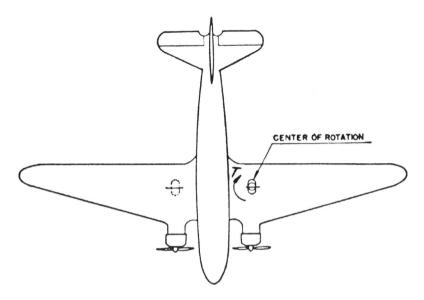

V_N and V_M are static ground reactions. For tail wheel type the airplane is in the three point attitude. Pivoting is assumed to take place about one main landing gear unit.

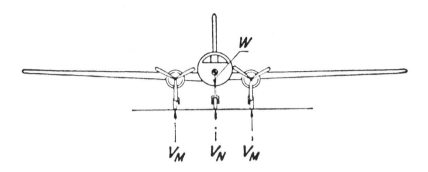

APPENDIX B TO PART 25

Appendix B

FIGURE 1—Pictorial definition of angles, dimensions, and directions on a seaplane

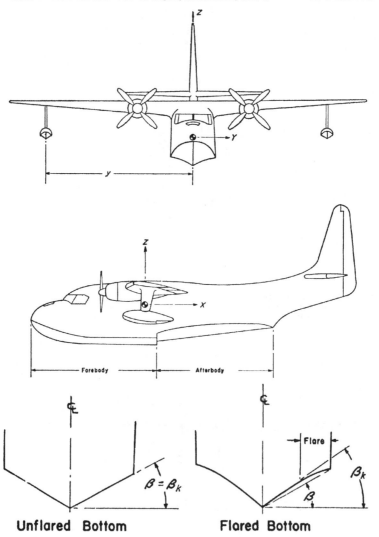

Unflared Bottom   **Flared Bottom**

FIGURE 2—Hull station weighing factor.

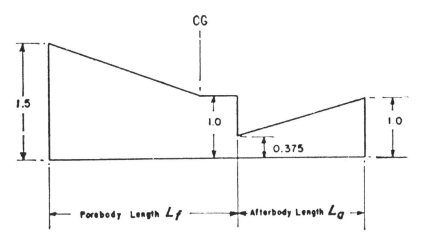

K_1 (Vertical Loads)

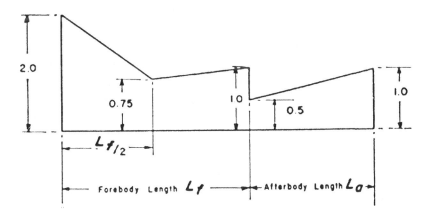

K_2 (Bottom Pressures)

FIGURE 3—Transverse pressure distributions.

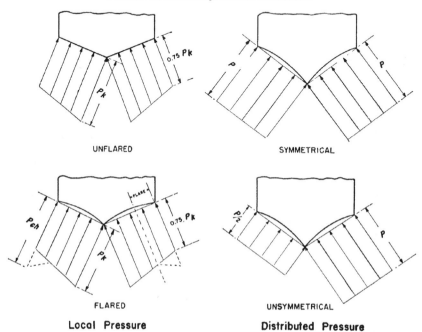

UNFLARED   SYMMETRICAL

FLARED   UNSYMMETRICAL

Local Pressure   **Distributed Pressure**

APPENDIX C TO PART 25

Part I—Atmospheric Icing Conditions

(a) *Continuous maximum icing.* The maximum continuous intensity of atmospheric icing conditions (continuous maximum icing) is defined by the variables of the cloud liquid water content, the mean effective diameter of the cloud droplets, the ambient air temperature, and the interrelationship of these three variables as shown in figure 1 of this appendix. The limiting icing envelope in terms of altitude and temperature is given in figure 2 of this appendix. The inter-relationship of cloud liquid water content with drop diameter and altitude is determined from figures 1 and 2. The cloud liquid water content for continuous maximum icing conditions of a horizontal extent, other than 17.4 nautical miles, is determined by the value of liquid water content of figure 1, multiplied by the appropriate factor from figure 3 of this appendix.

(b) *Intermittent maximum icing.* The intermittent maximum intensity of atmospheric icing conditions (intermittent maximum icing) is defined by the variables of the cloud liquid water content, the mean effective diameter of the cloud droplets, the ambient air temperature, and the interrelationship of these three variables as shown in figure 4 of this appendix. The limiting icing envelope in terms of altitude and temperature is given in figure 5 of this appendix. The inter-relationship of cloud liquid water content with drop diameter and altitude is determined from figures 4 and 5. The cloud liquid water content for intermittent maximum icing conditions of a horizontal extent, other than 2.6 nautical miles, is determined by the value of cloud liquid water content of figure 4 multiplied by the appropriate factor in figure 6 of this appendix.

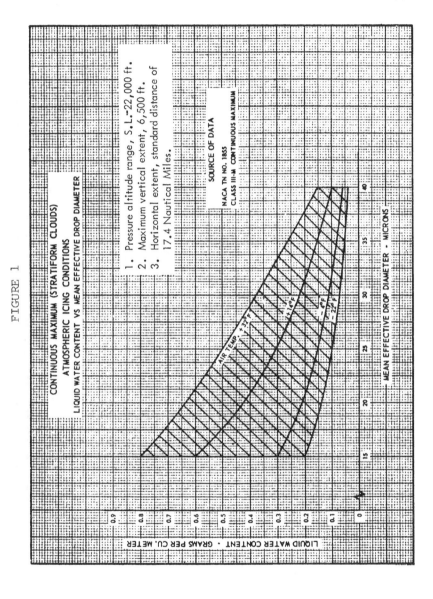

FIGURE 1

FIGURE 2

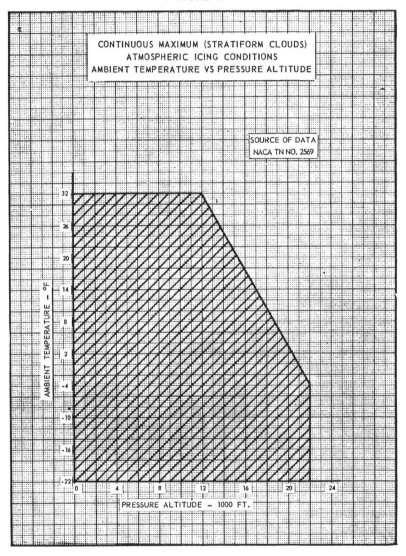

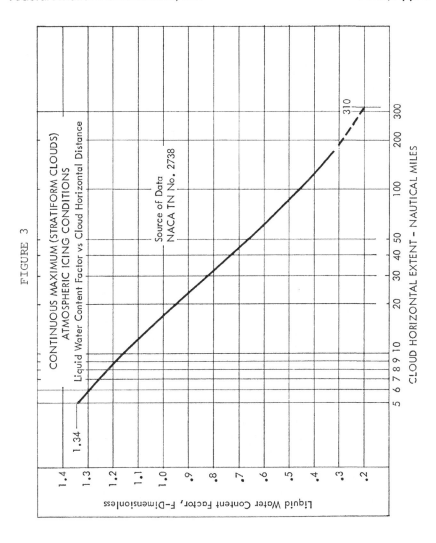

FIGURE 3

CONTINUOUS MAXIMUM (STRATIFORM CLOUDS)
ATMOSPHERIC ICING CONDITIONS
Liquid Water Content Factor vs Cloud Horizontal Distance

Source of Data
NACA TN No. 2738

FIGURE 4

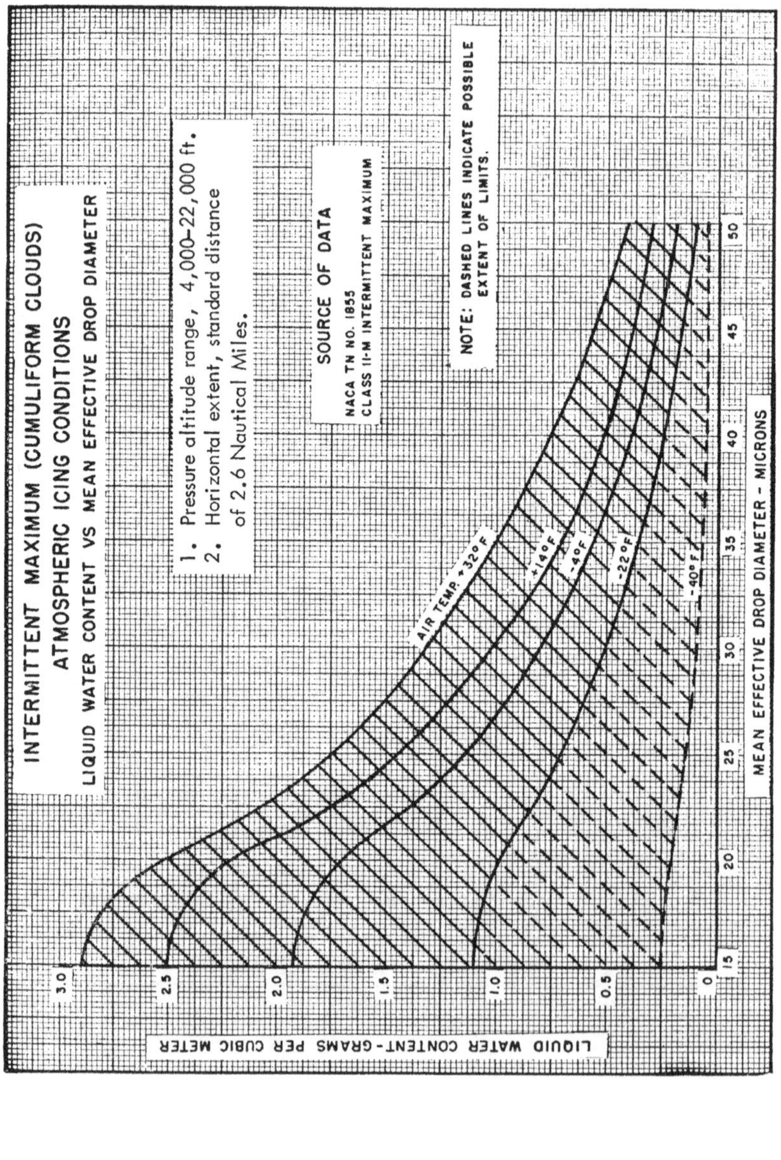
INTERMITTENT MAXIMUM (CUMULIFORM CLOUDS)
ATMOSPHERIC ICING CONDITIONS
LIQUID WATER CONTENT VS MEAN EFFECTIVE DROP DIAMETER
1. Pressure altitude range, 4,000–22,000 ft.
2. Horizontal extent, standard distance of 2.6 Nautical Miles.
SOURCE OF DATA
NACA TN NO. 1855
CLASS II-M INTERMITTENT MAXIMUM
NOTE: DASHED LINES INDICATE POSSIBLE EXTENT OF LIMITS.
AIR TEMP. +32° F
+14° F
-4° F
-22° F
-40° F
LIQUID WATER CONTENT-GRAMS PER CUBIC METER
3.0
2.5
2.0
1.5
1.0
0.5
0
MEAN EFFECTIVE DROP DIAMETER - MICRONS
15
20
25
30
35
40
45
50

FIGURE 5

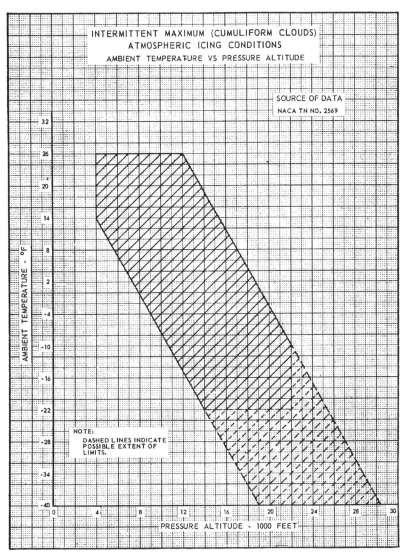

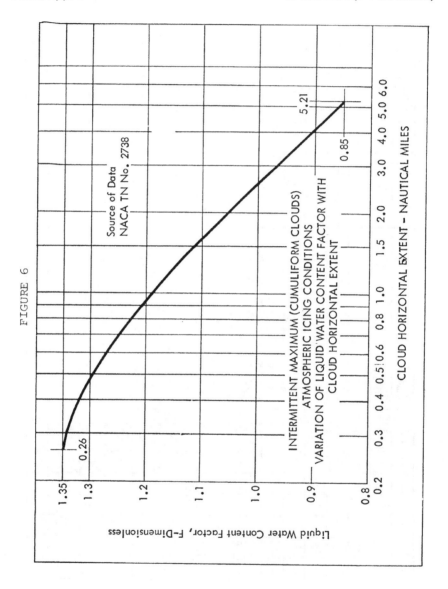

FIGURE 6

(c) *Takeoff maximum icing.* The maximum intensity of atmospheric icing conditions for takeoff (takeoff maximum icing) is defined by the cloud liquid water content of 0.35 g/m3, the mean effective diameter of the cloud droplets of 20 microns, and the ambient air temperature at ground level of minus 9 degrees Celsius (−9 °C). The takeoff maximum icing conditions extend from ground level to a height of 1,500 feet above the level of the takeoff surface.

Part II—Airframe Ice Accretions for Showing Compliance With Subpart B.

(a) *Ice accretions—General.* The most critical ice accretion in terms of airplane performance and handling qualities for each flight phase must be used to show compliance with the applicable airplane performance and handling requirements in icing conditions of subpart B of this part. Applicants must demonstrate that the full range of atmospheric icing conditions specified in part I of this appendix have been considered, including the mean effective drop diameter, liquid water content, and temperature appropriate to the flight conditions (for example, configuration, speed, angle-of-attack, and altitude). The ice accretions for each flight phase are defined as follows:

(1) *Takeoff ice* is the most critical ice accretion on unprotected surfaces and any ice accretion on the protected surfaces appropriate to normal ice protection system operation, occurring between the end of the takeoff distance and 400 feet above the takeoff surface, assuming accretion starts at the end of the takeoff distance in the takeoff maximum icing conditions defined in part I of this Appendix.

(2) *Final takeoff ice* is the most critical ice accretion on unprotected surfaces, and any ice accretion on the protected surfaces appropriate to normal ice protection system operation, between 400 feet and either 1,500 feet above the takeoff surface, or the height at which the transition from the takeoff to the en route configuration is completed and V_{FTO} is reached, whichever is higher. Ice accretion is assumed to start at the end of the takeoff distance in the takeoff maximum icing conditions of part I, paragraph (c) of this Appendix.

(3) *En route ice* is the critical ice accretion on the unprotected surfaces, and any ice accretion on the protected surfaces appropriate to normal ice protection system operation, during the en route phase.

(4) *Holding ice* is the critical ice accretion on the unprotected surfaces, and any ice accretion on the protected surfaces appropriate to normal ice protection system operation, during the holding flight phase.

(5) *Approach ice* is the critical ice accretion on the unprotected surfaces, and any ice accretion on the protected surfaces appropriate to normal ice protection system operation following exit from the holding flight phase and transition to the most critical approach configuration.

(6) *Landing ice* is the critical ice accretion on the unprotected surfaces, and any ice accretion on the protected surfaces appropriate to normal ice protection system operation following exit from the approach flight phase and transition to the final landing configuration.

(b) In order to reduce the number of ice accretions to be considered when demonstrating compliance with the requirements of §25.21(g), any of the ice accretions defined in paragraph (a) of this section may be used for any other flight phase if it is shown to be more critical than the specific ice accretion defined for that flight phase. Configuration differences and their effects on ice accretions must be taken into account.

(c) The ice accretion that has the most adverse effect on handling qualities may be used for airplane performance tests provided any difference in performance is conservatively taken into account.

(d) For both unprotected and protected parts, the ice accretion for the takeoff phase may be determined by calculation, assuming the takeoff maximum icing conditions defined in appendix C, and assuming that:

(1) Airfoils, control surfaces and, if applicable, propellers are free from frost, snow, or ice at the start of the takeoff;

(2) The ice accretion starts at the end of the takeoff distance.

(3) The critical ratio of thrust/power-to-weight;

(4) Failure of the critical engine occurs at V_{EF}; and

(5) Crew activation of the ice protection system is in accordance with a normal operating procedure provided in the Airplane Flight Manual, except that after beginning the takeoff roll, it must be assumed that the crew takes no action to activate the ice protection system until the airplane is at least 400 feet above the takeoff surface.

(e) The ice accretion before the ice protection system has been activated and is performing its intended function is the critical ice accretion formed on the unprotected and normally protected surfaces before activation and effective operation of the ice protection system in continuous maximum atmospheric icing conditions. This ice accretion only applies in showing compliance to §§25.143(j) and 25.207(h), and 25.207(i).

[Doc. No. 4080, 29 FR 17955, Dec. 18, 1964, as amended by Amdt. 25–121, 72 FR 44669, Aug. 8, 2007; 72 FR 50467, Aug. 31, 2007; Amdt. 25–129, 74 FR 38340, Aug. 3, 2009; Amdt. 25–140, 79 FR 65528, Nov. 4, 2014]

APPENDIX D TO PART 25

Criteria for determining minimum flight crew. The following are considered by the Agency in determining the minimum flight crew under §25.1523:

(a) *Basic workload functions.* The following basic workload functions are considered:

(1) Flight path control.

(2) Collision avoidance.

(3) Navigation.

(4) Communications.

(5) Operation and monitoring of aircraft engines and systems.

(6) Command decisions.

(b) *Workload factors.* The following workload factors are considered significant when analyzing and demonstrating workload for minimum flight crew determination:

(1) The accessibility, ease, and simplicity of operation of all necessary flight, power, and equipment controls, including emergency fuel shutoff valves, electrical controls, electronic controls, pressurization system controls, and engine controls.

(2) The accessibility and conspicuity of all necessary instruments and failure warning devices such as fire warning, electrical system malfunction, and other failure or caution indicators. The extent to which such instruments or devices direct the proper corrective action is also considered.

(3) The number, urgency, and complexity of operating procedures with particular consideration given to the specific fuel management schedule imposed by center of gravity, structural or other considerations of an airworthiness nature, and to the ability of each engine to operate at all times from a single tank or source which is automatically replenished if fuel is also stored in other tanks.

(4) The degree and duration of concentrated mental and physical effort involved in normal operation and in diagnosing and coping with malfunctions and emergencies.

(5) The extent of required monitoring of the fuel, hydraulic, pressurization, electrical, electronic, deicing, and other systems while en route.

(6) The actions requiring a crewmember to be unavailable at his assigned duty station, including: observation of systems, emergency operation of any control, and emergencies in any compartment.

(7) The degree of automation provided in the aircraft systems to afford (after failures or malfunctions) automatic crossover or isolation of difficulties to minimize the need for flight crew action to guard against loss of hydraulic or electric power to flight controls or to other essential systems.

(8) The communications and navigation workload.

(9) The possibility of increased workload associated with any emergency that may lead to other emergencies.

(10) Incapacitation of a flight crewmember whenever the applicable operating rule requires a minimum flight crew of at least two pilots.

(c) *Kind of operation authorized.* The determination of the kind of operation authorized requires consideration of the operating rules under which the airplane will be operated. Unless an applicant desires approval for a more limited kind of operation. It is assumed that each airplane certificated under this Part will operate under IFR conditions.

[Amdt. 25–3, 30 FR 6067, Apr. 29, 1965]

APPENDIX E TO PART 25

I—*Limited Weight Credit For Airplanes Equipped With Standby Power*

(a) Each applicant for an increase in the maximum certificated takeoff and landing weights of an airplane equipped with a type-certificated standby power rocket engine may obtain an increase as specified in paragraph (b) if—

(1) The installation of the rocket engine has been approved and it has been established by flight test that the rocket engine and its controls can be operated safely and reliably at the increase in maximum weight; and

(2) The Airplane Flight Manual, or the placard, markings or manuals required in place thereof, set forth in addition to any other operating limitations the Administrator may require, the increased weight approved under this regulation and a prohibition against the operation of the airplane at the approved increased weight when—

(i) The installed standby power rocket engines have been stored or installed in excess of the time limit established by the manufacturer of the rocket engine (usually stenciled on the engine casing); or

(ii) The rocket engine fuel has been expended or discharged.

(b) The currently approved maximum takeoff and landing weights at which an airplane is certificated without a standby power rocket engine installation may be increased by an amount that does not exceed any of the following:

(1) An amount equal in pounds to 0.014 IN, where I is the maximum usable impulse in pounds-seconds available from each standby power rocket engine and N is the number of rocket engines installed.

(2) An amount equal to 5 percent of the maximum certificated weight approved in accordance with the applicable airworthiness regulations without standby power rocket engines installed.

(3) An amount equal to the weight of the rocket engine installation.

(4) An amount that, together with the currently approved maximum weight, would equal the maximum structural weight established for the airplane without standby rocket engines installed.

II—*Performance Credit for Transport Category Airplanes Equipped With Standby Power*

The Administrator may grant performance credit for the use of standby power on transport category airplanes. However, the performance credit applies only to the maximum certificated takeoff and landing weights, the takeoff distance, and the takeoff paths, and may not exceed that found by the Administrator to result in an overall level of safety in the takeoff, approach, and landing regimes of flight equivalent to that prescribed in the regulations under which the airplane was originally certificated without standby power. For the purposes of this appendix, "standby power" is power or thrust, or both, obtained from rocket engines for a relatively short period and actuated only in cases of emergency. The following provisions apply:

(1) *Takeoff; general.* The takeoff data prescribed in paragraphs (2) and (3) of this appendix must be determined at all weights and altitudes, and at ambient temperatures if applicable, at which performance credit is to be applied.

(2) *Takeoff path.*

(a) The one-engine-inoperative takeoff path with standby power in use must be determined in accordance with the performance requirements of the applicable airworthiness regulations.

(b) The one-engine-inoperative takeoff path (excluding that part where the airplane is on or just above the takeoff surface) determined in accordance with paragraph (a) of this section must lie above the one-engine-inoperative takeoff path without standby power at the maximum takeoff weight at which all of the applicable air-worthiness requirements are met. For the purpose of this comparison, the flight path is considered to extend to at least a height of 400 feet above the takeoff surface.

(c) The takeoff path with all engines operating, but without the use of standby power, must reflect a conservatively greater overall level of performance than the one-engine-inoperative takeoff path established in accordance with paragraph (a) of this section. The margin must be established by the Administrator to insure safe day-to-day operations, but in no case may it be less than 15 percent. The all-engines-operating takeoff path must be determined by a procedure consistent with that established in complying with paragraph (a) of this section.

(d) For reciprocating-engine-powered airplanes, the takeoff path to be scheduled in the Airplane Flight Manual must represent the one-engine-operative takeoff path determined in accordance with paragraph (a) of this section and modified to reflect the procedure (see paragraph (6)) established by the applicant for flap retraction and attainment of the en route speed. The scheduled takeoff path must have a positive slope at all points of the airborne portion and at no point must it lie above the takeoff path specified in paragraph (a) of this section.

(3) *Takeoff distance.* The takeoff distance must be the horizontal distance along the one-engine-inoperative take off path determined in accordance with paragraph (2)(a)

from the start of the takeoff to the point where the airplane attains a height of 50 feet above the takeoff surface for reciprocating-engine-powered airplanes and a height of 35 feet above the takeoff surface for turbine-powered airplanes.

(4) *Maximum certificated takeoff weights.* The maximum certificated takeoff weights must be determined at all altitudes, and at ambient temperatures, if applicable, at which performance credit is to be applied and may not exceed the weights established in compliance with paragraphs (a) and (b) of this section.

(a) The conditions of paragraphs (2)(b) through (d) must be met at the maximum certificated takeoff weight.

(b) Without the use of standby power, the airplane must meet all of the en route requirements of the applicable airworthiness regulations under which the airplane was originally certificated. In addition, turbine-powered airplanes without the use of standby power must meet the final takeoff climb requirements prescribed in the applicable airworthiness regulations.

(5) *Maximum certificated landing weights.*

(a) The maximum certificated landing weights (one-engine-inoperative approach and all-engine-operating landing climb) must be determined at all altitudes, and at ambient temperatures if applicable, at which performance credit is to be applied and must not exceed that established in compliance with paragraph (b) of this section.

(b) The flight path, with the engines operating at the power or thrust, or both, appropriate to the airplane configuration and with standby power in use, must lie above the flight path without standby power in use at the maximum weight at which all of the applicable airworthiness requirements are met. In addition, the flight paths must comply with subparagraphs (i) and (ii) of this paragraph.

(i) The flight paths must be established without changing the appropriate airplane configuration.

(ii) The flight paths must be carried out for a minimum height of 400 feet above the point where standby power is actuated.

(6) *Airplane configuration, speed, and power and thrust; general.* Any change in the airplane's configuration, speed, and power or thrust, or both, must be made in accordance with the procedures established by the applicant for the operation of the airplane in service and must comply with paragraphs (a) through (c) of this section. In addition, procedures must be established for the execution of balked landings and missed approaches.

(a) The Administrator must find that the procedure can be consistently executed in service by crews of average skill.

(b) The procedure may not involve methods or the use of devices which have not been proven to be safe and reliable.

(c) Allowances must be made for such time delays in the execution of the procedures as may be reasonably expected to occur during service.

(7) *Installation and operation; standby power.* The standby power unit and its installation must comply with paragraphs (a) and (b) of this section.

(a) The standby power unit and its installation must not adversely affect the safety of the airplane.

(b) The operation of the standby power unit and its control must have proven to be safe and reliable.

[Amdt. 25–6, 30 FR 8468, July 2, 1965]

APPENDIX F TO PART 25

Part I—Test Criteria and Procedures for Showing Compliance With § 25.853 or § 25.855

(a) *Material test criteria*—(1) *Interior compartments occupied by crew or passengers.* (i) Interior ceiling panels, interior wall panels, partitions, galley structure, large cabinet walls, structural flooring, and materials used in the construction of stowage compartments (other than underseat stowage compartments and compartments for stowing small items such as magazines and maps) must be self-extinguishing when tested vertically in accordance with the applicable portions of part I of this appendix. The average burn length may not exceed 6 inches and the average flame time after removal of the flame source may not exceed 15 seconds. Drippings from the test specimen may not continue to flame for more than an average of 3 seconds after falling.

(ii) Floor covering, textiles (including draperies and upholstery), seat cushions, padding, decorative and non-decorative coated fabrics, leather, trays and galley furnishings, electrical conduit, air ducting, joint and edge covering, liners of Class B and E cargo or baggage compartments, floor panels of Class B, C, E, or F cargo or baggage compartments, cargo covers and transparencies, molded and thermoformed parts, air ducting joints, and trim strips (decorative and chafing), that are constructed of materials not covered in paragraph (a)(1)(iv) below, must be self-extinguishing when tested vertically in accordance with the applicable portions of part I of this appendix or other approved equivalent means. The average burn length may not exceed 8 inches, and the average flame time after removal of the flame source may not exceed 15 seconds. Drippings from the test specimen may not continue to flame for more than an average of 5 seconds after falling.

(iii) Motion picture film must be safety film meeting the Standard Specifications for Safety Photographic Film PHI.25 (available from the American National Standards Institute, 1430 Broadway, New York, NY 10018). If the film travels through ducts, the ducts must meet the requirements of subparagraph (ii) of this paragraph.

(iv) Clear plastic windows and signs, parts constructed in whole or in part of elastomeric materials, edge lighted instrument assemblies consisting of two or more instruments in a common housing, seat belts, shoulder harnesses, and cargo and baggage tiedown equipment, including containers, bins, pallets, etc., used in passenger or crew compartments, may not have an average burn rate greater than 2.5 inches per minute when tested horizontally in accordance with the applicable portions of this appendix.

(v) Except for small parts (such as knobs, handles, rollers, fasteners, clips, grommets, rub strips, pulleys, and small electrical parts) that would not contribute significantly to the propagation of a fire and for electrical wire and cable insulation, materials in items not specified in paragraphs (a)(1)(i), (ii), (iii), or (iv) of part I of this appendix may not have a burn rate greater than 4.0 inches per minute when tested horizontally in accordance with the applicable portions of this appendix.

(2) *Cargo and baggage compartments not occupied by crew or passengers.*

(i) [Reserved]

(ii) A cargo or baggage compartment defined in § 25.857 as Class B or E must have a liner constructed of materials that meet the requirements of paragraph (a)(1)(ii) of part I of this appendix and separated from the airplane structure (except for attachments). In addition, such liners must be subjected to the 45 degree angle test. The flame may not penetrate (pass through) the material during application of the flame or subsequent to its removal. The average flame time after removal of the flame source may not exceed 15 seconds, and the average glow time may not exceed 10 seconds.

(iii) A cargo or baggage compartment defined in § 25.857 as Class B, C, E, or F must have floor panels constructed of materials which meet the requirements of paragraph (a)(1)(ii) of part I of this appendix and which are separated from the airplane structure (except for attachments). Such panels must be subjected to the 45 degree angle test. The flame may not penetrate (pass through) the material during application of the flame or subsequent to its removal. The average flame time after removal of the flame source may not exceed 15 seconds, and the average glow time may not exceed 10 seconds.

(iv) Insulation blankets and covers used to protect cargo must be constructed of materials that meet the requirements of paragraph (a)(1)(ii) of part I of this appendix. Tiedown equipment (including containers, bins, and pallets) used in each cargo and baggage compartment must be constructed of materials that meet the requirements of paragraph (a)(1)(v) of part I of this appendix.

(3) *Electrical system components.* Insulation on electrical wire or cable installed in any area of the fuselage must be self-extinguishing when subjected to the 60 degree test specified in part I of this appendix. The average burn length may not exceed 3 inches, and the average flame time after removal of the flame source may not exceed 30 seconds. Drippings from the test specimen may not continue to flame for more than an average of 3 seconds after falling.

(b) *Test Procedures*—(1) *Conditioning.* Specimens must be conditioned to 70 ±5 F., and at 50 percent ±5 percent relative humidity until moisture equilibrium is reached or for 24 hours. Each specimen must remain in the conditioning environment until it is subjected to the flame.

(2) *Specimen configuration.* Except for small parts and electrical wire and cable insulation, materials must be tested either as section cut from a fabricated part as installed in the airplane or as a specimen simulating a cut section, such as a specimen cut from a flat sheet of the material or a model of the fabricated part. The specimen may be cut from any location in a fabricated part; however, fabricated units, such as sandwich panels, may not be separated for test. Except as noted below, the specimen thickness must be no thicker than the minimum thickness to be qualified for use in the airplane. Test specimens of thick foam parts, such as seat cushions, must be ½-inch in thickness. Test specimens of materials that must meet the requirements of paragraph (a)(1)(v) of part I of this appendix must be no more than ⅛-inch in thickness. Electrical wire and cable specimens must be the same size as used in the airplane. In the case of fabrics, both the warp and fill direction of the weave must be tested to determine the most critical flammability condition. Specimens must be mounted in a metal frame so that the two long edges and the upper edge are held securely during the vertical test prescribed in subparagraph (4) of this paragraph and the two long edges and the edge away from the flame are held securely during the horizontal test prescribed in subparagraph (5) of this paragraph. The exposed area of the specimen must be at least 2 inches wide and 12 inches long, unless the actual size used in the airplane is smaller. The edge to which the burner flame is applied must not consist of the finished or protected edge of the specimen but must be representative of the actual cross-section of the material or part as installed in the airplane. The specimen must be mounted in a metal frame so that all four edges are held securely and the exposed area of the specimen is at least 8 inches by 8 inches during the 45° test prescribed in subparagraph (6) of this paragraph.

(3) *Apparatus.* Except as provided in subparagraph (7) of this paragraph, tests must be conducted in a draft-free cabinet in accordance with Federal Test Method Standard 191 Model 5903 (revised Method 5902) for the vertical test, or Method 5906 for horizontal test (available from the General Services Administration, Business Service Center, Region 3, Seventh & D Streets SW., Washington, DC 20407). Specimens which are too large for the cabinet must be tested in similar draft-free conditions.

(4) *Vertical test.* A minimum of three specimens must be tested and results averaged. For fabrics, the direction of weave corresponding to the most critical flammability conditions must be parallel to the longest dimension. Each specimen must be supported vertically. The specimen must be exposed to a Bunsen or Tirrill burner with a nominal ⅜-inch I.D. tube adjusted to give a flame of 1½ inches in height. The minimum flame temperature measured by a calibrated thermocouple pyrometer in the center of the flame must be 1550 °F. The lower edge of the specimen must be ¾-inch above the top edge of the burner. The flame must be applied to the center line of the lower edge of the specimen. For materials covered by paragraph (a)(1)(i) of part I of this appendix, the flame must be applied for 60 seconds and then removed. For materials covered by paragraph (a)(1)(ii) of part I of this appendix, the flame must be applied for 12 seconds and then removed. Flame time, burn length, and flaming time of drippings, if any, may be recorded. The burn length determined in accordance with subparagraph (7) of this paragraph must be measured to the nearest tenth of an inch.

(5) *Horizontal test.* A minimum of three specimens must be tested and the results averaged. Each specimen must be supported horizontally. The exposed surface, when installed in the aircraft, must be face down for the test. The specimen must be exposed to a Bunsen or Tirrill burner with a nominal ⅜-inch I.D. tube adjusted to give a flame of 1½ inches in height. The minimum flame temperature measured by a calibrated thermocouple pyrometer in the center of the flame must be 1550 °F. The specimen must be positioned so that the edge being tested is centered ¾-inch above the top of the burner. The flame must be applied for 15 seconds and then removed. A minimum of 10 inches of specimen must be used for timing purposes, approximately 1½ inches must burn before the burning front reaches the timing zone, and the average burn rate must be recorded.

(6) *Forty-five degree test.* A minimum of three specimens must be tested and the results averaged. The specimens must be supported at an angle of 45° to a horizontal surface. The exposed surface when installed in the aircraft must be face down for the test. The specimens must be exposed to a Bunsen or Tirrill burner with a nominal ⅜-inch I.D. tube adjusted to give a flame of 1½ inches in height. The minimum flame temperature

measured by a calibrated thermocouple pyrometer in the center of the flame must be 1550 °F. Suitable precautions must be taken to avoid drafts. The flame must be applied for 30 seconds with one-third contacting the material at the center of the specimen and then removed. Flame time, glow time, and whether the flame penetrates (passes through) the specimen must be recorded.

(7) *Sixty degree test.* A minimum of three specimens of each wire specification (make and size) must be tested. The specimen of wire or cable (including insulation) must be placed at an angle of 60° with the horizontal in the cabinet specified in subparagraph (3) of this paragraph with the cabinet door open during the test, or must be placed within a chamber approximately 2 feet high by 1 foot by 1 foot, open at the top and at one vertical side (front), and which allows sufficient flow of air for complete combustion, but which is free from drafts. The specimen must be parallel to and approximately 6 inches from the front of the chamber. The lower end of the specimen must be held rigidly clamped. The upper end of the specimen must pass over a pulley or rod and must have an appropriate weight attached to it so that the specimen is held tautly throughout the flammability test. The test specimen span between lower clamp and upper pulley or rod must be 24 inches and must be marked 8 inches from the lower end to indicate the central point for flame application. A flame from a Bunsen or Tirrill burner must be applied for 30 seconds at the test mark. The burner must be mounted underneath the test mark on the specimen, perpendicular to the specimen and at an angle of 30° to the vertical plane of the specimen. The burner must have a nominal bore of 3/8-inch and be adjusted to provide a 3-inch high flame with an inner cone approximately one-third of the flame height. The minimum temperature of the hottest portion of the flame, as measured with a calibrated thermocouple pyrometer, may not be less than 1750 °F. The burner must be positioned so that the hottest portion of the flame is applied to the test mark on the wire. Flame time, burn length, and flaming time of drippings, if any, must be recorded. The burn length determined in accordance with paragraph (8) of this paragraph must be measured to the nearest tenth of an inch. Breaking of the wire specimens is not considered a failure.

(8) *Burn length.* Burn length is the distance from the original edge to the farthest evidence of damage to the test specimen due to flame impingement, including areas of partial or complete consumption, charring, or embrittlement, but not including areas sooted, stained, warped, or discolored, nor areas where material has shrunk or melted away from the heat source.

Part II—Flammability of Seat Cushions

(a) *Criteria for Acceptance.* Each seat cushion must meet the following criteria:

(1) At least three sets of seat bottom and seat back cushion specimens must be tested.

(2) If the cushion is constructed with a fire blocking material, the fire blocking material must completely enclose the cushion foam core material.

(3) Each specimen tested must be fabricated using the principal components (i.e., foam core, flotation material, fire blocking material, if used, and dress covering) and assembly processes (representative seams and closures) intended for use in the production articles. If a different material combination is used for the back cushion than for the bottom cushion, both material combinations must be tested as complete specimen sets, each set consisting of a back cushion specimen and a bottom cushion specimen. If a cushion, including outer dress covering, is demonstrated to meet the requirements of this appendix using the oil burner test, the dress covering of that cushion may be replaced with a similar dress covering provided the burn length of the replacement covering, as determined by the test specified in §25.853(c), does not exceed the corresponding burn length of the dress covering used on the cushion subjected to the oil burner test.

(4) For at least two-thirds of the total number of specimen sets tested, the burn length from the burner must not reach the side of the cushion opposite the burner. The burn length must not exceed 17 inches. Burn length is the perpendicular distance from the inside edge of the seat frame closest to the burner to the farthest evidence of damage to the test specimen due to flame impingement, including areas of partial or complete consumption, charring, or embrittlement, but not including areas sooted, stained, warped, or discolored, or areas where material has shrunk or melted away from the heat source.

(5) The average percentage weight loss must not exceed 10 percent. Also, at least two-thirds of the total number of specimen sets tested must not exceed 10 percent weight loss. All droppings falling from the cushions and mounting stand are to be discarded before the after-test weight is determined. The percentage weight loss for a specimen set is the weight of the specimen set before testing less the weight of the specimen set after testing expressed as the percentage of the weight before testing.

(b) *Test Conditions.* Vertical air velocity should average 25 fpm±10 fpm at the top of the back seat cushion. Horizontal air velocity should be below 10 fpm just above the bottom seat cushion. Air velocities should be measured with the ventilation hood operating and the burner motor off.

(c) *Test Specimens.* (1) For each test, one set of cushion specimens representing a seat bottom and seat back cushion must be used.

(2) The seat bottom cushion specimen must be 18 ±⅛ inches (457 ±3 mm) wide by 20 ±⅛ inches (508 ±3 mm) deep by 4 ±⅛ inches (102 ±3 mm) thick, exclusive of fabric closures and seam overlap.

(3) The seat back cushion specimen must be 18 ±⅛ inches (432 ±3 mm) wide by 25 ±⅛ inches (635 ±3 mm) high by 2 ±⅛ inches (51 ±3 mm) thick, exclusive of fabric closures and seam overlap.

(4) The specimens must be conditioned at 70 ±5 °F (21 ±2 °C) 55%±10% relative humidity for at least 24 hours before testing.

(d) *Test Apparatus.* The arrangement of the test apparatus is shown in Figures 1 through 5 and must include the components described in this section. Minor details of the apparatus may vary, depending on the model burner used.

(1) *Specimen Mounting Stand.* The mounting stand for the test specimens consists of steel angles, as shown in Figure 1. The length of the mounting stand legs is 12 ±⅛ inches (305 ±3 mm). The mounting stand must be used for mounting the test specimen seat bottom and seat back, as shown in Figure 2. The mounting stand should also include a suitable drip pan lined with aluminum foil, dull side up.

(2) *Test Burner.* The burner to be used in testing must—

(i) Be a modified gun type;

(ii) Have an 80-degree spray angle nozzle nominally rated for 2.25 gallons/hour at 100 psi;

(iii) Have a 12-inch (305 mm) burner cone installed at the end of the draft tube, with an opening 6 inches (152 mm) high and 11 inches (280 mm) wide, as shown in Figure 3; and

(iv) Have a burner fuel pressure regulator that is adjusted to deliver a nominal 2.0 gallon/hour of # 2 Grade kerosene or equivalent required for the test.

Burner models which have been used successfully in testing are the Lennox Model OB–32, Carlin Model 200 CRD, and Park Model DPL 3400. FAA published reports pertinent to this type of burner are: (1) Powerplant Enginering Report No. 3A, Standard Fire Test Apparatus and Procedure for Flexible Hose Assemblies, dated March 1978; and (2) Report No. DOT/FAA/RD/76/213, Reevaluation of Burner Characteristics for Fire Resistance Tests, dated January 1977.

(3) *Calorimeter.*

(i) The calorimeter to be used in testing must be a (0–15.0 BTU/ft²-sec. 0–17.0 W/cm²) calorimeter, accurate ±3%, mounted in a 6-inch by 12-inch (152 by 305 mm) by ¾-inch (19 mm) thick calcium silicate insulating board which is attached to a steel angle bracket for placement in the test stand during burner calibration, as shown in Figure 4.

(ii) Because crumbling of the insulating board with service can result in misalignment of the calorimeter, the calorimeter must be monitored and the mounting shimmed, as necessary, to ensure that the calorimeter face is flush with the exposed plane of the insulating board in a plane parallel to the exit of the test burner cone.

(4) *Thermocouples.* The seven thermocouples to be used for testing must be ¹⁄₁₆- to ⅛-inch metal sheathed, ceramic packed, type K, grounded thermocouples with a nominal 22 to 30 American wire gage (AWG)-size conductor. The seven thermocouples must be attached to a steel angle bracket to form a thermocouple rake for placement in the test stand during burner calibration, as shown in Figure 5.

(5) *Apparatus Arrangement.* The test burner must be mounted on a suitable stand to position the exit of the burner cone a distance of 4 ±⅛ inches (102 ±3 mm) from one side of the specimen mounting stand. The burner stand should have the capability of allowing the burner to be swung away from the specimen mounting stand during warmup periods.

(6) *Data Recording.* A recording potentiometer or other suitable calibrated instrument with an appropriate range must be used to measure and record the outputs of the calorimeter and the thermocouples.

(7) *Weight Scale.* Weighing Device—A device must be used that with proper procedures may determine the before and after test weights of each set of seat cushion specimens within 0.02 pound (9 grams). A continuous weighing system is preferred.

(8) *Timing Device.* A stopwatch or other device (calibrated to ±1 second) must be used to measure the time of application of the burner flame and self-extinguishing time or test duration.

(e) *Preparation of Apparatus.* Before calibration, all equipment must be turned on and the burner fuel must be adjusted as specified in paragraph (d)(2).

(f) *Calibration.* To ensure the proper thermal output of the burner, the following test must be made:

(1) Place the calorimeter on the test stand as shown in Figure 4 at a distance of 4 ±⅛ inches (102 ±3 mm) from the exit of the burner cone.

(2) Turn on the burner, allow it to run for 2 minutes for warmup, and adjust the burner air intake damper to produce a reading of 10.5 ±0.5 BTU/ft²-sec. (11.9 ±0.6 w/cm²) on the calorimeter to ensure steady state conditions have been achieved. Turn off the burner.

(3) Replace the calorimeter with the thermocouple rake (Figure 5).

(4) Turn on the burner and ensure that the thermocouples are reading 1900 ±100 °F (1038 ±38 °C) to ensure steady state conditions have been achieved.

(5) If the calorimeter and thermocouples do not read within range, repeat steps in paragraphs 1 through 4 and adjust the burner air intake damper until the proper readings are obtained. The thermocouple rake and the calorimeter should be used frequently to maintain and record calibrated test parameters. Until the specific apparatus has demonstrated consistency, each test should be calibrated. After consistency has been confirmed, several tests may be conducted with the pre-test calibration before and a calibration check after the series.

(g) *Test Procedure.* The flammability of each set of specimens must be tested as follows:

(1) Record the weight of each set of seat bottom and seat back cushion specimens to be tested to the nearest 0.02 pound (9 grams).

(2) Mount the seat bottom and seat back cushion test specimens on the test stand as shown in Figure 2, securing the seat back cushion specimen to the test stand at the top.

(3) Swing the burner into position and ensure that the distance from the exit of the burner cone to the side of the seat bottom cushion specimen is 4 ±⅛ inches (102 ±3 mm).

(4) Swing the burner away from the test position. Turn on the burner and allow it to run for 2 minutes to provide adequate warmup of the burner cone and flame stabilization.

(5) To begin the test, swing the burner into the test position and simultaneously start the timing device.

(6) Expose the seat bottom cushion specimen to the burner flame for 2 minutes and then turn off the burner. Immediately swing the burner away from the test position. Terminate test 7 minutes after initiating cushion exposure to the flame by use of a gaseous extinguishing agent (i.e., Halon or CO_2).

(7) Determine the weight of the remains of the seat cushion specimen left on the mounting stand to the nearest 0.02 pound (9 grams) excluding all droppings.

(h) *Test Report.* With respect to all specimen sets tested for a particular seat cushion for which testing of compliance is performed, the following information must be recorded:

(1) An identification and description of the specimens being tested.

(2) The number of specimen sets tested.

(3) The initial weight and residual weight of each set, the calculated percentage weight loss of each set, and the calculated average percentage weight loss for the total number of sets tested.

(4) The burn length for each set tested.

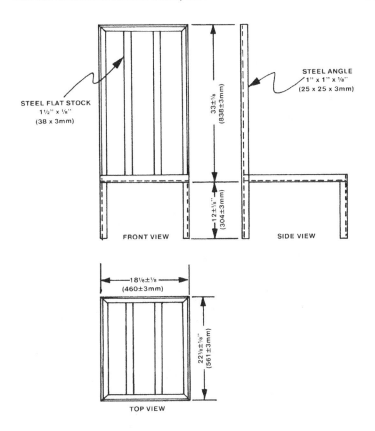

STEEL FLAT STOCK
1½" x ⅛"
(38 x 3mm)

STEEL ANGLE
1" x 1" x ⅛"
(25 x 25 x 3mm)

33±⅛
(838±3mm)

12±⅛"
(304±3mm)

FRONT VIEW

SIDE VIEW

18⅛±⅛
(460±3mm)

22⅛±⅛"
(561±3mm)

TOP VIEW

NOTE:
ALL JOINTS WELDED
FLAT STOCK BUTT WELDED
ALL MEASUREMENTS INSIDE

FIGURE 1

399

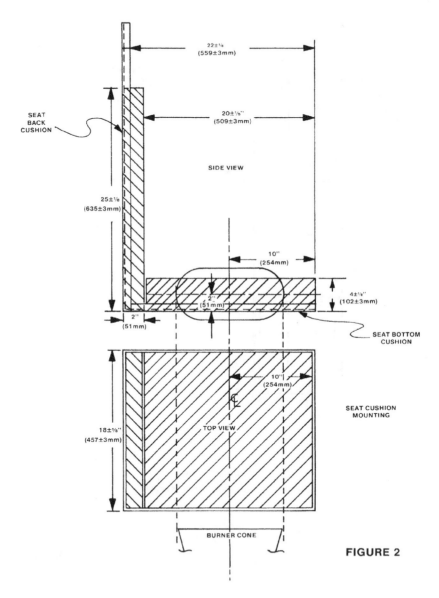

FIGURE 2

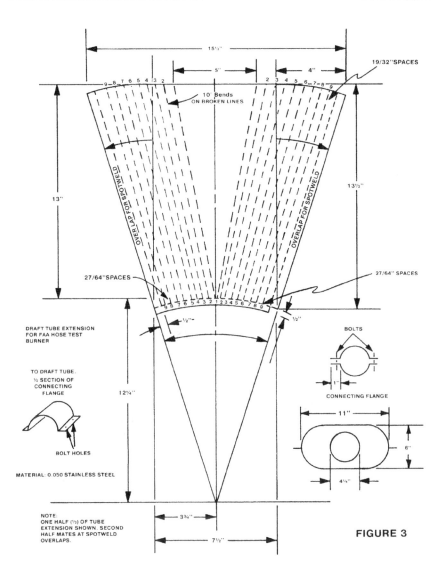

15½"

5"

4"

19/32"SPACES

9—8 7 6 5 4 3 2   2 3 4 5 6—7—8—9

10° Bends
ON BROKEN LINES

13½"

13"

OVER LAP FOR SPOTWELD

OVERLAP FOR SPOTWELD

27/64"SPACES

27/64" SPACES

9 8 7 6 5 4 3 2 1 2 3 4 5 6 7 8 9

½"

½"

BOLTS

DRAFT TUBE EXTENSION
FOR FAA HOSE TEST
BURNER

CONNECTING FLANGE

1"

TO DRAFT TUBE.
½ SECTION OF
CONNECTING
FLANGE

12¼"

11"

6"

4¼"

BOLT HOLES

MATERIAL: 0.050 STAINLESS STEEL

NOTE:
ONE HALF (½) OF TUBE
EXTENSION SHOWN. SECOND
HALF MATES AT SPOTWELD
OVERLAPS.

3¾"

7½"

FIGURE 3

401

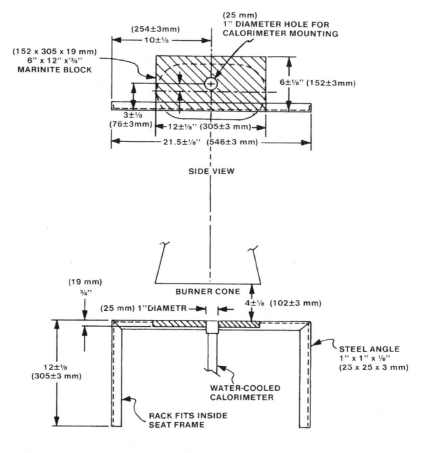

SIDE VIEW

BURNER CONE

TOP VIEW
CALORIMETER BRACKET

FIGURE 4

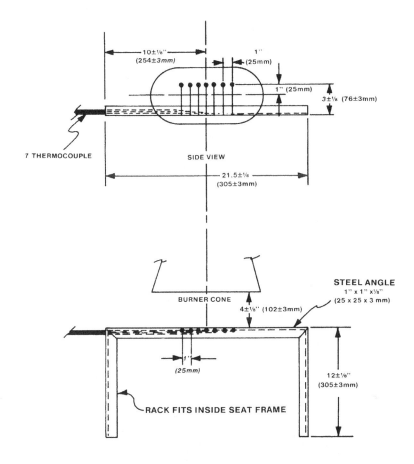

FIGURE 5

Part III—Test Method To Determine Flame Penetration Resistance of Cargo Compartment Liners.

(a) *Criteria for Acceptance.* (1) At least three specimens of cargo compartment sidewall or ceiling liner panels must be tested.

(2) Each specimen tested must simulate the cargo compartment sidewall or ceiling liner panel, including any design features, such as joints, lamp assemblies, etc., the failure of which would affect the capability of the liner to safely contain a fire.

(3) There must be no flame penetration of any specimen within 5 minutes after application of the flame source, and the peak temperature measured at 4 inches above the upper surface of the horizontal test sample must not exceed 400 °F.

(b) *Summary of Method.* This method provides a laboratory test procedure for measuring the capability of cargo compartment lining materials to resist flame penetration with a 2 gallon per hour (GPH) #2 Grade kerosene or equivalent burner fire source. Ceiling and sidewall liner panels may be tested individually provided a baffle is used to simulate the missing panel. Any specimen that passes the test as a ceiling liner panel may be used as a sidewall liner panel.

(c) *Test Specimens.* (1) The specimen to be tested must measure 16 ±⅛ inches (406 ±3 mm) by 24 + ⅛ inches (610 ±3 mm).

(2) The specimens must be conditioned at 70 °F.±5 °F. (21 °C. ±2 °C.) and 55%±5% humidity for at least 24 hours before testing.

(d) *Test Apparatus.* The arrangement of the test apparatus, which is shown in Figure 3 of Part II and Figures 1 through 3 of this part of appendix F, must include the components described in this section. Minor details of the apparatus may vary, depending on the model of the burner used.

(1) *Specimen Mounting Stand.* The mounting stand for the test specimens consists of steel angles as shown in Figure 1.

(2) *Test Burner.* The burner to be used in tesing must—

(i) Be a modified gun type.

(ii) Use a suitable nozzle and maintain fuel pressure to yield a 2 GPH fuel flow. For example: an 80 degree nozzle nominally rated at 2.25 GPH and operated at 85 pounds per square inch (PSI) gage to deliver 2.03 GPH.

(iii) Have a 12 inch (305 mm) burner extension installed at the end of the draft tube with an opening 6 inches (152 mm) high and 11 inches (280 mm) wide as shown in Figure 3 of Part II of this appendix.

(iv) Have a burner fuel pressure regulator that is adjusted to deliver a nominal 2.0 GPH of #2 Grade kerosene or equivalent.

Burner models which have been used successfully in testing are the Lenox Model OB–32, Carlin Model 200 CRD and Park Model DPL. The basic burner is described in FAA Powerplant Engineering Report No. 3A, Standard Fire Test Apparatus and Procedure for Flexible Hose Assemblies, dated March 1978; however, the test settings specified in this appendix differ in some instances from those specified in the report.

(3) *Calorimeter.* (i) The calorimeter to be used in testing must be a total heat flux Foil Type Gardon Gage of an appropriate range (approximately 0 to 15.0 British thermal unit (BTU) per ft.² sec., 0–17.0 watts/cm²). The calorimeter must be mounted in a 6 inch by 12 inch (152 by 305 mm) by ¾ inch (19 mm) thick insulating block which is attached to a steel angle bracket for placement in the test stand during burner calibration as shown in Figure 2 of this part of this appendix.

(ii) The insulating block must be monitored for deterioration and the mounting shimmed as necessary to ensure that the calorimeter face is parallel to the exit plane of the test burner cone.

(4) *Thermocouples.* The seven thermocouples to be used for testing must be ¹⁄₁₆ inch ceramic sheathed, type K, grounded thermocouples with a nominal 30 American wire gage (AWG) size conductor. The seven thermocouples must be attached to a steel angle bracket to form a thermocouple rake for placement in the test stand during burner calibration as shown in Figure 3 of this part of this appendix.

(5) *Apparatus Arrangement.* The test burner must be mounted on a suitable stand to position the exit of the burner cone a distance of 8 inches from the ceiling liner panel and 2 inches from the sidewall liner panel. The burner stand should have the capability of allowing the burner to be swung away from the test specimen during warm-up periods.

(6) *Instrumentation.* A recording potentiometer or other suitable instrument with an appropriate range must be used to measure and record the outputs of the calorimeter and the thermocouples.

(7) *Timing Device.* A stopwatch or other device must be used to measure the time of flame application and the time of flame penetration, if it occurs.

(e) *Preparation of Apparatus.* Before calibration, all equipment must be turned on and allowed to stabilize, and the burner fuel flow must be adjusted as specified in paragraph (d)(2).

(f) *Calibration.* To ensure the proper thermal output of the burner the following test must be made:

(1) Remove the burner extension from the end of the draft tube. Turn on the blower portion of the burner without turning the fuel or igniters on. Measure the air velocity using a hot wire anemometer in the center of the draft tube across the face of the opening. Adjust the damper such that the air velocity is in the range of 1550 to 1800 ft./min. If tabs are being used at the exit of the draft tube, they must be removed prior to this measurement. Reinstall the draft tube extension cone.

(2) Place the calorimeter on the test stand as shown in Figure 2 at a distance of 8 inches (203 mm) from the exit of the burner cone to simulate the position of the horizontal test specimen.

(3) Turn on the burner, allow it to run for 2 minutes for warm-up, and adjust the damper to produce a calorimeter reading of 8.0 ±0.5 BTU per ft.² sec. (9.1 ±0.6 Watts/cm²).

(4) Replace the calorimeter with the thermocouple rake (see Figure 3).

(5) Turn on the burner and ensure that each of the seven thermocouples reads 1700 °F. ±100 °F. (927 °C. ±38 °C.) to ensure steady state conditions have been achieved. If the temperature is out of this range, repeat steps 2 through 5 until proper readings are obtained.

(6) Turn off the burner and remove the thermocouple rake.

(7) Repeat (1) to ensure that the burner is in the correct range.

(g) *Test Procedure.* (1) Mount a thermocouple of the same type as that used for calibration at a distance of 4 inches (102 mm) above the horizontal (ceiling) test specimen. The thermocouple should be centered over the burner cone.

(2) Mount the test specimen on the test stand shown in Figure 1 in either the horizontal or vertical position. Mount the insulating material in the other position.

(3) Position the burner so that flames will not impinge on the specimen, turn the burner on, and allow it to run for 2 minutes. Rotate the burner to apply the flame to the specimen and simultaneously start the timing device.

(4) Expose the test specimen to the flame for 5 minutes and then turn off the burner. The test may be terminated earlier if flame penetration is observed.

(5) When testing ceiling liner panels, record the peak temperature measured 4 inches above the sample.

(6) Record the time at which flame penetration occurs if applicable.

(h) *Test Report.* The test report must include the following:

(1) A complete description of the materials tested including type, manufacturer, thickness, and other appropriate data.

(2) Observations of the behavior of the test specimens during flame exposure such as delamination, resin ignition, smoke, ect., including the time of such occurrence.

(3) The time at which flame penetration occurs, if applicable, for each of the three specimens tested.

(4) Panel orientation (ceiling or sidewall).

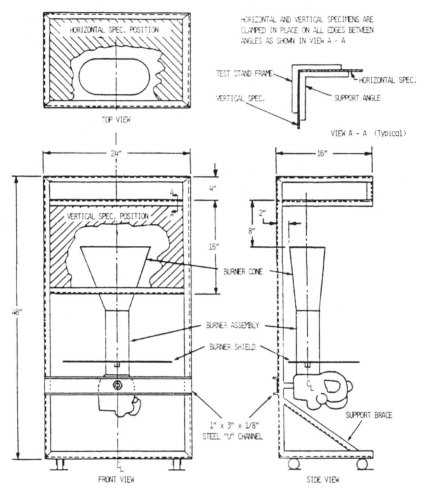

TEST STAND IS CONSTRUCTED WITH 1" x 1" x 1/8" STEEL ANGLES, ALL JOINTS WELDED
SUPPORT ANGLES ARE 1" x 1" x 1/8" CUT TO FIT

FIGURE 1.   TEST APPARATUS FOR HORIZONTAL AND VERTICAL MOUNTING

406

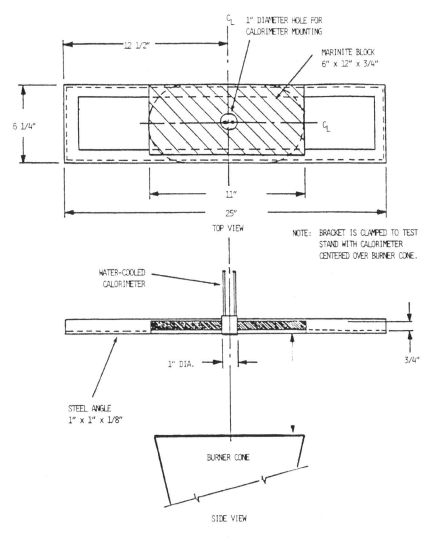

FIGURE 2. CALORIMETER BRACKET

407

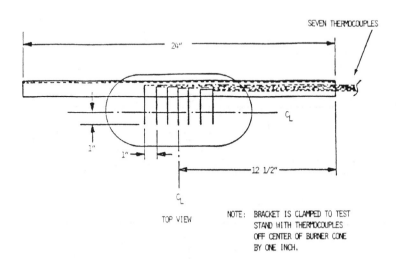

TOP VIEW

NOTE: BRACKET IS CLAMPED TO TEST
STAND WITH THERMOCOUPLES
OFF CENTER OF BURNER CONE
BY ONE INCH.

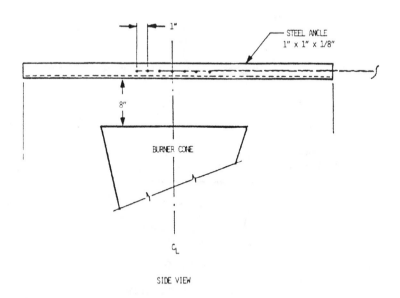

SIDE VIEW

FIGURE 3. THERMOCOUPLE RAKE BRACKET

408

Part IV—Test Method To Determine the Heat Release Rate From Cabin Materials Exposed to Radiant Heat.

(a) *Summary of Method.* Three or more specimens representing the completed aircraft component are tested. Each test specimen is injected into an environmental chamber through which a constant flow of air passes. The specimen's exposure is determined by a radiant heat source adjusted to produce, on the specimen, the desired total heat flux of 3.5 W/cm². The specimen is tested with the exposed surface vertical. Combustion is initiated by piloted ignition. The combustion products leaving the chamber are monitored in order to calculate the release rate of heat.

(b) *Apparatus.* The Ohio State University (OSU) rate of heat release apparatus, as described below, is used. This is a modified version of the rate of heat release apparatus standardized by the American Society of Testing and Materials (ASTM), ASTM E–906.

(1) This apparatus is shown in Figures 1A and 1B of this part IV. All exterior surfaces of the apparatus, except the holding chamber, must be insulated with 1 inch (25 mm) thick, low density, high temperature, fiberglass board insulation. A gasketed door, through which the sample injection rod slides, must be used to form an airtight closure on the specimen hold chamber.

(2) *Thermopile.* The temperature difference between the air entering the environmental chamber and that leaving must be monitored by a thermopile having five hot, and five cold, 24-guage Chromel-Alumel junctions. The hot junctions must be spaced across the top of the exhaust stack, .38 inches (10 mm) below the top of the chimney. The thermocouples must have a .050 ±.010 inch (1.3 ±.3mm) diameter, ball-type, welded tip. One thermocouple must be located in the geometric center, with the other four located 1.18 inch (30 mm) from the center along the diagonal toward each of the corners (Figure 5 of this part IV). The cold junctions must be located in the pan below the lower air distribution plate (see paragraph (b)(4) of this part IV). Thermopile hot junctions must be cleared of soot deposits as needed to maintain the calibrated sensitivity.

(3) *Radiation Source.* A radiant heat source incorporating four Type LL silicon carbide elements, 20 inches (508 mm) long by .63 inch (16 mm) O.D., must be used, as shown in Figures 2A and 2B of this part IV. The heat source must have a nominal resistance of 1.4 ohms and be capable of generating a flux up to 100 kW/m². The silicone carbide elements must be mounted in the stainless steel panel box by inserting them through .63 inch (16 mm) holes in .03 inch (1 mm) thick ceramic fiber or calcium-silicate millboard. Locations of the holes in the pads and stainless steel cover plates are shown in Figure 2B of this part IV. The truncated diamond-shaped mask of .042 ±.002 inch (1.07 ±.05mm) stainless steel must be added to provide uniform heat flux density over the area occupied by the vertical sample.

(4) *Air Distribution System.* The air entering the environmental chamber must be distributed by a .25 inch (6.3 mm) thick aluminum plate having eight No. 4 drill-holes, located 2 inches (51 mm) from sides on 4 inch (102 mm) centers, mounted at the base of the environmental chamber. A second plate of 18 guage stainless steel having 120, evenly spaced, No. 28 drill holes must be mounted 6 inches (152 mm) above the aluminum plate. A well-regulated air supply is required. The air-supply manifold at the base of the pyramidal section must have 48, evenly spaced, No. 26 drill holes located .38 inch (10 mm) from the inner edge of the manifold, resulting in an airflow split of approximately three to one within the apparatus.

(5) *Exhaust Stack.* An exhaust stack, 5.25 × 2.75 inches (133 × 70 mm) in cross section, and 10 inches (254 mm) long, fabricated from 28 guage stainless steel must be mounted on the outlet of the pyramidal section. A. 1.0 × 3.0 inch (25 × 76 mm) baffle plate of .018 ±.002 inch (.50 ±.05 mm) stainless steel must be centered inside the stack, perpendicular to the air flow, 3 inches (76 mm) above the base of the stack.

(6) *Specimen Holders.* (i) The specimen must be tested in a vertical orientation. The specimen holder (Figure 3 of this part IV) must incorporate a frame that touches the specimen (which is wrapped with aluminum foil as required by paragraph (d)(3) of this Part) along only the .25 inch (6 mm) perimeter. A "V" shaped spring is used to hold the assembly together. A detachable .50 × 50 × 5.91 inch (12 × 12 × 150 mm) drip pan and two .020 inch (.5 mm) stainless steel wires (as shown in Figure 3 of this part IV) must be used for testing materials prone to melting and dripping. The positioning of the spring and frame may be changed to accommodate different specimen thicknesses by inserting the retaining rod in different holes on the specimen holder.

(ii) Since the radiation shield described in ASTM E–906 is not used, a guide pin must be added to the injection mechanism. This fits into a slotted metal plate on the injection mechanism outside of the holding chamber. It can be used to provide accurate positioning of the specimen face after injection. The front surface of the specimen must be 3.9 inches (100 mm) from the closed radiation doors after injection.

(iii) The specimen holder clips onto the mounted bracket (Figure 3 of this part IV). The mounting bracket must be attached to the injection rod by three screws that pass through a wide-area washer welded onto a ½-inch (13 mm) nut. The end of the injection rod must be threaded to screw into the nut, and a .020 inch (5.1 mm) thick wide area washer must be held between two ½-inch (13 mm) nuts that are adjusted to tightly cover the hole in the radiation doors through which the injection rod or calibration calorimeter pass.

(7) *Calorimeter.* A total-flux type calorimeter must be mounted in the center of a ½-inch Kaowool "M" board inserted in the sample holder to measure the total heat flux. The calorimeter must have a view angle of 180 degrees and be calibrated for incident flux. The calorimeter calibration must be acceptable to the Administrator.

(8) *Pilot-Flame Positions.* Pilot ignition of the specimen must be accomplished by simultaneously exposing the specimen to a lower pilot burner and an upper pilot burner, as described in paragraph (b)(8)(i) and (b)(8)(ii) or (b)(8)(iii) of this part IV, respectively. Since intermittent pilot flame extinguishment for more than 3 seconds would invalidate the test results, a spark ignitor may be installed to ensure that the lower pilot burner remains lighted.

(i) *Lower Pilot Burner.* The pilot-flame tubing must be .25 inch (6.3 mm) O.D., .03 inch (0.8mm) wall, stainless steel tubing. A mixture of 120 cm³/min. of methane and 850 cm³/min. of air must be fed to the lower pilot flame burner. The normal position of the end of the pilot burner tubing is .40 inch (10 mm) from and perpendicular to the exposed vertical surface of the specimen. The centerline at the outlet of the burner tubing must intersect the vertical centerline of the sample at a point .20 inch (5 mm) above the lower exposed edge of the specimen.

(ii) *Standard Three-Hole Upper Pilot Burner.* The pilot burner must be a straight length of .25 inch (6.3 mm) O.D., .03 inch (0.8 mm) wall, stainless steel tubing that is 14 inches (360 mm) long. One end of the tubing must be closed, and three No. 40 drill holes must be drilled into the tubing, 2.38 inch (60 mm) apart, for gas ports, all radiating in the same direction. The first hole must be .19 inch (5 mm) from the closed end of the tubing. The tube must be positioned .75 inch (19 mm) above and .75 inch (19 mm) behind the exposed upper edge of the specimen. The middle hole must be in the vertical plane perpendicular to the exposed surface of the specimen which passes through its vertical centerline and must be pointed toward the radiation source. The gas supplied to the burner must be methane and must be adjusted to produce flame lengths of 1 inch (25 mm).

(iii) *Optional Fourteen-Hole Upper Pilot Burner.* This burner may be used in lieu of the standard three-hole burner described in paragraph (b)(8)(ii) of this part IV. The pilot burner must be a straight length of .25 inch (6.3 mm) O.D., .03 inch (0.8 mm) wall, stainless steel tubing that is 15.75 inches (400 mm) long. One end of the tubing must be closed, and 14 No. 59 drill holes must be drilled into

the tubing, .50 inch (13 mm) apart, for gas ports, all radiating in the same direction. The first hole must be .50 inch (13 mm) from the closed end of the tubing. The tube must be positioned above the specimen holder so that the holes are placed above the specimen as shown in Figure 1B of this part IV. The fuel supplied to the burner must be methane mixed with air in a ratio of approximately 50/50 by volume. The total gas flow must be adjusted to produce flame lengths of 1 inch (25 mm). When the gas/air ratio and the flow rate are properly adjusted, approximately .25 inch (6 mm) of the flame length appears yellow in color.

(c) *Calibration of Equipment—*(1) *Heat Release Rate.* A calibration burner, as shown in Figure 4, must be placed over the end of the lower pilot flame tubing using a gas tight connection. The flow of gas to the pilot flame must be at least 99 percent methane and must be accurately metered. Prior to usage, the wet test meter must be properly leveled and filled with distilled water to the tip of the internal pointer while no gas is flowing. Ambient temperature and pressure of the water are based on the internal wet test meter temperature. A baseline flow rate of approximately 1 liter/min. must be set and increased to higher preset flows of 4, 6, 8, 6 and 4 liters/min. Immediately prior to recording methane flow rates, a flow rate of 8 liters/min. must be used for 2 minutes to precondition the chamber. This is not recorded as part of calibration. The rate must be determined by using a stopwatch to time a complete revolution of the wet test meter for both the baseline and higher flow, with the flow returned to baseline before changing to the next higher flow. The thermopile baseline voltage must be measured. The gas flow to the burner must be increased to the higher preset flow and allowed to burn for 2.0 minutes, and the thermopile voltage must be measured. The sequence must be repeated until all five values have been determined. The average of the five values must be used as the calibration factor. The procedure must be repeated if the percent relative standard deviation is greater than 5 percent. Calculations are shown in paragraph (f) of this part IV.

(2) *Flux Uniformity.* Uniformity of flux over the specimen must be checked periodically and after each heating element change to determine if it is within acceptable limits of plus or minus 5 percent.

(3) As noted in paragraph (b)(2) of this part IV, thermopile hot junctions must be cleared of soot deposits as needed to maintain the calibrated sensitivity.

(d) *Preparation of Test Specimens.* (1) The test specimens must be representative of the aircraft component in regard to materials and construction methods. The standard size for the test specimens is 5.91 ±.03 × 5.91 ±.03 inches (149 ±1 × 149 ±1 mm). The thickness of the specimen must be the same as that of the aircraft component it represents up to a maximum thickness of 1.75 inches (45 mm). Test specimens representing thicker components must be 1.75 inches (45 mm).

(2) *Conditioning.* Specimens must be conditioned as described in Part 1 of this appendix.

(3) *Mounting.* Each test specimen must be wrapped tightly on all sides of the specimen, except for the one surface that is exposed with a single layer of .001 inch (.025 mm) aluminum foil.

(e) *Procedure.* (1) The power supply to the radiant panel must be set to produce a radiant flux of 3.5 ±.05 W/cm², as measured at the point the center of the specimen surface will occupy when positioned for the test. The radiant flux must be measured after the air flow through the equipment is adjusted to the desired rate.

(2) After the pilot flames are lighted, their position must be checked as described in paragraph (b)(8) of this part IV.

(3) Air flow through the apparatus must be controlled by a circular plate orifice located in a 1.5 inch (38.1 mm) I.D. pipe with two pressure measuring points, located 1.5 inches (38 mm) upstream and .75 inches (19 mm) downstream of the orifice plate. The pipe must be connected to a manometer set at a pressure differential of 7.87 inches (200 mm) of Hg. (See Figure 1B of this part IV.) The total air flow to the equipment is approximately .04 m³/seconds. The stop on the vertical specimen holder rod must be adjusted so that the exposed surface of the specimen is positioned 3.9 inches (100 mm) from the entrance when injected into the environmental chamber.

(4) The specimen must be placed in the hold chamber with the radiation doors closed. The airtight outer door must be secured, and the recording devices must be started. The specimen must be retained in the hold chamber for 60 seconds, plus or minus 10 seconds, before injection. The thermopile "zero" value must be determined during the last 20 seconds of the hold period. The sample must not be injected before completion of the "zero" value determination.

(5) When the specimen is to be injected, the radiation doors must be opened. After the specimen is injected into the environmental chamber, the radiation doors must be closed behind the specimen.

(6) [Reserved]

(7) Injection of the specimen and closure of the inner door marks time zero. A record of the thermopile output with at least one data point per second must be made during the time the specimen is in the environmental chamber.

(8) The test duration is five minutes. The lower pilot burner and the upper pilot burner must remain lighted for the entire duration of the test, except that there may be intermittent flame extinguishment for periods that do not exceed 3 seconds. Furthermore, if the optional three-hole upper burner is used, at least two flamelets must remain lighted for the entire duration of the test, except that there may be intermittent flame extinguishment of all three flamelets for periods that do not exceed 3 seconds.

(9) A minimum of three specimens must be tested.

(f) *Calculations.* (1) The calibration factor is calculated as follows:

$$K_h = \frac{(F_1 - F_O)}{(V_1 - V_O)} \times \frac{(210.8 - 22)k_{cal}}{mole} \times \frac{273}{T_a} \times \frac{P - P_v}{760} \times \frac{mole\ CH4STP}{22.41} \times \frac{WATT\ min}{.01433kcal} \times \frac{kw}{1000w}$$

F_o = flow of methane at baseline (1pm)
F_1 = higher preset flow of methane (1pm)
V_o = thermopile voltage at baseline (mv)
V_1 = thermopile voltage at higher flow (mv)
T_a = Ambient temperature (K)
P = Ambient pressure (mm Hg)
P_v = Water vapor pressure (mm Hg)

(2) Heat release rates may be calculated from the reading of the thermopile output voltage at any instant of time as:

$$HRR = \frac{(V_m - V_b)K_n}{.02323m^2}$$

HRR = heat release rate (kw/m²)
V_b = baseline voltage (mv)
V_m = measured thermopile voltage (mv)

K_h = calibration factor (kw/mv)

(3) The integral of the heat release rate is the total heat release as a function of time and is calculated by multiplying the rate by the data sampling frequency in minutes and summing the time from zero to two minutes.

(g) *Criteria.* The total positive heat release over the first two minutes of exposure for each of the three or more samples tested must be averaged, and the peak heat release rate for each of the samples must be averaged. The average total heat release must not exceed 65 kilowatt-minutes per square meter, and the average peak heat release rate must not exceed 65 kilowatts per square meter.

411

(h) *Report.* The test report must include the following for each specimen tested:

(1) Description of the specimen.

(2) Radiant heat flux to the specimen, expressed in W/cm².

(3) Data giving release rates of heat (in kW/m²) as a function of time, either graphically or tabulated at intervals no greater than 10 seconds. The calibration factor (k_n) must be recorded.

(4) If melting, sagging, delaminating, or other behavior that affects the exposed surface area or the mode of burning occurs, these behaviors must be reported, together with the time at which such behaviors were observed.

(5) The peak heat release and the 2-minute integrated heat release rate must be reported.

FIGURES TO PART IV OF APPENDIX F

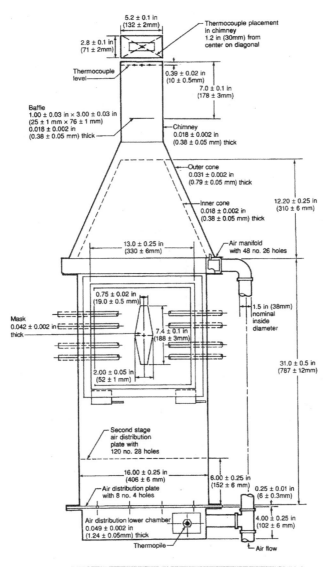

Figure 1A   Rate of Heat Release Apparatus

413

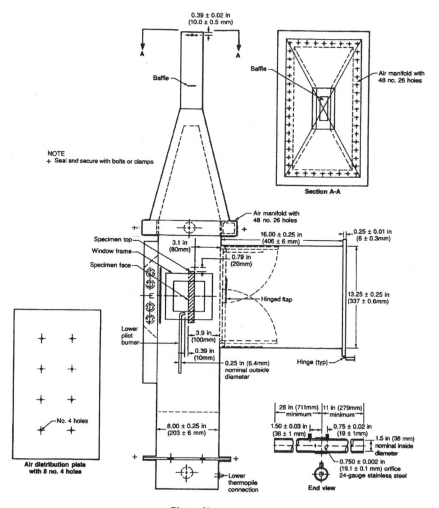

Figure 1B
Rate of Heat Release Apparatus

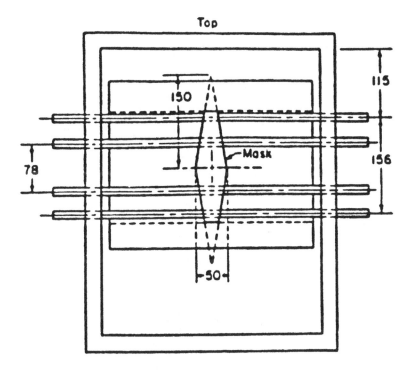

(Unless denoted otherwise all dimensions are in millimeters.)
Figure 2A. "Globar" Radiant Panel

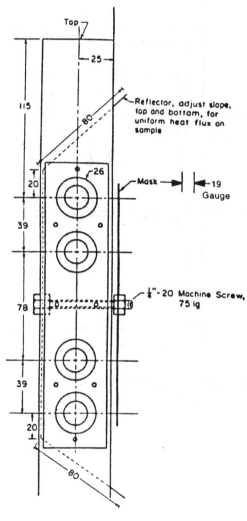

(Unless denoted otherwise all dimensions are in millimeters.)
Figure 2B. "Globar" Radiant Panel

416

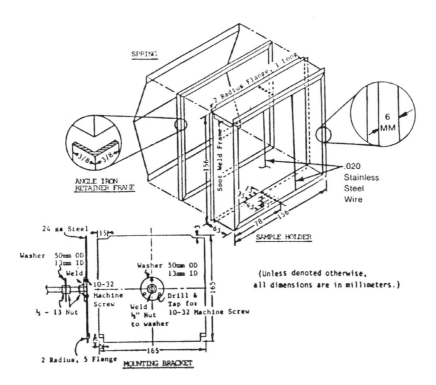

Figure 3.

417

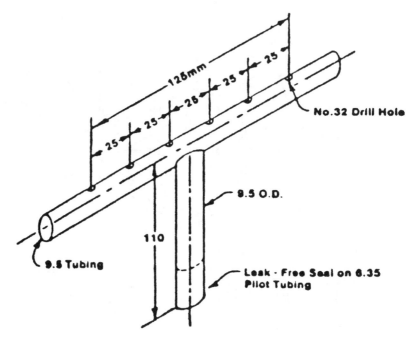

(Unless denoted otherwise, all dimensions are in millimeters.)

Figure 4.

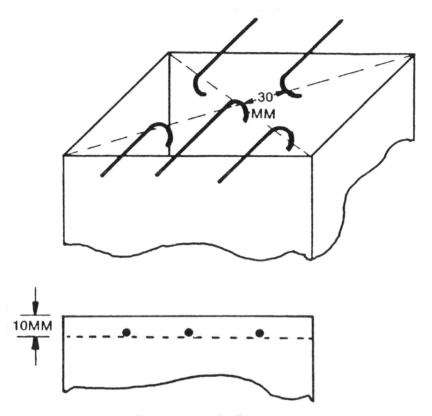

Figure 5. Thermocouple Position

Part V. Test Method To Determine the Smoke Emission Characteristics of Cabin Materials

(a) *Summary of Method.* The specimens must be constructed, conditioned, and tested in the flaming mode in accordance with American Society of Testing and Materials (ASTM) Standard Test Method ASTM F814–83.

(b) *Acceptance Criteria.* The specific optical smoke density (D$_s$), which is obtained by averaging the reading obtained after 4 minutes with each of the three specimens, shall not exceed 200.

Part VI—Test Method To Determine the Flammability and Flame Propagation Characteristics of Thermal/Acoustic Insulation Materials

Use this test method to evaluate the flammability and flame propagation characteristics of thermal/acoustic insulation when exposed to both a radiant heat source and a flame.

(a) *Definitions.*

"Flame propagation" means the furthest distance of the propagation of visible flame towards the far end of the test specimen, measured from the midpoint of the ignition source flame. Measure this distance after initially applying the ignition source and before all flame on the test specimen is extinguished. The measurement is not a determination of burn length made after the test.

"Radiant heat source" means an electric or air propane panel.

"Thermal/acoustic insulation" means a material or system of materials used to provide thermal and/or acoustic protection. Examples include fiberglass or other batting material encapsulated by a film covering and foams.

"Zero point" means the point of application of the pilot burner to the test specimen.

(b) *Test apparatus.*

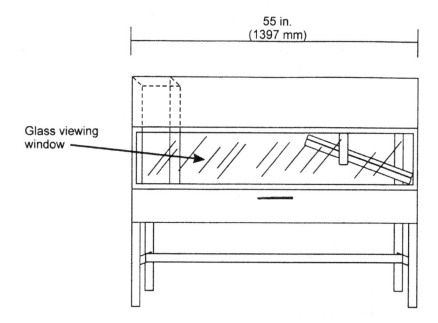

Figure 1 - Radiant Panel Test Chamber

(1) *Radiant panel test chamber.* Conduct tests in a radiant panel test chamber (see figure 1 above). Place the test chamber under an exhaust hood to facilitate clearing the chamber of smoke after each test. The radiant panel test chamber must be an enclosure 55 inches (1397 mm) long by 19.5 (495 mm) deep by 28 (710 mm) to 30 inches (maximum) (762 mm) above the test specimen. Insulate the sides, ends, and top with a fibrous ceramic insulation, such as Kaowool M™ board. On the front side, provide a 52 by 12-inch (1321 by 305 mm) draft-free, high-temperature, glass window for viewing the sample during testing. Place a door below the window to provide access to the movable specimen platform holder. The bottom of the test chamber must be a sliding steel platform that has provision for securing the test specimen holder in a fixed and level position. The chamber must have an internal chimney with exterior dimensions of 5.1 inches (129 mm) wide, by 16.2 inches (411 mm) deep by 13 inches (330 mm) high at the opposite end of the chamber from the radiant energy source. The interior dimensions must be 4.5 inches (114 mm) wide by 15.6 inches (395 mm) deep. The chimney must extend to the top of the chamber (see figure 2).

½ in. (13 mm) Kaowool M   board

16 gauge (1/16 in. 1.6mm) aluminum sheet metal

1/8 in. (3.2 mm) angle iron

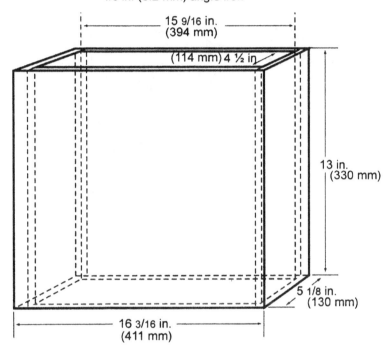

Figure 2 - Internal Chimney

(2) *Radiant heat source.* Mount the radiant heat energy source in a cast iron frame or equivalent. An electric panel must have six, 3-inch wide emitter strips. The emitter strips must be perpendicular to the length of the panel. The panel must have a radiation surface of 12⅞ by 18½ inches (327 by 470 mm). The panel must be capable of operating at temperatures up to 1300 °F (704 °C). An air propane panel must be made of a porous refractory material and have a radiation surface of 12 by 18 inches (305 by 457 mm). The panel must be capable of operating at temperatures up to 1,500 °F (816 °C). *See* figures 3a and 3b.

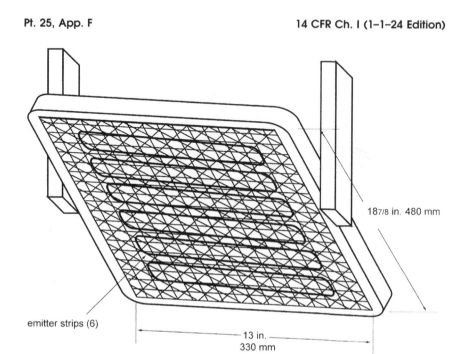

emitter strips (6)

13 in.
330 mm

18 7/8 in. 480 mm

Figure 3a – Electric Panel

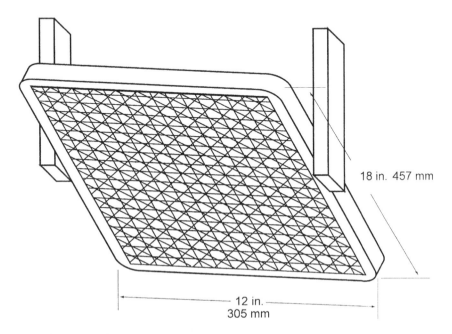

18 in. 457 mm

12 in.
305 mm

Figure 3b – Air Propane Radiant Panel

(i) *Electric radiant panel.* The radiant panel must be 3-phase and operate at 208 volts. A single-phase, 240 volt panel is also acceptable. Use a solid-state power controller and microprocessor-based controller to set the electric panel operating parameters.

(ii) *Gas radiant panel.* Use propane (liquid petroleum gas—2.1 UN 1075) for the radiant panel fuel. The panel fuel system must consist of a venturi-type aspirator for mixing gas and air at approximately atmospheric pressure. Provide suitable instrumentation for monitoring and controlling the flow of fuel and air to the panel. Include an air flow gauge, an air flow regulator, and a gas pressure gauge.

(iii) *Radiant panel placement.* Mount the panel in the chamber at 30° to the horizontal specimen plane, and 7½ inches above the zero point of the specimen.

(3) *Specimen holding system.* (i) The sliding platform serves as the housing for test specimen placement. Brackets may be attached (via wing nuts) to the top lip of the platform in order to accommodate various thicknesses of test specimens. Place the test specimens on a sheet of Kaowool M™ board or 1260 Standard Board (manufactured by Thermal Ceramics and available in Europe), or equivalent, either resting on the bottom lip of the sliding platform or on the base of the brackets. It may be necessary to use multiple sheets of material based on the thickness of the test specimen (to meet the sample height requirement). Typically, these non-combustible sheets of material are available in ¼ inch (6 mm) thicknesses. See figure 4. A sliding platform that is deeper than the 2-inch (50.8mm) platform shown in figure 4 is also acceptable as long as the sample height requirement is met.

423

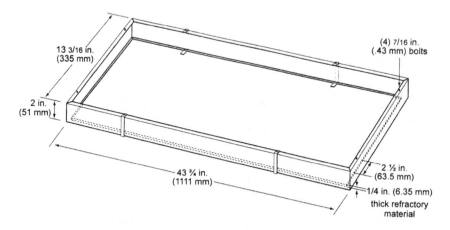

Figure 4 - Sliding Platform

(ii) Attach a ½ inch (13 mm) piece of Kaowool M™ board or other high temperature material measuring 41½ by 8¼ inches (1054 by 210 mm) to the back of the platform. This board serves as a heat retainer and protects the test specimen from excessive preheating. The height of this board must not impede the sliding platform movement (in and out of the test chamber). If the platform has been fabricated such that the back side of the platform is high enough to prevent excess preheating of the specimen when the sliding platform is out, a retainer board is not necessary.

(iii) Place the test specimen horizontally on the non-combustible board(s). Place a steel retaining/securing frame fabricated of mild steel, having a thickness of ⅛ inch (3.2 mm) and overall dimensions of 23 by 13⅛ inches (584 by 333 mm) with a specimen opening of 19 by 10¾ inches (483 by 273 mm) over the test specimen. The front, back, and right portions of the top flange of the frame must rest on the top of the sliding platform, and the bottom flanges must pinch all 4 sides of the test specimen. The right bottom flange must be flush with the sliding platform. See figure 5.

424

TOP VIEW

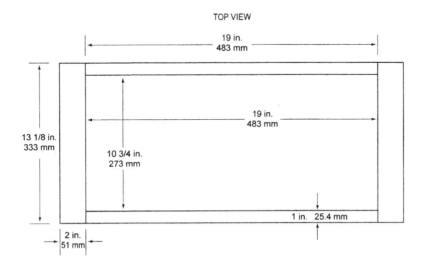

SIDE VIEW

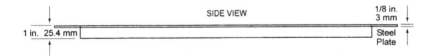

THREE DIMENSIONAL VIEW

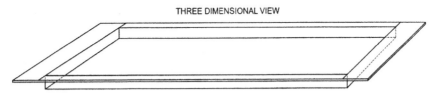

Figure 5: 3 views

(4) *Pilot Burner.* The pilot burner used to ignite the specimen must be a Bernzomatic™ commercial propane venturi torch with an axially symmetric burner tip and a propane supply tube with an orifice diameter of 0.006 inches (0.15 mm). The length of the burner tube must be 2⅞ inches (71 mm). The propane flow must be adjusted via gas pressure through an in-line regulator to produce a blue inner cone length of ¾ inch (19 mm). A ¾ inch (19 mm) guide (such as a thin strip of metal) may be soldered to the top of the burner to aid in setting the flame height. The overall flame length must be approximately 5 inches long (127 mm). Provide a way to move the burner out of the ignition position so that the flame is horizontal and at least 2 inches (50 mm) above the specimen plane. See figure 6.

425

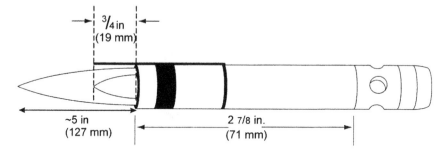

Figure 6 – Propane Pilot Burner

(5) *Thermocouples.* Install a 24 American Wire Gauge (AWG) Type K (Chromel-Alumel) thermocouple in the test chamber for temperature monitoring. Insert it into the chamber through a small hole drilled through the back of the chamber. Place the thermocouple so that it extends 11 inches (279 mm) out from the back of the chamber wall, 11½ inches (292 mm) from the right side of the chamber wall, and is 2 inches (51 mm) below the radiant panel. The use of other thermocouples is optional.

(6) *Calorimeter.* The calorimeter must be a one-inch cylindrical water-cooled, total heat flux density, foil type Gardon Gage that has a range of 0 to 5 BTU/ft²-second (0 to 5.7 Watts/cm²).

(7) *Calorimeter calibration specification and procedure.*

(i) *Calorimeter specification.*

(A) Foil diameter must be 0.25 ±0.005 inches (6.35 ±0.13 mm).

(B) Foil thickness must be 0.0005 ±0.0001 inches (0.013 ±0.0025 mm).

(C) Foil material must be thermocouple grade Constantan.

(D) Temperature measurement must be a Copper Constantan thermocouple.

(E) The copper center wire diameter must be 0.0005 inches (0.013 mm).

(F) The entire face of the calorimeter must be lightly coated with "Black Velvet" paint having an emissivity of 96 or greater.

(ii) *Calorimeter calibration.* (A) The calibration method must be by comparison to a like standardized transducer.

(B) The standardized transducer must meet the specifications given in paragraph VI(b)(6) of this appendix.

(C) Calibrate the standard transducer against a primary standard traceable to the National Institute of Standards and Technology (NIST).

(D) The method of transfer must be a heated graphite plate.

(E) The graphite plate must be electrically heated, have a clear surface area on each side of the plate of at least 2 by 2 inches (51 by 51 mm), and be ⅛ inch ±¹⁄₁₆ inch thick (3.2 ±1.6 mm).

(F) Center the 2 transducers on opposite sides of the plates at equal distances from the plate.

(G) The distance of the calorimeter to the plate must be no less than 0.0625 inches (1.6 mm), nor greater than 0.375 inches (9.5 mm).

(H) The range used in calibration must be at least 0–3.5 BTUs/ft² second (0–3.9 Watts/cm²) and no greater than 0–5.7 BTUs/ft² second (0–6.4 Watts/cm²).

(I) The recording device used must record the 2 transducers simultaneously or at least within ¹⁄₁₀ of each other.

(8) *Calorimeter fixture.* With the sliding platform pulled out of the chamber, install the calorimeter holding frame and place a sheet of non-combustible material in the bottom of the sliding platform adjacent to the holding frame. This will prevent heat losses during calibration. The frame must be 13⅛ inches (333 mm) deep (front to back) by 8 inches (203 mm) wide and must rest on the top of the sliding platform. It must be fabricated of ⅛ inch (3.2 mm) flat stock steel and have an opening that accommodates a ½ inch (12.7 mm) thick piece of refractory board, which is level with the top of the sliding platform. The board must have three 1-inch (25.4 mm) diameter holes drilled through the board for calorimeter insertion. The distance to the radiant panel surface from the centerline of the first hole ("zero" position) must be 7½ ±⅛ inches (191 ±3 mm). The distance between the centerline of the first hole to the centerline of the second hole must be 2 inches (51 mm). It must also be the same distance from the centerline of the second hole to the centerline of the third hole. *See* figure 7. A calorimeter holding frame that differs in construction is acceptable as long as the height from the centerline of the

first hole to the radiant panel and the distance between holes is the same as described in this paragraph.

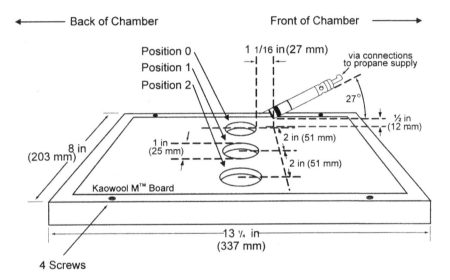

Figure 7 - Calorimeter Holding Frame

(9) *Instrumentation.* Provide a calibrated recording device with an appropriate range or a computerized data acquisition system to measure and record the outputs of the calorimeter and the thermocouple. The data acquisition system must be capable of recording the calorimeter output every second during calibration.

(10) *Timing device.* Provide a stopwatch or other device, accurate to ±1 second/hour, to measure the time of application of the pilot burner flame.

(c) *Test specimens.* (1) *Specimen preparation.* Prepare and test a minimum of three test specimens. If an oriented film cover material is used, prepare and test both the warp and fill directions.

(2) *Construction.* Test specimens must include all materials used in construction of the insulation (including batting, film, scrim, tape etc.). Cut a piece of core material such as foam or fiberglass, and cut a piece of film cover material (if used) large enough to cover the core material. Heat sealing is the preferred method of preparing fiberglass samples, since they can be made without compressing the fiberglass ("box sample"). Cover materials that are not heat sealable may be stapled, sewn, or taped as long as the cover material is over-cut enough to be drawn down the sides without compressing the core material. The fastening means should be as continuous as possible along the length of the seams. The specimen thickness must be of the same thickness as installed in the airplane.

(3) *Specimen Dimensions.* To facilitate proper placement of specimens in the sliding platform housing, cut non-rigid core materials, such as fiberglass, 12½ inches (318mm) wide by 23 inches (584mm) long. Cut rigid materials, such as foam, 11½ ±¼ inches (292 mm ±6mm) wide by 23 inches (584mm) long in order to fit properly in the sliding platform housing and provide a flat, exposed surface equal to the opening in the housing.

(d) *Specimen conditioning.* Condition the test specimens at 70 ±5 °F (21 ±2 °C) and 55% ±10% relative humidity, for a minimum of 24 hours prior to testing.

(e) *Apparatus Calibration.* (1) With the sliding platform out of the chamber, install the calorimeter holding frame. Push the platform back into the chamber and insert the calorimeter into the first hole ("zero" position). *See* figure 7. Close the bottom door located below the sliding platform. The distance from the centerline of the calorimeter to the radiant panel surface at this point must be 7.½ inches ±⅛ (191 mm ±3). Prior to igniting the radiant panel, ensure that the calorimeter face is clean and that there is water running through the calorimeter.

(2) Ignite the panel. Adjust the fuel/air mixture to achieve 1.5 BTUs/ft^2-second ±5% (1.7 Watts/cm^2 ±5%) at the "zero" position. If using an electric panel, set the power controller to achieve the proper heat flux. Allow the unit to reach steady state (this may take up to 1 hour). The pilot burner must be off and in the down position during this time.

(3) After steady-state conditions have been reached, move the calorimeter 2 inches (51 mm) from the "zero" position (first hole) to position 1 and record the heat flux. Move the calorimeter to position 2 and record the heat flux. Allow enough time at each position for the calorimeter to stabilize. Table 1 depicts typical calibration values at the three positions.

TABLE 1—CALIBRATION TABLE

Position	BTU's/ft^{2}sec	Watts/cm^2
"Zero" Position	1.5	1.7
Position 1	1.51–1.50–1.49	1.71–1.70–1.69
Position 2	1.43–1.44	1.62–1.63

(4) Open the bottom door, remove the calorimeter and holder fixture. Use caution as the fixture is very hot.

(f) *Test Procedure.* (1) Ignite the pilot burner. Ensure that it is at least 2 inches (51 mm) above the top of the platform. The burner must not contact the specimen until the test begins.

(2) Place the test specimen in the sliding platform holder. Ensure that the test sample surface is level with the top of the platform. At "zero" point, the specimen surface must be 7½ inches ±⅛ inch (191 mm ±3) below the radiant panel.

(3) Place the retaining/securing frame over the test specimen. It may be necessary (due to compression) to adjust the sample (up or down) in order to maintain the distance from the sample to the radiant panel (7½ inches ±⅛ inch (191 mm±3) at "zero" position). With film/fiberglass assemblies, it is critical to make a slit in the film cover to purge any air inside. This allows the operator to maintain the proper test specimen position (level with the top of the platform) and to allow ventilation of gases during testing. A longitudinal slit, approximately 2 inches (51mm) in length, must be centered 3 inches ±½ inch (76mm±13mm) from the left flange of the securing frame. A utility knife is acceptable for slitting the film cover.

(4) Immediately push the sliding platform into the chamber and close the bottom door.

(5) Bring the pilot burner flame into contact with the center of the specimen at the "zero" point and simultaneously start the timer. The pilot burner must be at a 27° angle with the sample and be approximately ½ inch (12 mm) above the sample. *See* figure 7. A stop, as shown in figure 8, allows the operator to position the burner correctly each time.

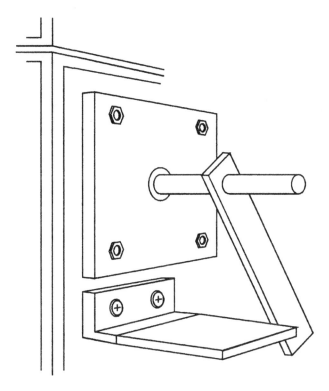

Figure 8 - Propane Burner Stop

(6) Leave the burner in position for 15 seconds and then remove to a position at least 2 inches (51 mm) above the specimen.

(g) *Report.* (1) Identify and describe the test specimen.

(2) Report any shrinkage or melting of the test specimen.

(3) Report the flame propagation distance. If this distance is less than 2 inches, report this as a pass (no measurement required).

(4) Report the after-flame time.

(h) *Requirements.* (1) There must be no flame propagation beyond 2 inches (51 mm) to the left of the centerline of the pilot flame application.

(2) The flame time after removal of the pilot burner may not exceed 3 seconds on any specimen.

Part VII—Test Method To Determine the Burnthrough Resistance of Thermal/Acoustic Insulation Materials

Use the following test method to evaluate the burnthrough resistance characteristics of aircraft thermal/acoustic insulation materials when exposed to a high intensity open flame.

(a) *Definitions.*

Burnthrough time means the time, in seconds, for the burner flame to penetrate the test specimen, and/or the time required for the heat flux to reach 2.0 Btu/ft²sec (2.27 W/cm²) on the inboard side, at a distance of 12 inches (30.5 cm) from the front surface of the insulation blanket test frame, whichever is sooner. The burnthrough time is measured at the inboard side of each of the insulation blanket specimens.

Insulation blanket specimen means one of two specimens positioned in either side of the test rig, at an angle of 30° with respect to vertical.

Specimen set means two insulation blanket specimens. Both specimens must represent the same production insulation blanket construction and materials, proportioned to correspond to the specimen size.

(b) *Apparatus.* (1) The arrangement of the test apparatus is shown in figures 1 and 2 and must include the capability of swinging the burner away from the test specimen during warm-up.

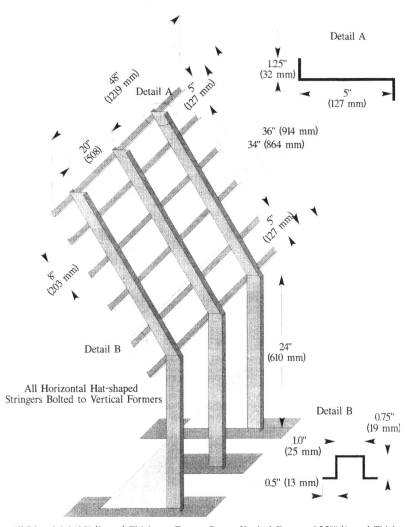

Detail A

1.25"
(32 mm)

5"
(127 mm)

48"
(1219 mm)

Detail A

5"
(127 mm)

36" (914 mm)
34" (864 mm)

20"
(508)

5"
(127 mm)

8"
(203 mm)

Detail B

24"
(610 mm)

All Horizontal Hat-shaped
Stringers Bolted to Vertical Formers

Detail B

0.75"
(19 mm)

1.0"
(25 mm)

0.5" (13 mm)

All Material 0.125" (3 mm) Thickness Except Center Vertical Former, 0.250" (6 mm) Thick

Figure 1 - Burnthrough Test Apparatus Specimen Holder

(2) *Test burner*. The test burner must be a modified gun-type such as the Park Model DPL 3400. Flame characteristics are highly dependent on actual burner setup. Param- eters such as fuel pressure, nozzle depth, stator position, and intake airflow must be properly adjusted to achieve the correct flame output.

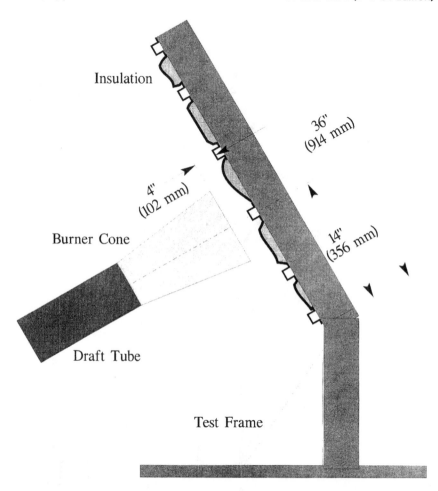

Figure 2 – Burnthrough Test Apparatus

(i) *Nozzle.* A nozzle must maintain the fuel pressure to yield a nominal 6.0 gal/hr (0.378 L/min) fuel flow. A Monarch-manufactured 80° PL (hollow cone) nozzle nominally rated at 6.0 gal/hr at 100 lb/in² (0.71 MPa) delivers a proper spray pattern.

(ii) *Fuel Rail.* The fuel rail must be adjusted to position the fuel nozzle at a depth of 0.3125 inch (8 mm) from the end plane of the exit stator, which must be mounted in the end of the draft tube.

(iii) *Internal Stator.* The internal stator, located in the middle of the draft tube, must

432

be positioned at a depth of 3.75 inches (95 mm) from the tip of the fuel nozzle. The stator must also be positioned such that the integral igniters are located at an angle midway between the 10 and 11 o'clock position, when viewed looking into the draft tube. Minor deviations to the igniter angle are acceptable if the temperature and heat flux requirements conform to the requirements of paragraph VII(e) of this appendix.

(iv) *Blower Fan.* The cylindrical blower fan used to pump air through the burner must measure 5.25 inches (133 mm) in diameter by 3.5 inches (89 mm) in width.

(v) *Burner cone.* Install a 12 + 0.125-inch (305 ±3 mm) burner extension cone at the end of the draft tube. The cone must have an opening 6 ±0.125-inch (152 ±3 mm) high and 11 ±0.125-inch (280 ±3 mm) wide (see figure 3).

(vi) *Fuel.* Use JP–8, Jet A, or their international equivalent, at a flow rate of 6.0 ±0.2 gal/hr (0.378 ±0.0126 L/min). If this fuel is unavailable, ASTM K2 fuel (Number 2 grade kerosene) or ASTM D2 fuel (Number 2 grade fuel oil or Number 2 diesel fuel) are acceptable if the nominal fuel flow rate, temperature, and heat flux measurements conform to the requirements of paragraph VII(e) of this appendix.

(vii) *Fuel pressure regulator.* Provide a fuel pressure regulator, adjusted to deliver a nominal 6.0 gal/hr (0.378 L/min) flow rate. An operating fuel pressure of 100 lb/in^2 (0.71 MPa) for a nominally rated 6.0 gal/hr 80° spray angle nozzle (such as a PL type) delivers 6.0 ±0.2 gal/hr (0.378 ±0.0126 L/min).

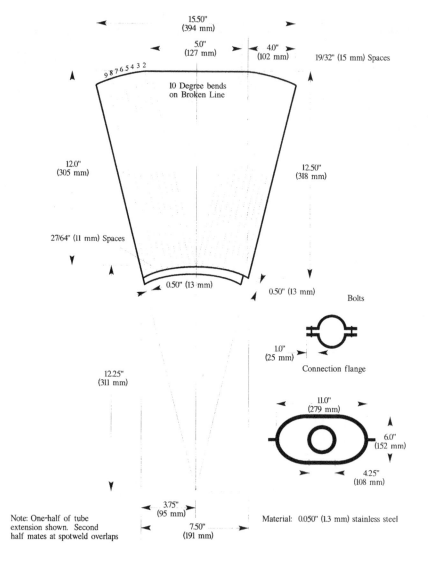

Figure 3 – Burner Draft Tube Extension Cone Diagram

(3) *Calibration rig and equipment.* (i) Construct individual calibration rigs to incorporate a calorimeter and thermocouple rake for the measurement of heat flux and temperature. Position the calibration rigs to allow movement of the burner from the test rig position to either the heat flux or temperature position with minimal difficulty.

(ii) *Calorimeter.* The calorimeter must be a total heat flux, foil type Gardon Gage of an appropriate range such as 0–20 Btu/ft²-sec (0–22.7 W/cm²), accurate to ±3% of the indicated reading. The heat flux calibration method must be in accordance with paragraph VI(b)(7) of this appendix.

(iii) *Calorimeter mounting.* Mount the calorimeter in a 6- by 12- ±0.125 inch (152- by 305- ±3 mm) by 0.75 ±0.125 inch (19 mm ±3 mm) thick insulating block which is attached to the heat flux calibration rig during calibration (figure 4). Monitor the insulating block for deterioration and replace it when necessary. Adjust the mounting as necessary to ensure that the calorimeter face is parallel to the exit plane of the test burner cone.

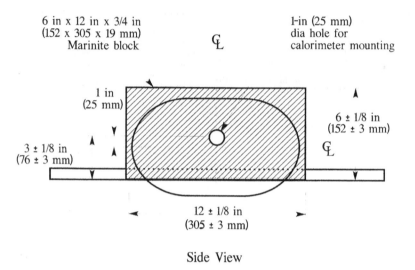

Side View

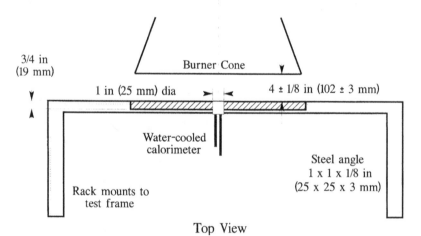

Top View

Figure 4 - Calorimeter Position Relative to Burner Cone

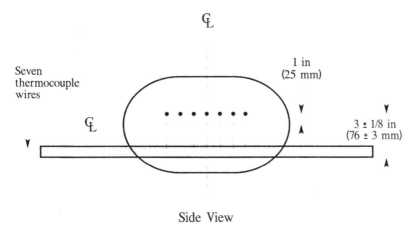

Side View

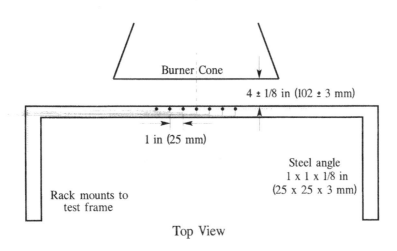

Top View

Figure 5 – Thermocouple Rake Position Relative to Burner Cone

(iv) *Thermocouples*. Provide seven ⅛-inch (3.2 mm) ceramic packed, metal sheathed, type K (Chromel-alumel), grounded junction thermocouples with a nominal 24 American Wire Gauge (AWG) size conductor for calibration. Attach the thermocouples to a steel angle bracket to form a thermocouple rake

437

for placement in the calibration rig during burner calibration (figure 5).

(v) *Air velocity meter.* Use a vane-type air velocity meter to calibrate the velocity of air entering the burner. An Omega Engineering Model HH30A is satisfactory. Use a suitable adapter to attach the measuring device to the inlet side of the burner to prevent air from entering the burner other than through the measuring device, which would produce erroneously low readings. Use a flexible duct, measuring 4 inches wide (102 mm) by 20 feet long (6.1 meters), to supply fresh air to the burner intake to prevent damage to the air velocity meter from ingested soot. An optional airbox permanently mounted to the burner intake area can effectively house the air velocity meter and provide a mounting port for the flexible intake duct.

(4) *Test specimen mounting frame.* Make the mounting frame for the test specimens of ⅛-inch (3.2 mm) thick steel as shown in figure 1, except for the center vertical former, which should be ¼-inch (6.4 mm) thick to minimize warpage. The specimen mounting frame stringers (horizontal) should be bolted to the test frame formers (vertical) such that the expansion of the stringers will not cause the entire structure to warp. Use the mounting frame for mounting the two insulation blanket test specimens as shown in figure 2.

(5) *Backface calorimeters.* Mount two total heat flux Gardon type calorimeters behind the insulation test specimens on the back side (cold) area of the test specimen mounting frame as shown in figure 6. Position the calorimeters along the same plane as the burner cone centerline, at a distance of 4 inches (102 mm) from the vertical centerline of the test frame.

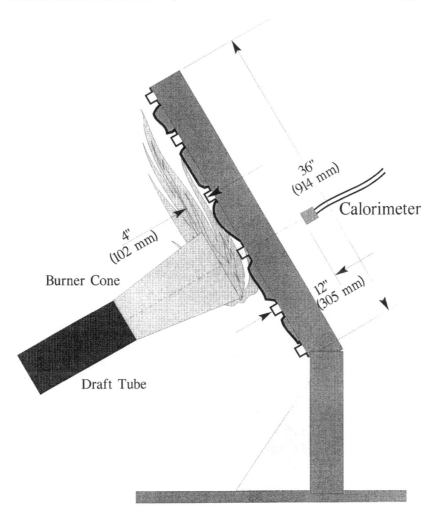

Figure 6 - . Position of Backface Calorimeters Relative to Test Specimen Frame

(i) The calorimeters must be a total heat flux, foil type Gardon Gage of an appropriate range such as 0–5 Btu/ft²-sec (0–5.7 W/cm²), accurate to ±3% of the indicated reading. The heat flux calibration method must comply with paragraph VI(b)(7) of this appendix.

(6) *Instrumentation.* Provide a recording potentiometer or other suitable calibrated instrument with an appropriate range to measure and record the outputs of the calorimeter and the thermocouples.

(7) *Timing device.* Provide a stopwatch or other device, accurate to ±1%, to measure the time of application of the burner flame and burnthrough time.

(8) *Test chamber.* Perform tests in a suitable chamber to reduce or eliminate the possibility of test fluctuation due to air movement. The chamber must have a minimum floor area of 10 by 10 feet (305 by 305 cm).

(i) *Ventilation hood.* Provide the test chamber with an exhaust system capable of removing the products of combustion expelled during tests.

(c) *Test Specimens.* (1) *Specimen preparation.* Prepare a minimum of three specimen sets of the same construction and configuration for testing.

(2) *Insulation blanket test specimen.*

(i) For batt-type materials such as fiberglass, the constructed, finished blanket specimen assemblies must be 32 inches wide by 36 inches long (81.3 by 91.4 cm), exclusive of heat sealed film edges.

(ii) For rigid and other non-conforming types of insulation materials, the finished test specimens must fit into the test rig in such a manner as to replicate the actual in-service installation.

(3) *Construction.* Make each of the specimens tested using the principal components (*i.e.,* insulation, fire barrier material if used, and moisture barrier film) and assembly processes (representative seams and closures).

(i) *Fire barrier material.* If the insulation blanket is constructed with a fire barrier material, place the fire barrier material in a manner reflective of the installed arrangement For example, if the material will be placed on the outboard side of the insulation material, inside the moisture film, place it the same way in the test specimen.

(ii) *Insulation material.* Blankets that utilize more than one variety of insulation (composition, density, etc.) must have specimen sets constructed that reflect the insulation combination used. If, however, several blanket types use similar insulation combinations, it is not necessary to test each combination if it is possible to bracket the various combinations.

(iii) *Moisture barrier film.* If a production blanket construction utilizes more than one type of moisture barrier film, perform separate tests on each combination. For example, if a polyimide film is used in conjunction with an insulation in order to enhance the burnthrough capabilities, also test the same insulation when used with a polyvinyl fluoride film.

(iv) *Installation on test frame.* Attach the blanket test specimens to the test frame using 12 steel spring type clamps as shown in figure 7. Use the clamps to hold the blankets in place in both of the outer vertical formers, as well as the center vertical former (4 clamps per former). The clamp surfaces should measure 1 inch by 2 inches (25 by 51 mm). Place the top and bottom clamps 6 inches (15.2 cm) from the top and bottom of the test frame, respectively. Place the middle clamps 8 inches (20.3 cm) from the top and bottom clamps.

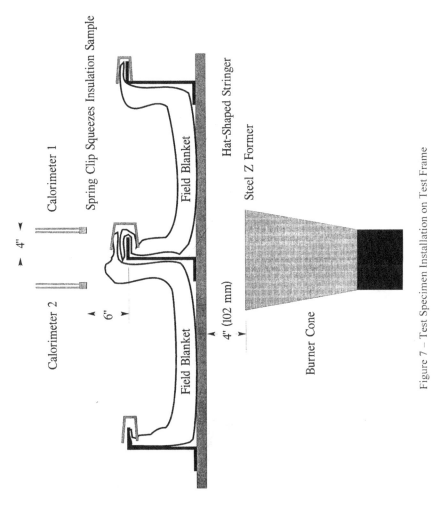

Figure 7 – Test Specimen Installation on Test Frame

(Note: For blanket materials that cannot be installed in accordance with figure 7 above, the blankets must be installed in a manner approved by the FAA.)

(v) *Conditioning.* Condition the specimens at 70° ±5 °F (21° ±2 °C) and 55% ±10% relative humidity for a minimum of 24 hours prior to testing.

(d) *Preparation of apparatus.* (1) Level and center the frame assembly to ensure alignment of the calorimeter and/or thermocouple rake with the burner cone.

(2) Turn on the ventilation hood for the test chamber. Do not turn on the burner blower. Measure the airflow of the test chamber using a vane anemometer or equivalent measuring device. The vertical air velocity just behind the top of the upper insulation blanket test specimen must be 100 ±50 ft/min (0.51 ±0.25 m/s). The horizontal air velocity at this point must be less than 50 ft/min (0.25 m/s).

(3) If a calibrated flow meter is not available, measure the fuel flow rate using a graduated cylinder of appropriate size. Turn on the burner motor/fuel pump, after insuring that the igniter system is turned off. Collect the fuel via a plastic or rubber tube into the

graduated cylinder for a 2-minute period. Determine the flow rate in gallons per hour. The fuel flow rate must be 6.0 ±0.2 gallons per hour (0.378 ±0.0126 L/min).

(e) *Calibration*. (1) Position the burner in front of the calorimeter so that it is centered and the vertical plane of the burner cone exit is 4 ±0.125 inches (102 ±3 mm) from the calo-

rimeter face. Ensure that the horizontal centerline of the burner cone is offset 1 inch below the horizontal centerline of the calorimeter (figure 8). Without disturbing the calorimeter position, rotate the burner in front of the thermocouple rake, such that the middle thermocouple (number 4 of 7) is centered on the burner cone.

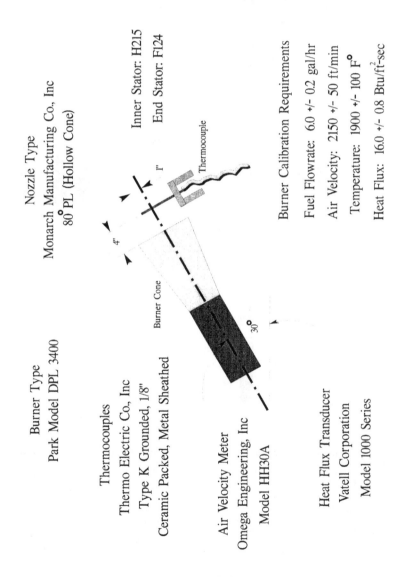

Figure 8 – Burner Information and Calibration Settings

Ensure that the horizontal centerline of the burner cone is also offset 1 inch below the horizontal centerline of the thermocouple tips. Re-check measurements by rotating the burner to each position to ensure proper alignment between the cone and the calorimeter and thermocouple rake. (Note: The test burner mounting system must incorporate "detents" that ensure proper centering of the burner cone with respect to both the calorimeter and the thermocouple rakes, so that rapid positioning of the burner can be achieved during the calibration procedure.)

(2) Position the air velocity meter in the adapter or airbox, making certain that no gaps exist where air could leak around the air velocity measuring device. Turn on the blower/motor while ensuring that the fuel solenoid and igniters are off. Adjust the air intake velocity to a level of 2150 ft/min, (10.92 m/s) then turn off the blower/motor. (Note: The Omega HH30 air velocity meter measures 2.625 inches in diameter. To calculate the intake airflow, multiply the cross-sectional area (0.03758 ft^2) by the air velocity (2150 ft/min) to obtain 80.80 ft^3/min. An air velocity meter other than the HH30 unit can be used, provided the calculated airflow of 80.80 ft^3/min (2.29 m^3/min) is equivalent.)

(3) Rotate the burner from the test position to the warm-up position. Prior to lighting the burner, ensure that the calorimeter face is clean of soot deposits, and there is water running through the calorimeter. Examine and clean the burner cone of any evidence of buildup of products of combustion, soot, etc. Soot buildup inside the burner cone may affect the flame characteristics and cause calibration difficulties. Since the burner cone may distort with time, dimensions should be checked periodically.

(4) While the burner is still rotated to the warm-up position, turn on the blower/motor, igniters and fuel flow, and light the burner. Allow it to warm up for a period of 2 minutes. Move the burner into the calibration position and allow 1 minute for calorimeter stabilization, then record the heat flux once every second for a period of 30 seconds. Turn off burner, rotate out of position, and allow to cool. Calculate the average heat flux over this 30-second duration. The average heat flux should be 16.0 ±0.8 Btu/ft^2 sec (18.2 ±0.9 W/cm^2).

(5) Position the burner in front of the thermocouple rake. After checking for proper alignment, rotate the burner to the warm-up position, turn on the blower/motor, igniters and fuel flow, and light the burner. Allow it to warm up for a period of 2 minutes. Move the burner into the calibration position and allow 1 minute for thermocouple stabilization, then record the temperature of each of the 7 thermocouples once every second for a period of 30 seconds. Turn off burner, rotate out of position, and allow to cool. Calculate the average temperature of each thermocouple over this 30-second period and record. The average temperature of each of the 7 thermocouples should be 1900 °F ±100 °F (1038 ±56 °C).

(6) If either the heat flux or the temperatures are not within the specified range, adjust the burner intake air velocity and repeat the procedures of paragraphs (4) and (5) above to obtain the proper values. Ensure that the inlet air velocity is within the range of 2150 ft/min ±50 ft/min (10.92 ±0.25 m/s).

(7) Calibrate prior to each test until consistency has been demonstrated. After consistency has been confirmed, several tests may be conducted with calibration conducted before and after a series of tests.

(f) *Test procedure.* (1) Secure the two insulation blanket test specimens to the test frame. The insulation blankets should be attached to the test rig center vertical former using four spring clamps positioned as shown in figure 7 (according to the criteria of paragraph paragraph (c)(3)(iv) of this part of this appendix).

(2) Ensure that the vertical plane of the burner cone is at a distance of 4 ±0.125 inch (102 ±3 mm) from the outer surface of the horizontal stringers of the test specimen frame, and that the burner and test frame are both situated at a 30° angle with respect to vertical.

(3) When ready to begin the test, direct the burner cone away from the test position to the warm-up position so that the flame will not impinge on the specimens prematurely. Turn on and light the burner and allow it to stabilize for 2 minutes.

(4) To begin the test, rotate the burner into the test position and simultaneously start the timing device.

(5) Expose the test specimens to the burner flame for 4 minutes and then turn off the burner. Immediately rotate the burner out of the test position.

(6) Determine (where applicable) the burnthrough time, or the point at which the heat flux exceeds 2.0 Btu/ft^2-sec (2.27 W/cm^2).

(g) *Report.* (1) Identify and describe the specimen being tested.

(2) Report the number of insulation blanket specimens tested.

(3) Report the burnthrough time (if any), and the maximum heat flux on the back face of the insulation blanket test specimen, and the time at which the maximum occurred.

(h) *Requirements.* (1) Each of the two insulation blanket test specimens must not allow fire or flame penetration in less than 4 minutes.

(2) Each of the two insulation blanket test specimens must not allow more than 2.0 Btu/

ft²-sec (2.27 W/cm²) on the cold side of the insulation specimens at a point 12 inches (30.5 cm) from the face of the test rig.

[Amdt. 25–32, 37 FR 3972, Feb. 24, 1972]

EDITORIAL NOTE: For FEDERAL REGISTER citations affecting appendix F to Part 25, see the List of CFR Sections Affected, which appears in the Finding Aids section of the printed volume and at *www.govinfo.gov.*

APPENDIX H TO PART 25—INSTRUCTIONS FOR CONTINUED AIRWORTHINESS

H25.1 *General.*

(a) This appendix specifies requirements for preparation of Instructions for Continued Airworthiness as required by §§ 25.1529, 25.1729, and applicable provisions of parts 21 and 26 of this chapter.

(b) The Instructions for Continued Airworthiness for each airplane must include the Instructions for Continued Airworthiness for each engine and propeller (hereinafter designated "products"), for each appliance required by this chapter, and any required information relating to the interface of those appliances and products with the airplane. If Instructions for Continued Airworthiness are not supplied by the manufacturer of an appliance or product installed in the airplane, the Instructions for Continued Airworthiness for the airplane must include the information essential to the continued airworthiness of the airplane.

(c) The applicant must submit to the FAA a program to show how changes to the Instructions for Continued Airworthiness made by the applicant or by the manufacturers or products and appliances installed in the airplane will be distributed.

H25.2 *Format.*

(a) The Instructions for Continued Airworthiness must be in the form of a manual or manuals as appropriate for the quantity of data to be provided.

(b) The format of the manual or manuals must provide for a practical arrangement.

H25.3 *Content.*

The contents of the manual or manuals must be prepared in the English language. The Instructions for Continued Airworthiness must contain the following manuals or sections, as appropriate, and information:

(a) *Airplane maintenance manual or section.* (1) Introduction information that includes an explanation of the airplane's features and data to the extent necessary for maintenance or preventive maintenance.

(2) A description of the airplane and its systems and installations including its engines, propellers, and appliances.

(3) Basic control and operation information describing how the airplane components and systems are controlled and how they operate, including any special procedures and limitations that apply.

(4) Servicing information that covers details regarding servicing points, capacities of tanks, reservoirs, types of fluids to be used, pressures applicable to the various systems, location of access panels for inspection and servicing, locations of lubrication points, lubricants to be used, equipment required for servicing, tow instructions and limitations, mooring, jacking, and leveling information.

(b) *Maintenance instructions.* (1) Scheduling information for each part of the airplane and its engines, auxiliary power units, propellers, accessories, instruments, and equipment that provides the recommended periods at which they should be cleaned, inspected, adjusted, tested, and lubricated, and the degree of inspection, the applicable wear tolerances, and work recommended at these periods. However, the applicant may refer to an accessory, instrument, or equipment manufacturer as the source of this information if the applicant shows that the item has an exceptionally high degree of complexity requiring specialized maintenance techniques, test equipment, or expertise. The recommended overhaul periods and necessary cross references to the Airworthiness Limitations section of the manual must also be included. In addition, the applicant must include an inspection program that includes the frequency and extent of the inspections necessary to provide for the continued airworthiness of the airplane.

(2) Troubleshooting information describing probable malfunctions, how to recognize those malfunctions, and the remedial action for those malfunctions.

(3) Information describing the order and method of removing and replacing products and parts with any necessary precautions to be taken.

(4) Other general procedural instructions including procedures for system testing during ground running, symmetry checks, weighing and determining the center of gravity, lifting and shoring, and storage limitations.

(c) Diagrams of structural access plates and information needed to gain access for inspections when access plates are not provided.

(d) Details for the application of special inspection techniques including radiographic and ultrasonic testing where such processes are specified.

(e) Information needed to apply protective treatments to the structure after inspection.

(f) All data relative to structural fasteners such as identification, discard recommendations, and torque values.

(g) A list of special tools needed.

H25.4 *Airworthiness Limitations section.*

(a) The Instructions for Continued Airworthiness must contain a section titled Airworthiness Limitations that is segregated and clearly distinguishable from the rest of the document. This section must set forth—

(1) Each mandatory modification time, replacement time, structural inspection interval, and related structural inspection procedure approved under § 25.571.

(2) Each mandatory replacement time, inspection interval, related inspection procedure, and all critical design configuration control limitations approved under § 25.981 for the fuel tank system.

(3) Any mandatory replacement time of EWIS components as defined in section 25.1701.

(4) A limit of validity of the engineering data that supports the structural maintenance program (LOV), stated as a total number of accumulated flight cycles or flight hours or both, approved under § 25.571. Until the full-scale fatigue testing is completed and the FAA has approved the LOV, the number of cycles accumulated by the airplane cannot be greater than ½ the number of cycles accumulated on the fatigue test article.

(5) Each mandatory replacement time, inspection interval, and related inspection and test procedure, and each critical design configuration control limitation for each lightning protection feature approved under § 25.954.

(b) If the Instructions for Continued Airworthiness consist of multiple documents, the section required by this paragraph must be included in the principal manual. This section must contain a legible statement in a prominent location that reads: "The Airworthiness Limitations section is FAA-approved and specifies maintenance required under §§ 43.16 and 91.403 of the Federal Aviation Regulations, unless an alternative program has been FAA approved."

H25.5 *Electrical Wiring Interconnection System (EWIS) Instructions for Continued Airworthiness.*

(a) The applicant must prepare Instructions for Continued Airworthiness (ICA) applicable to EWIS as defined by § 25.1701 that are approved by the FAA and include the following:

(1) Maintenance and inspection requirements for the EWIS developed with the use of an enhanced zonal analysis procedure that includes:

(i) Identification of each zone of the airplane.

(ii) Identification of each zone that contains EWIS.

(iii) Identification of each zone containing EWIS that also contains combustible materials.

(iv) Identification of each zone in which EWIS is in close proximity to both primary and back-up hydraulic, mechanical, or electrical flight controls and lines.

(v) Identification of—

(A) Tasks, and the intervals for performing those tasks, that will reduce the likelihood of ignition sources and accumulation of combustible material, and

(B) Procedures, and the intervals for performing those procedures, that will effectively clean the EWIS components of combustible material if there is not an effective task to reduce the likelihood of combustible material accumulation.

(vi) Instructions for protections and caution information that will minimize contamination and accidental damage to EWIS, as applicable, during performance of maintenance, alteration, or repairs.

(2) Acceptable EWIS maintenance practices in a standard format.

(3) Wire separation requirements as determined under § 25.1707.

(4) Information explaining the EWIS identification method and requirements for identifying any changes to EWIS under § 25.1711.

(5) Electrical load data and instructions for updating that data.

(b) The EWIS ICA developed in accordance with the requirements of H25.5(a)(1) must be in the form of a document appropriate for the information to be provided, and they must be easily recognizable as EWIS ICA. This document must either contain the required EWIS ICA or specifically reference other portions of the ICA that contain this information.

[Amdt. 25–54, 45 FR 60177, Sept. 11, 1980, as amended by Amdt. 25–68, 54 FR 34329, Aug. 18, 1989; Amdt. 25–102, 66 FR 23130, May 7, 2001; Amdt. 25–123, 72 FR 63408, Nov. 8, 2007; Amdt. 25–132, 75 FR 69782, Nov. 15, 2010; Doc. No. FAA–2014–1027, Amdt. No. 25–146, 83 FR 47557, Sept. 20, 2018]

APPENDIX I TO PART 25—INSTALLATION OF AN AUTOMATIC TAKEOFF THRUST CONTROL SYSTEM (ATTCS)

I25.1 *General.*

(a) This appendix specifies additional requirements for installation of an engine power control system that automatically resets thrust or power on operating engine(s) in the event of any one engine failure during takeoff.

(b) With the ATTCS and associated systems functioning normally as designed, all applicable requirements of Part 25, except as provided in this appendix, must be met without requiring any action by the crew to increase thrust or power.

I25.2 *Definitions.*

(a) *Automatic Takeoff Thrust Control System (ATTCS).* An ATTCS is defined as the entire automatic system used on takeoff, including all devices, both mechanical and electrical, that sense engine failure, transmit signals, actuate fuel controls or power levers or increase engine power by other means on operating engines to achieve scheduled thrust or power increases, and furnish cockpit information on system operation.

(b) *Critical Time Interval*. When conducting an ATTCS takeoff, the critical time interval is between V_1 minus 1 second and a point on the minimum performance, all-engine flight path where, assuming a simultaneous occurrence of an engine and ATTCS failure, the resulting minimum flight path thereafter intersects the Part 25 required actual flight path at no less than 400 feet above the takeoff surface. This time interval is shown in the following illustration:

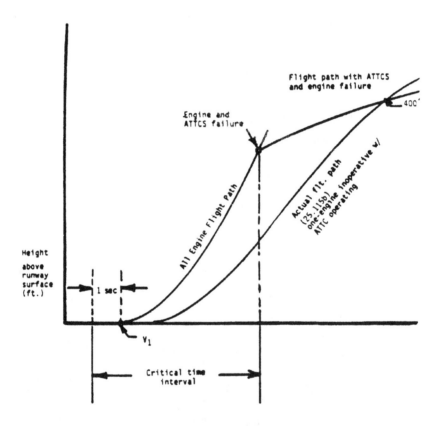

I25.3 *Performance and System Reliability Requirements.*

The applicant must comply with the performance and ATTCS reliability requirements as follows:

(a) An ATTCS failure or a combination of failures in the ATTCS during the critical time interval:

(1) Shall not prevent the insertion of the *maximum approved takeoff* thrust or power, or must be shown to be an improbable event.

(2) Shall not result in a significant loss or reduction in thrust or power, or must be shown to be an extremely improbable event.

(b) The concurrent existence of an ATTCS failure and an engine failure during the critical time interval must be shown to be extremely improbable.

(c) All applicable performance requirements of Part 25 must be met with an engine failure occurring at the most critical point

446

during takeoff with the ATTCS system functioning.

125.4 *Thrust Setting.*

The initial takeoff thrust or power setting on each engine at the beginning of the takeoff roll may not be less than any of the following:

(a) Ninety (90) percent of the thrust or power set by the ATTCS (the maximum takeoff thrust or power approved for the airplane under existing ambient conditions);

(b) That required to permit normal operation of all safety-related systems and equipment dependent upon engine thrust or power lever position; or

(c) That shown to be free of hazardous engine response characteristics when thrust or power is advanced from the initial takeoff thrust or power to the maximum approved takeoff thrust or power.

125.5 *Powerplant Controls.*

(a) In addition to the requirements of § 25.1141, no single failure or malfunction, or probable combination thereof, of the ATTCS, including associated systems, may cause the failure of any powerplant function necessary for safety.

(b) The ATTCS must be designed to:

(1) Apply thrust or power on the operating engine(s), following any one engine failure during takeoff, to achieve the maximum approved takeoff thrust or power without exceeding engine operating limits;

(2) Permit manual decrease or increase in thrust or power up to the maximum takeoff thrust or power approved for the airplane under existing conditions through the use of the power lever. For airplanes equipped with limiters that automatically prevent engine operating limits from being exceeded under existing ambient conditions, other means may be used to increase the thrust or power in the event of an ATTCS failure provided the means is located on or forward of the power levers; is easily identified and operated under all operating conditions by a single action of either pilot with the hand that is normally used to actuate the power levers; and meets the requirements of § 25.777 (a), (b), and (c);

(3) Provide a means to verify to the flightcrew before takeoff that the ATTCS is in a condition to operate; and

(4) Provide a means for the flightcrew to deactivate the automatic function. This means must be designed to prevent inadvertent deactivation.

125.6 *Powerplant Instruments.*

In addition to the requirements of § 25.1305:

(a) A means must be provided to indicate when the ATTCS is in the armed or ready condition; and

(b) If the inherent flight characteristics of the airplane do not provide adequate warning that an engine has failed, a warning system that is independent of the ATTCS must be provided to give the pilot a clear warning of any engine failure during takeoff.

[Amdt. 25–62, 52 FR 43156, Nov. 9, 1987]

APPENDIX J TO PART 25—EMERGENCY
EVACUATION

The following test criteria and procedures must be used for showing compliance with § 25.803:

(a) The emergency evacuation must be conducted with exterior ambient light levels of no greater than 0.3 foot-candles prior to the activation of the airplane emergency lighting system. The source(s) of the initial exterior ambient light level may remain active or illuminated during the actual demonstration. There must, however, be no increase in the exterior ambient light level except for that due to activation of the airplane emergency lighting system.

(b) The airplane must be in a normal attitude with landing gear extended.

(c) Unless the airplane is equipped with an off-wing descent means, stands or ramps may be used for descent from the wing to the ground. Safety equipment such as mats or inverted life rafts may be placed on the floor or ground to protect participants. No other equipment that is not part of the emergency evacuation equipment of the airplane may be used to aid the participants in reaching the ground.

(d) Except as provided in paragraph (a) of this appendix, only the airplane's emergency lighting system may provide illumination.

(e) All emergency equipment required for the planned operation of the airplane must be installed.

(f) Each internal door or curtain must be in the takeoff configuration.

(g) Each crewmember must be seated in the normally assigned seat for takeoff and must remain in the seat until receiving the signal for commencement of the demonstration. Each crewmember must be a person having knowledge of the operation of exits and emergency equipment and, if compliance with § 121.291 is also being demonstrated, each flight attendant must be a member of a regularly scheduled line crew.

(h) A representative passenger load of persons in normal health must be used as follows:

(1) At least 40 percent of the passenger load must be female.

(2) At least 35 percent of the passenger load must be over 50 years of age.

(3) At least 15 percent of the passenger load must be female and over 50 years of age.

(4) Three life-size dolls, not included as part of the total passenger load, must be carried by passengers to simulate live infants 2 years old or younger.

(5) Crewmembers, mechanics, and training personnel, who maintain or operate the airplane in the normal course of their duties, may not be used as passengers.

(i) No passenger may be assigned a specific seat except as the Administrator may require. Except as required by subparagraph (g) of this paragraph, no employee of the applicant may be seated next to an emergency exit.

(j) Seat belts and shoulder harnesses (as required) must be fastened.

(k) Before the start of the demonstration, approximately one-half of the total average amount of carry-on baggage, blankets, pillows, and other similar articles must be distributed at several locations in aisles and emergency exit access ways to create minor obstructions.

(l) No prior indication may be given to any crewmember or passenger of the particular exits to be used in the demonstration.

(m) The applicant may not practice, rehearse, or describe the demonstration for the participants nor may any participant have taken part in this type of demonstration within the preceding 6 months.

(n) Prior to entering the demonstration aircraft, the passengers may also be advised to follow directions of crewmembers but may not be instructed on the procedures to be followed in the demonstration, except with respect to safety procedures in place for the demonstration or which have to do with the demonstration site. Prior to the start of the demonstration, the pre-takeoff passenger briefing required by § 121.571 may be given. Flight attendants may assign demonstration subjects to assist persons from the bottom of a slide, consistent with their approved training program.

(o) The airplane must be configured to prevent disclosure of the active emergency exits to demonstration participants in the airplane until the start of the demonstration.

(p) Exits used in the demonstration must consist of one exit from each exit pair. The demonstration may be conducted with the escape slides, if provided, inflated and the exits open at the beginning of the demonstration. In this case, all exits must be configured such that the active exits are not disclosed to the occupants. If this method is used, the exit preparation time for each exit utilized must be accounted for, and exits that are not to be used in the demonstration must not be indicated before the demonstration has started. The exits to be used must be representative of all of the emergency exits on the airplane and must be designated by the applicant, subject to approval by the Administrator. At least one floor level exit must be used.

(q) Except as provided in paragraph (c) of this section, all evacuees must leave the airplane by a means provided as part of the airplane's equipment.

(r) The applicant's approved procedures must be fully utilized, except the flightcrew must take no active role in assisting others inside the cabin during the demonstration.

(s) The evacuation time period is completed when the last occupant has evacuated the airplane and is on the ground. Provided that the acceptance rate of the stand or ramp is no greater than the acceptance rate of the means available on the airplane for descent from the wing during an actual crash situation, evacuees using stands or ramps allowed by paragraph (c) of this appendix are considered to be on the ground when they are on the stand or ramp.

[Amdt. 25–72, 55 FR 29788, July 20, 1990, as amended by Amdt. 25–79, Aug. 26, 1993; Amdt. 25–117, 69 FR 67499, Nov. 17, 2004]

APPENDIX K TO PART 25—EXTENDED OPERATIONS (ETOPS)

This appendix specifies airworthiness requirements for the approval of an airplane-engine combination for extended operations (ETOPS). For two-engine airplanes, the applicant must comply with sections K25.1 and K25.2 of this appendix. For airplanes with more than two engines, the applicant must comply with sections K25.1 and K25.3 of this appendix.

K25.1 *Design requirements.*

K25.1.1 *Part 25 compliance.*

The airplane-engine combination must comply with the requirements of part 25 considering the maximum flight time and the longest diversion time for which the applicant seeks approval.

K25.1.2 *Human factors.*

An applicant must consider crew workload, operational implications, and the crew's and passengers' physiological needs during continued operation with failure effects for the longest diversion time for which it seeks approval.

K25.1.3 *Airplane systems.*

(a) *Operation in icing conditions.*

(1) The airplane must be certificated for operation in icing conditions in accordance with § 25.1419.

(2) The airplane must be able to safely conduct an ETOPS diversion with the most critical ice accretion resulting from:

(i) Icing conditions encountered at an altitude that the airplane would have to fly following an engine failure or cabin decompression.

(ii) A 15-minute hold in the continuous maximum icing conditions specified in Appendix C of this part with a liquid water content factor of 1.0.

(iii) Ice accumulated during approach and landing in the icing conditions specified in Appendix C of this part.

(b) *Electrical power supply.* The airplane must be equipped with at least three independent sources of electrical power.

(c) *Time limited systems.* The applicant must define the system time capability of each ETOPS significant system that is time-limited.

K25.1.4 *Propulsion systems.*

(a) *Fuel system design.* Fuel necessary to complete an ETOPS flight (including a diversion for the longest time for which the applicant seeks approval) must be available to the operating engines at the pressure and fuel-flow required by §25.955 under any airplane failure condition not shown to be extremely improbable. Types of failures that must be considered include, but are not limited to: crossfeed valve failures, automatic fuel management system failures, and normal electrical power generation failures.

(1) If the engine has been certified for limited operation with negative engine-fuel-pump-inlet pressures, the following requirements apply:

(i) Airplane demonstration-testing must cover worst case cruise and diversion conditions involving:

(A) Fuel grade and temperature.

(B) Thrust or power variations.

(C) Turbulence and negative G.

(D) Fuel system components degraded within their approved maintenance limits.

(ii) Unusable-fuel quantity in the suction feed configuration must be determined in accordance with §25.959.

(2) For two-engine airplanes to be certificated for ETOPS beyond 180 minutes, one fuel boost pump in each main tank and at least one crossfeed valve, or other means for transferring fuel, must be powered by an independent electrical power source other than the three power sources required to comply with section K25.1.3(b) of this appendix. This requirement does not apply if the normal fuel boost pressure, crossfeed valve actuation, or fuel transfer capability is not provided by electrical power.

(3) An alert must be displayed to the flightcrew when the quantity of fuel available to the engines falls below the level required to fly to the destination. The alert must be given when there is enough fuel remaining to safely complete a diversion. This alert must account for abnormal fuel management or transfer between tanks, and possible loss of fuel. This paragraph does not apply to airplanes with a required flight engineer.

(b) *APU design.* If an APU is needed to comply with this appendix, the applicant must demonstrate that:

(1) The reliability of the APU is adequate to meet those requirements; and

(2) If it is necessary that the APU be able to start in flight, it is able to start at any altitude up to the maximum operating altitude of the airplane, or 45,000 feet, whichever is lower, and run for the remainder of any flight .

(c) *Engine oil tank design.* The engine oil tank filler cap must comply with §33.71(c)(4) of this chapter.

K25.1.5 *Engine-condition monitoring.*

Procedures for engine-condition monitoring must be specified and validated in accordance with Part 33, Appendix A, paragraph A33.3(c) of this chapter.

K25.1.6 *Configuration, maintenance, and procedures.*

The applicant must list any configuration, operating and maintenance requirements, hardware life limits, MMEL constraints, and ETOPS approval in a CMP document.

K25.1.7 *Airplane flight manual.*

The airplane flight manual must contain the following information applicable to the ETOPS type design approval:

(a) Special limitations, including any limitation associated with operation of the airplane up to the maximum diversion time being approved.

(b) Required markings or placards.

(c) The airborne equipment required for extended operations and flightcrew operating procedures for this equipment.

(d) The system time capability for the following:

(1) The most limiting fire suppression system for Class C cargo or baggage compartments.

(2) The most limiting ETOPS significant system other than fire suppression systems for Class C cargo or baggage compartments.

(e) This statement: "The type-design reliability and performance of this airplane-engine combination has been evaluated under 14 CFR 25.1535 and found suitable for (identify maximum approved diversion time) extended operations (ETOPS) when the configuration, maintenance, and procedures standard contained in (identify the CMP document) are met. The actual maximum approved diversion time for this airplane may be less based on its most limiting system time capability. This finding does not constitute operational approval to conduct ETOPS."

K25.2. *Two-engine airplanes.*

An applicant for ETOPS type design approval of a two-engine airplane must use one of the methods described in section K25.2.1, K25.2.2, or K25.2.3 of this appendix.

K25.2.1 *Service experience method.*

An applicant for ETOPS type design approval using the service experience method must comply with sections K25.2.1(a) and K25.2.1(b) of this appendix before conducting the assessments specified in sections K25.2.1(c) and K25.2.1(d) of this appendix, and the flight test specified in section K25.2.1(e) of this appendix.

(a) *Service experience.* The world fleet for the airplane-engine combination must accumulate a minimum of 250,000 engine-hours. The FAA may reduce this number of hours if the applicant identifies compensating factors that are acceptable to the FAA. The compensating factors may include experience on another airplane, but experience on the candidate airplane must make up a significant portion of the total service experience.

(b) *In-flight shutdown (IFSD) rates.* The demonstrated 12-month rolling average IFSD rate for the world fleet of the airplane-engine combination must be commensurate with the level of ETOPS approval being sought.

(1) For type design approval up to and including 120 minutes: An IFSD rate of 0.05 or less per 1,000 world-fleet engine-hours, unless otherwise approved by the FAA. Unless the IFSD rate is 0.02 or less per 1,000 world-fleet engine-hours, the applicant must provide a list of corrective actions in the CMP document specified in section K25.1.6 of this appendix, that, when taken, would result in an IFSD rate of 0.02 or less per 1,000 fleet engine-hours.

(2) For type design approval up to and including 180 minutes: An IFSD rate of 0.02 or less per 1,000 world-fleet engine-hours, unless otherwise approved by the FAA. If the airplane-engine combination does not meet this rate by compliance with an existing 120-minute CMP document, then new or additional CMP requirements that the applicant has demonstrated would achieve this IFSD rate must be added to the CMP document.

(3) For type design approval beyond 180 minutes: An IFSD rate of 0.01 or less per 1,000 fleet engine-hours unless otherwise approved by the FAA. If the airplane-engine combination does not meet this rate by compliance with an existing 120-minute or 180-minute CMP document, then new or additional CMP requirements that the applicant has demonstrated would achieve this IFSD rate must be added to the CMP document.

(c) *Propulsion system assessment.* (1) The applicant must conduct a propulsion system assessment based on the following data collected from the world-fleet of the airplane-engine combination:

(i) A list of all IFSD's, unplanned ground engine shutdowns, and occurrences (both ground and in-flight) when an engine was not shut down, but engine control or the desired thrust or power level was not achieved, including engine flameouts. Planned IFSD's performed during flight training need not be included. For each item, the applicant must provide—

(A) Each airplane and engine make, model, and serial number;

(B) Engine configuration, and major alteration history;

(C) Engine position;

(D) Circumstances leading up to the engine shutdown or occurrence;

(E) Phase of flight or ground operation;

(F) Weather and other environmental conditions; and

(G) Cause of engine shutdown or occurrence.

(ii) A history of unscheduled engine removal rates since introduction into service (using 6- and 12-month rolling averages), with a summary of the major causes for the removals.

(iii) A list of all propulsion system events (whether or not caused by maintenance or flightcrew error), including dispatch delays, cancellations, aborted takeoffs, turnbacks, diversions, and flights that continue to destination after the event.

(iv) The total number of engine hours and cycles, the number of hours for the engine with the highest number of hours, the number of cycles for the engine with the highest number of cycles, and the distribution of hours and cycles.

(v) The mean time between failures (MTBF) of propulsion system components that affect reliability.

(vi) A history of the IFSD rates since introduction into service using a 12-month rolling average.

(2) The cause or potential cause of each item listed in K25.2.1(c)(1)(i) must have a corrective action or actions that are shown to be effective in preventing future occurrences. Each corrective action must be identified in the CMP document specified in section K25.1.6. A corrective action is not required:

(i) For an item where the manufacturer is unable to determine a cause or potential cause.

(ii) For an event where it is technically unfeasible to develop a corrective action.

(iii) If the world-fleet IFSD rate—

(A) Is at or below 0.02 per 1,000 world-fleet engine-hours for approval up to and including 180-minute ETOPS; or

(B) Is at or below 0.01 per 1,000 world-fleet engine-hours for approval greater than 180-minute ETOPS.

(d) *Airplane systems assessment.* The applicant must conduct an airplane systems assessment. The applicant must show that the airplane systems comply with § 25.1309(b) using available in-service reliability data for ETOPS significant systems on the candidate airplane-engine combination. Each cause or potential cause of a relevant design, manufacturing, operational, and maintenance problem occurring in service must have a corrective action or actions that are shown to be effective in preventing future occurrences. Each corrective action must be identified in the CMP document specified in section K25.1.6 of this appendix. A corrective action is not required if the problem would not significantly impact the safety or reliability of the airplane system involved. A relevant problem is a problem with an ETOPS group 1 significant system that has or could result in, an IFSD or diversion. The applicant must include in this assessment relevant problems with similar or identical equipment installed on other types of airplanes to the extent such information is reasonably available.

(e) *Airplane flight test.* The applicant must conduct a flight test to validate the flightcrew's ability to safely conduct an ETOPS diversion with an inoperative engine and worst-case ETOPS Significant System failures and malfunctions that could occur in service. The flight test must validate the airplane's flying qualities and performance with the demonstrated failures and malfunctions.

K25.2.2 *Early ETOPS method.*

An applicant for ETOPS type design approval using the Early ETOPS method must comply with the following requirements:

(a) *Assessment of relevant experience with airplanes previously certificated under part 25.* The applicant must identify specific corrective actions taken on the candidate airplane to prevent relevant design, manufacturing, operational, and maintenance problems experienced on airplanes previously certificated under part 25 manufactured by the applicant. Specific corrective actions are not required if the nature of a problem is such that the problem would not significantly impact the safety or reliability of the airplane system involved. A relevant problem is a problem with an ETOPS group 1 significant system that has or could result in an IFSD or diversion. The applicant must include in this assessment relevant problems of supplier-provided ETOPS group 1 significant systems and similar or identical equipment used on airplanes built by other manufacturers to the extent such information is reasonably available.

(b) *Propulsion system design.* (1) The engine used in the applicant's airplane design must be approved as eligible for Early ETOPS in accordance with § 33.201 of this chapter.

(2) The applicant must design the propulsion system to preclude failures or malfunctions that could result in an IFSD. The applicant must show compliance with this requirement by analysis, test, in-service experience on other airplanes, or other means acceptable to the FAA. If analysis is used, the applicant must show that the propulsion system design will minimize failures and malfunctions with the objective of achieving the following IFSD rates:

(i) An IFSD rate of 0.02 or less per 1,000 world-fleet engine-hours for type design approval up to and including 180 minutes.

(ii) An IFSD rate of 0.01 or less per 1,000 world-fleet engine-hours for type design approval beyond 180 minutes.

(c) *Maintenance and operational procedures.* The applicant must validate all maintenance and operational procedures for ETOPS significant systems. The applicant must identify, track, and resolve any problems found during the validation in accordance with the problem tracking and resolution system specified in section K25.2.2(h) of this appendix.

(d) *Propulsion system validation test.* (1) The installed engine configuration for which approval is being sought must comply with § 33.201(c) of this chapter. The test engine must be configured with a complete airplane nacelle package, including engine-mounted equipment, except for any configuration differences necessary to accommodate test stand interfaces with the engine nacelle package. At the conclusion of the test, the propulsion system must be—

(i) Visually inspected according to the applicant's on-wing inspection recommendations and limits; and

(ii) Completely disassembled and the propulsion system hardware inspected to determine whether it meets the service limits specified in the Instructions for Continued Airworthiness submitted in compliance with § 25.1529.

(2) The applicant must identify, track, and resolve each cause or potential cause of IFSD, loss of thrust control, or other power loss encountered during this inspection in accordance with the problem tracking and resolution system specified in section K25.2.2(h) of this appendix.

(e) *New technology testing.* Technology new to the applicant, including substantially new manufacturing techniques, must be tested to substantiate its suitability for the airplane design.

(f) *APU validation test.* If an APU is needed to comply with this appendix, one APU of the type to be certified with the airplane must be tested for 3,000 equivalent airplane operational cycles. Following completion of the test, the APU must be disassembled and inspected. The applicant must identify, track, and resolve each cause or potential cause of an inability to start or operate the APU in flight as intended in accordance with the problem tracking and resolution system specified in section K25.2.2(h) of this appendix.

(g) *Airplane demonstration.* For each airplane-engine combination to be approved for ETOPS, the applicant must flight test at least one airplane to demonstrate that the airplane, and its components and equipment are capable of functioning properly during ETOPS flights and diversions of the longest duration for which the applicant seeks approval. This flight testing may be performed in conjunction with, but may not substitute for the flight testing required by § 21.35(b)(2) of this chapter.

(1) The airplane demonstration flight test program must include:

(i) Flights simulating actual ETOPS, including flight at normal cruise altitude, step climbs, and, if applicable, APU operation.

(ii) Maximum duration flights with maximum duration diversions.

(iii) Maximum duration engine-inoperative diversions distributed among the engines installed on the airplanes used for the airplane demonstration flight test program. At least two one-engine-inoperative diversions must be conducted at maximum continuous thrust or power using the same engine.

(iv) Flights under non-normal conditions to demonstrate the flightcrew's ability to safely conduct an ETOPS diversion with worst-case ETOPS significant system failures or malfunctions that could occur in service.

(v) Diversions to airports that represent airports of the types used for ETOPS diversions.

(vi) Repeated exposure to humid and inclement weather on the ground followed by a long-duration flight at normal cruise altitude.

(2) The airplane demonstration flight test program must validate the adequacy of the airplane's flying qualities and performance, and the flightcrew's ability to safely conduct an ETOPS diversion under the conditions specified in section K25.2.2(g)(1) of this appendix.

(3) During the airplane demonstration flight test program, each test airplane must be operated and maintained using the applicant's recommended operating and maintenance procedures.

(4) At the completion of the airplane demonstration flight test program, each ETOPS significant system must undergo an on-wing inspection or test in accordance with the tasks defined in the proposed Instructions for Continued Airworthiness to establish its condition for continued safe operation. Each engine must also undergo a gas path inspection. These inspections must be conducted in a manner to identify abnormal conditions that could result in an IFSD or diversion. The applicant must identify, track and resolve any abnormal conditions in accordance with the problem tracking and resolution system specified in section K25.2.2(h) of this appendix.

(h) *Problem tracking and resolution system.* (1) The applicant must establish and maintain a problem tracking and resolution system. The system must:

(i) Contain a process for prompt reporting to the FAA office responsible for the design approval of each occurrence reportable under §21.4(a)(6) encountered during the phases of airplane and engine development used to assess Early ETOPS eligibility.

(ii) Contain a process for notifying the FAA office responsible for the design approval of each proposed corrective action that the applicant determines necessary for each problem identified from the occurrences reported under section K25.2.2. (h)(1)(i) of this appendix. The timing of the notification must permit appropriate FAA review before taking the proposed corrective action.

(2) If the applicant is seeking ETOPS type design approval of a change to an airplane-engine combination previously approved for ETOPS, the problem tracking and resolution system need only address those problems specified in the following table, provided the applicant obtains prior authorization from the FAA:

If the change does not require a new airplane type certificiate and . . .	Then the Problem Tracking and Resolution System must address . . .
(i) Requires a new engine type certificate	All problems applicable to the new engine installation, and for the remainder of the airplane, problems in changed systems only.
(ii) Does not require a new engine type certificate	Problems in changed systems only.

(i) *Acceptance criteria.* The type and frequency of failures and malfunctions on ETOPS significant systems that occur during the airplane flight test program and the airplane demonstration flight test program specified in section K25.2.2(g) of this appendix must be consistent with the type and frequency of failures and malfunctions that would be expected to occur on currently certificated airplanes approved for ETOPS.

K25.2.3. *Combined service experience and Early ETOPS method.*

An applicant for ETOPS type design approval using the combined service experience and Early ETOPS method must comply with the following requirements.

(a) A service experience requirement of not less than 15,000 engine-hours for the world fleet of the candidate airplane-engine combination.

(b) The Early ETOPS requirements of K25.2.2, except for the airplane demonstration specified in section K25.2.2(g) of this appendix; and

(c) The flight test requirement of section K25.2.1(e) of this appendix.

K25.3. *Airplanes with more than two engines.*

An applicant for ETOPS type design approval of an airplane with more than two engines must use one of the methods described in section K25.3.1, K25.3.2, or K25.3.3 of this appendix.

K25.3.1 *Service experience method.*

An applicant for ETOPS type design approval using the service experience method must comply with section K25.3.1(a) of this appendix before conducting the airplane systems assessment specified in K25.3.1(b), and the flight test specified in section K25.3.1(c) of this appendix.

(a) *Service experience.* The world fleet for the airplane-engine combination must accumulate a minimum of 250,000 engine-hours. The FAA may reduce this number of hours if the applicant identifies compensating factors that are acceptable to the FAA. The compensating factors may include experience on another airplane, but experience on the candidate airplane must make up a significant portion of the total required service experience.

(b) *Airplane systems assessment.* The applicant must conduct an airplane systems assessment. The applicant must show that the airplane systems comply with the §25.1309(b) using available in-service reliability data for ETOPS significant systems on the candidate airplane-engine combination. Each cause or potential cause of a relevant design, manufacturing, operational or maintenance problem occurring in service must have a corrective action or actions that are shown to be effective in preventing future occurrences. Each corrective action must be identified in the CMP document specified in section K25.1.6 of this appendix. A corrective action is not required if the problem would not significantly impact the safety or reliability of the airplane system involved. A relevant problem is a problem with an ETOPS group 1 significant system that has or could result in an IFSD or diversion. The applicant must include in this assessment relevant problems with similar or identical equipment installed on other types of airplanes to the extent such information is reasonably available.

(c) *Airplane flight test.* The applicant must conduct a flight test to validate the flightcrew's ability to safely conduct an ETOPS diversion with an inoperative engine and worst-case ETOPS significant system failures and malfunctions that could occur in service. The flight test must validate the airplane's flying qualities and performance with the demonstrated failures and malfunctions.

K25.3.2   Early ETOPS method.

An applicant for ETOPS type design approval using the Early ETOPS method must comply with the following requirements:

(a) *Maintenance and operational procedures.* The applicant must validate all maintenance and operational procedures for ETOPS significant systems. The applicant must identify, track and resolve any problems found during the validation in accordance with the problem tracking and resolution system specified in section K25.3.2(e) of this appendix.

(b) *New technology testing.* Technology new to the applicant, including substantially new manufacturing techniques, must be tested to

substantiate its suitability for the airplane design.

(c) *APU validation test.* If an APU is needed to comply with this appendix, one APU of the type to be certified with the airplane must be tested for 3,000 equivalent airplane operational cycles. Following completion of the test, the APU must be disassembled and inspected. The applicant must identify, track, and resolve each cause or potential cause of an inability to start or operate the APU in flight as intended in accordance with the problem tracking and resolution system specified in section K25.3.2(e) of this appendix.

(d) *Airplane demonstration.* For each airplane-engine combination to be approved for ETOPS, the applicant must flight test at least one airplane to demonstrate that the airplane, and its components and equipment are capable of functioning properly during ETOPS flights and diversions of the longest duration for which the applicant seeks approval. This flight testing may be performed in conjunction with, but may not substitute for the flight testing required by §21.35(b)(2).

(1) The airplane demonstration flight test program must include:

(i) Flights simulating actual ETOPS including flight at normal cruise altitude, step climbs, and, if applicable, APU operation.

(ii) Maximum duration flights with maximum duration diversions.

(iii) Maximum duration engine-inoperative diversions distributed among the engines installed on the airplanes used for the airplane demonstration flight test program. At least two one engine-inoperative diversions must be conducted at maximum continuous thrust or power using the same engine.

(iv) Flights under non-normal conditions to validate the flightcrew's ability to safely conduct an ETOPS diversion with worst-case ETOPS significant system failures or malfunctions that could occur in service.

(v) Diversions to airports that represent airports of the types used for ETOPS diversions.

(vi) Repeated exposure to humid and inclement weather on the ground followed by a long duration flight at normal cruise altitude.

(2) The airplane demonstration flight test program must validate the adequacy of the airplane's flying qualities and performance, and the flightcrew's ability to safely conduct an ETOPS diversion under the conditions specified in section K25.3.2(d)(1) of this appendix.

(3) During the airplane demonstration flight test program, each test airplane must be operated and maintained using the applicant's recommended operating and maintenance procedures.

(4) At the completion of the airplane demonstration, each ETOPS significant system must undergo an on-wing inspection or test

in accordance with the tasks defined in the proposed Instructions for Continued Airworthiness to establish its condition for continued safe operation. Each engine must also undergo a gas path inspection. These inspections must be conducted in a manner to identify abnormal conditions that could result in an IFSD or diversion. The applicant must identify, track and resolve any abnormal conditions in accordance with the problem tracking and resolution system specified in section K25.3.2(e) of this appendix.

(e) *Problem tracking and resolution system.* (1) The applicant must establish and maintain a problem tracking and resolution system. The system must:

(i) Contain a process for prompt reporting to the FAA office responsible for the design approval of each occurrence reportable under § 21.4(a)(6) encountered during the phases of

airplane and engine development used to assess Early ETOPS eligibility.

(ii) Contain a process for notifying the FAA office responsible for the design approval of each proposed corrective action that the applicant determines necessary for each problem identified from the occurrences reported under section K25.3.2(h)(1)(i) of this appendix. The timing of the notification must permit appropriate FAA review before taking the proposed corrective action.

(2) If the applicant is seeking ETOPS type design approval of a change to an airplane-engine combination previously approved for ETOPS, the problem tracking and resolution system need only address those problems specified in the following table, provided the applicant obtains prior authorization from the FAA:

If the change does not require a new airplane type certificate and . . .	Then the Problem Tracking and Resolution System must address . . .
(i) Requires a new engine type certificate	All problems applicable to the new engine installation, and for the remainder of the airplane, problems in changed systems only.
(ii) Does not require a new engine type certificate	Problems in changed systems only.

(f) *Acceptance criteria.* The type and frequency of failures and malfunctions on ETOPS significant systems that occur during the airplane flight test program and the airplane demonstration flight test program specified in section K25.3.2(d) of this appendix must be consistent with the type and frequency of failures and malfunctions that would be expected to occur on currently certificated airplanes approved for ETOPS.

K25.3.3 *Combined service experience and Early ETOPS method.*

An applicant for ETOPS type design approval using the Early ETOPS method must comply with the following requirements:

(a) A service experience requirement of less than 15,000 engine-hours for the world fleet of the candidate airplane-engine combination;

(b) The Early ETOPS requirements of section K25.3.2 of this appendix, except for the airplane demonstration specified in section K25.3.2(d) of this appendix; and

(c) The flight test requirement of section K25.3.1(c) of this appendix.

[Doc. No. FAA–2002–6717, 72 FR 1873, Jan. 16, 2007, as amended by Doc. No. FAA–2018–0119, Amdt. 25–145, 83 FR 9169, Mar. 5, 2018]

APPENDIX L TO PART 25—HIRF ENVIRONMENTS AND EQUIPMENT HIRF TEST LEVELS

This appendix specifies the HIRF environments and equipment HIRF test levels for electrical and electronic systems under § 25.1317. The field strength values for the

HIRF environments and equipment HIRF test levels are expressed in root-mean-square units measured during the peak of the modulation cycle.

(a) HIRF environment I is specified in the following table:

TABLE I.—HIRF ENVIRONMENT I

Frequency	Field strength (volts/meter)	
	Peak	Average
10 kHz–2 MHz	50	50
2 MHz–30 MHz	100	100
30 MHz–100 MHz	50	50
100 MHz–400 MHz	100	100
400 MHz–700 MHz	700	50
700 MHz–1 GHz	700	100
1 GHz–2 GHz	2,000	200
2 GHz–6 GHz	3,000	200
6 GHz–8 GHz	1,000	200
8 GHz–12 GHz	3,000	300
12 GHz–18 GHz	2,000	200
18 GHz–40 GHz	600	200

In this table, the higher field strength applies at the frequency band edges.

(b) HIRF environment II is specified in the following table:

TABLE II.—HIRF ENVIRONMENT II

Frequency	Field strength (volts/meter)	
	Peak	Average
10 kHz–500 kHz	20	20
500 kHz–2 MHz	30	30
2 MHz–30 MHz	100	100

TABLE II.–HIRF ENVIRONMENT II—Continued

Frequency	Field strength (volts/meter)	
	Peak	Average
30 MHz–100 MHz	10	10
100 MHz–200 MHz	30	10
200 MHz–400 MHz	10	10
400 MHz–1 GHz	700	40
1 GHz–2 GHz	1,300	160
2 GHz–4 GHz	3,000	120
4 GHz–6 GHz	3,000	160
6 GHz–8 GHz	400	170
8 GHz–12 GHz	1,230	230
12 GHz–18 GHz	730	190
18 GHz–40 GHz	600	150

In this table, the higher field strength applies at the frequency band edges.

(c) *Equipment HIRF Test Level 1.* (1) From 10 kilohertz (kHz) to 400 megahertz (MHz), use conducted susceptibility tests with continuous wave (CW) and 1 kHz square wave modulation with 90 percent depth or greater. The conducted susceptibility current must start at a minimum of 0.6 milliamperes (mA) at 10 kHz, increasing 20 decibels (dB) per frequency decade to a minimum of 30 mA at 500 kHz.

(2) From 500 kHz to 40 MHz, the conducted susceptibility current must be at least 30 mA.

(3) From 40 MHz to 400 MHz, use conducted susceptibility tests, starting at a minimum of 30 mA at 40 MHz, decreasing 20 dB per frequency decade to a minimum of 3 mA at 400 MHz.

(4) From 100 MHz to 400 MHz, use radiated susceptibility tests at a minimum of 20 volts per meter (V/m) peak with CW and 1 kHz square wave modulation with 90 percent depth or greater.

(5) From 400 MHz to 8 gigahertz (GHz), use radiated susceptibility tests at a minimum of 150 V/m peak with pulse modulation of 4 percent duty cycle with a 1 kHz pulse repetition frequency. This signal must be switched on and off at a rate of 1 Hz with a duty cycle of 50 percent.

(d) *Equipment HIRF Test Level 2.* Equipment HIRF test level 2 is HIRF environment II in table II of this appendix reduced by acceptable aircraft transfer function and attenuation curves. Testing must cover the frequency band of 10 kHz to 8 GHz.

(e) *Equipment HIRF Test Level 3.* (1) From 10 kHz to 400 MHz, use conducted susceptibility tests, starting at a minimum of 0.15 mA at 10 kHz, increasing 20 dB per frequency decade to a minimum of 7.5 mA at 500 kHz.

(2) From 500 kHz to 40 MHz, use conducted susceptibility tests at a minimum of 7.5 mA.

(3) From 40 MHz to 400 MHz, use conducted susceptibility tests, starting at a minimum of 7.5 mA at 40 MHz, decreasing 20 dB per frequency decade to a minimum of 0.75 mA at 400 MHz.

(4) From 100 MHz to 8 GHz, use radiated susceptibility tests at a minimum of 5 V/m.

[Doc. No. FAA–2006–23657, 72 FR 44026, Aug. 6, 2007]

APPENDIX M TO PART 25—FUEL TANK SYSTEM FLAMMABILITY REDUCTION MEANS

M25.1 *Fuel tank flammability exposure requirements.*

(a) The Fleet Average Flammability Exposure of each fuel tank, as determined in accordance with Appendix N of this part, may not exceed 3 percent of the Flammability Exposure Evaluation Time (FEET), as defined in Appendix N of this part. As a portion of this 3 percent, if flammability reduction means (FRM) are used, each of the following time periods may not exceed 1.8 percent of the FEET:

(1) When any FRM is operational but the fuel tank is not inert and the tank is flammable; and

(2) When any FRM is inoperative and the tank is flammable.

(b) The Fleet Average Flammability Exposure, as defined in Appendix N of this part, of each fuel tank may not exceed 3 percent of the portion of the FEET occurring during either ground or takeoff/climb phases of flight during warm days. The analysis must consider the following conditions.

(1) The analysis must use the subset of those flights that begin with a sea level ground ambient temperature of 80 °F (standard day plus 21 °F atmosphere) or above, from the flammability exposure analysis done for overall performance.

(2) For the ground and takeoff/climb phases of flight, the average flammability exposure must be calculated by dividing the time during the specific flight phase the fuel tank is flammable by the total time of the specific flight phase.

(3) Compliance with this paragraph may be shown using only those flights for which the airplane is dispatched with the flammability reduction means operational.

M25.2 *Showing compliance.*

(a) The applicant must provide data from analysis, ground testing, and flight testing, or any combination of these, that:

(1) Validate the parameters used in the analysis required by paragraph M25.1 of this appendix;

(2) Substantiate that the FRM is effective at limiting flammability exposure in all compartments of each tank for which the FRM is used to show compliance with paragraph M25.1 of this appendix; and

(3) Describe the circumstances under which the FRM would not be operated during each phase of flight.

(b) The applicant must validate that the FRM meets the requirements of paragraph

M25.1 of this appendix with any airplane or engine configuration affecting the performance of the FRM for which approval is sought.

M25.3 *Reliability indications and maintenance access.*

(a) Reliability indications must be provided to identify failures of the FRM that would otherwise be latent and whose identification is necessary to ensure the fuel tank with an FRM meets the fleet average flammability exposure requirements listed in paragraph M25.1 of this appendix, including when the FRM is inoperative.

(b) Sufficient accessibility to FRM reliability indications must be provided for maintenance personnel or the flightcrew.

(c) The access doors and panels to the fuel tanks with FRMs (including any tanks that communicate with a tank via a vent system), and to any other confined spaces or enclosed areas that could contain hazardous atmosphere under normal conditions or failure conditions, must be permanently stenciled, marked, or placarded to warn maintenance personnel of the possible presence of a potentially hazardous atmosphere.

M25.4 *Airworthiness limitations and procedures.*

(a) If FRM is used to comply with paragraph M25.1 of this appendix, Airworthiness Limitations must be identified for all maintenance or inspection tasks required to identify failures of components within the FRM that are needed to meet paragraph M25.1 of this appendix.

(b) Maintenance procedures must be developed to identify any hazards to be considered during maintenance of the FRM. These procedures must be included in the instructions for continued airworthiness (ICA).

M25.5 *Reliability reporting.*

The effects of airplane component failures on FRM reliability must be assessed on an on-going basis. The applicant/holder must do the following:

(a) Demonstrate effective means to ensure collection of FRM reliability data. The means must provide data affecting FRM reliability, such as component failures.

(b) Unless alternative reporting procedures are approved by the responsible Aircraft Certification Service office, as defined in part 26 of this subchapter, provide a report to the FAA every six months for the first five years after service introduction. After that period, continued reporting every six months may be replaced with other reliability tracking methods found acceptable to the FAA or eliminated if it is established that the reliability of the FRM meets, and will continue to meet, the exposure requirements of paragraph M25.1 of this appendix.

(c) Develop service instructions or revise the applicable airplane manual, according to a schedule approved by the responsible Aircraft Certification Service office, as defined in part 26 of this subchapter, to correct any failures of the FRM that occur in service that could increase any fuel tank's Fleet Average Flammability Exposure to more than that required by paragraph M25.1 of this appendix.

[Doc. No. FAA–2005–22997, 73 FR 42494, July 21, 2008, as amended by Doc. No. FAA–2018–0119, Amdt. 25–145, 83 FR 9169, Mar. 5, 2018]

APPENDIX N TO PART 25—FUEL TANK FLAMMABILITY EXPOSURE AND RELIABILITY ANALYSIS

N25.1 *General.*

(a) This appendix specifies the requirements for conducting fuel tank fleet average flammability exposure analyses required to meet § 25.981(b) and Appendix M of this part. For fuel tanks installed in aluminum wings, a qualitative assessment is sufficient if it substantiates that the tank is a conventional unheated wing tank.

(b) This appendix defines parameters affecting fuel tank flammability that must be used in performing the analysis. These include parameters that affect all airplanes within the fleet, such as a statistical distribution of ambient temperature, fuel flash point, flight lengths, and airplane descent rate. Demonstration of compliance also requires application of factors specific to the airplane model being evaluated. Factors that need to be included are maximum range, cruise mach number, typical altitude where the airplane begins initial cruise phase of flight, fuel temperature during both ground and flight times, and the performance of a flammability reduction means (FRM) if installed.

(c) The following definitions, input variables, and data tables must be used in the program to determine fleet average flammability exposure for a specific airplane model.

N25.2 *Definitions.*

(a) *Bulk Average Fuel Temperature* means the average fuel temperature within the fuel tank or different sections of the tank if the tank is subdivided by baffles or compartments.

(b) *Flammability Exposure Evaluation Time (FEET).* The time from the start of preparing the airplane for flight, through the flight and landing, until all payload is unloaded, and all passengers and crew have disembarked. In the Monte Carlo program, the flight time is randomly selected from the Flight Length Distribution (Table 2), the pre-flight times are provided as a function of the flight time, and the post-flight time is a constant 30 minutes.

(c) *Flammable.* With respect to a fluid or gas, flammable means susceptible to igniting readily or to exploding (14 CFR Part 1, Definitions). A non-flammable ullage is one where the fuel-air vapor is too lean or too rich to burn or is inert as defined below. For the purposes of this appendix, a fuel tank that is not inert is considered flammable when the bulk average fuel temperature within the tank is within the flammable range for the fuel type being used. For any fuel tank that is subdivided into sections by baffles or compartments, the tank is considered flammable when the bulk average fuel temperature within any section of the tank, that is not inert, is within the flammable range for the fuel type being used.

(d) *Flash Point*. The flash point of a flammable fluid means the lowest temperature at which the application of a flame to a heated sample causes the vapor to ignite momentarily, or "flash." Table 1 of this appendix provides the flash point for the standard fuel to be used in the analysis.

(e) *Fleet average flammability exposure* is the percentage of the flammability exposure evaluation time (FEET) each fuel tank ullage is flammable for a fleet of an airplane type operating over the range of flight lengths in a world-wide range of environmental conditions and fuel properties as defined in this appendix.

(f) *Gaussian Distribution* is another name for the normal distribution, a symmetrical frequency distribution having a precise mathematical formula relating the mean and standard deviation of the samples. Gaussian distributions yield bell-shaped frequency curves having a preponderance of values around the mean with progressively fewer observations as the curve extends outward.

(g) *Hazardous atmosphere*. An atmosphere that may expose maintenance personnel, passengers or flight crew to the risk of death, incapacitation, impairment of ability to self-rescue (that is, escape unaided from a confined space), injury, or acute illness.

(h) *Inert*. For the purpose of this appendix, the tank is considered inert when the bulk average oxygen concentration within each compartment of the tank is 12 percent or less from sea level up to 10,000 feet altitude, then linearly increasing from 12 percent at 10,000 feet to 14.5 percent at 40,000 feet altitude, and extrapolated linearly above that altitude.

(i) *Inerting*. A process where a noncombustible gas is introduced into the ullage of a fuel tank so that the ullage becomes nonflammable.

(j) *Monte Carlo Analysis*. The analytical method that is specified in this appendix as the compliance means for assessing the fleet average flammability exposure time for a fuel tank.

(k) *Oxygen evolution* occurs when oxygen dissolved in the fuel is released into the ullage as the pressure and temperature in the fuel tank are reduced.

(l) *Standard deviation* is a statistical measure of the dispersion or variation in a distribution, equal to the square root of the arithmetic mean of the squares of the deviations from the arithmetic means.

(m) *Transport Effects*. For purposes of this appendix, transport effects are the change in fuel vapor concentration in a fuel tank caused by low fuel conditions and fuel condensation and vaporization.

(n) *Ullage*. The volume within the fuel tank not occupied by liquid fuel.

N25.3 *Fuel tank flammability exposure analysis.*

(a) A flammability exposure analysis must be conducted for the fuel tank under evaluation to determine fleet average flammability exposure for the airplane and fuel types under evaluation. For fuel tanks that are subdivided by baffles or compartments, an analysis must be performed either for each section of the tank, or for the section of the tank having the highest flammability exposure. Consideration of transport effects is not allowed in the analysis. The analysis must be done in accordance with the methods and procedures set forth in the Fuel Tank Flammability Assessment Method User's Manual, dated May 2008, document number DOT/FAA/AR–05/8 (incorporated by reference, see §25.5). The parameters specified in sections N25.3(b) and (c) of this appendix must be used in the fuel tank flammability exposure "Monte Carlo" analysis.

(b) The following parameters are defined in the Monte Carlo analysis and provided in paragraph N25.4 of this appendix:

(1) Cruise Ambient Temperature, as defined in this appendix.

(2) Ground Ambient Temperature, as defined in this appendix.

(3) Fuel Flash Point, as defined in this appendix.

(4) Flight Length Distribution, as defined in Table 2 of this appendix.

(5) Airplane Climb and Descent Profiles, as defined in the Fuel Tank Flammability Assessment Method User's Manual, dated May 2008, document number DOT/FAA/AR–05/8 (incorporated by reference in §25.5).

(c) Parameters that are specific to the particular airplane model under evaluation that must be provided as inputs to the Monte Carlo analysis are:

(1) Airplane cruise altitude.

(2) Fuel tank quantities. If fuel quantity affects fuel tank flammability, inputs to the Monte Carlo analysis must be provided that represent the actual fuel quantity within the fuel tank or compartment of the fuel tank throughout each of the flights being evaluated. Input values for this data must be obtained from ground and flight test data or the approved FAA fuel management procedures.

(3) Airplane cruise mach number.

(4) Airplane maximum range.

(5) *Fuel tank thermal characteristics.* If fuel temperature affects fuel tank flammability, inputs to the Monte Carlo analysis must be provided that represent the actual bulk average fuel temperature within the fuel tank at each point in time throughout each of the flights being evaluated. For fuel tanks that are subdivided by baffles or compartments, bulk average fuel temperature inputs must be provided for each section of the tank. Input values for these data must be obtained from ground and flight test data or a thermal model of the tank that has been validated by ground and flight test data.

(6) Maximum airplane operating temperature limit, as defined by any limitations in the airplane flight manual.

(7) *Airplane Utilization.* The applicant must provide data supporting the number of flights per day and the number of hours per flight for the specific airplane model under evaluation. If there is no existing airplane fleet data to support the airplane being evaluated, the applicant must provide substantiation that the number of flights per day and the number of hours per flight for that airplane model is consistent with the existing fleet data they propose to use.

(d) *Fuel Tank FRM Model.* If FRM is used, an FAA approved Monte Carlo program must be used to show compliance with the flammability requirements of § 25.981 and Appendix M of this part. The program must determine the time periods during each flight phase when the fuel tank or compartment with the FRM would be flammable. The following factors must be considered in establishing these time periods:

(1) Any time periods throughout the flammability exposure evaluation time and under the full range of expected operating conditions, when the FRM is operating properly but fails to maintain a non-flammable fuel tank because of the effects of the fuel tank vent system or other causes,

(2) If dispatch with the system inoperative under the Master Minimum Equipment List (MMEL) is requested, the time period assumed in the reliability analysis (60 flight hours must be used for a 10-day MMEL dispatch limit unless an alternative period has been approved by the Administrator),

(3) Frequency and duration of time periods of FRM inoperability, substantiated by test or analysis acceptable to the FAA, caused by latent or known failures, including airplane system shut-downs and failures that could cause the FRM to shut down or become inoperative.

(4) Effects of failures of the FRM that could increase the flammability exposure of the fuel tank.

(5) If an FRM is used that is affected by oxygen concentrations in the fuel tank, the time periods when oxygen evolution from the fuel results in the fuel tank or compartment exceeding the inert level. The applicant must include any times when oxygen evolution from the fuel in the tank or compartment under evaluation would result in a flammable fuel tank. The oxygen evolution rate that must be used is defined in the Fuel Tank Flammability Assessment Method User's Manual, dated May 2008, document number DOT/FAA/AR–05/8 (incorporated by reference in § 25.5).

(6) If an inerting system FRM is used, the effects of any air that may enter the fuel tank following the last flight of the day due to changes in ambient temperature, as defined in Table 4, during a 12-hour overnight period.

(e) The applicant must submit to the responsible Aircraft Certification Service officefor approval the fuel tank flammability analysis, including the airplane-specific parameters identified under paragraph N25.3(c) of this appendix and any deviations from the parameters identified in paragraph N25.3(b) of this appendix that affect flammability exposure, substantiating data, and any airworthiness limitations and other conditions assumed in the analysis.

N25.4 *Variables and data tables.*

The following data must be used when conducting a flammability exposure analysis to determine the fleet average flammability exposure. Variables used to calculate fleet flammability exposure must include atmospheric ambient temperatures, flight length, flammability exposure evaluation time, fuel flash point, thermal characteristics of the fuel tank, overnight temperature drop, and oxygen evolution from the fuel into the ullage.

(a) *Atmospheric Ambient Temperatures and Fuel Properties.*

(1) In order to predict flammability exposure during a given flight, the variation of ground ambient temperatures, cruise ambient temperatures, and a method to compute the transition from ground to cruise and back again must be used. The variation of the ground and cruise ambient temperatures and the flash point of the fuel is defined by a Gaussian curve, given by the 50 percent value and a ±1-standard deviation value.

(2) *Ambient Temperature:* Under the program, the ground and cruise ambient temperatures are linked by a set of assumptions on the atmosphere. The temperature varies with altitude following the International Standard Atmosphere (ISA) rate of change from the ground ambient temperature until the cruise temperature for the flight is reached. Above this altitude, the ambient temperature is fixed at the cruise ambient temperature. This results in a variation in the upper atmospheric temperature. For cold days, an inversion is applied up to 10,000 feet, and then the ISA rate of change is used.

(3) *Fuel properties:*

(i) For Jet A fuel, the variation of flash point of the fuel is defined by a Gaussian curve, given by the 50 percent value and a ±1-standard deviation, as shown in Table 1 of this appendix.

(ii) The flammability envelope of the fuel that must be used for the flammability exposure analysis is a function of the flash point of the fuel selected by the Monte Carlo for a given flight. The flammability envelope for the fuel is defined by the upper flammability limit (UFL) and lower flammability limit (LFL) as follows:

(A) LFL at sea level = flash point temperature of the fuel at sea level minus 10 °F. LFL

decreases from sea level value with increasing altitude at a rate of 1 °F per 808 feet.

(B) UFL at sea level = flash point temperature of the fuel at sea level plus 63.5 °F. UFL decreases from the sea level value with increasing altitude at a rate of 1 °F per 512 feet.

(4) For each flight analyzed, a separate random number must be generated for each of the three parameters (ground ambient temperature, cruise ambient temperature, and fuel flash point) using the Gaussian distribution defined in Table 1 of this appendix.

TABLE 1.—GAUSSIAN DISTRIBUTION FOR GROUND AMBIENT TEMPERATURE, CRUISE AMBIENT TEMPERATURE, AND FUEL FLASH POINT

Parameter	Temperature in deg F		
	Ground ambient temperature	Cruise ambient temperature	Fuel flash point (FP)
Mean Temp	59.95	−70	120
Neg 1 std dev	20.14	8	8
Pos 1 std dev	17.28	8	8

(b) The Flight Length Distribution defined in Table 2 must be used in the Monte Carlo analysis.

TABLE 2.—FLIGHT LENGTH DISTRIBUTION

Flight length (NM)		Airplane maximum range—nautical miles (NM)									
From	To	1000	2000	3000	4000	5000	6000	7000	8000	9000	10000
		Distribution of flight lengths (percentage of total)									
0	200	11.7	7.5	6.2	5.5	4.7	4.0	3.4	3.0	2.6	2.3
200	400	27.3	19.9	17.0	15.2	13.2	11.4	9.7	8.5	7.5	6.7
400	600	46.3	40.0	35.7	32.6	28.5	24.9	21.2	18.7	16.4	14.8
600	800	10.3	11.6	11.0	10.2	9.1	8.0	6.9	6.1	5.4	4.8
800	1000	4.4	8.5	8.6	8.2	7.4	6.6	5.7	5.0	4.5	4.0
1000	1200	0.0	4.8	5.3	5.3	4.8	4.3	3.8	3.3	3.0	2.7
1200	1400	0.0	3.6	4.4	4.5	4.2	3.8	3.3	3.0	2.7	2.4
1400	1600	0.0	2.2	3.3	3.5	3.3	3.1	2.7	2.4	2.2	2.0
1600	1800	0.0	1.2	2.3	2.6	2.5	2.4	2.1	1.9	1.7	1.6
1800	2000	0.0	0.7	2.2	2.6	2.6	2.5	2.2	2.0	1.8	1.7
2000	2200	0.0	0.0	1.6	2.1	2.2	2.1	1.9	1.7	1.6	1.4
2200	2400	0.0	0.0	1.1	1.6	1.7	1.7	1.6	1.4	1.3	1.2
2400	2600	0.0	0.0	0.7	1.2	1.4	1.4	1.3	1.2	1.1	1.0
2600	2800	0.0	0.0	0.4	0.9	1.0	1.1	1.0	0.9	0.9	0.8
2800	3000	0.0	0.0	0.2	0.6	0.7	0.8	0.7	0.7	0.6	0.6
3000	3200	0.0	0.0	0.0	0.6	0.8	0.8	0.8	0.8	0.7	0.7
3200	3400	0.0	0.0	0.0	0.7	1.1	1.2	1.2	1.1	1.1	1.0
3400	3600	0.0	0.0	0.0	0.7	1.3	1.6	1.6	1.5	1.5	1.4
3600	3800	0.0	0.0	0.0	0.9	2.2	2.7	2.8	2.7	2.6	2.5
3800	4000	0.0	0.0	0.0	0.5	2.0	2.6	2.8	2.8	2.7	2.6
4000	4200	0.0	0.0	0.0	0.0	2.1	3.0	3.2	3.3	3.2	3.1
4200	4400	0.0	0.0	0.0	0.0	1.4	2.2	2.5	2.6	2.6	2.5
4400	4600	0.0	0.0	0.0	0.0	1.0	2.0	2.3	2.5	2.5	2.4
4600	4800	0.0	0.0	0.0	0.0	0.6	1.5	1.8	2.0	2.0	2.0
4800	5000	0.0	0.0	0.0	0.0	0.2	1.0	1.4	1.5	1.6	1.5
5000	5200	0.0	0.0	0.0	0.0	0.0	0.8	1.1	1.3	1.3	1.3
5200	5400	0.0	0.0	0.0	0.0	0.0	0.8	1.2	1.5	1.6	1.6
5400	5600	0.0	0.0	0.0	0.0	0.0	0.9	1.7	2.1	2.2	2.3
5600	5800	0.0	0.0	0.0	0.0	0.0	0.6	1.6	2.2	2.4	2.5
5800	6000	0.0	0.0	0.0	0.0	0.0	0.2	1.8	2.4	2.8	2.9
6000	6200	0.0	0.0	0.0	0.0	0.0	0.0	1.7	2.6	3.1	3.3
6200	6400	0.0	0.0	0.0	0.0	0.0	0.0	1.4	2.4	2.9	3.1
6400	6600	0.0	0.0	0.0	0.0	0.0	0.0	0.9	1.8	2.2	2.5
6600	6800	0.0	0.0	0.0	0.0	0.0	0.0	0.5	1.2	1.6	1.9
6800	7000	0.0	0.0	0.0	0.0	0.0	0.0	0.2	0.8	1.1	1.3

TABLE 2.—FLIGHT LENGTH DISTRIBUTION—Continued

Flight length (NM)		Airplane maximum range—nautical miles (NM)									
From	To	1000	2000	3000	4000	5000	6000	7000	8000	9000	10000
7000	7200	0.0	0.0	0.0	0.0	0.0	0.0	0.0	0.4	0.7	0.8
7200	7400	0.0	0.0	0.0	0.0	0.0	0.0	0.0	0.3	0.5	0.7
7400	7600	0.0	0.0	0.0	0.0	0.0	0.0	0.0	0.2	0.5	0.6
7600	7800	0.0	0.0	0.0	0.0	0.0	0.0	0.0	0.1	0.5	0.7
7800	8000	0.0	0.0	0.0	0.0	0.0	0.0	0.0	0.1	0.6	0.8
8000	8200	0.0	0.0	0.0	0.0	0.0	0.0	0.0	0.0	0.5	0.8
8200	8400	0.0	0.0	0.0	0.0	0.0	0.0	0.0	0.0	0.5	1.0
8400	8600	0.0	0.0	0.0	0.0	0.0	0.0	0.0	0.0	0.6	1.3
8600	8800	0.0	0.0	0.0	0.0	0.0	0.0	0.0	0.0	0.4	1.1
8800	9000	0.0	0.0	0.0	0.0	0.0	0.0	0.0	0.0	0.2	0.8
9000	9200	0.0	0.0	0.0	0.0	0.0	0.0	0.0	0.0	0.0	0.5
9200	9400	0.0	0.0	0.0	0.0	0.0	0.0	0.0	0.0	0.0	0.2
9400	9600	0.0	0.0	0.0	0.0	0.0	0.0	0.0	0.0	0.0	0.1
9600	9800	0.0	0.0	0.0	0.0	0.0	0.0	0.0	0.0	0.0	0.1
9800	10000	0.0	0.0	0.0	0.0	0.0	0.0	0.0	0.0	0.0	0.1

(c) Overnight Temperature Drop. For airplanes on which FRM is installed, the overnight temperature drop for this appendix is defined using:

(1) A temperature at the beginning of the overnight period that equals the landing temperature of the previous flight that is a random value based on a Gaussian distribution; and

(2) An overnight temperature drop that is a random value based on a Gaussian distribution.

(3) For any flight that will end with an overnight ground period (one flight per day out of an average number of flights per day, depending on utilization of the particular airplane model being evaluated), the landing outside air temperature (OAT) is to be chosen as a random value from the following Gaussian curve:

TABLE 3.—LANDING OUTSIDE AIR TEMPERATURE

Parameter	Landing outside air temperature °F
Mean Temperature	58.68
negative 1 std dev	20.55
positive 1 std dev	13.21

(4) The outside ambient air temperature (OAT) overnight temperature drop is to be chosen as a random value from the following Gaussian curve:

TABLE 4.—OUTSIDE AIR TEMPERATURE (OAT) DROP

Parameter	OAT drop temperature °F
Mean Temp	12.0
1 std dev	6.0

(d) Number of Simulated Flights Required in Analysis. In order for the Monte Carlo analysis to be valid for showing compliance

with the fleet average and warm day flammability exposure requirements, the applicant must run the analysis for a minimum number of flights to ensure that the fleet average and warm day flammability exposure for the fuel tank under evaluation meets the applicable flammability limits defined in Table 5 of this appendix.

TABLE 5.—FLAMMABILITY EXPOSURE LIMIT

Minimum number of flights in Monte Carlo analysis	Maximum acceptable Monte Carlo average fuel tank flammability exposure (percent) to meet 3 percent requirements	Maximum acceptable Monte Carlo average fuel tank flammability exposure (percent) to meet 7 percent part 26 requirements
10,000	2.91	6.79
100,000	2.98	6.96
1,000,000	3.00	7.00

[Doc. No. FAA–2005–22997, 73 FR 42495, July 21, 2008, as amended by Doc. No. FAA–2018–0119, Amdt. 25–145, 83 FR 9169, Mar. 5, 2018]

APPENDIX O TO PART 25—SUPERCOOLED LARGE DROP ICING CONDITIONS

This Appendix consists of two parts. Part I defines this Appendix as a description of supercooled large drop icing conditions in which the drop median volume diameter (MVD) is less than or greater than 40 µm, the maximum mean effective drop diameter (MED) of Appendix C of this part continuous maximum (stratiform clouds) icing conditions. For this Appendix, supercooled large drop icing conditions consist of freezing drizzle and freezing rain occurring in and/or below stratiform clouds. Part II defines ice accretions used to show compliance with the airplane performance and handling qualities requirements of subpart B of this part.

PART I—METEOROLOGY

In this Appendix icing conditions are defined by the parameters of altitude, vertical and horizontal extent, temperature, liquid water content, and water mass distribution as a function of drop diameter distribution.

(a) Freezing Drizzle (Conditions with spectra maximum drop diameters from 100µm to 500 µm):

(1) Pressure altitude range: 0 to 22,000 feet MSL.

(2) Maximum vertical extent: 12,000 feet.

(3) Horizontal extent: Standard distance of 17.4 nautical miles.

(4) Total liquid water content.

NOTE: Liquid water content (LWC) in grams per cubic meter (g/m^3) based on horizontal extent standard distance of 17.4 nautical miles.

(5) Drop diameter distribution: Figure 2.

(6) Altitude and temperature envelope: Figure 3.

(b) Freezing Rain (Conditions with spectra maximum drop diameters greater than 500 µm):

(1) Pressure altitude range: 0 to 12,000 ft MSL.

(2) Maximum vertical extent: 7,000 ft.

(3) Horizontal extent: Standard distance of 17.4 nautical miles.

(4) Total liquid water content.

NOTE: LWC in grams per cubic meter (g/m^3) based on horizontal extent standard distance of 17.4 nautical miles.

(5) Drop Diameter Distribution: Figure 5.

(6) Altitude and temperature envelope: Figure 6.

(c) Horizontal extent.

The liquid water content for freezing drizzle and freezing rain conditions for horizontal extents other than the standard 17.4 nautical miles can be determined by the value of the liquid water content determined from Figure 1 or Figure 4, multiplied by the factor provided in Figure 7, which is defined by the following equation:

$$S = 1.266 - 0.213 \log 10(H)$$

Where:

S = Liquid Water Content Scale Factor (dimensionless) and

H = horizontal extent in nautical miles

FIGURE 1 — Appendix O, Freezing Drizzle, Liquid Water Content

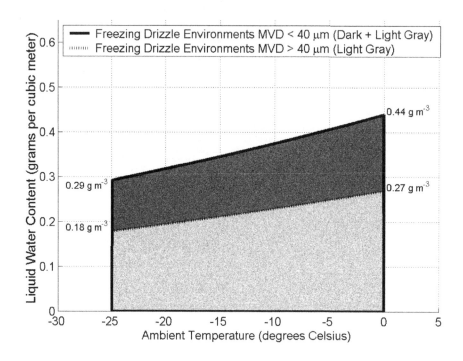

FIGURE 2 — Appendix O, Freezing Drizzle, Drop Diameter Distribution

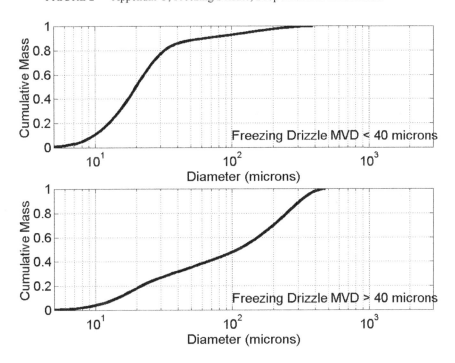

FIGURE 3 — Appendix O, Freezing Drizzle, Temperature and Altitude

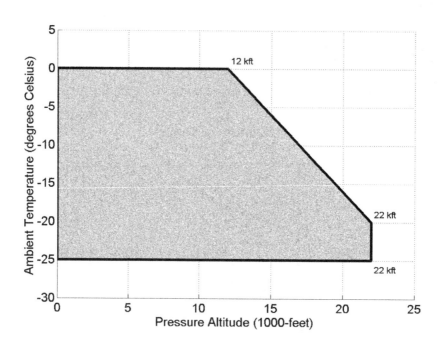

FIGURE 4 — Appendix O, Freezing Rain, Liquid Water Content

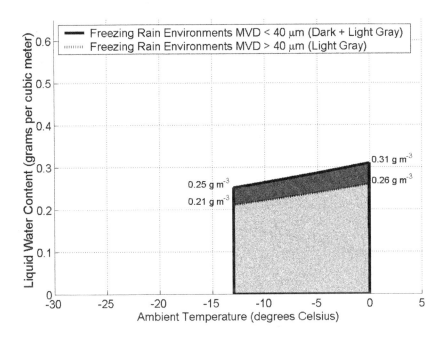

FIGURE 5 — Appendix O, Freezing Rain, Drop Diameter Distribution

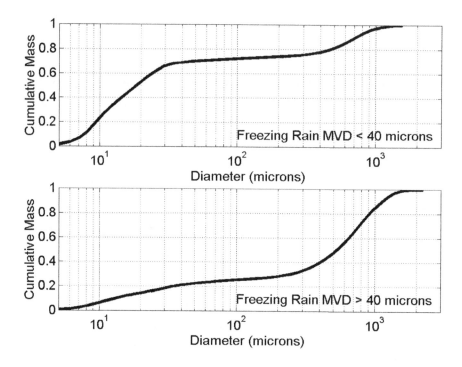

FIGURE 6 — Appendix O, Freezing Rain, Temperature and Altitude

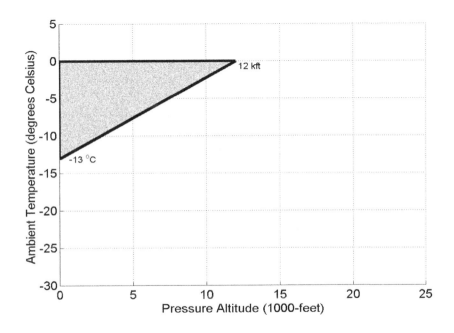

FIGURE 7 — Appendix O, Horizontal Extent, Freezing Drizzle and Freezing Rain

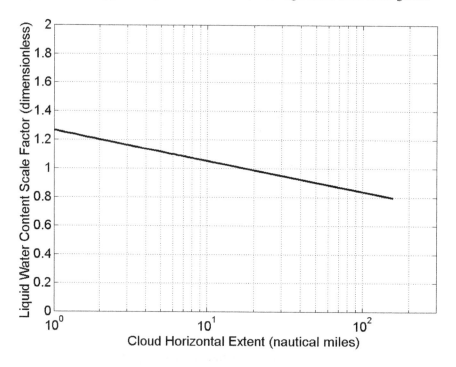

PART II—AIRFRAME ICE ACCRETIONS FOR SHOWING COMPLIANCE WITH SUBPART B OF THIS PART

(a) *General.* The most critical ice accretion in terms of airplane performance and handling qualities for each flight phase must be used to show compliance with the applicable airplane performance and handling qualities requirements for icing conditions contained in subpart B of this part. Applicants must demonstrate that the full range of atmospheric icing conditions specified in part I of this Appendix have been considered, including drop diameter distributions, liquid water content, and temperature appropriate to the flight conditions (for example, configuration, speed, angle of attack, and altitude).

(1) For an airplane certified in accordance with § 25.1420(a)(1), the ice accretions for each flight phase are defined in part II, paragraph (b) of this Appendix.

(2) For an airplane certified in accordance with § 25.1420(a)(2), the most critical ice accretion for each flight phase defined in part II, paragraphs (b) and (c) of this Appendix, must be used. For the ice accretions defined in part II, paragraph (c) of this Appendix, only the portion of part I of this Appendix in which the airplane is capable of operating safely must be considered.

(3) For an airplane certified in accordance with § 25.1420(a)(3), the ice accretions for each flight phase are defined in part II, paragraph (c) of this Appendix.

(b) Ice accretions for airplanes certified in accordance with § 25.1420(a)(1) or (2).

(1) *En route ice* is the en route ice as defined by part II, paragraph (c)(3), of this Appendix, for an airplane certified in accordance with § 25.1420(a)(2), or defined by part II, paragraph (a)(3), of Appendix C of this part, for an airplane certified in accordance with § 25.1420(a)(1), plus:

(i) Pre-detection ice as defined by part II, paragraph (b)(5), of this Appendix; and

(ii) The ice accumulated during the transit of one cloud with a horizontal extent of 17.4 nautical miles in the most critical of the icing conditions defined in part I of this Appendix and one cloud with a horizontal extent of 17.4 nautical miles in the continuous maximum icing conditions defined in Appendix C of this part.

(2) *Holding ice* is the holding ice defined by part II, paragraph (c)(4), of this Appendix, for an airplane certified in accordance with §25.1420(a)(2), or defined by part II, paragraph (a)(4), of Appendix C of this part, for an airplane certified in accordance with §25.1420(a)(1), plus:

(i) Pre-detection ice as defined by part II, paragraph (b)(5), of this Appendix; and

(ii) The ice accumulated during the transit of one cloud with a 17.4 nautical miles horizontal extent in the most critical of the icing conditions defined in part I of this Appendix and one cloud with a horizontal extent of 17.4 nautical miles in the continuous maximum icing conditions defined in Appendix C of this part.

(iii) Except the total exposure to holding ice conditions does not need to exceed 45 minutes.

(3) *Approach ice* is the more critical of the holding ice defined by part II, paragraph (b)(2), of this Appendix, or the ice calculated in the applicable paragraphs (b)(3)(i) or (ii) of part II, of this Appendix:

(i) For an airplane certified in accordance with §25.1420(a)(2), the ice accumulated during descent from the maximum vertical extent of the icing conditions defined in part I of this Appendix to 2,000 feet above the landing surface in the cruise configuration, plus transition to the approach configuration, plus:

(A) Pre-detection ice, as defined by part II, paragraph (b)(5), of this Appendix; and

(B) The ice accumulated during the transit at 2,000 feet above the landing surface of one cloud with a horizontal extent of 17.4 nautical miles in the most critical of the icing conditions defined in part I of this Appendix and one cloud with a horizontal extent of 17.4 nautical miles in the continuous maximum icing conditions defined in Appendix C of this part.

(ii) For an airplane certified in accordance with §25.1420(a)(1), the ice accumulated during descent from the maximum vertical extent of the maximum continuous icing conditions defined in part I of Appendix C to 2,000 feet above the landing surface in the cruise configuration, plus transition to the approach configuration, plus:

(A) Pre-detection ice, as defined by part II, paragraph (b)(5), of this Appendix; and

(B) The ice accumulated during the transit at 2,000 feet above the landing surface of one cloud with a horizontal extent of 17.4 nautical miles in the most critical of the icing conditions defined in part I of this Appendix and one cloud with a horizontal extent of 17.4 nautical miles in the continuous maximum icing conditions defined in Appendix C of this part.

(4) *Landing ice* is the more critical of the holding ice as defined by part II, paragraph (b)(2), of this Appendix, or the ice calculated in the applicable paragraphs (b)(4)(i) or (ii) of part II of this Appendix:

(i) For an airplane certified in accordance with §25.1420(a)(2), the ice accretion defined by part II, paragraph (c)(5)(i), of this Appendix, plus a descent from 2,000 feet above the landing surface to a height of 200 feet above the landing surface with a transition to the landing configuration in the icing conditions defined in part I of this Appendix, plus:

(A) Pre-detection ice, as defined in part II, paragraph (b)(5), of this Appendix; and

(B) The ice accumulated during an exit maneuver, beginning with the minimum climb gradient required by §25.119, from a height of 200 feet above the landing surface through one cloud with a horizontal extent of 17.4 nautical miles in the most critical of the icing conditions defined in part I of this Appendix and one cloud with a horizontal extent of 17.4 nautical miles in the continuous maximum icing conditions defined in Appendix C of this part.

(ii) For an airplane certified in accordance with §25.1420(a)(1), the ice accumulated in the maximum continuous icing conditions defined in Appendix C of this part, during a descent from the maximum vertical extent of the icing conditions defined in Appendix C of this part, to 2,000 feet above the landing surface in the cruise configuration, plus transition to the approach configuration and flying for 15 minutes at 2,000 feet above the landing surface, plus a descent from 2,000 feet above the landing surface to a height of 200 feet above the landing surface with a transition to the landing configuration, plus:

(A) Pre-detection ice, as described by part II, paragraph (b)(5), of this Appendix; and

(B) The ice accumulated during an exit maneuver, beginning with the minimum climb gradient required by §25.119, from a height of 200 feet above the landing surface through one cloud with a horizontal extent of 17.4 nautical miles in the most critical of the icing conditions defined in part I of this Appendix and one cloud with a horizontal extent of 17.4 nautical miles in the continuous maximum icing conditions defined in Appendix C of this part.

(5) *Pre-detection ice* is the ice accretion before detection of flight conditions in this Appendix that require exiting per §25.1420(a)(1) and (2). It is the pre-existing ice accretion that may exist from operating in icing conditions in which the airplane is approved to operate prior to encountering the icing conditions requiring an exit, plus the ice accumulated during the time needed to detect the icing conditions, followed by two minutes of further ice accumulation to take into account the time for the flightcrew to take action to exit the icing conditions, including coordination with air traffic control.

469

(i) For an airplane certified in accordance with §25.1420(a)(1), the pre-existing ice accretion must be based on the icing conditions defined in Appendix C of this part.

(ii) For an airplane certified in accordance with §25.1420(a)(2), the pre-existing ice accretion must be based on the more critical of the icing conditions defined in Appendix C of this part, or the icing conditions defined in part I of this Appendix in which the airplane is capable of safely operating.

(c) *Ice accretions for airplanes certified in accordance with §§25.1420(a)(2) or (3).* For an airplane certified in accordance with §25.1420(a)(2), only the portion of the icing conditions of part I of this Appendix in which the airplane is capable of operating safely must be considered.

(1) *Takeoff ice* is the most critical ice accretion on unprotected surfaces, and any ice accretion on the protected surfaces, occurring between the end of the takeoff distance and 400 feet above the takeoff surface, assuming accretion starts at the end of the takeoff distance in the icing conditions defined in part I of this Appendix.

(2) *Final takeoff ice* is the most critical ice accretion on unprotected surfaces, and any ice accretion on the protected surfaces appropriate to normal ice protection system operation, between 400 feet and either 1,500 feet above the takeoff surface, or the height at which the transition from the takeoff to the en route configuration is completed and V_{FTO} is reached, whichever is higher. Ice accretion is assumed to start at the end of the takeoff distance in the icing conditions defined in part I of this Appendix.

(3) *En route ice* is the most critical ice accretion on the unprotected surfaces, and any ice accretion on the protected surfaces appropriate to normal ice protection system operation, during the en route flight phase in the icing conditions defined in part I of this Appendix.

(4) *Holding ice* is the most critical ice accretion on the unprotected surfaces, and any ice accretion on the protected surfaces appropriate to normal ice protection system operation, resulting from 45 minutes of flight within a cloud with a 17.4 nautical miles horizontal extent in the icing conditions defined in part I of this Appendix, during the holding phase of flight.

(5) *Approach ice* is the ice accretion on the unprotected surfaces, and any ice accretion on the protected surfaces appropriate to normal ice protection system operation, resulting from the more critical of the:

(i) Ice accumulated in the icing conditions defined in part I of this Appendix during a descent from the maximum vertical extent of the icing conditions defined in part I of this Appendix, to 2,000 feet above the landing surface in the cruise configuration, plus transition to the approach configuration and

flying for 15 minutes at 2,000 feet above the landing surface; or

(ii) Holding ice as defined by part II, paragraph (c)(4), of this Appendix.

(6) *Landing ice* is the ice accretion on the unprotected surfaces, and any ice accretion on the protected surfaces appropriate to normal ice protection system operation, resulting from the more critical of the:

(i) Ice accretion defined by part II, paragraph (c)(5)(i), of this Appendix, plus ice accumulated in the icing conditions defined in part I of this Appendix during a descent from 2,000 feet above the landing surface to a height of 200 feet above the landing surface with a transition to the landing configuration, followed by a go-around at the minimum climb gradient required by §25.119, from a height of 200 feet above the landing surface to 2,000 feet above the landing surface, flying for 15 minutes at 2,000 feet above the landing surface in the approach configuration, and a descent to the landing surface (touchdown) in the landing configuration; or

(ii) Holding ice as defined by part II, paragraph (c)(4), of this Appendix.

(7) For both unprotected and protected parts, the ice accretion for the takeoff phase must be determined for the icing conditions defined in part I of this Appendix, using the following assumptions:

(i) The airfoils, control surfaces, and, if applicable, propellers are free from frost, snow, or ice at the start of takeoff;

(ii) The ice accretion starts at the end of the takeoff distance;

(iii) The critical ratio of thrust/power-to-weight;

(iv) Failure of the critical engine occurs at V_{EF}; and

(v) Crew activation of the ice protection system is in accordance with a normal operating procedure provided in the airplane flight manual, except that after beginning the takeoff roll, it must be assumed that the crew takes no action to activate the ice protection system until the airplane is at least 400 feet above the takeoff surface.

(d) The ice accretion before the ice protection system has been activated and is performing its intended function is the critical ice accretion formed on the unprotected and normally protected surfaces before activation and effective operation of the ice protection system in the icing conditions defined in part I of this Appendix. This ice accretion only applies in showing compliance to §§25.143(j) and 25.207(h).

(e) In order to reduce the number of ice accretions to be considered when demonstrating compliance with the requirements of §25.21(g), any of the ice accretions defined in this Appendix may be used for any other flight phase if it is shown to be at least as critical as the specific ice accretion defined for that flight phase. Configuration differences and their effects on ice accretions must be taken into account.

(f) The ice accretion that has the most adverse effect on handling qualities may be used for airplane performance tests provided any difference in performance is conservatively taken into account.

[Amdt. 25–140, 79 FR 65528, Nov. 4, 2014]

PART 26—CONTINUED AIRWORTHINESS AND SAFETY IMPROVEMENTS FOR TRANSPORT CATEGORY AIRPLANES

AUTHORITY: 49 U.S.C. 106(g), 40113, 44701, 44702 and 44704.

SOURCE: Docket No. FAA–2004–18379, 72 FR 63409, Nov. 8, 2007, unless otherwise noted.

Subpart A—General

§ 26.1 Purpose and scope.

(a) This part establishes requirements for support of the continued airworthiness of and safety improvements for transport category airplanes. These requirements may include performing assessments, developing design changes, developing revisions to Instructions for Continued Airworthiness (ICA), and making necessary documentation available to affected persons. Requirements of this part that establish standards for design changes and revisions to the ICA are considered airworthiness requirements.

(b) Except as provided in paragraph (c) of this section, this part applies to the following persons, as specified in each subpart of this part:

(1) Holders of type certificates and supplemental type certificates.

(2) Applicants for type certificates and supplemental type certificates and changes to those certificates (including service bulletins describing design changes).

(3) Persons seeking design approval for airplane repairs, alterations, or modifications that may affect airworthiness.

(4) Holders of type certificates and their licensees producing new airplanes.

(c) An applicant for approval of a design change is not required to comply with any applicable airworthiness requirement of this part if the applicant elects or is required to comply with a corresponding amendment to part 25 of this chapter that is adopted concurrently or after that airworthiness requirement.

(d) For the purposes of this part, the word "type certificate" does not include supplemental type certificates.

§ 26.3 [Reserved]

§ 26.5 Applicability table.

Table 1 of this section provides an overview of the applicability of this part. It provides guidance in identifying what sections apply to various types of entities. The specific applicability of each subpart and section is specified in the regulatory text.

TABLE 1—APPLICABILITY OF PART 26 RULES

	Applicable sections			
	Subpart B EAPAS/FTS	Subpart C widespread fatigue damage	Subpart D fuel tank flammability	Subpart E damage tolerance data
Effective date of rule	December 10, 2007	January 14, 2011	December 26, 2008	January 11, 2008
Existing [1] TC Holders	26.11	26.21	26.33	26.43, 26.45, 26.49
Pending [1] TC Applicants	26.11	26.21	26.37	26.43, 26.45
Future [2] TC applicants	N/A	N/A	N/A	26.43
Existing [1] STC Holders	N/A	26.21	26.35	26.47, 26.49
Pending [1] STC/ATC applicants ..	26.11	26.21	26.35	26.45, 26.47, 26.49
Future [2] STC/ATC applicants	26.11	26.21	26.35	26.45, 26.47, 26.49
Manufacturers	N/A	N/A	26.39	N/A

[1] As of the effective date of the identified rule.
[2] Application made after the effective date of the identified rule.

[Doc. No. FAA–2006–24281, 75 FR 69782, Nov. 15, 2010]

Subpart B—Enhanced Airworthiness Program for Airplane Systems

§ 26.11 Electrical wiring interconnection systems (EWIS) maintenance program.

(a) Except as provided in paragraph (g) of this section, this section applies to transport category, turbine-powered airplanes with a type certificate issued after January 1, 1958, that, as a result of the original certification, or later increase in capacity, have—

(1) A maximum type-certificated passenger capacity of 30 or more or

(2) A maximum payload capacity of 7,500 pounds or more.

(b) Holders of, and applicants for, type certificates, as identified in paragraph (d) of this section must develop Instructions for Continued Airworthiness (ICA) for the representative airplane's EWIS in accordance with part 25, Appendix H paragraphs H25.5(a)(1) and (b) of this subchapter in effect on December 10, 2007 for each affected type design, and submit those ICA for review and approval by the responsible Aircraft Certification Service office. For purposes of this section, the "representative airplane" is the configuration of each model series airplane that incorporates all variations of EWIS used in production on that series airplane, and all TC-holder-designed modifications mandated by airworthiness directive as of the effective date of this rule. Each person specified in paragraph (d) of this section must also review any fuel tank system ICA developed by that person to comply with SFAR 88 to ensure compatibility with the EWIS ICA, including minimizing redundant requirements.

(c) Applicants for amendments to type certificates and supplemental type certificates, as identified in paragraph (d) of this section, must:

(1) Evaluate whether the design change for which approval is sought necessitates a revision to the ICA required by paragraph (b) of this section to comply with the requirements of Appendix H, paragraphs H25.5(a)(1) and (b). If so, the applicant must develop and submit the necessary revisions for review and approval by the responsible Aircraft Certification Service office.

(2) Ensure that any revised EWIS ICA remain compatible with any fuel tank system ICA previously developed to comply with SFAR 88 and any redundant requirements between them are minimized.

(d) The following persons must comply with the requirements of paragraph (b) or (c) of this section, as applicable, before the dates specified.

(1) Holders of type certificates (TC): December 10, 2009.

(2) Applicants for TCs, and amendments to TCs (including service bulletins describing design changes), if the date of application was before December 10, 2007 and the certificate was

issued on or after December 10, 2007: December 10, 2009 or the date the certificate is issued, whichever occurs later.

(3) Unless compliance with §25.1729 of this subchapter is required or elected, applicants for amendments to TCs, if the application was filed on or after December 10, 2007: December 10, 2009, or the date of approval of the certificate, whichever occurs later.

(4) Applicants for supplemental type certificates (STC), including changes to existing STCs, if the date of application was before December 10, 2007 and the certificate was issued on or after December 10, 2007: June 7, 2010, or the date of approval of the certificate, whichever occurs later.

(5) Unless compliance with §25.1729 of this subchapter is required or elected, applicants for STCs, including changes to existing STCs, if the application was filed on or after December 10, 2007, June 7, 2010, or the date of approval of the certificate, whichever occurs later.

(e) Each person identified in paragraphs (d)(1), (d)(2), and (d)(4) of this section must submit to the responsible Aircraft Certification Service office for approval a compliance plan by March 10, 2008. The compliance plan must include the following information:

(1) A proposed project schedule, identifying all major milestones, for meeting the compliance dates specified in paragraph (d) of this section.

(2) A proposed means of compliance with this section, identifying all required submissions, including all compliance items as mandated in part 25, Appendix H paragraphs H25.5(a)(1) and (b) of this subchapter in effect on December 10, 2007, and all data to be developed to substantiate compliance.

(3) A proposal for submitting a draft of all compliance items required by paragraph (e)(2) of this section for review by the responsible Aircraft Certification Service office not less than 60 days before the compliance time specified in paragraph (d) of this section.

(4) A proposal for how the approved ICA will be made available to affected persons.

(f) Each person specified in paragraph (e) must implement the compliance plan, or later approved revisions, as approved in compliance with paragraph (e) of this section.

(g) This section does not apply to the following airplane models:

(1) Lockheed L–188
(2) Bombardier CL–44
(3) Mitsubishi YS–11
(4) British Aerospace BAC 1–11
(5) Concorde
(6) deHavilland D.H. 106 Comet 4C
(7) VFW—Vereinigte Flugtechnische Werk VFW–614
(8) Illyushin Aviation IL 96T
(9) Bristol Aircraft Britannia 305
(10) Handley Page Herald Type 300
(11) Avions Marcel Dassault—Breguet Aviation Mercure 100C
(12) Airbus Caravelle
(13) Lockheed L–300

[Amdt. 26–0, 72 FR 63409, Nov. 8, 2007; 72 FR 68618, Dec. 5, 2007, as amended by Doc. No. FAA–2018–0119, Amdt. 26–7, 83 FR 9170, Mar. 5, 2018]

Subpart C—Aging Airplane Safety—Widespread Fatigue Damage

SOURCE: Docket No. FAA–2006–24281, 75 FR 69782, Nov. 15, 2010, unless otherwise noted.

§26.21 Limit of validity.

(a) *Applicability*. Except as provided in paragraph (g) of this section, this section applies to transport category, turbine-powered airplanes with a maximum takeoff gross weight greater than 75,000 pounds and a type certificate issued after January 1, 1958, regardless of whether the maximum takeoff gross weight is a result of an original type certificate or a later design change. This section also applies to transport category, turbine-powered airplanes with a type certificate issued after January 1, 1958, if a design change approval for which application is made after January 14, 2011 has the effect of reducing the maximum takeoff gross weight from greater than 75,000 pounds to 75,000 pounds or less.

(b) *Limit of validity*. Each person identified in paragraph (c) of this section must comply with the following requirements:

(1) Establish a limit of validity of the engineering data that supports the

structural maintenance program (hereafter referred to as LOV) that corresponds to the period of time, stated as a number of total accumulated flight cycles or flight hours or both, during which it is demonstrated that widespread fatigue damage will not occur in the airplane. This demonstration must include an evaluation of airplane structural configurations and be supported by test evidence and analysis at a minimum and, if available, service experience, or service experience and teardown inspection results, of high-time airplanes of similar structural design, accounting for differences in operating conditions and procedures. The airplane structural configurations to be evaluated include—

(i) All model variations and derivatives approved under the type certificate; and

(ii) All structural modifications to and replacements for the airplane structural configurations specified in paragraph (b)(1)(i) of this section, mandated by airworthiness directives as of January 14, 2011.

(2) If the LOV depends on performance of maintenance actions for which service information has not been mandated by airworthiness directive as of January 14, 2011, submit the following to the responsible Aircraft Certification Service office:

(i) For those maintenance actions for which service information has been issued as of the applicable compliance date specified in paragraph (c) of this section, a list identifying each of those actions.

(ii) For those maintenance actions for which service information has not been issued as of the applicable compliance date specified in paragraph (c) of this section, a list identifying each of those actions and a binding schedule for providing in a timely manner the necessary service information for those actions. Once the responsible Aircraft Certification Service office approves this schedule, each person identified in paragraph (c) of this section must comply with that schedule.

(3) Unless previously accomplished, establish an Airworthiness Limitations section (ALS) for each airplane structural configuration evaluated under paragraph (b)(1) of this section.

(4) Incorporate the applicable LOV established under paragraph (b)(1) of this section into the ALS for each airplane structural configuration evaluated under paragraph (b)(1) and submit it to the responsible Aircraft Certification Service office for approval.

(c) *Persons who must comply and compliance dates.* The following persons must comply with the requirements of paragraph (b) of this section by the specified date.

(1) Holders of type certificates (TC) of airplane models identified in Table 1 of this section: No later than the applicable date identified in Table 1 of this section.

(2) Applicants for TCs, if the date of application was before January 14, 2011: No later than the latest of the following dates:

(i) January 14, 2016;

(ii) The date the certificate is issued; or

(iii) The date specified in the plan approved under § 25.571(b) for completion of the full-scale fatigue testing and demonstrating that widespread fatigue damage will not occur in the airplane structure.

(3) Applicants for amendments to TCs, with the exception of amendments to TCs specified in paragraphs (c)(6) or (c)(7) of this section, if the original TC was issued before January 14, 2011: No later than the latest of the following dates:

(i) January 14, 2016;

(ii) The date the amended certificate is issued; or

(iii) The date specified in the plan approved under § 25.571(b) for completion of the full-scale fatigue testing and demonstrating that widespread fatigue damage will not occur in the airplane structure.

(4) Applicants for amendments to TCs, with the exception of amendments to TCs specified in paragraphs (c)(6) or (c)(7) of this section, if the application for the original TC was made before January 14, 2011 but the TC was not issued before January 14, 2011: No later than the latest of the following dates:

(i) January 14, 2016;

(ii) The date the amended certificate is issued; or

(iii) The date specified in the plan approved under § 25.571(b) for completion of the full-scale fatigue testing and demonstrating that widespread fatigue damage will not occur in the airplane structure.

(5) Holders of either supplemental type certificates (STCs) or amendments to TCs that increase maximum takeoff gross weights from 75,000 pounds or less to greater than 75,000 pounds: No later than July 14, 2012.

(6) Applicants for either STCs or amendments to TCs that increase maximum takeoff gross weights from 75,000 pounds or less to greater than 75,000 pounds: No later than the latest of the following dates:

(i) July 14, 2012;

(ii) The date the certificate is issued; or

(iii) The date specified in the plan approved under §25.571(b) for completion of the full-scale fatigue testing and demonstrating that widespread fatigue damage will not occur in the airplane structure.

(7) Applicants for either STCs or amendments to TCs that decrease maximum takeoff gross weights from greater than 75,000 pounds to 75,000 pounds or less, if the date of application was after January 14, 2011: No later than the latest of the following dates:

(i) July 14, 2012;

(ii) The date the certificate is issued; or

(iii) The date specified in the plan approved under §25.571(b) for completion of the full-scale fatigue testing and demonstrating that widespread fatigue damage will not occur in the airplane structure.

(d) *Compliance plan.* Each person identified in paragraph (e) of this section must submit a compliance plan consisting of the following:

(1) A proposed project schedule, identifying all major milestones, for meeting the compliance dates specified in paragraph (c) of this section.

(2) A proposed means of compliance with paragraphs (b)(1) through (b)(4) of this section.

(3) A proposal for submitting a draft of all compliance items required by paragraph (b) of this section for review by the responsible Aircraft Certification Service office not less than 60 days before the compliance date speci-

fied in paragraph (c) of this section, as applicable.

(4) A proposal for how the LOV will be distributed.

(e) *Compliance dates for compliance plans.* The following persons must submit the compliance plan described in paragraph (d) of this section to the responsible Aircraft Certification Service office by the specified date.

(1) Holders of type certificates: No later than April 14, 2011.

(2) Applicants for TCs and amendments to TCs, with the exception of amendments to TCs specified in paragraphs (e)(4), (e)(5), or (e)(6) of this section, if the date of application was before January 14, 2011 but the TC or TC amendment was not issued before January 14, 2011: No later than April 14, 2011.

(3) Holders of either supplemental type certificates or amendments to TCs that increase maximum takeoff gross weights from 75,000 pounds or less to greater than 75,000 pounds: No later than April 14, 2011.

(4) Applicants for either STCs or amendments to TCs that increase maximum takeoff gross weights from 75,000 pounds or less to greater than 75,000 pounds, if the date of application was before January 14, 2011: No later than April 14, 2011.

(5) Applicants for either STCs or amendments to TCs that increase maximum takeoff gross weights from 75,000 pounds or less to greater than 75,000 pounds, if the date of application is on or after January 14, 2011: Within 90 days after the date of application.

(6) Applicants for either STCs or amendments to TCs that decrease maximum takeoff gross weights from greater than 75,000 pounds to 75,000 pounds or less, if the date of application is on or after January 14, 2011: Within 90 days after the date of application.

(f) *Compliance plan implementation.* Each affected person must implement the compliance plan as approved in compliance with paragraph (d) of this section.

(g) *Exceptions.* This section does not apply to the following airplane models:

(1) Bombardier BD–700.

(2) Bombardier CL–44.

(3) Gulfstream GV.

(4) Gulfstream GV–SP.

(5) British Aerospace, Aircraft Group, and Societe Nationale Industrielle Aerospatiale Concorde Type 1.

(6) British Aerospace (Commercial Aircraft) Ltd., Armstrong Whitworth Argosy A.W. 650 Series 101.

(7) British Aerospace Airbus, Ltd., BAC 1–11.

(8) BAE Systems (Operations) Ltd., BAe 146.

(9) BAE Systems (Operations) Ltd., Avro 146.

(10) Lockheed 300–50A01 (USAF C141A).

(11) Boeing 707.

(12) Boeing 720.

(13) deHavilland D.H. 106 Comet 4C.

(14) Ilyushin Aviation IL–96T.

(15) Bristol Aircraft Britannia 305.

(16) Avions Marcel Dassault-Breguet Aviation Mercure 100C.

(17) Airbus Caravelle.

(18) D & R Nevada, LLC, Convair Model 22.

(19) D & R Nevada, LLC, Convair Model 23M.

TABLE 1—COMPLIANCE DATES FOR AFFECTED AIRPLANES

Airplane model (all existing [1] models)	Compliance date— (months after January 14, 2011)
Airbus:	
A300 Series	18
A310 Series, A300–600 Series	48
A318 Series	48
A319 Series	48
A320 Series	48
A321 Series	48
A330–200, –200 Freighter, –300 Series	48
A340–200, –300, –500, –600 Series	48
A380–800 Series	60
Boeing:	
717	48
727 (all series)	18
737 (Classics): 737–100, –200, –200C, –300, –400, –500	18
737 (NG): 737–600, –700, –700C, –800, –900, –900ER	48
747 (Classics): 747–100, –100B, –100B SUD, –200B, –200C, –200F, –300, 747SP, 747SR	18
747–400: 747–400, –400D, –400F	48
757	48
767	48
777–200, –300	48
777–200LR, 777–300ER, 777F	60
Bombardier:	
CL–600: 2D15 (Regional Jet Series 705), 2D24 (Regional Jet Series 900)	60
Embraer:	
ERJ 170	60
ERJ 190	60
Fokker:	
F.28 Mark 0070, Mark 0100	18
Lockheed:	
L–1011	18
188	18
382 (all series)	18
McDonnell Douglas:	
DC–8, –8F	18
DC–9	18
MD–80 (DC–9–81, –82, –83, –87, MD–88)	18
MD–90	48
DC–10	18
MD–10	48
MD–11, –11F	48
All Other Airplane Models Listed on a Type Certificate as of January 14, 2011	60

[1] Type certificated as of January 14, 2011.

[Doc. No. FAA–2006–24281, 75 FR 69782, Nov. 15, 2010, as amended at 77 FR 30878, May 24, 2012; Doc. No. FAA–2018–0119, Amdt. 26–7, 83 FR 9169, Mar. 5, 2018]

§ 26.23   Extended limit of validity.

(a) *Applicability.* Any person may apply to extend a limit of validity of the engineering data that supports the structural maintenance program (hereafter referred to as LOV) approved under § 25.571 of this subchapter, § 26.21, or this section. Extending an LOV is a major design change. The applicant must comply with the relevant provisions of subparts D or E of part 21 of this subchapter and paragraph (b) of this section.

(b) *Extended limit of validity.* Each person applying for an extended LOV must comply with the following requirements:

(1) Establish an extended LOV that corresponds to the period of time, stated as a number of total accumulated flight cycles or flight hours or both, during which it is demonstrated that widespread fatigue damage will not occur in the airplane. This demonstration must include an evaluation of airplane structural configurations and be supported by test evidence and analysis at a minimum and, if available, service experience, or service experience and teardown inspection results, of high-time airplanes of similar structural design, accounting for differences in operating conditions and procedures. The airplane structural configurations to be evaluated include—

(i) All model variations and derivatives approved under the type certificate for which approval for an extension is sought; and

(ii) All structural modifications to and replacements for the airplane structural configurations specified in paragraph (b)(1)(i) of this section, mandated by airworthiness directive, up to the date of approval of the extended LOV.

(2) Establish a revision or supplement, as applicable, to the Airworthiness Limitations section (ALS) of the Instructions for Continued Airworthiness required by § 25.1529 of this subchapter, and submit it to the responsible Aircraft Certification Service office for approval. The revised ALS or supplement to the ALS must include the applicable extended LOV established under paragraph (b)(1) of this section.

(3) Develop the maintenance actions determined by the WFD evaluation performed in paragraph (b)(1) of this section to be necessary to preclude WFD from occurring before the airplane reaches the proposed extended LOV. These maintenance actions must be documented as airworthiness limitation items in the ALS and submitted to the responsible Aircraft Certification Service office for approval.

[Docket No. FAA–2006–24281, 75 FR 69782, Nov. 15, 2010, as amended by Doc. No. FAA–2018–0119, Amdt. 26–7, 83 FR 9169, Mar. 5, 2018]

Subpart D—Fuel Tank Flammability

SOURCE: Docket No. FAA–2005–22997, 73 FR 42499, July 21, 2008, unless otherwise noted.

§ 26.31 Definitions.

For purposes of this subpart—

(a) *Fleet Average Flammability Exposure* has the meaning defined in Appendix N of part 25 of this chapter.

(b) *Normally Emptied* means a fuel tank other than a Main Fuel Tank. Main Fuel Tank is defined in 14 CFR 25.981(b).

§ 26.33 Holders of type certificates: Fuel tank flammability.

(a) *Applicability.* This section applies to U.S. type certificated transport category, turbine-powered airplanes, other than those designed solely for all-cargo operations, for which the State of Manufacture issued the original certificate of airworthiness or export airworthiness approval on or after January 1, 1992, that, as a result of original type certification or later increase in capacity have:

(1) A maximum type-certificated passenger capacity of 30 or more, or

(2) A maximum payload capacity of 7,500 pounds or more.

(b) *Flammability Exposure Analysis.* (1) *General.* Within 150 days after December 26, 2008, holders of type certificates must submit for approval to the responsible Aircraft Certification Service office a flammability exposure analysis of all fuel tanks defined in the type design, as well as all design variations approved under the type certificate that affect flammability exposure. This analysis must be conducted in accordance with Appendix N of part 25 of this chapter.

(2) *Exception.* This paragraph (b) does not apply to—

(i) Fuel tanks for which the type certificate holder has notified the FAA under paragraph (g) of this section that it will provide design changes and service instructions for Flammability Reduction Means or an Ignition Mitigation Means (IMM) meeting the requirements of paragraph (c) of this section.

(ii) Fuel tanks substantiated to be conventional unheated aluminum wing tanks.

(c) *Design Changes.* For fuel tanks with a Fleet Average Flammability Exposure exceeding 7 percent, one of the following design changes must be made.

(1) *Flammability Reduction Means (FRM).* A means must be provided to reduce the fuel tank flammability.

(i) Fuel tanks that are designed to be Normally Emptied must meet the flammability exposure criteria of Appendix M of part 25 of this chapter if any portion of the tank is located within the fuselage contour.

(ii) For all other fuel tanks, the FRM must meet all of the requirements of Appendix M of part 25 of this chapter, except, instead of complying with paragraph M25.1 of this appendix, the Fleet Average Flammability Exposure may not exceed 7 percent.

(2) *Ignition Mitigation Means (IMM).* A means must be provided to mitigate the effects of an ignition of fuel vapors within the fuel tank such that no damage caused by an ignition will prevent continued safe flight and landing.

(d) *Service Instructions.* No later than December 27, 2010, holders of type certificates required by paragraph (c) of this section to make design changes must meet the requirements specified in either paragraph (d)(1) or (d)(2) of this section. The required service instructions must identify each airplane subject to the applicability provisions of paragraph (a) of this section.

(1) *FRM.* The type certificate holder must submit for approval by the responsible Aircraft Certification Service office design changes and service instructions for installation of fuel tank flammability reduction means (FRM) meeting the criteria of paragraph (c) of this section.

(2) *IMM.* The type certificate holder must submit for approval by the responsible Aircraft Certification Service

office design changes and service instructions for installation of fuel tank IMM that comply with 14 CFR 25.981(c) in effect on December 26, 2008.

(e) *Instructions for Continued Airworthiness (ICA).* No later than December 27, 2010, holders of type certificates required by paragraph (c) of this section to make design changes must submit for approval by the responsible Aircraft Certification Service office, critical design configuration control limitations (CDCCL), inspections, or other procedures to prevent increasing the flammability exposure of any tanks equipped with FRM above that permitted under paragraph (c)(1) of this section and to prevent degradation of the performance of any IMM provided under paragraph (c)(2) of this section. These CDCCL, inspections, and procedures must be included in the Airworthiness Limitations Section (ALS) of the ICA required by 14 CFR 25.1529 or paragraph (f) of this section. Unless shown to be impracticable, visible means to identify critical features of the design must be placed in areas of the airplane where foreseeable maintenance actions, repairs, or alterations may compromise the critical design configuration limitations. These visible means must also be identified as a CDCCL.

(f) *Airworthiness Limitations.* Unless previously accomplished, no later than December 27, 2010, holders of type certificates affected by this section must establish an ALS of the maintenance manual or ICA for each airplane configuration evaluated under paragraph (b)(1) of this section and submit it to the responsible Aircraft Certification Service office for approval. The ALS must include a section that contains the CDCCL, inspections, or other procedures developed under paragraph (e) of this section.

(g) *Compliance Plan for Flammability Exposure Analysis.* Within 90 days after December 26, 2008, each holder of a type certificate required to comply with paragraph (b) of this section must submit to the responsible Aircraft Certification Service office a compliance plan consisting of the following:

(1) A proposed project schedule for submitting the required analysis, or a determination that compliance with

paragraph (b) of this section is not required because design changes and service instructions for FRM or IMM will be developed and made available as required by this section.

(2) A proposed means of compliance with paragraph (b) of this section, if applicable.

(h) *Compliance Plan for Design Changes and Service Instructions.* Within 210 days after December 26, 2008, each holder of a type certificate required to comply with paragraph (d) of this section must submit to the responsible Aircraft Certification Service office a compliance plan consisting of the following:

(1) A proposed project schedule, identifying all major milestones, for meeting the compliance dates specified in paragraphs (d), (e) and (f) of this section.

(2) A proposed means of compliance with paragraphs (d), (e) and (f) of this section.

(3) A proposal for submitting a draft of all compliance items required by paragraphs (d), (e) and (f) of this section for review by the responsible Aircraft Certification Service office not less than 60 days before the compliance times specified in those paragraphs.

(4) A proposal for how the approved service information and any necessary modification parts will be made available to affected persons.

(i) Each affected type certificate holder must implement the compliance plans, or later revisions, as approved under paragraph (g) and (h) of this section.

[Doc. No. FAA–2005–22997, 73 FR 42499, July 21, 2008, as amended by Amdt. 26–3, 74 FR 31619, July 2, 2009; Doc. No. FAA–2018–0119, Amdt. 26–7, 83 FR 9169, Mar. 5, 2018]

§26.35 **Changes to type certificates affecting fuel tank flammability.**

(a) *Applicability.* This section applies to holders and applicants for approvals of the following design changes to any airplane subject to 14 CFR 26.33(a):

(1) Any fuel tank designed to be Normally Emptied if the fuel tank installation was approved pursuant to a supplemental type certificate or a field approval before December 26, 2008;

(2) Any fuel tank designed to be Normally Emptied if an application for a

supplemental type certificate or an amendment to a type certificate was made before December 26, 2008 and if the approval was not issued before December 26, 2008; and

(3) If an application for a supplemental type certificate or an amendment to a type certificate is made on or after December 26, 2008, any of the following design changes:

(i) Installation of a fuel tank designed to be Normally Emptied,

(ii) Changes to existing fuel tank capacity, or

(iii) Changes that may increase the flammability exposure of an existing fuel tank for which FRM or IMM is required by §26.33(c).

(b) *Flammability Exposure Analysis*—(1) *General.* By the times specified in paragraphs (b)(1)(i) and (b)(1)(ii) of this section, each person subject to this section must submit for approval a flammability exposure analysis of the auxiliary fuel tanks or other affected fuel tanks, as defined in the type design, to the responsible Aircraft Certification Service office. This analysis must be conducted in accordance with Appendix N of part 25 of this chapter.

(i) Holders of supplemental type certificates and field approvals: Within 12 months of December 26, 2008,

(ii) Applicants for supplemental type certificates and for amendments to type certificates: Within 12 months after December 26, 2008, or before the certificate is issued, whichever occurs later.

(2) *Exception.* This paragraph does not apply to—

(i) Fuel tanks for which the type certificate holder, supplemental type certificate holder, or field approval holder has notified the FAA under paragraph (f) of this section that it will provide design changes and service instructions for an IMM meeting the requirements of §25.981(c) in effect December 26, 2008; and

(ii) Fuel tanks substantiated to be conventional unheated aluminum wing tanks.

(c) *Impact Assessment.* By the times specified in paragraphs (c)(1) and (c)(2) of this section, each person subject to paragraph (a)(1) of this section holding an approval for installation of a Normally Emptied fuel tank on an airplane model listed in Table 1 of this section, and each person subject to paragraph (a)(3)(iii) of this section, must submit for approval to the responsible Aircraft Certification Service office an assessment of the fuel tank system, as modified by their design change. The assessment must identify any features of the design change that compromise any critical design configuration control limitation (CDCCL) applicable to any airplane on which the design change is eligible for installation.

(1) Holders of supplemental type certificates and field approvals: Before June 26, 2011.

(2) Applicants for supplemental type certificates and for amendments to type certificates: Before June 26, 2011 or before the certificate is issued, whichever occurs later.

TABLE 1

Model—Boeing
747 Series
737 Series
777 Series
767 Series
757 Series

Model—Airbus
A318, A319, A320, A321 Series
A300, A310 Series
A330, A340 Series

(d) *Design Changes and Service Instructions.* By the times specified in paragraph (e) of this section, each person subject to this section must meet the requirements of paragraphs (d)(1) or (d)(2) of this section, as applicable.

(1) For holders and applicants subject to paragraph (a)(1) or (a)(3)(iii) of this section, if the assessment required by paragraph (c) of this section identifies any features of the design change that compromise any CDCCL applicable to any airplane on which the design change is eligible for installation, the holder or applicant must submit for approval by the responsible Aircraft Certification Service office design changes and service instructions for Flammability Impact Mitigation Means (FIMM) that would bring the design change into compliance with the

CDCCL. Any fuel tank modified as required by this paragraph must also be evaluated as required by paragraph (b) of this section.

(2) Applicants subject to paragraph (a)(2), or (a)(3)(i) of this section must comply with the requirements of 14 CFR 25.981, in effect on December 26, 2008.

(3) Applicants subject to paragraph (a)(3)(ii) of this section must comply with the requirements of 14 CFR 26.33.

(e) *Compliance Times for Design Changes and Service Instructions.* The following persons subject to this section must comply with the requirements of paragraph (d) of this section at the specified times.

(1) Holders of supplemental type certificates and field approvals: Before December 26, 2012.

(2) Applicants for supplemental type certificates and for amendments to type certificates: Before December 26, 2012, or before the certificate is issued, whichever occurs later.

(f) *Compliance Planning.* By the applicable date specified in Table 2 of this section, each person subject to paragraph (a)(1) of this section must submit for approval by the responsible Aircraft Certification Service office compliance plans for the flammability exposure analysis required by paragraph (b) of this section, the impact assessment required by paragraph (c) of this section, and the design changes and service instructions required by paragraph (d) of this section. Each person's compliance plans must include the following:

(1) A proposed project schedule for submitting the required analysis or impact assessment.

(2) A proposed means of compliance with paragraph (d) of this section.

(3) For the requirements of paragraph (d) of this section, a proposal for submitting a draft of all design changes, if any are required, and Airworthiness Limitations (including CDCCLs) for review by the responsible Aircraft Certification Service office not less than 60 days before the compliance time specified in paragraph (e) of this section.

(4) For the requirements of paragraph (d) of this section, a proposal for how the approved service information and

480

any necessary modification parts will be made available to affected persons.

TABLE 2—COMPLIANCE PLANNING DATES

	Flammability exposure analysis plan	Impact assessment plan	Design changes and service instructions plan
STC and Field Approval Holders.	March 26, 2009	February 26, 2011	August 26, 2011.

(g) Each person subject to this section must implement the compliance plans, or later revisions, as approved under paragraph (f) of this section.

[Doc. No. FAA–2005–22997, 73 FR 42499, July 21, 2008, as amended by Amdt. 26–3, 74 FR 31619, July 2, 2009; Doc. No. FAA–2018–0119, Amdt. 26–7, 83 FR 9170, Mar. 5, 2018]

§26.37 Pending type certification projects: Fuel tank flammability.

(a) *Applicability.* This section applies to any new type certificate for a transport category airplane, if the application was made before December 26, 2008, and if the certificate was not issued before December 26, 2008. This section applies only if the airplane would have—

(1) A maximum type-certificated passenger capacity of 30 or more, or

(2) A maximum payload capacity of 7,500 pounds or more.

(b) If the application was made on or after June 6, 2001, the requirements of 14 CFR 25.981 in effect on December 26, 2008, apply.

[Doc. No. FAA–2005–22997, 73 FR 42499, July 21, 2008, as amended by Amdt. 26–3, 74 FR 31619, July 2, 2009]

§26.39 Newly produced airplanes: Fuel tank flammability.

(a) *Applicability:* This section applies to Boeing model airplanes specified in Table 1 of this section, including passenger and cargo versions of each model, when application is made for original certificates of airworthiness or export airworthiness approvals after December 27, 2010.

TABLE 1

Model—Boeing
747 Series
737 Series
777 Series
767 Series

(b) Any fuel tank meeting all of the criteria stated in paragraphs (b)(1), (b)(2) and (b)(3) of this section must have flammability reduction means (FRM) or ignition mitigation means (IMM) that meet the requirements of 14 CFR 25.981 in effect on December 26, 2008.

(1) The fuel tank is Normally Emptied.

(2) Any portion of the fuel tank is located within the fuselage contour.

(3) The fuel tank exceeds a Fleet Average Flammability Exposure of 7 percent.

(c) All other fuel tanks that exceed an Fleet Average Flammability Exposure of 7 percent must have an IMM that meets 14 CFR 25.981(d) in effect on December 26, 2008, or an FRM that meets all of the requirements of Appendix M to this part, except instead of complying with paragraph M25.1 of that appendix, the Fleet Average Flammability Exposure may not exceed 7 percent.

[Doc. No. FAA–2005–22997, 73 FR 42499, July 21, 2008, as amended by Amdt. 26–3, 74 FR 31619, July 2, 2009]

Subpart E—Aging Airplane Safety—Damage Tolerance Data for Repairs and Alterations

SOURCE: Docket No. FAA–2005–21693, 72 FR 70505, Dec. 12, 2007, unless otherwise noted.

§26.41 Definitions.

Affects (or Affected) means structure has been physically repaired, altered, or modified, or the structural loads acting on the structure have been increased or redistributed.

Baseline structure means structure that is designed under the original type certificate or amended type certificate for that airplane model.

Damage Tolerance Evaluation (DTE) means a process that leads to a determination of maintenance actions necessary to detect or preclude fatigue cracking that could contribute to a catastrophic failure. As applied to repairs and alterations, a DTE includes the evaluation both of the repair or alteration and of the fatigue critical structure affected by the repair or alteration.

Damage Tolerance Inspection (DTI) means the inspection developed as a result of a DTE. A DTI includes the areas to be inspected, the inspection method, the inspection procedures, including acceptance and rejection criteria, the threshold, and any repeat intervals associated with those inspections. The DTI may specify a time limit when a repair or alteration needs to be replaced or modified. If the DTE concludes that DT-based supplemental structural inspections are not necessary, the DTI contains a statement to that effect.

DT data mean DTE documentation and the DTI.

DTE documentation means data that identify the evaluated fatigue critical structure, the basic assumptions applied in a DTE, and the results of a DTE.

Fatigue critical structure means airplane structure that is susceptible to fatigue cracking that could contribute to a catastrophic failure, as determined in accordance with § 25.571 of this chapter. Fatigue critical structure includes structure, which, if repaired or altered, could be susceptible to fatigue cracking and contribute to a catastrophic failure. Such structure may be part of the baseline structure or part of an alteration.

Implementation schedule consists of documentation that establishes the timing for accomplishing the necessary actions for developing DT data for repairs and alterations, and for incorporating those data into an operator's continuing airworthiness maintenance program. The documentation must identify times when actions must be taken as specific numbers of airplane flight hours, flight cycles, or both.

Published repair data mean instructions for accomplishing repairs, which are published for general use in structural repair manuals and service bulletins (or equivalent types of documents).

§ 26.43 Holders of and applicants for type certificates—Repairs.

(a) *Applicability.* Except as specified in paragraph (g) of this section, this section applies to transport category, turbine powered airplane models with a type certificate issued after January 1, 1958, that as a result of original type certification or later increase in capacity have—

(1) A maximum type certificated passenger seating capacity of 30 or more; or

(2) A maximum payload capacity of 7,500 pounds or more.

(b) *List of fatigue critical baseline structure.* For airplanes specified in paragraph (a) of this section, the holder of or applicant for a type certificate must—

(1) Identify fatigue critical baseline structure for all airplane model variations and derivatives approved under the type certificate; and

(2) Develop and submit to the responsible Aircraft Certification Service office for review and approval, a list of the structure identified under paragraph (b)(1) of this section and, upon approval, make the list available to persons required to comply with § 26.47 and §§ 121.1109 and 129.109 of this chapter.

(c) *Existing and future published repair data.* For repair data published by a holder of a type certificate that is current as of January 11, 2008 and for all later published repair data, the holder of a type certificate must—

(1) Review the repair data and identify each repair specified in the data that affects fatigue critical baseline structure identified under paragraph (b)(1) of this section;

(2) Perform a DTE and develop the DTI for each repair identified under paragraph (c)(1) of this section, unless previously accomplished;

(3) Submit the DT data to the responsible Aircraft Certification Service office or its properly authorized designees for review and approval; and

(4) Upon approval, make the DTI available to persons required to comply

with §§121.1109 and 129.109 of this chapter.

(d) *Future repair data not published.* For repair data developed by a holder of a type certificate that are approved after January 11, 2008 and are not published, the type certificate holder must accomplish the following for repairs specified in the repair data that affect fatigue critical baseline structure:

(1) Perform a DTE and develop the DTI.

(2) Submit the DT data required in paragraph (d)(1) of this section for review and approval by the responsible Aircraft Certification Service office or its properly authorized designees.

(3) Upon approval, make the approved DTI available to persons required to comply with §§121.1109 and 129.109 of this chapter.

(e) *Repair evaluation guidelines.* Except for airplane models whose type certificate is issued after January 11, 2008, holders of a type certificate for each airplane model subject to this section must—

(1) Develop repair evaluation guidelines for operators' use that include—

(i) A process for conducting surveys of affected airplanes that will enable identification and documentation of all existing repairs that affect fatigue critical baseline structure identified under paragraph (b)(1) of this section and §26.45(b)(2);

(ii) A process that will enable operators to obtain the DTI for repairs identified under paragraph (e)(1)(i) of this section; and

(iii) An implementation schedule for repairs covered by the repair evaluation guidelines. The implementation schedule must identify times when actions must be taken as specific numbers of airplane flight hours, flight cycles, or both.

(2) Submit the repair evaluation guidelines to the responsible Aircraft Certification Service office for review and approval.

(3) Upon approval, make the guidelines available to persons required to comply with §§121.1109 and 129.109 of this chapter.

(4) If the guidelines direct the operator to obtain assistance from the holder of a type certificate, make such

assistance available in accordance with the implementation schedule.

(f) *Compliance times.* Holders of type certificates must submit the following to the responsible Aircraft Certification Service office or its properly authorized designees for review and approval by the specified compliance time:

(1) The identified list of fatigue critical baseline structure required by paragraph (b)(2) of this section must be submitted no later than 180 days after January 11, 2008 or before issuance of the type certificate, whichever occurs later.

(2) For published repair data that are current as of January 11, 2008, the DT data required by paragraph (c)(3) of this section must be submitted by June 30, 2009.

(3) For repair data published after January 11, 2008, the DT data required by paragraph (c)(3) of this section must be submitted before FAA approval of the repair data.

(4) For unpublished repair data developed after January 11, 2008, the DT data required by paragraph (d)(1) of this section must be submitted within 12 months of the airplane's return to service or in accordance with a schedule approved by the responsible Aircraft Certification Service office.

(5) The repair evaluation guidelines required by paragraph (e)(1) of this section must be submitted by December 30, 2009.

(g) *Exceptions.* The requirements of this section do not apply to the following transport category airplane models:

(1) Convair CV–240, 340, 440, if modified to include turbine engines.

(2) Vickers Armstrong Viscount, TCDS No. A–814.

(3) Douglas DC–3, if modified to include turbine engines, TCDS No. A–618.

(4) Bombardier CL–44, TCDS No. 1A20.

(5) Mitsubishi YS–11, TCDS No. A1PC.

(6) British Aerospace BAC 1–11, TCDS No. A5EU.

(7) Concorde, TCDS No. A45EU.

(8) deHavilland D.H. 106 Comet 4C, TCDS No. 7A10.

(9) deHavilland DHC–7, TCDS No. A20EA.

(10) VFW-Vereinigte Flugtechnische Werk VFW-614, TCDS No. A39EU.

(11) Illyushin Aviation IL 96T, TCDS No. A54NM.

(12) Bristol Aircraft Britannia 305, TCDS No. 7A2.

(13) Handley Page Herald Type 300, TCDS No. A21N.

(14) Avions Marcel Dassault—Breguet Aviation Mercure 100C, TCDS No. A40EU.

(15) Airbus Caravelle, TCDS No. 7A6.

(16) Lockheed L-300, TCDS No. A2S0.

(17) Boeing 707-100/-200, TCDS No. 4A21.

(18) Boeing 707-300/-400, TCDS No. 4A26.

(19) Boeing 720, TCDS No. 4A28.

[Doc. No. FAA-2005-21693, 72 FR 70505, Dec. 12, 2007, as amended by Amdt. 26-4, 75 FR 11734, Mar. 12, 2010; Doc. No. FAA-2018-0119, Amdt. 26-7, 83 FR 9170, Mar. 5, 2018]

§ 26.45 Holders of type certificates—Alterations and repairs to alterations.

(a) *Applicability.* This section applies to transport category airplanes subject to § 26.43.

(b) *Fatigue critical alteration structure.* For existing and future alteration data developed by the holder of a type certificate, the holder must—

(1) Review alteration data and identify all alterations that affect fatigue critical baseline structure identified under § 26.43(b)(1);

(2) For each alteration identified under paragraph (b)(1) of this section, identify any fatigue critical alteration structure;

(3) Develop and submit to the responsible Aircraft Certification Service office for review and approval a list of the structure identified under paragraph (b)(2) of this section; and

(4) Upon approval, make the list required in paragraph (b)(3) of this section available to persons required to comply with §§ 121.1109 and 129.109 of this chapter.

(c) *DT Data.* For existing and future alteration data developed by the holder of a type certificate that affect fatigue critical baseline structure identified under § 26.43(b)(1), unless previously accomplished, the holder must—

(1) Perform a DTE and develop the DTI for the alteration and fatigue crit-

ical baseline structure that is affected by the alteration;

(2) Submit the DT data developed in accordance with paragraphs (c)(1) of this section to the responsible Aircraft Certification Service office or its properly authorized designees for review and approval; and

(3) Upon approval, make the DTI available to persons required to comply with §§ 121.1109 and 129.109 of this chapter.

(d) *DT Data for Repairs Made to Alterations.* For existing and future repair data developed by a holder of a type certificate, the type certificate holder must—

(1) Review the repair data, and identify each repair that affects any fatigue critical alteration structure identified under paragraph (b)(2) of this section;

(2) For each repair identified under paragraph (d)(1) of this section, unless previously accomplished, perform a DTE and develop DTI;

(3) Submit the DT data developed in accordance with paragraph (d)(2) of this section to the responsible Aircraft Certification Service office or its properly authorized designees for review and approval; and

(4) Upon approval, make the DTI available to persons required to comply with §§ 121.1109 and 129.109 of this chapter.

(e) *Compliance times.* Holders of type certificates must submit the following to the responsible Aircraft Certification Service office or its properly authorized designees for review and approval by the specified compliance time:

(1) The list of fatigue critical alteration structure identified under paragraph (b)(3) of this section must be submitted—

(i) No later than 360 days after January 11, 2008, for alteration data approved before January 11, 2008.

(ii) No later than 30 days after March 12, 2010 or before initial approval of the alteration data, whichever occurs later, for alteration data approved on or after January 11, 2008.

(2) For alteration data developed and approved before January 11, 2008, the DT data required by paragraph (c)(2) of

this section must be submitted by June 30, 2009.

(3) For alteration data approved on or after January 11, 2008, DT data required by paragraph (c)(2) of this section must be submitted before initial approval of the alteration data.

(4) For repair data developed and approved before January 11, 2008, the DT data required by paragraph (d)(2) of this section must be submitted by June 30, 2009.

(5) For repair data developed and approved after January 11, 2008, the DT data required by paragraph (d)(2) of this section must be submitted within 12 months after initial approval of the repair data and before making the DT data available to persons required to comply with §§ 121.1109 and 129.109 of this chapter.

[Doc. No. FAA–2005–21693, 72 FR 70505, Dec. 12, 2007, as amended by Amdt. 26–4, 75 FR 11734, Mar. 12, 2010; Doc. No. FAA–2018–0119, Amdt. 26–7, 83 FR 9170, Mar. 5, 2018]

§ 26.47 Holders of and applicants for a supplemental type certificate—Alterations and repairs to alterations.

(a) *Applicability*. This section applies to transport category airplanes subject to § 26.43.

(b) *Fatigue critical alteration structure.* For existing structural alteration data approved under a supplemental certificate, the holder of the supplemental certificate must—

(1) Review the alteration data and identify all alterations that affect fatigue critical baseline structure identified under § 26.43(b)(1);

(2) For each alteration identified under paragraph (b)(1) of this section, identify any fatigue critical alteration structure;

(3) Develop and submit to the responsible Aircraft Certification Service office for review and approval a list of the structure identified under paragraph (b)(2) of this section; and

(4) Upon approval, make the list required in paragraph (b)(3) of this section available to persons required to comply with §§ 121.1109 and 129.109 of this chapter.

(c) *DT Data.* For existing and future alteration data developed by the holder of a supplemental type certificate that affect fatigue critical baseline struc-

ture identified under § 26.43(b)(1), unless previously accomplished, the holder of a supplemental type certificate must—

(1) Perform a DTE and develop the DTI for the alteration and fatigue critical baseline structure that is affected by the alteration;

(2) Submit the DT data developed in accordance with paragraphs (c)(1) of this section to the responsible Aircraft Certification Service office or its properly authorized designees for review and approval; and

(3) Upon approval, make the DTI available to persons required to comply with §§ 121.1109 and 129.109 of this chapter.

(d) *DT Data for Repairs Made to Alterations.* For existing and future repair data developed by the holder of a supplemental holder of a supplemental type certificate, the holder of a supplemental type certificate must—

(1) Review the repair data, and identify each repair that affects any fatigue critical alteration structure identified under paragraph (b)(2) of this section;

(2) For each repair identified under paragraph (d)(1) of this section, unless previously accomplished, perform a DTE and develop DTI;

(3) Submit the DT data developed in accordance with paragraph (d)(2) of this section to the responsible Aircraft Certification Service office or its properly authorized designees for review and approval; and

(4) Upon approval, make the DTI available to persons required to comply with §§ 121.1109 and 129.109 of this chapter.

(e) *Compliance times.* Holders of supplemental type certificates must submit the following to the responsible Aircraft Certification Service office or its properly authorized designees for review and approval by the specified compliance time:

(1) The list of fatigue critical alteration structure required by paragraph (b)(3) of this section must be submitted no later than 360 days after January 11, 2008.

(2) For alteration data developed and approved before January 11, 2008, the DT data required by paragraph (c)(2) of this section must be submitted by June 30, 2009.

(3) For alteration data developed after January 11, 2008, the DT data required by paragraph (c)(2) of this section must be submitted before approval of the alteration data and making it available to persons required to comply with §§ 121.1109 and 129.109 of this chapter.

(4) For repair data developed and approved before January 11, 2008, the DT data required by paragraph (d)(2) of this section must be submitted by June 30, 2009.

(5) For repair data developed and approved after January 11, 2008, the DT data required by paragraph (d)(2) of this section, must be submitted within 12 months after initial approval of the repair data and before making the DT data available to persons required to comply with §§ 121.1109 and 129.109 of this chapter.

[Docket No. FAA–2005–21693, 72 FR 70505, Dec. 12, 2007, as amended by Doc. No. FAA–2018–0119, Amdt. 26–7, 83 FR 9170, Mar. 5, 2018]

§ 26.49 Compliance plan.

(a) *Compliance plan.* Except for applicants for type certificates and supplemental type certificates whose applications are submitted after January 11, 2008, each person identified in §§ 26.43, 26.45, and 26.47, must submit a compliance plan consisting of the following:

(1) A project schedule identifying all major milestones for meeting the compliance times specified in §§ 26.43(f), 26.45(e), and 26.47(e), as applicable.

(2) A proposed means of compliance with §§ 26.43, 26.45, and 26.47, as applicable.

(3) A plan for submitting a draft of all compliance items required by this subpart for review by the responsible Aircraft Certification Service office not less than 60 days before the applicable compliance date.

(b) *Compliance dates for compliance plans.* The following persons must submit the compliance plan described in paragraph (a) of this section to the responsible Aircraft Certification Service office for approval on the following schedule:

(1) For holders of type certificates, no later than 90 days after January 11, 2008.

(2) For holders of supplemental type certificates no later than 180 days after January 11, 2008.

(3) For applicants for changes to type certificates whose application are submitted before January 11, 2008, no later than 180 days after January 11, 2008.

(c) *Compliance Plan Implementation.* Each affected person must implement the compliance plan as approved in compliance with paragraph (a) of this section.

[Docket No. FAA–2005–21693, 72 FR 70505, Dec. 12, 2007, as amended by Doc. No. FAA–2018–0119, Amdt. 26–7, 83 FR 9170, Mar. 5, 2018]

PART 27—AIRWORTHINESS STANDARDS: NORMAL CATEGORY ROTORCRAFT

Subpart A—General

Subpart B—Flight

Made in the USA
Las Vegas, NV
23 April 2025

21276466R00267